Insect Diapause

Our highly seasonal world restricts insect activity to brief portions of the year. This feature necessitates a sophisticated interpretation of seasonal changes and enactment of mechanisms for bringing development to a halt and then reinitiating it when the inimical season is past. The dormant state of diapause serves to bridge the unfavorable seasons, and its timing provides a powerful mechanism for synchronizing insect development. This book explores how seasonal signals are monitored and used by insects to enact specific molecular pathways that generate the diapause phenotype. The broad perspective offered here scales from the ecological to the molecular and thus provides a comprehensive view of this exciting and vibrant research field, offering insights on topics ranging from pest management, evolution, speciation, climate change, and disease transmission, to human health, as well as analogies with other forms of invertebrate dormancy and mammalian hibernation.

David L. Denlinger is one of the world's leading researchers on insect diapause. He is a Distinguished University Professor and Professor Emeritus of Entomology at The Ohio State University, USA. He is a member of the National Academy of Sciences, a Fellow of the Entomological Society of America, and an Honorary Fellow of the Royal Entomological Society. Professor Denlinger's current laboratory research focuses primarily on molecular mechanisms involved in insect overwintering. His interests range from the use of clock genes to perceive environmental signals through the endocrine and molecular events that result in the expression of the diapause phenotype. He has received numerous awards for his research, including the Gregor Mendel Medal from the Czech Academy of Sciences (2006), the Antarctic Service Medal (2006), and the International Centre of Insect Physiology and Ecology (ICIPE) Achievement Award (2020).

Insect Diapause

DAVID L. DENLINGER
The Ohio State University

Shaftesbury Road, Cambridge CB2 8EA, United Kingdom

One Liberty Plaza, 20th Floor, New York, NY 10006, USA

477 Williamstown Road, Port Melbourne, VIC 3207, Australia

314–321, 3rd Floor, Plot 3, Splendor Forum, Jasola District Centre, New Delhi – 110025, India

103 Penang Road, #05–06/07, Visioncrest Commercial, Singapore 238467

Cambridge University Press is part of Cambridge University Press & Assessment, a department of the University of Cambridge.

We share the University's mission to contribute to society through the pursuit of education, learning and research at the highest international levels of excellence.

www.cambridge.org
Information on this title: www.cambridge.org/9781108497527

DOI: 10.1017/9781108609364

First published 2022

A catalogue record for this publication is available from the British Library

ISBN   978-1-108-49752-7   Hardback

To Else, Esben, Solomon, Jude, and Liv

Contents

Color plates can be found between pages 214 and 215.

Preface

To every thing there is a season

Ecclesiastes 3:1

The surge of new plant growth in the spring is followed quickly by a flush of insects that exploit those plants for food. And then, just as quickly, as summer wanes and winter approaches, plants die back and insects disappear until the cycle repeats the following year. Ours is a highly seasonal world. The tilt of the Earth as it circles the Sun dictates the intensity of heat impinging on the Earth's surface, thus driving the striking seasonal rhythms of life. Temperature contrasts between winter and summer are dramatic, and even in the tropics, where temperatures remain more constant, wet and dry seasons pose a conspicuous seasonal pattern. The impact this has on all life forms is consequential, but the challenge is especially pronounced for insects and other arthropods that, as ectotherms, can perform only within a narrow seasonal window. Escaping in time thus becomes an essential and crucial feature of the insect life cycle, allowing insects to survive seasons that are too cold, too dry, or lacking essential food resources required for development and reproduction.

Diapause, a dormant stage akin to mammalian hibernation, thus emerges as a key feature that may encompass the major portion of an insect's life. Development is shut down or dramatically retarded, allowing the insect to bridge inimical seasons. The ability of insects to avoid such unfavorable seasons by entering diapause has made it possible for this group of animals to not only invade all seven continents but also become the dominant life form in most of Earth's terrestrial habitats. Insects, of course, are not alone in their ability to respond to seasonal changes in the environment, but the diversity of insects, our ability to experimentally manipulate their environment, and our growing understanding of insect molecular processes have made insects a particularly rich taxon for probing seasonal responses. How an insect knows when bad times are coming, why and how it responds to evoke the diapause response, and the consequences of diapause for post-diapause development is the story I hope to relay.

Several wonderful reviews greatly influenced my understanding of diapause down through the years. My first encounters as a student with the diapause literature were delightful treatises by A.D. Lees (1955) entitled *The Physiology of Diapause in Arthropods* and by A.S. Danilevskii (1965 translation from Russian) entitled

Photoperiodism and Seasonal Development of Insects. This was followed by books that I eagerly devoured by S.D. Beck (*Insect Photoperiodism*, 1968); D.S. Saunders (*Insect Clocks*, 1976 [2002]); M.J. Tauber, C.A. Tauber, and S. Masaki (*Seasonal Adaptations of Insects*, 1986); and H.V. Danks (*Insect Dormancy: An Ecological Perspective*, 1987). The quality of these remarkable books makes it a bit daunting to consider adding my own perspective, but do so, I will! These previous books on insect diapause contain great background information on diapause as well as lists of diapausing insects that I will not repeat here, but instead, I hope to highlight major features of diapause with special emphasis on recent developments in the field.

In spite of the diversity of insects, quite a few common themes unite diapause responses observed across a range of species. I offer some hints on the diversity of responses, but I err on the side of giving more depth to some of the best-studied models at the expense of portraying all the variants. I place special emphasis on flies and moths, a bias that comes from my life experiences with these two prominent groups.

I am enormously grateful to my mentors who introduced me to diapause and encouraged me along the way. My graduate advisors, Gottfried Fraenkel and Judith Willis, introduced me to flesh fly diapause and the broader realm of diapause physiology; my postdoctoral advisor Jan de Wilde shared his diapause expertise and dared to send this Pennsylvania Dutchman to Kenya as a Dutch representative to explore diapause in the tropics; and Carroll M. Williams provided an engaging laboratory, replete with a rich history of diapause experimentation, when I returned to North America. I am also enormously grateful to all the students, postdocs, and visiting scholars who have been through my laboratory at Ohio State and contributed so much to my own understanding of diapause. While I often got the credit, it was their hard work that allowed us to deepen our understanding of diapause mechanisms. And, more broadly, the entire diapause community has attracted a great cadre of people with whom I have enjoyed interacting. Thanks for all you have taught me. I hope I have captured your insights correctly.

Thanks to Dominic Lewis, Aleksandra Serocka, and Jenny van der Meijden from Cambridge University Press for encouraging me to embark on this project and for providing guidance along the way. Select portions of the book benefited from thoughtful comments provided by David Saunders (University of Edinburgh), Peter Cherbas (Indiana University), Dan Hahn and Clancy Short (University of Florida), William Bradshaw and Christina Holzapfel (University of Oregon), Peter Armbruster (Georgetown University), Mariana Wolfner (Cornell University), and Megan Meuti (The Ohio State University). Their helpful critiques, especially when adopted, resulted in a better book. Thanks to Jonathan Denlinger for assistance in preparing several figures, and a special thanks to my wife, Judy, for supporting my devotion to the preparation of this book over the past few years.

1 Confronting the Challenges of a Seasonal Environment

Our Earth imposes challenging seasonal obstacles for insect development, with few sites capable of continuously sustaining activity throughout the year. Low winter temperatures at temperate and polar latitudes are a conspicuous obstacle to continuous growth and reproduction. Not only is the cold a significant threat to insect survival, but the accompanying disappearance of plants, on which most insects depend for food, generates a seasonally depauperated environment not suitable for continuous development. Equally challenging are tropical dry seasons that lack critical food and water resources.

But, the advantages of diapause go well beyond avoidance of adverse seasons. A finely tuned diapause is critical for the precise timing of adult emergence in the spring. The fact that many insects are host specialists means that diapause is an essential element for coordinating the timing of adult emergence with the growth or flowering of favored host plants. By becoming dormant, insects also mitigate risks of predation, and periodic disappearance into diapause may help reduce biotic challenges from parasites and microbes.

Diapause occurring during the winter has the advantage of concomitant low temperatures that, for an ectotherm, help suppress metabolic rate and conserve energy reserves. Summer diapauses and diapauses in tropical insects lack this advantage and thus confront the extra challenge of conserving energy reserves in a warm environment. The ability to suppress metabolism without the aid of low temperature thus becomes a special challenge in select environments.

High latitudes pose yet another different challenge. Short growing seasons restrict development to narrow seasonal windows that may be inadequate for the completion of a single generation. Such conditions may require multiple years for completion of the life cycle, thus making it necessary for more than one developmental stage to enter diapause. In some cases, a developmental arrest may be repeatedly induced and broken to capitalize on brief favorable seasons.

Just as seasons have different profiles in different regions of the Earth, the role of diapause, the environmental regulators that are essential for coordinating the response, and the features of the diapause response can be expected to vary across the globe. Diapause thus emerges as a fascinating and plastic life-history trait whose evolution reflects adaptations essential for confronting challenges of life in a wide range of specific environments as well as life in a changing environment.

1.1   What Is Diapause?

1.1.1   Origin of the Term Diapause

The Greek word "diapause" means rest or interruption of work. The usage of this word has an interesting history (Lees 1955). As first used by Wheeler (1893) in his work with the grasshopper *Xiphidium ensiferum*, diapause defined the brief halt in development and embryonic transition between anatrepsis (when the embryo of a hemimetabolous insect moves tail-first through the yolk, away from the pole) and katatrepsis (when the embryo then migrates to either the ventral or dorsal surface, depending on the species). Its meaning was expanded by Henneguy (1904) to refer not to a *stage* of morphogenesis but the *condition* of arrested development, but Henneguy failed to make a distinction between the simple arrest prompted by cold and the type of arrest that is stage-specific and developmentally programmed. Shelford (1929) recognized the need to make such a distinction and used the term quiescence for the interruption of growth due directly to an unfavorable condition, while he used the term diapause to refer to "spontaneous arrest," an arrest that we now refer to as being developmentally programmed and stage-specific. This same sort of distinction was implied in the terms "diapause vrai" and "pseudodiapause" used by Roubaud (1930). Numerous and subtle distinctions can be made when arrested development is described for diverse species (Danks 2002), and distinctions between species have led to rather complex classifications of insect dormancy (e.g., Müller 1970, Mansingh 1971, Ushatinskaya 1976a), resulting in the introduction of terms such as oligopause, parapause, pluvipause, and others to describe developmental arrests with slightly different features. My own bias is that the terms diapause and quiescence are useful distinctions, but I see no need for further subdivisions, assuming we recognize that some features of diapause and quiescence may vary among species.

The term diapause extends beyond the insect literature and includes numerous examples among other arthropods, especially ticks and mites (e.g., Belozerov 2008, Lohmeyer et al. 2009, Bryon et al. 2017a). Diapause in Crustacea is well documented among copepods and brine shrimp (Marcus and Scheef 2010, Hand et al. 2016, Baumgartner and Tarrant 2017) and is prevalent in rotifers (Garcia-Rogers et al. 2019, Tarazona et al. 2020), as well as a few species of mollusks (Numata and Udaka 2010). Tardigrades, invertebrates best known for their capacity to enter a cryptobiotic state, also include some species that diapause as embryos (Guidetti et al. 2011). Diapause is also widely used to refer to delayed implantation of embryos in marsupials, polar bears, badgers, mink, anteaters, mice, gerbils, and over 130 other species of mammals (Renfree and Fenelon 2017) and is routinely used to refer to embryonic dormancies noted in non-mammalian taxa ranging from lizards to some of the bony fish species (Rafferty and Reina 2012), sharks and rays (Waltrick et al. 2012), and some of the bony fish species such as annual killifish (Martin and Podrabsky 2017, Hu et al. 2020).

Although developmental arrest in nematodes is a similar form of dormancy, the arrest in nematodes is more commonly referred to as the **dauer state** (Riddle 1997). Mammalian **hibernation** shares many features with insect diapause (Andrews 2019), but the unique physiological features of endotherms and ectotherms justify making a distinction between these two forms of dormancy. Though we may use different terms for these forms of animal dormancy, there is growing evidence that certain common themes operate across the animal kingdom and much can be gained from a quest for universal principles.

In most cases diapause is one of two developmental alternatives produced by the same genotype, thus it fits the classic definition of **polyphenism**, coined by Ernst Mayr (1963). Insects of the same genotype can either enter diapause or develop without diapause, a developmental decision that depends on the seasonal environmental cues received. The term polyphenism is thus distinct from the common usage of polymorphism, a term usually restricted to differences that have a genetic basis. The features of diapause, in which the environment dictates the phenotype (diapause or nondiapause), conform to the classic definition of **phenotypic plasticity** or a plastic response to environmental conditions.

1.1.2   Diapause Is a Programmed Event

Diapause is a programmed, stage-specific arrest or retardation of development commonly used to circumvent an adverse season. Upon entry into diapause, insects remain in this arrested state for some time, even if prevailing environmental conditions are favorable for development.

"Programmed" implies that it is not an immediate response to environmental adversity but a response to environmental signals that have been received in advance of the actual onset of diapause or, alternatively, is hard-wired genetically to occur at a specific stage. The distinction between an environmentally programmed diapause and one that is hard-wired genetically is captured in the terms "facultative" and "obligate" diapause. **Facultative diapause** implies a plastic response dependent upon receipt of specific environmental cues commonly received well in advance of diapause onset. By contrast, **obligate diapause** refers to a genetically programmed diapause that occurs at a specific stage regardless of the prevailing environmental input. While an insect with an obligate diapause is likely to complete only one generation a year, an insect with a facultative diapause has the flexibility to complete multiple generations, commonly several nondiapausing generations during the summer, followed by an overwintering generation that enters diapause.

Both obligate and facultative diapauses can sometimes be seen within the same family, such as the stink bugs Pentatomidae (Musolin and Saulich 2018). Certain trends exist within taxa such as hymenopteran parasitoids: parasitoids that attack univoltine (one generation/year) hosts tend to have an obligate diapause, while those attacking polyvoltine (multiple generations/year) hosts usually have a facultative diapause (Polgar and Hardie 2000). Although rare, one species may exhibit both an obligate and facultative diapause: certain populations of the European spruce bark beetle *Ips typographus* harbor both of these diapause phenotypes (Schebeck et al. 2017). Some species that appear to have an obligate diapause may, under close scrutiny, avert diapause under specific circumstances. For example, the saturniid moth

Hyalophora cecropia, a species that always enters pupal diapause when reared outside, can actually be enticed to develop without entering diapause when exposed to artificially long daylengths (Waldbauer 1996); hence this species, long thought to be an iconic example of a species with an obligate diapause, is more correctly recognized as a species with a facultative diapause, albeit a species that normally completes only a single generation each year and appears to seldom exploit its potential to skip diapause.

The beauty of the programming aspect of diapause is that it provides a preparatory phase that allows sequestration of additional energy reserves, augmentation of cuticular waterproofing, changes in color to match the winter or dry season habitat, migration, and selection of a well-protected hibernaculum (overwintering refuge).

1.1.3   Making a Distinction between Diapause and Quiescence

The programmed feature of diapause is in contrast to **quiescence**, a form of dormancy that is an immediate response to adversity and includes no preparatory phase. Unlike diapause, quiescence is broken immediately when favorable conditions return. Quiescence offers the capacity to quickly stop and restart development multiple times and at any stage in response to certain environmental challenges. For example, an insect placed in a refrigerator will halt development and become inactive but will resume development almost immediately when retrieved from the cold environment. Similarly, an insect denied food may enter an arrest that is terminated immediately when food again becomes available. This sort of rapid entry into and recovery from a dormant state distinguishes quiescence from diapause. Distinct endocrine signatures underlie these two forms of developmental arrest, as noted in adult females of the linden bug *Pyrrhocoris apterus* (Hodková and Okuda 2019). The corpora allata (CA) cease producing the juvenile hormone needed for vitellogenesis in both starvation-induced quiescence and diapause, and when food again becomes available the CA is immediately activated in starved bugs but not in diapausing bugs.

One of the interesting current debates is whether the well-studied adult dormancy in *Drosophila melanogaster* is a programmed arrest (i.e., a diapause) or a simple quiescence that is temperature-induced. Although the term diapause is frequently used in this context (e.g., Zonato et al. 2017), strong arguments counter that quiescence is a more appropriate term for the dormancy of *D. melanogaster* (e.g., Emerson et al. 2009b,c, Saunders 2020b) due to the immediacy of the response. One strong argument that the dormancy of *D. melanogaster* should not be regarded as diapause is the fact that egg chambers of the female initiate yolk uptake but then, in response to the low temperatures or other stresses that evoke the dormancy response, degenerate, an energy-consuming event not normally observed in a diapausing insect (Lirakis et al. 2018). The fact that this same response can be evoked in tropical populations of *D. melanogaster* has been used to suggest that the dormancy is not a diapause, but many tropical species do indeed have a diapause (see Section 2.3), thus that argument is not compelling. That egg chambers degenerate, that other stresses such as starvation elicit an identical response, and that the response is so volatile support the argument that dormancy noted in *D. melanogaster* represents a general stress response in this species, rather than a strictly binary trait. Yet, an examination of latitudinal clines suggests that certain populations of *D. melanogaster* do indeed possess a photoperiodically programmed diapause while others have features more akin to quiescence (Tatar et al. 2001, Anduaga et al. 2018).

A distinction between diapause and quiescence can sometimes be challenging to discern in other species as well. The invasive species *Drosophila suzukii* seems to show a slight photoperiodic response but, like *D. melanogaster*, its winter dormancy is more akin to quiescence (Everman et al. 2018). The mountain pine beetle *Dendroctonus ponderosae* was initially reported to overwinter in quiescence, but more recent experimental evidence shows a period of developmental latency, suggesting that the overwintering stage of the larvae can more accurately be described as diapause (Bentz and Hansen 2017). It is still unclear whether the midsummer disappearance of the yellow dung fly *Scathophaga stercoraria* in Central Europe can be attributed to diapause or quiescence (Blanckenhorn et al. 2001). Although less common, quiescence is also a viable option for overwintering, as noted in the diamondback moth *Plutella xylostella*, a species found in various life stages during winter in eastern North America (Dancau et al. 2018). Perhaps as a bet-hedging strategy, the European water strider *Velia caprai* can overwinter both as an adult in quiescence and as an embryo in diapause (Ditrich and Koštál 2011).

Though diapause, with its opportunity for preparation, would appear to be the ideal option for bridging unfavorable seasons, there are certain advantages to quiescence. Not only is it an option available to any developmental stage, but it also offers flexibility for rapidly responding to environmental conditions. For example, when temporary rock pools inhabited by the midge *Polypedilum vanderplanki* dry up in Nigeria, midge larvae enter a cryptobiotic state, a form of quiescence that is quickly broken when the rains return (Hinton 1951, Cornette and Kikawada 2011). This form of dormancy offers extreme flexibility, for both time of entry as well as exit from the dormant state. Quiescence may also be the preferable state for aquatic insects residing in cool, high-latitude streams (Danks 1987). Unlike the habitat of the African midge, cool streams represent a stable habitat, and quiescence enables swift transition between an active and inactive lifestyle, making it possible to maximally exploit periods when temperatures exceed a certain threshold.

1.1.4   Stage Specificity

The adjective "stage-specific" implies that diapause capacity is restricted to a single stage of development for each species. Although diapause can occur in embryos, larvae, pupae, or adults, for any one species the capacity for diapause is usually restricted to a single stage. There are, however, good examples of species, especially those confronting short growing seasons at high latitudes, that have the capacity for diapause at two stages of development. Northern populations of the spruce budworm

Choristoneura biennis first overwinter in a second-instar larval diapause and spend the second winter in a final larval-instar diapause (Nealis 2005), northern populations of the blow fly *Calliphora vicina* diapause both as third-instar larvae and as adults (Vinogradova and Reznik 2013), and the bruchid beetle *Bruchidius dorsalis* diapauses in cooler areas as a final-instar larva (Kurota and Shimada 2003a) and as an adult in warmer regions (Kurota and Shimada 2003b). Larval diapause in the longicorn beetle *Psacothea hilaris* is unusual in that it can occur across a wide developmental range from the fifth to ninth instar, although it most commonly occurs in the sixth or seventh instar (Asano et al. 2004). In central Sweden, the butterfly *Parage aegeria* diapauses as a pupa, but both larval and pupal diapause are common in southern Sweden (van Dyck and Wiklund 2002). *Lycaena hippothoe*, a butterfly with a wide Eurasian range, is capable of entering larval diapause as either a third or fourth instar (Fischer and Fiedler 2002). However, in the majority of insect species, the capacity for diapause resides in only one stage. This sort of stage specificity is not true for quiescence. Insects can become quiescent in any developmental stage, although the threshold for entering quiescence may differ among stages.

1.1.5   Diapause Development

In the strictest sense, development is not completely arrested during diapause. A slow progression of developmental processes occurs during diapause, culminating eventually in termination of diapause, a progression that Andrewartha (1952) referred to as **diapause development**. Diapause is best viewed not as a "stop/start" event but as a dynamic developmental trajectory operating on a much slower time scale than the developmental events observed in nondiapausing insects. Although covert molecular events are occurring as diapause progresses in all species, in most cases development appears at a standstill. In a few cases, however, overt changes can be noted during diapause, generating a response more akin to a slowing of the developmental rate rather than a complete developmental arrest. For example, during adult diapause in females of the mosquito *Culex pipiens*, ovarian follicles progressively enlarge, but it takes more than 20 weeks for the follicles to attain the same size attained by a nondiapausing female in 3 days (Readio et al. 1999). Similarly, the sexually produced embryos of the pea aphid *Acyrthosiphon pisum* overwinter in a diapause that can best be described as extremely slow, albeit progressive development during the interval between anatrepsis and katatrepsis (Shingleton et al. 2003). Legs and body organs continue to grow but at a rate that is largely temperature-independent, a contrast to the temperature-dependent growth rate observed after katatrepsis. A similar slow progression of development is noted for diapausing embryos of aphids *Cinara cupressi* and *Cinara juniperi* (Durak et al. 2020): Mitotic cell division continues at a slow pace, resulting in a doubling of body size from day 16 to day 70 after oviposition. In some Orthoptera, for example the cricket *Modicogryllus siamensis*, diapause is manifested not as a complete halt in development but as a much slower rate of development that incorporates additional nymphal molts (Miki et al. 2020).

1.2   Who Does It and in What Stage?

Diapause has evolved numerous times, as discussed in Chapter 11. In some taxa, diapause characteristically occurs at one certain stage, but that is not universally true. In some families and genera, the stage of diapause is not highly conserved. Among the many species of flesh flies (family Sarcophagidae) that have been examined from around the world, those from South America appear to lack the capacity for diapause (Denlinger et al. 1988a), but all species examined from North America, Europe, Asia, and Africa have a pupal diapause. Ladybird beetles (family Coccinellidae), due to their importance as biological control agents, have been examined extensively, and reports on species from Europe, North America, and Asia consistently reveal the capacity for an adult diapause (Hodek 2012).

At the other extreme, the genus *Drosophila* contains species that diapause as larvae (*D. deflexa*), pupae (*D. obscura*), or adults (e.g., *D. virilis* and *D. obscura* groups) (Lumme 1978). The diapause stage differs in two closely related Swiss dung flies that occupy a similar niche: *Scathophaga stercoraria* diapauses as a pupa, while *Sepsis cynipsea* diapauses as an adult (Blanckenhorn 1998a). Scandinavian butterflies within the Nymphalini tribe show remarkably different overwintering strategies (Wiklund et al. 2019). One of the nymphalids enters pupal diapause, two migrate south for the winter, four are univoltine and have an obligate adult diapause, and three have a partial second generation (two species have a facultative adult diapause determined by larval daylength, and one species shunts half the population into an obligate adult diapause while the other half enters a facultative adult diapause dependent on larval daylength). Stink bugs, family Pentatomidae, are known to diapause as embryos, nymphs, and adults (Musolin and Saulich 2018). Most bruchid beetles diapause as adults but at least two temperate species, *Kytorhinus sharpianus* and *Callosobruchus ademptus*, diapause as larvae (Kurota and Shimada 2001). Two closely related weevils differ considerably in their diapause attributes: *Exapion ulicis* diapauses as an adult and lays eggs in the spring, while *Exapion lemovicinum* lays eggs in the autumn and diapauses overwinter as a larva, responses driven largely by different fruiting times of their host plants (Barat et al. 2010). Among mosquitoes, certain trends emerge (Denlinger and Armbruster 2016, Diniz et al. 2017), as shown in Figure 1.1. While embryonic and larval stages predominate as the diapause stage in the genus *Aedes*, most mosquitoes in the genus *Culex* diapause as adults. In other genera, such as *Anopheles*, diapause is reported in embryonic, larval, and adult stages. No mosquitoes appear to use the pupal stage for diapause. Among crickets, embryonic diapause is the most prevalent stage of diapause, represented by over 80% of the species in Japan and the eastern United States (Masaki and Walker 1987). Among 200 species of mirid bugs from the Netherlands, 85% diapause as embryos (Cobben 1968), and among 14 species representing 5 genera of plant bugs in the subfamily Mirinae, 4 genera consistently diapause as embryos and members of another genus, *Lygus*, consistently diapause as adults (Saulich and Musolin 2020). So, some trends are noted within taxonomic

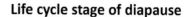

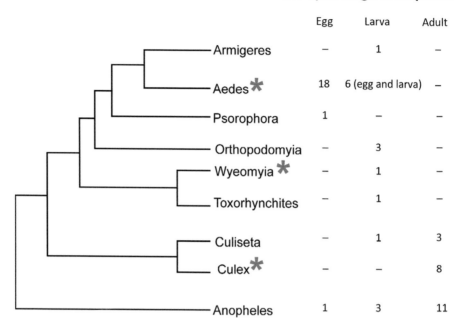

Figure 1.1 Phylogenetic distribution of life cycle stages of diapause in nine mosquito genera. *
indicates genera containing at least one species for which considerable information exists on the
transcriptional basis for diapause.
From Denlinger and Armbruster (2016).

groups, but many exceptions are evident. Diapause is a highly variable trait,
not only among closely related species but also among populations, as further
discussed in Chapter 3.

1.2.1   Embryonic Diapause

Embryonic diapause is especially well documented in Orthoptera (crickets, grasshoppers, walking sticks), true bugs in the order Heteroptera, and butterflies and moths in
the order Lepidoptera, and among the lower Diptera, such as mosquitoes and midges.
It occurs in some Coleoptera but appears to be less common in that order. Diapause is
not restricted to a specific stage of embryonic development but has been noted across a
wide range of embryonic stages. During embryonic development, the zygote divides
(cleavage), forms a layer of cells surrounding the yolk (blastoderm formation), and in
hemimetabolous insects, the blastoderm then invaginates slightly and migrates (blastokinesis). The embryo first migrates posteriorly (anatrepsis) and then dorsally (katatrepsis), movements that are somewhat simplified in holometabolous insects (Panfilio
2008). Diapause can apparently intercede at almost any point during embryogenesis,
from the early stages of blastoderm formation through to the completion of pharate
larval development (at which point the first-instar larva is completely formed but has

not yet hatched). A study on the crane fly *Tipula simplex* suggests that diapause in this species occurs prior to cleavage (Hartman and Hynes 1980), although additional studies at the electron microscope level would be helpful in validating this report. One particularly well-documented case of a species with an early-stage diapause is the band-legged ground cricket *Dianemobius nigrofasciatus* (Tanigawa et al. 2009). This diapause occurs prior to germ band formation when the embryo is merely a cellular blastoderm, a continuous layer of single cells. Diverse stages of embryonic diapauses have been documented for other species. Diapause occurs in a pre-anatrepsis stage in the Australian plague locust *Chortoicetes terminifera* (Deveson and Woodman 2014) and balsam twig aphid *Mindarus abietinus* (Doherty et al. 2018). The well-studied commercial silk moth Bombyx mori also enters diapause quite early, when the blastoderm is in the form of an unsegmented dumbbell (Yamashita and Hasegawa 1985). Embryonic diapause in the pea aphid *A. pisum* occurs at a slightly later stage when the embryo is fully segmented and the limb buds are clearly defined (Shingleton et al. 2003). Embryonic development in a Tibetan population of *Locusta migratoria* is classified into 27 stages, with diapause interceding at stage 19, a stage prior to katatrepsis, occurring 9 days after oviposition at 30°C (Su et al. 2019, Wang et al. 2021). Two Japanese cicadas diapause at distinctly different developmental stages prior to katatrepsis: *Cryptotympana facialis* diapauses upon completion of body segmentation when the antennal buds are segregated from the cephalic lobes, while *Graptopsaltria nigrofuscata* diapauses at a slightly later stage when the antennae, maxillary buds, and thoracic appendages have already differentiated (Moriyama and Numata 2008). At the other extreme, the katydid *Eobiana engelhardti subtropica* enters diapause as an almost fully developed embryo (Higaki and Ando 2005), and many Lepidoptera, such as the gypsy moth *Lymantria dispar*, enter diapause at the completion of embryogenesis when the embryo is more appropriately termed a pharate first instar (Leonard 1968). When diapause in the gypsy moth is broken, the fully developed pharate first-instar larva consumes the final remaining yolk, breaks through the chorion, and begins its life as a first-instar larva.

Different stages of embryonic diapause may exist in closely related species. The Heteroptera present a fascinating array of diapausing embryonic stages (Cobben 1968). Nearly all conceivable stages of embryonic development, ranging from the early blastoderm through to pharate first-instar nymph, are used by this taxon as the diapausing stage (Figure 1.2). This observation underscores the idea that natural selection can capture different embryonic stages among closely related species. One caveat is that the array of diapause stages depicted in Figure 1.2 is based on single snapshots in time, thus any sort of progression as noted in pea aphid embryos (Shingleton et al. 2003) would be obscured.

Early studies forming the foundation for embryonic diapause include extensive work on *B. mori,* initiated by Fukuda (1951) and Hasegawa (1951), followed by beautiful studies from Yamashita (1996) and his colleagues, experiments on the gypsy moth *L. dispar* (Leonard 1968, Hoy 1977), detailed series of experiments in Japan on diapause in several species of crickets by Masaki (1967) and his students, and experiments with French populations of *L. migratoria* (Le Berre 1953). More recently,

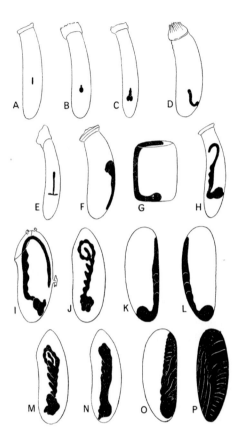

Figure 1.2 Stages of embryonic diapause in different species of Heteroptera. A, *Plagiognathus arbustorum*; B, *Megalocoleus molliculus*; C, *Pantilius tunicatus*; D, *Loricula elegantula*; E, *Leptopterna ferrugata*; F, *Coranus subapterus*; G, *Picromerus bidens*; H, *Himacerus apterus*; I, *Myrmus miriformis*; J, *Chiloxanthus pilosus*; K, *Notonecta lutea*; L, *Notonecta maculata*; M, *Chiloxanthus pilosus*; N, *Salda littoralis*; O, *Notonecta reuteri*; P, *Nysius thymi*. From Cobben (1968), with permission from Wageningen University.

embryonic diapause of the mosquito *Aedes albopictus* is attracting considerable attention as it invades new territories in Europe and North America (Armbruster 2016).

1.2.2   Larval Diapause

Larval (or nymphal) diapause is common among Lepidoptera, certain groups of Diptera, Hymenoptera, Coleoptera, Neuroptera, Odonata, Orthoptera, Hemiptera, and Plecoptera. Among the Lepidoptera, some species diapause in early larval instars, such as the first overwintering generation of the northern population of the spruce budworm *Choristonerura biennis* (Nealis 2005). But, more commonly diapause is entered at the end of the final larval instar when larvae have attained their full size and have ceased feeding. Although less common, a few Lepidoptera diapause midway through larval development. For example, the copper butterfly *Lycaena tityrus* diapauses as a third-instar larva before molting in the spring to a final, fourth-instar larval stage (Fischer and Fiedler 2001). *L. hippothoe*, a species with five larval instars, overwinters in either a third or fourth larval instar (Fischer and Fiedler 2002), and the large copper butterfly *L. dispar batavus*, diapauses as a second instar (Nicholls and Pullin 2003). The European corn borer *Ostrinia nubilalis* (Beck and Hanec 1960) and the southwestern corn borer *Diatraea grandiosella* (Chippendale and Reddy 1972) are among the best-studied examples providing early insights into larval diapause in Lepidoptera.

Among the lower Diptera, the pitcher plant mosquito *Wyeomyia smithii* offers the most robust and extensive body of literature on larval diapause, thanks mainly to the dedicated work from the Bradshaw-Holzapfel Laboratory and that of their colleagues (e.g., Bradshaw and Lounibos 1972). Larval diapause of *W. smithii* can intercede during either the third or fourth instar in certain populations (Lounibos and Bradshaw 1975), but a third-instar diapause is most prevalent. The mosquito *Anopheles barberi* most commonly enters diapause in the second instar, but some individuals diapause in the third instar as well (Copeland and Craig 1989).

Among the higher Diptera, I am not aware of larval diapause occurring in stages earlier than the fully grown, post-feeding third (final) larval instar, also known as the wandering stage. A well-studied example of such a third-instar fly diapause is the blow fly *C. vicina* (Vinogradova and Zinovjeva 1972, Saunders 2000). More recently, the drosophilid *Chymomyza costata* (Shimada and Riihimaa 1990, Koštál et al. 2011) joins *C. vicina* as a powerful model for larval diapause in Diptera. The final larval instar also dominates as the diapausing stage for larval diapause in the Hymenoptera. Diapause in the ectoparasitoid *Nasonia vitripennis* is perhaps the most thoroughly studied larval diapause among the Hymenoptera, prompted by several key early papers (Schneiderman and Horwitz 1958, Saunders 1965). More recent studies on the fifth-instar larval diapause of the alfalfa leafcutting bee *Megachile rotundata* adds this economically important species to the list of well-studied examples of larval diapause among the Hymenoptera (Yocum et al. 2006).

1.2.3 Pupal Diapause

Pupal diapause, best known for Lepidoptera and Diptera, is also noted in a few Hymenoptera but appears to be completely absent in Coleoptera, the largest insect order. The arrest usually occurs in the true pupal stage, that is, before the onset of adult differentiation. However, in a few species of moths (Sahota et al. 1982, 1985, Monro 1972), diapause occurs in pharate adults, that is, when adult development has been completed but the insect is still within the pupal cuticle. Thus, development can be halted either before differentiation begins, the most common stopping point, or after it is completed, but not at some midpoint between these two extremes. This is in marked contrast to embryonic diapauses that can intercede at nearly any point during the course of embryonic development.

Pioneering studies on the saturniid moth *H. cecropia* (Williams 1946), the tobacco hornworm *Manduca sexta* (Rabb 1966), the flesh fly *Sarcophaga argyrostoma*

(Fraenkel and Hsiao 1968), and fruit flies in the genus *Rhagoletis* (Bush 1969) provide the backdrop for much recent work on pupal diapause and for understanding seasonal patterns of distribution. Experiments using species from the agriculturally important *Heliothis/Helicoverpa* complex of moths add a more recent model for understanding pupal diapause (Meola and Adkisson 1977, Xu and Denlinger 2003).

1.2.4   Adult Diapause

Adult diapause is characterized by the arrested development of the ovaries, testes, accessory glands and related reproductive structures, suppressed feeding, locomotor and mating behavior, and in some cases, degeneration of the flight muscles. Adult diapause is common among Coleoptera, Lepidoptera, Diptera, Hemiptera, Orthoptera, Neuroptera, Trichoptera, Thysanoptera, as well as the Acarina. It is most frequently manifested in newly emerged adults that are not yet reproductively active, and characteristically the adults remain reproductively inactive until diapause is terminated. But, there are exceptions where diapause is entered following a bout of reproductive activity. For example, the leaf beetle *Diorhabda elongata* when transferred from long days to short days, can switch to diapause after a period of egg-laying (Bean et al. 2007a), and before entering summer diapause, females of the Colorado potato beetle *Leptinotarsa decemlineata* may lay a few eggs (Tauber et al. 1988). As discussed in Section 3.2, a few species exit and enter adult diapause multiple times.

A short period of **feeding** by newly emerged adults is sometimes essential for entry into diapause. Without feeding the twospotted spider mite *Tetranychus urticae* fails to assume the bright orange coloration of diapause and fewer individuals enter diapause (Kawaguchi et al. 2016). Although feeding may continue in some diapausing adults, feeding intensity is greatly suppressed, as reported for several species of *Drosophila* (Matsunaga et al. 1995), the blow fly *Phormia regina* (Stoffolano 1975), and the ladybird beetle *Harmonia axyridis* (Gao et al. 2019), among others. In *P. regina*, like other flies, the proboscis is extended when tarsi of the adult detect a food source. Interestingly, the threshold for this tarsal response differs little between diapausing and nondiapausing flies, suggesting that the feeding inhibition during diapause is controlled by the central nervous system rather than the peripheral system (Stoffolano 1975). Distinctions may also occur in the food source used by diapausing and nondiapausing adults. This is especially evident in the northern house mosquito *Culex pipiens*. While nondiapausing females are avid blood-feeders, females programmed for diapause feed exclusively on nectar (Bowen 1992). Only at the completion of diapause do females regain an interest in seeking a blood meal from their avian hosts (Faraji and Gaugler 2015).

Diapause in males has been examined less extensively than in females (Pener 1992), but in many cases, the same sort of reproductive shut-down is seen in both sexes (Kubrak et al. 2016, Urbanová et al. 2016, Ala-Honkola et al. 2018). **Mating behavior** is curtailed. Accessory glands remain undeveloped. Spermatogenesis ceases and most undifferentiated sperm degenerate. The only cells within the testes that appear to be affected by diapause in larvae of the waxworm *Galleria mellonella* are spermatocytes in proximal regions of the testicular follicles, cells that undergo apoptotic degeneration (Bebas et al. 2018). In diapausing males of the leaf beetle *Gastrophysa atrocyanea*, the testes slowly increase in size during the first two to three months of diapause and then regress and remain small for the remaining three to four months of diapause (Ojima et al. 2015). During diapause in the nymphalid butterfly *Polygonia c-aureum*, and presumably in other Lepidoptera that characteristically produce both nucleated eupyrene sperm (fertile) and anucleated apyrene sperm (infertile), production of both types of sperm is synchronously reinitiated at diapause termination (Hiroyoshi et al. 2017).

Diapausing females are characteristically not attractive to either diapausing or nondiapausing males, and diapausing males usually will not attempt to mate with sexually attractive females. Males of *Drosophila montana* that are reproductively active will court reproducing, post-diapause females but completely ignore diapausing females (Ala-Honkola et al. 2018). The same is true for *D. melanogaster* (Kubrak et al. 2016). Diapausing males of the grasshopper *Anacridium aegyptium* fail to display mating behavior until several months after reaching the adult stage (Greenfield and Pener 1992). Late in diapause, males of the fungus beetle *Stenotarsus rotundus* will attempt to mate if they are physically handled, but mating is not normally observed until the environmental signal for diapause termination has been received (Tanaka et al. 1987b).

Foundational experiments on the Colorado potato beetle *L. decemlineata* (de Wilde and Bonga 1958), the grasshopper *Oedipoda miniata* (Pener and Broza 1971), linden bug *P. apterus* (Hodek 1968), lacewing *Chrysopa carnea* (Tauber and Tauber 1970a,b), and mosquito *Culex pipiens* (Spielman and Wong 1973) offer a rich launching point for more recent studies on adult diapause. Though *D. melanogaster* has a rather weak adult dormancy (Saunders et al. 1989), probably most correctly termed a quiescence (Section 1.1.3), the powerful genetic tools available for this model species have prompted a surge of recent interest in this species, as well as in other members of the genus.

In some cases, **only adult females** enter diapause. Males of the northern house mosquito *Culex pipiens* inseminate females in the autumn and then die without entering diapause, leaving only inseminated, diapausing females, to survive the winter (Spielman 1964). Similarly, only females of the twospotted spider mite *Tetranychus urticae* (Veerman 1985, Suzuki and Takeda 2009), anthocorid bugs in the genus *Orius* (Musolin and Ito 2008), eastern yellow-jacket *Vespula maculifrons* (Kovacs and Goodisman 2012), bumble bees *Bombus impatiens* (Amsalem et al. 2015), and *B. terrestris* (Colgan et al. 2019), among others, overwinter in diapause, after mating in the autumn. In the autumn, overwintering females of the hover fly *Episyrphus balteatus* accumulate impressive fat reserves, while males fail to do so and all males succumb by the end of December, leaving only the inseminated females to survive until spring in northern Germany (Hondelmann and Poehling 2007). In tropical Australia, only females of the nymphalid butterfly *Hypolimnas bolina* appear to use diapause to bridge the dry season (Pieloor and Seymour 2001). Males of this species survive the dry season and have sperm present throughout the year, suggesting they are consistently able to take advantage of unpredictable female activity.

In other species such as the ladybird beetle *Coccinella septempunctata* (Hodek 2012), the Colorado potato beetle *L. decemlineata* (de Wilde et al. 1959), and the handsome fungus beetle *S. rotundus* (Wolda and Denlinger 1984) both sexes diapause and mating occurs only when diapause is terminated. But, there are exceptions. A high proportion (40–60%) of central European populations of *C. septempunctata* mate prior to entering diapause, while *Ceratomegilla undecimnotata*, a ladybird beetle that hibernates in the same area, delays mating until spring (Hodek and Ceryngier 2000). Most females of the linden bug *P. apterus* mate in the spring after diapause has been completed, but a few females (up to 7% in South Bohemia) mate prior to overwintering and retain viable sperm throughout the winter (Socha 2010). Similarly, both sexes of the kudzu bug *Megacopta cribraria* overwinter in adult diapause, and most females delay mating until spring, but approximately 15% of overwintering females mate before the onset of diapause and thus store sperm for up to seven months (Golec and Hu 2015). Female dung flies (family Sepsidae) store sperm over winter, but fertility increases when they also have an opportunity to mate in the spring (Zeender et al. 2019). It seems unlikely there are cases in which only adult males diapause.

The fate and survival of stored sperm within diapausing females have received little attention in insects. However, in the marine copepod *Neocalanus flemingeri*, sperm within fertilized females remain quiescent during diapause and complete maturation inside the female, as indicated by upregulation of spermatogenesis toward the end of diapause (Roncalli et al. 2018). Similar scenarios may operate in insects, but such results have not been documented.

Mating prior to diapause sometimes alters the overwintering female's physiology through unexpected mechanisms. Pre-diapause mating boosts the abundance of anti-microbial hemolymph peptides in the buff-tailed bumble bee *B. terrestris*, an effect that is sustained throughout diapause and likely enhances survival of the diapausing queen (Colgan et al. 2019).

Little is known about the physiology of males that die rather than entering diapause. Do the environmental factors that trigger diapause in females also impact male physiology? Sperm produced by the male are transferred to the female's spermatheca in the autumn and remained stored there until the following spring. Do they have special properties or are they indistinguishable from sperm produced by their nondiapausing male counterparts that mature under spring or summer conditions?

1.3   Do Colonies of Social Insects Diapause?

Overwintering survival of many social insects is vested in a queen that overwinters and restarts the colony again in the spring, but many social insects including honey bees, termites, and ants maintain their colony structure throughout the winter. Are these overwintering colonies in diapause? The literature usually avoids this terminology, and instead opts for terms such as **summer bees and winter bees**. Yet, when viewed as a superorganism, the winter phases of colony life share many attributes of diapause.

Ant colonies that Kipyatkov and his colleagues (e.g., Kipyatkov and Lopatina 1999, Kipyatkov 2006) have studied in northern Russia are considered to be in diapause during the northern winter. Like most diapausing insects, ants in several genera have been documented to respond to short daylength by entering a nonreproductive phase. In some species, only adults and workers are capable of entering diapause (tribe Formicini), but in several other species, larvae also have the capacity for diapause. Which larval stage is used for diapause varies with genus: Diapause occurs in early instars (1–3) in *Lepisiota, Plagiolepis, Tapinoma*, and some *Camponotus*, in middle instars (2–4) in some *Camponotus*, in late instars (3–4) in *Harpagoxenus, Leptothorax, Temnothorax* and *Messor*, and in the final larval instar in *Manica, Diplorhoptrum, Leptanilla, Monomorium, Myrmica* and *Tetramorium*, and in all six instars in *Aphaenogaster, Crematogaster, Lasius, Paratrechina,* and some species of *Camponotus*. In southern Russia, larvae overwinter only once, but as the growing season shortens in the north, more and more larvae overwinter twice or more. In many of these taxa, overwintering larvae give rise to alates (winged adults that are reproductively active) that emerge in the spring to establish new colonies. Few physiological properties have been evaluated in these overwintering colonies, but the cessation of reproduction, the response to daylength, and the stage specificity of the overwintering stage share many of the attributes one expects to see in diapause. In the fire ant, *Solenopsis invicta*, a species recently introduced into the southern portion of the United States, clear behavioral and physiological attributes distinguish summer and winter colonies (Tschinkel 1993, Cook et al. 2016): Summer colonies have faster turnover rates, higher rates of metabolism, and nutrient utilization patterns that are distinct from patterns noted in winter colonies. What prompts these seasonal shifts, however, remains undefined and can be challenging to uncover in this and other social species.

In **honey bees** (*Apis mellifera*), winter workers tend the queen and continue to regulate hive temperatures, but they remain rather idle, they no longer forage and have reduced duties in brood rearing. There is a striking difference in their longevity compared to summer bees. Although summer bees live for 15–38 days, the lifespans of winter bees average 140 days, a tenfold increase over summer bees (Winston 1987). In addition, winter bees have a more extensively developed fat body than their summer counterparts and display a more robust immune response (Dostálková et al. 2021). Brood-rearing activity declines in response to short days and is accelerated by long days (Kefuss 1978). Thus, the winter bee is in many ways similar to other insects in diapause. The transcriptional profile of winter bees, based on 14 candidate genes, differs from that of summer bees: Abdominal tissues resemble profiles of nurse bees and thoracic tissues resemble foragers (Bresnahan et al. 2021), thus the winter bee appears to be a distinct phenotype. Thermoregulatory activities that invoke the use of thoracic flight muscles in winter bees are presumably the functional link with summer-active forager bees. A full transcriptomic analysis will be useful in searching more broadly for expression patterns that may be shared with other insects in diapause, especially bumble bee queens and other Hymenoptera that overwinter in diapause. It is unclear which seasonal environmental features are responsible for the switch to the production of winter bees. Whether the winter bee should be considered to be in diapause is a semantic issue. Although it is reasonable to view winter bees as being in diapause, the more important issue is to at least recognize that winter bees and diapausing insects do share some common attributes.

The same can be said for many species of **termites.** Workers of the eastern subterranean termite *Reticulitermes flavipes* survive low temperatures better when subjected to a decreasing thermoperiod and photoperiod prior to the onset of winter, a response that appears to drive the workers to a greater depth where temperatures are not as severe (Cabrera and Kamble 2001). Activity levels remain low during winter, but workers do not appear to sequester additional fat reserves. During the dry season in the African savanna, termite densities are low and activity is suppressed (Davies et al. 2015), but when the rains arrive, the next generation of reproductives, the alates, emerge from the ground in enormous numbers to initiate the next cycle of nest founding and reproduction. The pause in development that precedes the emergence of reproductive adults is in many ways akin to post-diapause quiescence. The alates are fully capable of progressing to the next phase of their life cycle, but they fail to do so until the proper environmental stimulus (rain) arrives. Again, termite researchers do not refer to this waiting period as diapause and post-diapause quiescence, but this life cycle does indeed share features in common with diapause.

1.4   Phases of Diapause

The impact of diapause on the insect life cycle is profound, extending from the perception of environmental signals used to program diapause, through diapause itself, to alteration of traits manifested after diapause has ended. In some cases the influence is also transgenerational, initiated by the parents and extending to subsequent generations. It is this pervasive impact of diapause across a broad temporal and developmental scale that justifies identifying diapause as an **alternative life cycle**, a syndrome that encompasses more than just the diapausing stage.

Although numerous terms have been used to define different phases of diapause in the older literature, the terminology championed by Koštál (2006) and articulated also by others (e.g., Tauber and Tauber 1976, Tauber et al. 1986, Danks 1987, Denlinger 2002) broadly divides diapause into three phases: pre-diapause, diapause, and post-diapause (Figure 1.3).

Pre-diapause includes the environmental programming events (also known as diapause induction) as well as the preparative steps that the insect undergoes prior to diapause. These two phases of pre-diapause are usually separate in time but may overlap, depending on the species. Diapause itself encompasses the total period when the insect is refractory to the progression of development, even though favorable environmental conditions are present. Diapause can be divided into three distinct phases, including **diapause initiation** (or onset), **maintenance**, and **termination**. It is during the initiation phase that development ceases or dramatically slows, and the metabolic rate usually drops, although some feeding and accumulation of energy

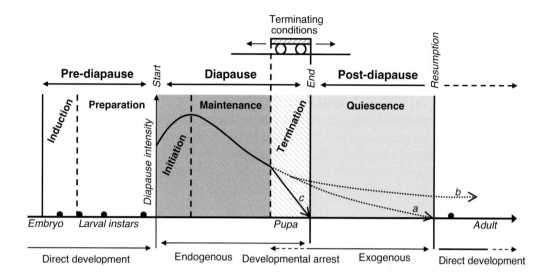

Figure 1.3 Schematic depiction of three major phases of diapause (pre-diapause, diapause, post-diapause) and finer subphases including environmental induction, preparation, initiation, maintenance, termination, and post-diapause quiescence. Changes in diapause intensity are represented by dotted lines (a and b) under two hypothetical constant conditions, while the solid line (c) indicates a change in environmental conditions that favors diapause termination. From Koštál (2006), with permission from Elsevier.

stores may continue. The intensity of diapause usually increases during this period, that is, it becomes increasingly more difficult to reverse the diapause decision by external stimuli.

The period of diapause maintenance is the longest, extending from diapause initiation to diapause termination. It is during this period that the insect gradually acquires the capacity to bring diapause to an end. Diapause termination marks the time when the actual diapause period is completed and the insect can resume development if the correct conditions are present. This point marks a rapid developmental transition defining the end of diapause. Diapause termination usually occurs long before the insect actually resumes development.

Post-diapause includes all developmental events that ensue following diapause termination. **Post-diapause quiescence** refers to the period between diapause termination and the actual resumption of development, an interval that frequently encompasses many months in temperate latitudes. For example, in northern latitudes, diapause may be terminated in early winter, but development may not be reinitiated until the arrival of warm days in spring because cold conditions in late winter are not permissive for development. The distinction between diapause and post-diapause quiescence is frequently overlooked because the two phases are morphologically indistinguishable. The only obvious distinction is whether the insect is capable of reinitiating development if favorable conditions are present. However, from a developmental perspective and for projecting the timing of spring emergence, the distinction between diapause and post-diapause quiescence is extremely meaningful. Following a period of post-diapause quiescence, development resumes when favorable conditions are present, and the ensuing development follows a developmental trajectory that usually closely matches that of its nondiapausing counterparts. It is not uncommon for post-diapausing insects to exhibit fitness consequences that result from having experienced diapause.

In subsequent chapters, I use this framework to discuss more fully the environmental programming events involved in diapause induction (Chapter 5), preparation (Chapter 6), diapause (Chapter 7), termination of diapause, and resumption of development (Chapter 8), as well as consequences of diapause (Chapter 4).

2   Which Seasons Are Being Avoided?

Seasonal patterns of temperature and rainfall, largely dictated by geography, define annual periods suitable for continuous insect growth and development and conversely define times that are unsuitable or suboptimal. Winter has received the most attention as the season for diapause, but summer diapauses are also well documented, and diapause is common in the tropics as well. Receiving less attention are the dormancy patterns operating at high latitudes where seasonal patterns are most extreme. This chapter introduces distinctions noted in diapauses used for different seasons. **Aestivation** is commonly used to describe dormancy that bridges dry seasons, but the aestivation literature often does not distinguish the programmed diapause response from the nonprogrammed response of quiescence. Aestivation can be found in both summer diapauses that occur in temperate latitudes and tropical diapauses, and not all summer and tropical diapauses are driven by dry seasons. I have thus elected to incorporate the aestivation literature within the categories of summer and tropical diapause, rather than as a distinct category.

2.1   Winter Diapause

Winter is an obvious obstacle for most insects living in temperate latitudes, and indeed the overwhelming body of diapause literature focuses on diapause that is used to bridge this unfavorable season. This book also reflects that popular winter bias, and hence features of winter diapause, from the ecological to the molecular, prevail throughout the text and will be given less specific attention in this section.

Insect performance curves plotting activity, growth, development, or reproduction as a function of temperature clearly show optimal temperatures, as well as the extremes that limit activity. Some insects have performance curves shifted to maximize activity at low temperatures, but in most cases, optimal activity curves do not include the sorts of temperatures typically encountered during the winter months. As ectotherms with limited ability to physiologically control their body temperature, insects are forced to escape in time or space, and usually, this means becoming dormant to circumvent the impact of winter. Diapause by itself does not imply cold hardiness as will be discussed later (Section 7.13.6), but it brings a halt to development and sets the stage for evoking cold hardening mechanisms.

But, low temperatures are likely only one of the driving evolutionary forces leading to diapause in temperate latitudes. The onset of winter brings with it the demise of host plants exploited by insects for food. Without a food source, insects cannot continue to develop and reproduce. The lack of host plants has a cascading effect, also impacting the life cycles of insect predators and parasitoids. When the food resource is gone, the insect must adjust its seasonal cycle to accommodate the lack of food. For most insects that confront winter, diapause is the answer. The active phase of the life cycle coincides with the favorable summer season (Figure 2.1A and B).

As discussed in Chapter 5, the overwhelming majority of insects that enter a facultative winter diapause do so in response to short daylengths that prevail in late summer and early autumn, a response that is frequently enhanced by low temperatures. Yet, as reported later, diapause at other times of the year and in regions beyond temperate latitudes is common, albeit understudied.

2.2   Summer Diapause

It may seem a bit counterintuitive that an insect would enter diapause in summer, a season when temperatures are clearly favorable for development, yet the presence of summer diapause underscores the fact that diapause is not just about winter survival but has much to do with the synchrony of insects with their food sources, as Masaki (1980) and Saulich and Musolin (2018) nicely outline in their reviews of summer diapause. Escape from parasitoids can be another good reason to avoid activity during the summer, as proposed for the white butterfly *Pieris brassicae* (Spieth et al. 2011). In the early literature, summer diapause was thought to be restricted to arid or Mediterranean climates, but Masaki's elegant review makes it clear that summer diapause is widely distributed geographically, from Finland, Alaska, and Canada to the lower latitudes and has been documented in nearly all major insect orders and mites (Ridsdill–Smith et al. 2005).

Many insects that overwinter in diapause enter diapause in mid-to-late summer (Figure 2.1B) or early autumn (Figure 2.1A), but this situation is not referred to as a summer diapause because that diapause persists through the winter. By contrast, a summer diapause is frequently terminated in the cooling days of autumn (Figure 2.1C). Summer diapause, which occurs during the hottest time of the year, can also be distinguished from tropical diapause that commonly is used at low latitudes to coordinate activity with predictable alternation between seasonal dry and wet seasons.

A common pattern of summer diapause is for the insect to be **active in spring**, enabling it to feed on buds, flowers, young shoots and leaves, or to prey on other insects if it is a predatory species and then become dormant during the summer. As temperatures drop and/or photoperiod declines in the autumn, summer diapause is commonly broken and the insect again becomes active (Figure 2.1C). The reproductive phase may occur either before or after summer diapause or both times. The insect may then enter a winter diapause, often in a different developmental stage, or remain

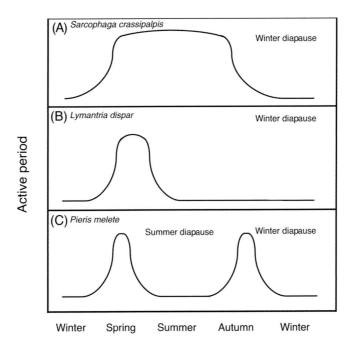

Figure 2.1 Schematic representation of the active (nondiapause) life phases of (A) an insect that overwinters in diapause and is active during most of the summer (e.g., *Sarcophaga crassipalpis*); (B) an insect that is active for a short interval in early summer but is in diapause for the remainder of the year (e.g., *Lymantria dispar*); and (C) an insect that is active in spring and autumn but enters both a summer and winter diapause (e.g., *Pieris melete*).

active during the winter. For example, the bug *Poecilocoris lewisi* enters an overwintering diapause as a fifth-instar nymph in response to short daylengths but then responds to long daylength in the spring by entering a summer adult diapause. This summer diapause and delay in adult activity is used to synchronize adult activity with the July appearance of seeds from its dogwood host *Cornus controversa* (Tanaka et al. 2002). In Greece, adults of the ladybird beetle *Hippodamia undecimnotata* feed voraciously on aphids in the spring and early summer in the lowlands, but in June, they head for the mountains where they enter a summer diapause in large aggregations (Katsoyannos et al. 2005). The beetles display classic attributes of diapause, failing to reproduce for several months, even when provided an optimal environment. Though they remain in the mountains throughout the winter, by late autumn they are capable of reproducing when collected and returned to optimal conditions in the lowlands. Thus, what starts as a summer diapause transitions into post-diapause quiescence during autumn, and the beetles remain in that stage until they return to the lowlands in the spring and again become reproductively active.

Environmental cues programming summer diapause are frequently a mirror image of cues used to program winter diapause, as illustrated for summer and winter pupal diapauses of the noctuid *Mamestra brassicae* in Figure 2.2 (Grüner and Masaki 1994). Rather than using short days and low temperatures to enter diapause, common signals for winter diapause, insects entering summer diapause frequently do so in response to long days and high temperatures. Interestingly, critical daylengths for both summer and winter diapauses in *P. lewisi* are between 14 and 15 hours of light/day, but the winter nymphal diapause requires daylengths shorter than 14–15 hours for diapause induction, while the summer adult diapause requires daylengths greater than 14–15 hours (Tanaka et al. 2002). In *M. brassicae*, winter diapause is induced by a range of daylengths shorter than 13 hours, while summer diapause is induced by daylengths that exceed 14.5 hours (Grüner and Masaki 1994). Nondiapausing pupae are generated only within the narrow range between 13 and 14.5 hours of daylength (Figure 2.2). A similar response is observed in the cabbage butterfly *Pieris melete*: Long daylengths prompt a summer pupal diapause, while short daylengths lead to a winter pupal diapause, and most pupae develop without diapause at intermediate daylengths between 11.5 and 13.5 hours of light (Xue et al. 1997, Xiao et al. 2008a). In the moth *Pseudopidorus fasciata*, an overwintering prepupal diapause is triggered by daylengths exceeding 14 hours of light, but a small portion of the population enters summer diapause during the fourth instar in response to daylengths shorter than 13.5 hours (Wu et al. 2006). In the leaf beetle *Phaedon brassicae*, long days result in an adult summer diapause, while short days elicit an adult winter diapause (Wang et al. 2007a). The winter embryonic diapause of the stink bug *Picromerus bidens* is obligate, but summer adult diapause in this species is facultative and relies on the long-day signals of early summer (Musolin and Saulich 2000). Daylength may impact not only the decision to enter summer diapause but also diapause duration, as in the Mediterranean tiger moth *Cymbalophora pudica* (Koštál and Hodek 1997). This species enters summer diapause under both long and short daylengths, but switching from one daylength to the other either shortens or lengthens diapause duration, demonstrating that the duration of diapause is responsive to change in daylength, rather than absolute daylength.

In the cabbage beetle *Colaphellus bowringi*, photoperiod seems not to be particularly important (Xue et al. 2002): Low temperatures in early spring induce summer diapause, high temperatures of summer promote nondiapause, and low temperatures of autumn induce winter diapause. The onion maggot *Delia antiqua* enters summer diapause in response to temperature ($>24°C$), but its winter diapause is evoked by both temperature ($<18°C$) and photoperiod (Ishikawa et al. 2000). Long daylength, coupled with high temperature, prompts summer diapause in the burying beetle *Nicrophorus quadripunctatus*, but high temperature by itself will also bring about a reproductive halt (Nisimura et al. 2002). Summer diapause in larvae of the hemlock woolly adelgid *Adelges tsugae* also seems to be evoked strictly by high temperature (Sussky and Elkinton 2015).

A single species may use summer diapause in some portions of its range and winter diapause in others. In Spain and Northern Africa, populations of *Pieris brassicae* respond to long daylengths by entering a summer pupal diapause, while relying on short daylength and low temperature to program diapause at higher latitudes of France and Denmark (Held and Spieth 1999, Spieth 2002). Likewise, within Japan, adults of

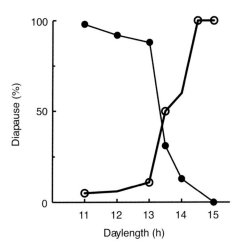

Figure 2.2 Photoperiodic response curves for induction of winter diapause (solid circles) and summer diapause (open circles) in pupae of a southern Japanese strain of *Mamestra brassicae*. Adapted from Grüner and Masaki (1994), with permission from Kluwer Academic.

the coccinellid beetle *Epilachna admirabilis* have a summer diapause in warm Kyoto (35° N) but not in much cooler Sapporo (43° N) (Imai 2004). Adults of the coccinellid *Coccinella septempunctata* fail to enter summer diapause in the north of Japan but do so in central and southern Japan, although the incidence of summer diapause varies not only geographically but annually as well, presumably a consequence of annual temperature variation; the incidence of summer diapause is high when July temperatures are high (Ohashi et al. 2003). As these few examples suggest, the incidence of summer diapause does indeed seem to be greater at lower, warmer latitudes, but numerous examples demonstrate summer diapause at high latitudes as well.

Although limitations of a **seasonal food resource** are likely the most common driving force for the evolution of summer diapause, related features of competition for a dwindling food source may also contribute. Experiments by Tougeron et al. (2018, 2019) suggest that increased competition between the braconid parasitoids *Aphidius avenae* and *A. rhopalosiphi* affects diapause incidence in the parasitoid's progeny. Reducing the abundance of host aphids does not induce the summer diapause, but increasing parasitoid abundance leads to a boost in summer diapause incidence. How such a fascinating mechanism works remains unknown. In populations of the white butterfly *Pieris brassicae* from the southern Iberian Peninsula, summer pupal diapause is possibly driven by an evolutionary attempt to escape parasitism (Spieth and Schwarzer 2001). While nearly all individuals of *P. brassicae* enter a three- to four-month diapause in the summer, the major parasitoid that attacks this butterfly, *Cotesia glomerata*, lacks the capacity for summer diapause and is forced to switch to less suitable hosts for the summer months, thus diminishing its reproductive success.

Physiological features of summer diapause have not been examined as extensively as winter diapause, but there seem to be no major differences. Body mass and energy resources such as lipid and glycogen reserves are commonly elevated in both summer and winter diapause, and metabolic activity is depressed in both cases (Ding et al. 2003, Liu et al. 2006, 2016, Weyda et al. 2015, Saulich and Musolin 2018). Net transpiration rates and critical transition temperatures, a reflection of cuticular hydrocarbon composition, are nearly identical for summer and winter diapauses in pupae of the Hessian fly *Mayetiola destructor* but differ considerably from values noted in nondiapausing fly pupae (Benoit et al. 2010b). Larvae of the stem borer *Chilo partellus* have a winter diapause in northern India but a summer diapause in southern India; one conspicuous difference in these larvae is that those in summer diapause are heavier than those in winter diapause (Dhillon et al. 2020). Another distinction apparent in some species such as the cabbage armyworm *M. brassicae*, is that pupae in winter diapause have a greater capacity for cold tolerance than their counterparts in summer diapause, a feature linked to higher trehalose levels evoked by winter diapause (Goto et al. 2001, Ding et al. 2003). Both summer and winter diapauses appear to have the same endocrine basis, as outlined in Chapter 9.

Molecular signatures of diapause in the onion maggot *Delia antiqua*, such as an elevated expression of genes encoding heat shock proteins, are evident in both summer and winter diapauses (Chen et al. 2005b, 2006, Ren et al. 2018). Genes involved in lipid metabolism, immune responses, and cytoskeletal modifications are also differentially expressed in both types of diapause in this species, but certain transcripts are more highly expressed in one or the other: Genes associated with lipid metabolism are more highly upregulated in winter diapause, while genes associated with immune responses are more highly upregulated in the summer diapause of *D. antiqua* (Hao et al. 2012). Though the transcript encoding a trypsin-like enzyme is upregulated in both summer and winter diapauses in *D. antiqua*, the level of upregulation is greater in winter diapause (Chen et al. 2005a). To date, *D. antiqua* is the only species in which both summer and winter pupal diapauses have been examined at the molecular level, and thus it is too early to be certain if differences noted in this species are common to other species as well. The fact that different temperatures are required to trigger summer and winter diapauses in *D. antiqua* makes it challenging to determine whether the molecular responses are due to temperature or the diapause program (Ren et al. 2018). Summer diapauses in adults of the chrysomelid beetle *Galeruca daurica* share many of the proteomic (Ma et al. 2019), microRNA (Duan et al. 2021), and endocrine (Ma et al. 2021) signatures seen in winter diapauses, suggesting that at the molecular level there is little to distinguish a summer diapause from a winter diapause.

2.3   Tropical Diapause

The common assumption that insects living in the tropics do not diapause is not well founded. A compilation of tropical insects with a diapause or at least some form of dormant state includes 73 species, spanning all major insect orders and representing diapause in all life stages (Denlinger 1986). A number of examples, primarily Hymenoptera from Hawaii, are also cited in a review (Nishida 1955). Since then, diapause in additional tropical species has been noted, as recorded for examples in Table 2.1.

Additional examples are abundant from subtropical areas, such as larvae of the braconid *Microplitis demolitor* (Seymour and Jones 2000) and adults of *Drosophila ananassae* (Lambhod et al. 2017). Circumstantial evidence for diapause can also be gleaned from natural history observations. For example, huge numbers of insects, representing 33 families in diverse orders, are reported during the dry season in ball moss, an epiphyte in the dry forest of Central Mexico (Luna–Cozar et al. 2020). The fact that most of the insects in the moss are inactive and in non-feeding stages suggests they are in diapause and are using the site as a dry season refuge. There is every reason to believe that diapause is widespread within the tropical and subtropical world, although some species clearly lack this capacity, as noted in a brief, but not exhaustive, listing of 16 tropical species that appear to lack a diapause response (Denlinger 1986).

Temperature variation, the conspicuous driver of seasonality in temperate latitudes, is evident in many tropical environments but is considerably more muted than at higher latitudes. Alternating **rainy and dry seasons** dictate the most obvious seasonal environmental pattern. Rainy seasons alternate with one or two dry seasons each year, depending on locality. Seasonal patterns of rainfall and other environmental variables range widely within the tropics, generating a variety of seasonal patterns of insect abundance in different tropical regions (Wolda 1988). Seasonal patterns of insects collected in traps have been well documented in the tropics, including Brazil (Davis 1945), Panama (Wolda 1978), and Kenya (Denlinger 1980). Pronounced **seasonality** is evident for most taxa examined. For example, a five-year study in the Nairobi National Park, Kenya (Denlinger 1980) reveals a huge single peak abundance of lacewings (Chrysopidae) each August, a peak of psocids (Psocidae) in November and December, a peak of horse flies (Tabanidae) in May, and bimodal patterns of abundance for robber flies (Asilidae) in November and April and wood nymph butterflies (Satyridae) in December and May. Nairobi annually experiences two rainy and two dry seasons, but the insect distribution pattern noted in Malaise trap catches indicates there is no single period universally favored by all taxa. Some are abundant in the rainy season, others in the dry season. A striking peak in abundance of adults of the seed bug *Jadera aeola* appears late in the dry season on Barro Colorado Island, Panama, but the bugs are nearly absent for the remainder of the year (Tanaka et al. 1987c). Even in regions of nearly constant temperature and rainfall, such as Fortuna, Panama, two species of dobsonflies (*Chloronia* sp. and *Platyneuromus* sp.) are highly seasonal, emerging during a two-month window each year (Wolda 1988).

Seasonal differences of insect abundance do not, of course, immediately imply diapause during the nadir of seasonal abundance. Tropical species such as the tsetse flies *Glossina morsitans* and *G. swynnertoni* display highly seasonal abundance patterns in trap catches in Tanzania, with peak abundance in July (Nnko 2017), yet,

Table 2.1 Recent examples of diapause in tropical insect species.

Species	Stage	Latitude	Country	Reference
Orthoptera				
Stictophaula armata	Embryo	19° N	Thailand	Ingrisch (1996)
Oedaleus senegalensis	Embryo	17° N	Mali	Gehrken and Doumbia (1996)
Hemiptera				
Deois flavopicta	Embryo	15° S	Brazil	Pires et al. (2000)
Encosternum delegorguei	Adult	22° S	S. Africa	Dzerefos et al. (2015)
Coleoptera				
Pachymerus nucleorum	Embryo	10° S	Brazil	Benton (2015)
Bruchidius atrolineatus	Adult	9° N	Togo	Glitho et al. (1996)
Callosobruchus rhodesianus	Adult	9° N	Togo	Amevoin et al. (2005)
C. subinnotatus	Adult	9° N	Ghana	Appleby and Credland (2007)
Lepidoptera				
Omphisa fuscidentalis	Embryo	19° N	Thailand	Singtripop et al. (1999)
Chlosyne lacinia	Embryo	16° S	Brazil	Moreira et al. (2012)
Coniesta ignefusalis	Embryo	11° N	Ghana	Tanzubil et al. (2002)
Papilio demoleus	Pupa	15° N	India	Singh (1993)
P. polytes	Pupa	15° N	India	Singh (1993)
Hypolimnas bolina	Adult	17° S	Australia	Kemp and Jones (2001)
Diptera				
Procontarinia mangiferae	Embryo	21° S	Réunion Isl.	Amouroux et al. (2014)
Anopheles coluzzii	Adult	14° N	Mali	Dao et al. (2014)
Hymenoptera				
Eurytoma sp.	Embryo	20° S	Brazil	Barbosa and Fernandes (2019)

despite my own extensive experimental attempts to find diapause in tsetse, there is no evidence for any such arrest. High seasonal mortality most likely explains the decline in tsetse populations during dry seasons. Local and long-distance **migration from drier to wetter sites** are presumed to account for many seasonal patterns of tropical insect abundance (Janzen and Schoener 1968), including the dramatic seasonal cycles noted in the malaria mosquitoes *Anopheles arabiensis* and *An. gambiae sensu stricto* in the Sahel (Dao et al. 2014). Remarkably, the dry season in the Sahelian region of Senegal generates extreme fluctuations in populations of *An. arabiensis*, yet there is no evidence for a dry season bottleneck, indicating that the mosquito maintains a large permanent deme spread across a broad geographic area (Simard et al. 2000). Likewise, the eastern highlands and coastal regions of Kenya appear to provide habitat for continuous breeding of the African armyworm *Spodoptera exempta*, thus offering a dry season refuge from the more seasonal grasslands (Dewhurst et al. 2001). An abundance of the Panamanian frog-biting midge (*Corethrella* spp.) is high during the rainy season but drops during the dry season, but again this seasonal difference does not appear to be linked to diapause but to the abundance of host frogs (Legett et al. 2018). During the rainy season these blood-feeding midges cue in on the calls of an abundant host, the túngara frog, but during the dry season, túngara frogs cease chorusing and midges then switch their response to identifying a less abundant host, the pug-nosed tree frog, a species that calls only during the dry season. However, in addition to differential seasonal mortality, migration, and host shifts, diapause is likely responsible for the seasonal patterns noted in many tropical species.

The striking differences between tropical diapauses and diapauses occurring at higher latitudes are the **environmental regulators** that preside over the diapause decision (see Section 5.14). In brief, the role of photoperiod, so dominant at high latitudes, is frequently supplanted near the equator by environmental regulators that rely on wet/dry season cues, including not only rainfall patterns but the downstream effects of rainfall on plant growth, flowering, seed set, and the nutritional status of food plants, as well as subtle seasonal changes in temperature (Denlinger 1986).

Another attribute of tropical seasons is that they are less predictable than winter in temperate latitudes. Timing of the onset and end of the rainy seasons can vary considerably, and in some years, the rains may not come at all. The inherent lack of seasonal predictability is possibly why only a portion of a population is likely to enter diapause. For example, approximately 18% of the population of the nymphalid butterfly *Hypolimnas bolina* fails to enter adult diapause even under optimal diapause conditions in tropical Australia (Pieloor and Seymour 2001). Similarly, among the sarcophagids from East Africa, low levels of diapause are consistently found in pupae, even when reared under the strongest diapause-inducing conditions (diapause incidences range from 7% to 79% among seven species under the extreme conditions of 12°C) (Denlinger 1979). In Nigeria, a portion of the maize stem borer *Busseola fusca* population enters larval diapause each month regardless of the condition of the maize host plant (Usua 1970).

As with summer diapause, diapause in the tropics requires an impressive mechanism for **suppressing metabolism** at high temperatures, a challenge not faced by insects overwintering at high latitudes. The stage of diapause (pupa) is the same for flesh flies (Sarcophagidae) from equatorial Africa as it is for their relatives living in temperate regions (Denlinger 1974, 1978). Similar levels of metabolic suppression, patterns of oxygen consumption cycles, the influence of water content in their food (higher water content yields higher diapause incidence), hormonal regulation by ecdysteroids and juvenile hormone, and the ability of the same select agents (e.g., hexane, acetone) to artificially break diapause all underscore the fact that the basic physiological properties of diapause in flesh flies from tropical Africa and temperate latitudes are remarkably similar (Denlinger 1979). The big differences in winter and tropical diapauses in flesh flies are the environmental features that characterize diapause: seasons of diapause (winter vs cool African dry season), duration of the dormancy (nine versus one to two months), and the triggers for diapause onset (photoperiod versus low daytime temperatures). Among other species that have been examined at the physiological level, few distinctions between tropical and temperate latitude diapauses are evident, such as tropical adult diapauses in both the beetle *Stenotarsus rotundus* and the seed bug *J. aeola* appear to be the consequence of a shut-down in juvenile hormone production, and both species degenerate their flight muscles and suppress gonadal development (Tanaka et al. 1987c), as noted in adult insects that diapause at temperate latitudes. Diapausing larvae of *C. partellus* and *C. orichalcociliella*, stem borers from the Kenyan coast, show similar levels of metabolic depression and reduction in the rate of the heartbeat as seen in temperate species during diapause (Scheltes 1978). Metabolic depression of approximately 28% is noted in diapausing adults of the Australian tropical butterflies *Euploea core* and *E. sylvester* (Canzano et al. 2006); this is a modest decline but not too different from that of temperate species that diapause as mobile adults. Like many temperate latitude species, diapausing adults of the tropical butterfly *H. bolina* are larger and darker than their nondiapausing counterparts (Kemp and Jones 2001). Enhanced desiccation resistance is associated with embryonic diapause in the grasshopper *Oedaleus senegalensis* from Mali (Gehrken and Doumbia 1996). Citrate synthase activity and other enzymes involved in oxidative metabolism are depressed in diapausing larvae of the tropical butterfly *Chlosyne lacinia* (Moreira et al. 2012), again suggesting that physiological features of tropical diapause, including metabolic suppression, are indistinguishable from features characterizing diapause in temperate species.

At the molecular level, many of the same patterns of gene expression observed for diapausing insects from temperate latitudes are also noted for diapause in tropical species, such as the tropical solitary bee *Tetrapedia diversipes* (Santos et al. 2018), further underscoring the fact that molecular and physiological attributes of diapause are shared in spite of differences in environmental regulation. A shared molecular underpinning for diapause in tropical and temperate species is also indicated by a large number of diapause-associated single nucleotide polymorphisms that are common to populations of *Drosophila melanogaster* from Zambia and populations from temperate latitudes (Erickson et al. 2020).

The more intriguing question is why insects would diapause in the tropics. Drought resistance is likely one of the major driving forces, especially in regions with pronounced dry seasons, but of parallel importance for many insects is the ability to coordinate development with highly seasonal patterns of plant development, growth, fruiting, and seed production. Like many tropical species, the butterfly *H. bolina* breaks diapause in tropical Australia at the onset of the rainy season to exploit new growth of its larval food plant, in this case, *Synedrella nodiflora* (Pieloor and Seymour 2001). Similarly, the onset of the rainy season in Panama terminates diapause and prompts a mating frenzy in the endomychid beetle *S. rotundus* (Wolda and Denlinger 1984, Denlinger 1994), enabling it to disperse from its diapause aggregation and lay eggs, presumably on fungi that appear only at that time of year. Seed production in the Sapindaceae is a dry season event in Panama, and accordingly, the seed bug *J. aeola* breaks its diapause late in the dry season in concordance with the mass production of seeds from its host plant (Tanaka et al. 1987c). The absence of suitable aquatic larval habitats during the dry season in the Sahel likely drives the cycle of adult dormancy in the mosquito *Anopheles coluzzii* (Yaro et al. 2012, Dao et al. 2014). Less obvious is why a portion of flesh fly (Sarcophagidae) populations living near the equator in Africa should enter pupal diapause during July and August, the coolest time of the year. Although temperatures at this high-altitude site (1,700 m) near Nairobi, Kenya are low by African standards (mean temperatures of 16–22°C during those two months), such temperatures are certainly not lethal and would not be assumed to significantly impair development. These flies are carrion and dung feeders, and such resources appear to be equally abundant throughout the year, yet some of the flies enter a short diapause each year at this time (Denlinger 1974, 1978). **Biotic factors** may be more important than abiotic factors in driving seasonality in certain species. Entering diapause to escape predators or parasites, avoiding entomophagous fungi or other pathogens during rainy seasons, avoiding competition, and other biotic features could be just as important driving forces for tropical insects as seasonal features of the abiotic world. Although one can imagine tropical diapause as a relict from the insect's evolutionary past, this seems unlikely. Diapause responses are lost quickly in laboratory cultures, and some species, including flesh flies, are thought to have evolved in the Old World tropics (Denlinger 1979).

Clearly, no single tropical season is bad for all species. Many tropical plants flower and bear fruit during the dry season, and thus provide seasonal resources for insect pollinators, frugivores, and seed predators. Not all features of the rainy season are positive: Rains can wash out stores of nectar and pollen, and destroy nesting and oviposition sites of soil-dwelling insects (Denlinger 1986). Despite the vagaries of tropical seasons and the generally mild conditions that prevail, there are still needs that can be met by diapause. Diapause remains a potent mechanism for tropical insects to synchronize mating, to optimize the timing of their development with a specialized and sometimes highly seasonal food source, and to mount protective measures for avoiding drought or other challenges of the unfavorable season. Despite the adaptive advantages of diapause, mortality can be high during diapause, such as in northern Ghana, mortality toward the end of the dry season exceeds 50% in diapausing larvae of the millet stem borer *Coniesta ignefusalis* (Tanzubil et al. 2002).

The fact that diapause is already present in many tropical species suggests that such species are primed to invade temperate latitudes. What is required is not the evolution of diapause capacity but simply the modification of the signaling system that responds to environmental cues. The sarcophagid flies, with a pupal diapause already intact in tropical Africa, appear to have moved from Africa to temperate regions of Europe, Asia, and North America, a migration that required a shift from temperature as the sole diapause determinant to the reliance of both photoperiod and temperature (Denlinger 1974, 1979). Similarly, as tropical and subtropical populations of the cotton bollworm *Helicoverpa armigera* invaded temperate regions of Japan, they intensified their reliance on both photoperiod and temperature as pupal diapause regulators (Shimizu and Fujisaki 2006), a response also reflected in populations of *D. melanogaster* that have migrated from Africa to Europe (Zonato et al. 2017).

2.4   What Happens at High Latitudes?

The challenges confronting insects during winter in temperate climates are exacerbated at high latitudes (Danks 2004a). Temperatures during winter can be incredibly low for an ectotherm, and even summer temperatures are cool by temperate zone standards. Impressive cold tolerance mechanisms are in place, and thermal thresholds for activity are adjusted downward. Humidity is typically low, with water resources frequently locked in a frozen state, making high latitude sites the equivalent of **polar deserts**. The seasonal cycle of daylength swings dramatically, yielding unusually short days during winter and long periods of continuous or nearly continuous daylight during the summer. The net effect is a brief, and sometimes highly variable, growing season for plant life, accompanied by a short season that can be exploited by insects for their development and reproduction. Despite these harsh conditions, an impressive number of insects thrive in the **Arctic**. As many as 2,200 species of insects and their relatives have been reported above the tree line in North America (Danks 2004a). Survival in such environments requires not only physiological and morphological adaptations to combat the cold, but behavioral responses to capitalize on the limited energy from sunlight, and temporal developmental patterns that allow the insect to maximize the short growing season. Insects living in high alpine regions at lower latitudes are confronted with many of these same constraints (Convey 2010).

These extreme challenges suggest that some form of dormancy is essential. Frequently, the favorable season is simply not sufficiently long to permit completion of the full life cycle, thus several years, with intermittent periods of dormancy, may be required for completion of a single generation. The brevity of the favorable season demands that the insect precisely synchronize its development to fully exploit the summer. But, is it diapause that is used to bridge these unfavorable times, or is it a form of quiescence that simply turns development on and off in direct response to the prevailing conditions? Very few species have actually been subjected to the type of rigorous investigation that enables us to distinguish between these two forms of dormancy. The synchronous response to favorable summer conditions could reflect either diapause or quiescence. Diapause is likely in at least some Arctic species, especially those such as the geometrid moth *Psychophora sabini* that molts each spring to the next instar and then halts development and overwinters at the end of the instar (Danks 2004a). Mitochondria in the lymantrid moth *Gynaephora groenlandica* degrade in winter (Levin et al. 2003), and overwintering larvae show metabolic depression characteristic of diapause (Bennett et al. 1999), but this feature alone does not distinguish diapause from quiescence. Diapauses occurring in larval, pupal, and adult stages have been nicely documented for species of drosophilids that live beyond 65° N (Lumme 1978). The malt fly *Drosophila montana* has been especially well studied (e.g., Tyukmaeva et al. 2011, Salminen et al. 2015). This species, found as far north as 67° N in Finland, exhibits a classic photoperiodically induced adult diapause, albeit a diapause that relies on an extremely long critical daylength (daily light:dark cycle of 19 hours:5 hours) for induction. Populations of the blow fly *Calliphora vicina* that live above the Arctic Circle (Lofoten Islands, northern Norway), retain responsiveness to photoperiod for induction of larval diapause, but some will enter diapause at low temperatures (4–10°C) regardless of photoperiod (Vinogradova and Reznik 2015). A population of the yellow dung fly *Scathophaga stercoraria* from Tromsø, Norway (69.7° N) enters pupal diapause at 12°C and will do so under either a short or long daylength, whereas more southerly populations enter diapause at that temperature only under short daylengths (Scharf et al. 2010).

Antarctica and the sub-Antarctic islands lack the species richness seen in the Arctic. In fact, only one insect species is endemic to the Antarctic continent, a flightless midge *Belgica antarctica* (Lee and Denlinger 2015). A few additional species have more recently invaded the northern tip of the Antarctic Peninsula, and a number of mites, ticks, and Collembola are common in this region and on the sub-Antarctic islands (Convey 2010). Little is known about the actual overwintering mechanisms used by these arthropods, but our experiments with *B. antarctica* are suggestive of diapause. Measurements of metabolic rate show a characteristic drop in metabolism in mid-February (near the end of the Antarctic summer), regardless of temperature, photoperiod, humidity, or other environmental signals to which they are exposed (Spacht et al. 2020). These results suggest an obligate diapause of some sort or, alternatively, that the environmental regulator has yet to be discovered. Laboratory conditions to which the larvae were exposed approximate summer conditions (4°C and long daylength), thus they should be able to continue developing, but they fail to do so. Larvae are capable of initiating metamorphosis and emerging as adults only after a few months at low temperatures have elapsed. A metabolic depression of this sort, occurring in the absence of any obvious environmental cue, is indicative of diapause, as is the synchrony of adult emergence the following summer. During the brief austral summer, the midge is continuously active, seemingly oblivious to the daily environmental cues that normally dictate activity patterns. Although the canonical clock genes are present, the cyclic patterns of expression seen in temperate species are not observed (Kobelkova et al. 2015). This lack of rhythmicity in the midge is similar to summer observations of reindeer (Lu et al. 2010) and Svalbard ptarmigan (Reierth and Stokkan 1998, but see Appenroth et al. 2020) living in the high Arctic.

From the limited data available, it is apparent that diapause does indeed occur in some species living at high latitudes, but dormancies that are more akin to quiescence are also likely to be prevalent. The quick response time for entering and exiting quiescence would seem to be an attractive option for an insect living in an extreme and highly variable environment. The flexibility of being able to rapidly become active when conditions are favorable and then just as quickly become quiescent at any life stage would appear to be a highly adaptive response to exploit the short intervals that favor growth and development at high latitudes.

3 Variation in the Diapause Response

Seldom is the diapause response absolute. This chapter explores variation in the response within and among populations, as well as some of the oddities that have evolved such as repeated diapause and prolonged diapause. Though most insects diapause only once during their life, some species do so repeatedly, and some have the capacity to extend their diapause for multiple years. This type of rich variation inherent in populations provides the grist for selection and thus is highly relevant for evolutionary change and adaptations, allowing a successful invasion of new niches.

Though there is usually, but not always, consistency in the stage used for diapause within a species, there can be a huge variation in the incidence of diapause within and between populations, as well as variation in the timing of diapause onset and termination. Individuals entering prolonged diapause frequently represent only a small portion of the overall population, and some species, such as flesh flies in the Old World tropics, consistently display a low incidence of diapause. Species with an obligate diapause are more uniform in their response, but even here it is not unusual for a few individuals to deviate from the norm by avoiding diapause. Diapause responses that appear fairly uniform under defined laboratory conditions can show considerable variation in the field (Danks 2007, Yocum et al. 2011).

3.1   Variation within and among Populations

Variation is especially pronounced in species with a facultative diapause. The incidence of diapause is rarely 100% under any given set of conditions. For example, the flesh fly *Sarcophaga bullata* reared under natural conditions in central Illinois yields a low incidence of diapause (<2%) in June and July, 20% diapause in August, and 98% diapause in September (Denlinger 1972a). What is striking about this result, and this is similar to what is observed with other species as well, is that a few individuals enter diapause during the peak summer season and, at the other extreme, a few pupae fail to enter diapause in late autumn. Likewise, under carefully controlled laboratory conditions *S. bullata* varies in its response to both photoperiod and temperature (Denlinger 1972b). Although a critical photoperiod of 13.5 hours of daylength can be calculated for the population, a few individuals enter diapause at daily light: dark cycle of 14 hours: 10 hours, while shortening the daylength to LD 13:9 is required to generate the maximum incidence of diapause. At 25°C, approximately 70% of the individuals enter diapause at short daylengths, but the diapause incidence jumps to nearly 100% when flies are reared at 17°C. A similar pattern of variability is noted for a population of blow flies *Calliphora vicina* monitored in the field in St. Petersburg, Russia (Vinogradova and Reznik 2013): A few larvae (20%) enter diapause in early August, but it is late September before nearly all larvae enter diapause. This is reflected in critical daylengths that vary from 16.5 L:D to 12:12 LD for different cohorts from the same location. In a population of the mosquito *Culex tarsalis* monitored in the Sacramento Valley of California, the incidence of adult diapause jumps from 41% in October to 90% in November (Nelms et al. 2013), again demonstrating differences in individual thresholds for diapause induction. The incidence of dormancy in adults of *Drosophila melanogaster* is remarkably variable throughout the season in North America (Erickson et al. 2020). Though the incidence of arrested ovarian development is highest in December and lowest in July, surprisingly high incidences of diapause-like ovaries are noted as well in flies collected during the intervening summer and autumn months.

Environmental conditions differ from year to year, impacting the proportion of individuals in the population entering diapause. This is evident from field observations with the hemlock woolly adelgid *Adelges tsugae*. This species most commonly completes two generations annually in western North America, but an unusually hot period coinciding with the settlement of the crawlers in April 2014 prompted entry into diapause, resulting in the completion of only a single generation that year (Weed et al. 2016). These sorts of differences in the diapause threshold or prevailing environmental conditions are also reflected in the seasonal appearance of molecular markers for diapause in field populations of the Colorado potato beetle *Leptinotarsa decemlineata* observed in the Red River Valley of North Dakota and Minnesota (Yocum et al. 2011). Such observations underscore the inherent variability of the diapause response within a single population. Even more variation is noted when populations from different geographic regions are compared, as discussed in Chapter 4.

Distinctions can also be noted in closely related populations that feed on **different host plants**. Rice and water-oat populations of the striped stem borer *Chilo suppressalis* from the same region of Wuhan, China differ in critical daylength and duration of the sensitive period that programs larval diapause (Zhou et al. 2018). Similarly, black-headed and red-headed forms of the moth *Hyphantria cunea*, sibling species from the same location in Missouri, feed on different host plants and have critical daylengths that differ by 30 minutes (Takeda 2005). Such differences are not necessarily innate but may be induced by feeding on different host plants. For example, when the ragweed beetle *Ophraella communa* feeds on *Ambrosia artemisiifolia*, it has a higher diapause incidence than when it feeds on a closely related species *A. trifida* (Tanaka and Murata 2016).

Variation is evident even **within a single clutch** of eggs. For example, most females of the Australian plague locust *Chortoicetes terminifera* lay eggs that are of one type, either diapause or nondiapause, but as many as 20–30% of egg pods contain a mix (Deveson and Woodman 2014), a situation likely to pose a challenge for nondiapausing hatchlings as they attempt to escape from an egg pod containing diapausing siblings that block their passage. Similar variation within a single clutch is evident in embryonic diapause of the cricket *Gryllus firmus*, a species in which the mix of the two phenotypes varies seasonally from less than 5% diapause in March–June to 50% in November–December (Walker 1980), in embryonic diapause of the striped ground cricket *Allonemobius socius* (Huestis and Marshall 2006), in larval diapauses of the pitcher plant mosquito *Wyeomyia smithii* (Bradshaw 1976) and the Swede midge *Contarinia nasturtii* (Des Marteaux and Hallett 2019), in pupal diapause of flesh flies in the genus *Sarcophaga* (Denlinger 1972b), as well as in many other species.

Among species with an obligate diapause, the incidence of diapause is commonly 100%, but even in such species, a few individuals may develop without diapause. For example, the vast majority of embryos of the gypsy moth *Lymantria dispar*, enter an obligate embryonic diapause, but a few eggs may hatch immediately without entering diapause (Hoy 1977), an aberrancy that has been exploited to establish a nondiapausing, laboratory strain of gypsy moths. Similarly, most embryos of the hemlock looper *Lambdina fiscellaria* enter an obligate overwintering diapause in eastern Canada, but a few individuals (8 out of 16,000) hatch in early autumn, suggesting they did not enter diapause or were arrested only briefly (Delisle et al. 2009). Most tephritid flies enter an obligate pupal diapause, but a few (<2%) develop without interruption, as noted in populations of *Rhagoletis cingulata* and *R. turpiniae* from high elevations in Central and Northeastern Mexico (Rull et al. 2017). A population of the eastern larch beetle *Dendroctonus simplex* was thought to overwinter in an obligate adult diapause, but closer examination reveals that at least a few females can reproduce without first entering diapause (McKee and Aukema 2015).

Distinct populations can be expected to differ in diapause attributes, as evidenced in the butterfly *Pieris napi*, a species that, in the warm climate of northern Spain, has a facultative pupal diapause that responds to photoperiod for induction, allowing it to complete three to four generations each year, while a population from northern Sweden, above the Arctic Circle, shows little response to photoperiod, suggesting an obligate diapause that allows completion of only a single generation annually, although a few individuals (<10%) from the northern population fail to diapause across a range of photoperiods (Pruisscher et al. 2017). Similarly, a southern subspecies of the eastern tiger swallowtail *Papilio glaucus* has a facultative pupal diapause, while diapause in the northern subspecies is obligate (Rockey et al. 1987). Or, diapause may be absent in some populations and present in others. For example, embryonic diapause in the mosquito *Aedes aegypti* is not seen in most geographic locations but is documented from Buenos Aires, Argentina (Fischer et al. 2019).

One aspect of variation that seems to differ between tropical and temperate latitude species is that, when strong diapause-inducing signals are presented, nearly all temperate latitude species display a high incidence of diapause, often near 100%, a feature that does not appear to be true for many tropical species. The capacity for diapause exists in **tropical species**, as noted in Chapter 2, but levels of diapause within a population frequently remain low. For example, among eight species of flesh flies (Sarcophagidae) originating from within 9° N or S of the equator in Africa, maximum diapause incidences observed under a range of laboratory environmental conditions ranges from 79% for *Poecilometopa spilogaster* to 7% for *Sarcophaga par* (Denlinger 1979), and under natural field conditions in Nairobi, Kenya (1° S), the highest pupal diapause incidence for *P. spilogaster* was a meager 4.7% (Denlinger 1978). Similar low diapause levels are noted for larvae of the tropical stem borers *Chilo zonellus* and *C. argyrolepia*: Maximum diapause incidences in three experimental fields on the Kenyan coast range between 20 and 60% (Scheltes 1976), although higher incidences of diapause likely occur in non-irrigated fields (Scheltes 1978). In another stem borer *Busseola fusca*, a few individuals enter larval diapause each generation in Nigeria, with incidences of diapause in field-collected larvae ranging from 3.4% in April to a high of 86.1% in December (Usua 1970). A low diapause incidence prevails among embryos of the Asian tiger mosquito *Aedes albopictus* in southern Florida (Lounibos et al. 2011, Urbanski et al. 2012), and individual females subjected to an unambiguous short-day photoperiod under laboratory conditions produce eggs with a highly variable diapause incidence, ranging from 0% to over 90% (Peter Armbruster, in preparation). On Reunion Island (20° 58′ S), a portion of the population of the mango blossom gall midge *Procontarinia mangiferae* can be found in larval diapause throughout the year, with incidences ranging from 3% (September) to 31% (February). Consequently, a large portion of the population is continuously active, a feature facilitated by the presence year-round of leaves and inflorescences of the mango tree (Amouroux et al. 2014). These observations suggest that levels of diapause in species living in moderate climates, where selective forces promoting diapause may not be as intense, still vary but oscillate at lower levels. But, there are exceptions, such as nearly all individuals of the tropical beetle *Stenotarsus rotundus* likely enter adult diapause (Denlinger 1994), presumably because the beetle feeds on a fungus that is restricted to a narrow seasonal window.

Natural variation is evident not only for the timing and incidence of diapause onset but also for **diapause duration** (Masaki 2002). Although the timing of spring emergence is notable for its amazing synchrony, as discussed in Chapter 8, there is again some variation in this response. A notable feature of diapause in the subtropical cockroach *Symploce japonica* is the high variability in the timing of diapause termination: While the majority of nymphs break diapause within a brief developmental window, a small portion remains in diapause much longer than their counterparts (Tanaka and Zhu 2003). Adult emergence of the silk moth *Hyalophora cecropia* from overwintering pupal diapause in central Illinois occurs over a broad and disjunct seasonal window, resulting in a bimodal emergence pattern (May 19–25 median emergence date over a nine-year period for the first emergence and June 17–26 for the second emergence) (Waldbauer 1978). Natural variation in the timing of diapause termination has led to distinct early and late-emerging genotypes of the cabbage root fly *Delia radicum* (Lepage et al. 2014) and the turnip root fly *D. floralis* (Biron et al. 2003), as well as the remarkable segregation of host races in fruit flies in the genus *Rhagoletis* (Lyons-Sobaski and Berlocher 2009, Egan et al. 2015, Meyers et al. 2016), as discussed further in Chapter 8. Variation is also evident in the timing of the **transition between diapause and post-diapause quiescence** (when the insect becomes competent to terminate diapause if exposed to suitable conditions).

A population of the wheat blossom midge *Sitodiplosis mosellana* from northern China requires a cold treatment of 60–90 days before making the transition, while a Canadian population requires greater than 112 days of chilling (Cheng et al. 2017). Under field conditions, a single population may show considerable variation in the timing of this transition, as demonstrated in overwintering larvae of the peach fruit moth, *Carposina sasakii* (Zhang et al. 2016a). Differences in diapause duration are evident even between individual orchards and among almond cultivars for the almond seed wasp, *Eurytoma amygdali* in northern Greece (Margaritopoulos and Tzanakakis 2006).

Though the **stage of diapause** is usually rather rigidly fixed within a species, that is not always so. The third instar is the predominant stage for larval diapause in the burnet moth *Zygaena trifolii*, but in natural populations diapause can also occur in later instars (Wipking 1988). Within the range of daylengths that are diapause-inductive for *Z. trifolii*, longer daylengths within this range shift the stage of diapause to later instars, and in a laboratory setting, it is possible to quickly select for a specific diapause stage (Wipking and Kurtz 2000). Larvae of the European butterfly *Lycaena hippothoe* diapause in either the third or fourth instar (Fischer and Fiedler 2002). Populations of the bruchid beetle *Bruchidius dorsalis* living in the cold, far northern region of Japan diapause mainly as fourth-instar larvae, populations in central Japan diapause both as fourth-instar larvae and adults, while those in even warmer regions overwinter not only as diapausing larvae or adults but also as nondiapausing larvae (Kurota and Shimada 2007).

Variation is also evident in **molecular profiles** observed during diapause (Yocum et al. 2018). Gene expression patterns observed during the course of diapause do not yield a simple ordered sequence. Instead, the environmental history of individual insects has a profound impact on the patterns observed, as noted in diapausing larvae of the alfalfa leafcutting bee *Megachile rotundata* reared under laboratory and various field conditions.

These few examples suggest there is sufficient variation within a natural population to allow selection to modify diverse features of the diapause response. This variation underscores the potential of diapause as a **bet-hedging** strategy if environmental conditions change. If not all individuals respond in exactly the same way, the ability of a species to respond through natural selection to shifting climates is maximized, and variation allows rare phenotypes to successfully invade new geographic areas or switch seasonal developmental patterns to accommodate changes in resource avail-ability. The rich variations of the diapause response also offer powerful tools for probing the genetic basis of diapause, as discussed in Chapter 10.

3.2   Repeated Diapauses

Another variation in the diapause response is the ability to enter diapause more than once. The common pattern is for diapause to be experienced only once during an insect's life cycle, but there are clear exceptions. At higher latitudes where the growing season is too short to permit completion of a whole generation in a single year, an insect may need to overwinter more than once before reaching adulthood, but repeated diapauses can occur in more moderate climates as well.

Northern populations of the spruce budworm *Choristoneura biennis* overwinter first in a second-instar diapause, then a second time as a third-instar larva, and, at least under laboratory conditions, a significant number of individuals can be environmentally triggered to enter a third diapause as a fourth-instar larva (Nealis 2005). In Korea, the katydid *Paratlanticus ussuriensis* first overwinters in a facultative embryonic diapause soon after blastoderm formation and then enters an obligate embryonic diapause the following winter as a nearly fully developed embryo (Shim et al. 2013). The predatory stink bug *Picromerus bidens* overwinters in an obligate embryonic diapause and then enters a long-day induced summer diapause as an adult (Musolin and Saulich 2000). The zygaenid moth *Pseudopidorus fasciata*, completes winter diapause as a fourth instar, and then a portion of the population enters summer diapause as a post-feeding larva (prepupa) within a cocoon (Wu et al. 2006). The burnet moth *Z. trifolii* shows remarkable flexibility in its stage for diapause: Near the northern limits of its range in Germany, the larva enters an obligate third-instar diapause, but in the following winter it may enter a facultative diapause in any of the subsequent six to seven larval instar (Wipking 1988). The spruce beetle *Dendroctonus rufipennis*, a major forest pest in North America, spends its first winter in a facultative larval (prepupal) diapause and the next winter enters an obligate adult diapause (Schebeck et al. 2017). Several British cockroaches in the genus *Ectobius* diapause in two different developmental stages, embryos and nymphs (Brown 1983). An even more extreme case is seen in a subtropical cockroach *S. japonica*; this Japanese species diapauses in three distinct stages: a winter diapause in mid-nymphal instars, a summer diapause in later nymphal instars, followed by an adult diapause (Tanaka and Zhu 2003). Another subtropical species, the white grub *Dasylepida ishigakiensis* has a two-year life cycle in southern Japan; it first enters a long (one year) third (final) instar larval diapause, followed by a somewhat shorter (two months) adult diapause before emerging from the soil as a reproductively active adult (Tanaka et al. 2008). The pitcher plant mosquito *Wyeomyia smithii* most commonly diapauses as a third-instar larva but can also repeat diapause as a fourth instar (Lounibos and Bradshaw 1975). *Ochlerotatus triseriatus* and *Aedes sierrensis* can diapause in either embryonic or larval stages, although the same individual would not normally enter diapause twice (Jordan 1980, Holzapfel and Bradshaw 1981). The leaf beetle *Timarcha tenebricosa*, a Mediterranean species, first enters an obligate embryonic diapause and later can enter either summer or winter diapause as an adult (Kutcherov et al. 2018).

Species living in extreme constraints of **polar environments** may require multiple years to complete larval development. The Antarctic midge *Belgica antarctica* requires at least two years to complete larval development (Lee and Denlinger 2015). An even more prolonged life cycle is noted in the high Arctic woolly bear caterpillar *Gynaephora groenlandica*, a species that is thought to require at least seven years to complete its life cycle (Morewood and Ring 1998). But, for these polar species, we do not know for certain whether the overwintering larval dormancies are diapause or some form of quiescence. Even some species from lower latitudes have life cycles that exceed one year. For example, in China, the seabuckthorn carpenter-worm *Holcocerus hippophaecolus* has a four-year life cycle extending over 16 larval instars; instars 5, 10, and 15 enter some sort of dormancy each winter (Zhou et al. 2016).

In addition to these examples in which a single individual enters diapause at more than one developmental stage, some species can enter and exit diapause in the same stage (adult) more than once. For example, males of the grasshopper *Oedipoda miniata*, from Israel, are capable of repeatedly going in and out of diapause, depending on the prevailing photoregime (Orshan and Pener 1976, Pener and Orshan 1983). The predaceous mite *Amblyseius potentillae* enters and exits diapause multiple times in direct response to photoperiod (van Houten 1989). The bug *Aelia acuminata* over-winters in adult diapause but if exposed to short days the following spring will again enter diapause after a few weeks despite other aspects of spring conditions being permissive (Hodek 1971a). Certain populations of the Colorado potato beetle *L. decemlineata* reproduce, at least briefly, before entering diapause (Tauber et al. 1988). In the Transcarpathian region of Russia, up to a third of females enter the soil to overwinter for a second diapause, although very few females survive the second winter (Minder and Petrova 1976). Repeated entry and exit from diapause, although rather rare, has been observed in quite a few species that diapause as adults. As one might expect, repeated diapauses are most prevalent in long-lived species.

3.3   Prolonged Diapause

Another variation in the diapause response is the ability to remain in diapause for longer than one year. Though diapause bridging a single season is most common, a few species remain in diapause for multiple years or at least channel a portion of the population into diapause for more than one year. Such a diapause, referred to as prolonged diapause, implies that the insect remains dormant through a season that would normally be considered suitable for development and has the capacity to delay its development for one or more years until the "right" conditions return. A prolonged diapause is challenging energetically because the insect must sequester energy reserves sufficient to bridge an unusually long time span, including summers that are especially expensive for an ectotherm due to prevailing high temperatures. In many ways, prolonged diapause is the entomological equivalent of plant seed banks containing stored seeds that germinate over multiple years.

Prolonged diapause is best documented in species living in drought-prone environments and among species that feed on seeds, cones, or plants that are sporadically produced (Ushatinskaya 1984, Powell 1987, Hanski 1988, Saulich 2010, Higaki 2016). Erratic rainfall patterns and seed production generate an environment where a **bet-hedging** strategy of distributing emergence over multiple years could offer an obvious selective advantage by spreading the risk over time. Additionally, prolonged diapause may provide a temporal escape from parasitoids. For example, some larvae of the gall midge *Contarinia vincetoxici* remain in diapause for up to 13 years (Solbreck and Widenfalk 2012). Although the two major parasitoids of the gall midge have also evolved the capacity for prolonged diapause, the parasitoid diapause is much shorter, two to four years, thus suggesting an evolutionary cat-and-mouse game currently being won by the gall midge. A similar disconnect is seen for the papilionid butterfly *Papilio machaon* and its ichneumonid *Trogus* parasitoid: The butterfly can remain in pupal diapause for up to three years in Canada, but two years seems to be the maximum for its parasitoid (Dupuis et al. 2016).

Among moths and butterflies, prolonged diapause is thought to be more prevalent in **larger species** (Powell 1987), although this could be a sampling bias. Prolonged diapause is usually not observed in all individuals within a population. Commonly, the majority of individuals from a single clutch of eggs break diapause after a single season, while a smaller portion of their siblings enters a prolonged diapause. While most diapausing adults (97.7%) of the Colorado potato beetle *L. decemlineata* emerge after a single winter in upstate New York, some adults remain in diapause for three to seven years, and one beetle emerged after nine years (Tauber and Tauber 2002). The incidence of prolonged diapause is highly variable within species as well as among species. In a comparison of seven closely related mushroom-breeding flies (*Pegomya* sp.) in Lapland, Finland, the frequency of prolonged diapause varies significantly among species, from 4 to 68% (Hanski 1988). An unusually high incidence of prolonged larval diapause (32–56%) is noted in a French population of the chestnut weevil *Curculio elephas* (Menu and Debouzie 1993), and similarly high incidences of prolonged diapause are reported for Greek populations of the European cherry fruit fly *Rhagoletis cerasi* (Moraiti and Papadopoulos 2017).

Prolonged diapause is easily overlooked. Its very nature and low incidence likely have led to an underestimate of its prevalence. When efforts are made to detect prolonged diapause, a surprisingly high number of examples are found. In his survey of Lepidoptera, Powell (1987) records prolonged diapause in 90 species within 10 superfamilies. He notes its prevalence among Prodoxidae, Saturniidae, Pieridae, and Papilionidae, but reports it among a range of other families of Lepidoptera as well, including Pyralidae, Geometridae, and Noctuidae. A listing of prolonged diapause in desert-dwelling solitary bees, representing several families (Danforth et al. 2019), includes 15 species known to diapause for two or more years and one Australian species *Amegilla dawsoni* that can remain in diapause for up to 10 years. Additional compilations of prolonged diapause are provided by Sunose (1983) and Ushatinskaya (1984), and each such list contains unique entries, leading further credence to the idea that prolonged diapause is not particularly rare.

Within the Lepidoptera, prolonged diapause appears to be restricted to fully developed larvae or pupae (Powell 1987). The desert bee *Perdita portalis* (Danforth 1999) and other species of solitary bees (Danforth et al. 2019) enter prolonged diapause as larvae (prepupae), just prior to pupation. Prolonged diapause in the larval stage is also well documented for several beetles, such as chestnut weevil *Curculio sikkimensis* (Higaki 2006, Higaki et al. 2010), acorn weevil *C. robustus* (Higaki 2016), and seed-predatory weevil *Exechesops leucopis* (Matsuo 2006). But, at least some beetles such as the northern corn rootworm *Diabrotica barberi* enter prolonged diapause as an embryo (Krysan et al. 1986), and the prolonged diapause of the Colorado potato beetle *L. decemlineata* is noted in adults (Tauber and Tauber 2002). Among Diptera, there are several species with prolonged diapause in the larval stage, such as the wheat blossom midge *S. mosellana* (Cheng et al. 2017) and sugarbeet root maggot *Tetanops myopaeformis* (Chirumamilla et al. 2008), or in the pupal stage, such as the European cherry fruit fly *R. cerasi* (Moraiti et al. 2012) and other species of *Rhagoletis*. The Mormon cricket *Anabrus simplex* (Srygley 2019, 2020) and European species of Tettigoniidae (Ingrisch 1986) and other Orthoptera (Hanski 1988) have prolonged diapauses in the embryonic stage.

Prolonged diapauses can be of **varying duration**, even within a single species (Danks 1987, Saulich 2010). The current record for the duration of a prolonged diapause is held by the Yucca moth *Prodoxus y-inversus*, a species that inhabits the desert of the American Southwest (Powell 2001). Diapausing larvae are encased in a hardened cyst within fruits of the Yucca plant, and if the larvae receive a cold winter typical of their Nevada environment, they end dormancy after a single winter, but if they remain in a warm environment, they are capable of surviving in diapause for up to 30 years! Such a long diapause is unlikely to occur in the field, but it underscores the capacity of this moth to extend its diapause for an extremely long time. Among other species, especially under field conditions, a one-year delay is much more common, and among species with a multiple-year prolonged diapause, the number of individuals remaining in diapause diminishes progressively with time. For example, the majority of *C. sikkimensis* weevils in a Japanese population (Maeto and Ozaki 2003) emerge in year 2, but a small portion delay emergence until year 3 (approximately 10%), and even fewer emerge in year 4 (approximately 5%). The two-year prolonged diapause noted in *C. sikkimensis* and two close relatives nicely coincides with the pattern of alternate year acorn production in the host oak *Quercus crispula*. In a Mediterranean population of the lycaenid butterfly *Tomares ballus*, 45% of pupae break diapause after one year, but the remaining pupae may remain in diapause for up to 8 years (Obregon et al. 2017). Most individuals of the tropical palm nut bruchid *Pachymerus nucleorum* break larval diapause in 15–48 months, but a few remain in prolonged diapause for as long as 5 years (Benton 2015). The median duration of prolonged diapause in larvae of the gall midge *Contarinia vincetoxici* is 6 years, but some individuals remain in diapause for up to 13 years (Solbreck and Widenfalk 2012).

What determines which individuals are destined for prolonged diapause remains a fascinating but unresolved question. Presumably, individuals that have sequestered minimal energy reserves would not be good candidates. This is possibly reflected in the larger adult size noted in individuals of the European cherry fruit fly *R. cerasi* that emerge from prolonged pupal diapause in comparison to their counterparts that emerge after a single winter (Moraiti et al. 2012). A similar correlation is seen in adults of the seed-predatory weevil *E. leucopis*, a species that diapauses as a larva (Matsuo 2006). Larval diapause in *P. portalis*, a ground-nesting bee native to the Chihuahuan desert in the American Southwest, can remain in diapause for as long as

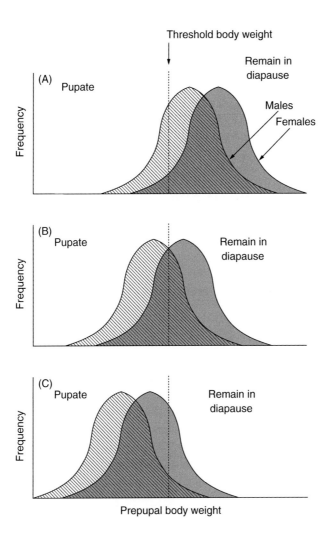

Figure 3.1 The incidence of prolonged diapause in larvae of the desert bee *Perdita portalis* is dependent on the body weight of the larva (prepupa). Females, who are heavier, are more highly represented in the portion remaining in prolonged diapause. With the passage of time, body weight declines, making larvae more likely to respond to the rainfall that triggers diapause termination (pupation). (A–C) represent groups with three different hypothetical weight distributions. From Danforth (1999), with permission from Royal Society, London.

three years, and within a sampling of field-collected bees of different weights, it is the lighter bees that are the first to break diapause (Danforth 1999). Females are consistently heavier than males and are more highly represented in the portion remaining in prolonged diapause (Figure 3.1). Larvae of the chestnut weevil *C. elephas* that enter prolonged diapause have higher lipid reserves (20.5 mg) than their counterparts that diapause for only one year (8.7 mg), and metabolic rates during prolonged diapause drop to levels lower than rates seen during a single-season winter diapause (Soula and Menu 2005), underscoring the importance of energy management for a successful

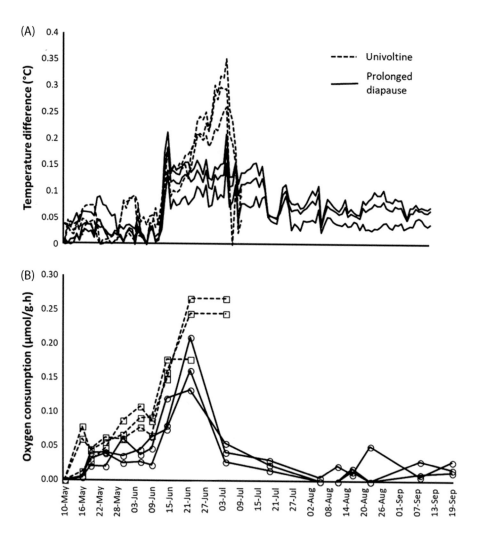

Figure 3.2 Seasonal changes in (A) body temperature and (B) rate of oxygen consumption in diapausing pupae of the processionary moth *Thaumetopoea pityocampa* that either terminate diapause after one year, a univoltine life cycle (dashed line), or enter prolonged diapause (solid line).

From Salman et al. (2019b), with permission from John Wiley & Sons.

prolonged diapause. Similarly, adults of the Colorado potato beetle *L. decemlineata* in prolonged diapause are heavier than those entering a single winter diapause, metabolic rates are three- to five-fold lower, and periods between bouts of discontinuous gas exchange (see Section 7.5) are extended (Ploomi et al. 2018).

In the processionary moth *Thaumetopoea pityocampa*, a spike in both body temperature and oxygen consumption is noted in early June for both those with a one-year pupal diapause and those in prolonged diapause (Figure 3.2). For those with a one-year diapause, the rise culminates in adult eclosion by mid-June. For pupae in prolonged diapause, the rise in body temperature and oxygen consumption rate begins at the same time, remains elevated until late June, and then declines by early July to the previously low-level characteristic of diapause (Salman et al. 2019b). Possibly a critical assessment is made during this interval, enabling the pupa to decide to either immediately proceed with development or remain in diapause for another year. A similar decision point seems to be present in *C. elephas*, suggesting that the decision to enter prolonged diapause is made at the end of the first winter of diapause rather than at the very onset of diapause (Soula and Menu 2005). The fact that one seldom knows *a priori* whether an individual is destined for prolonged diapause makes work on this topic particularly challenging. But, careful monitoring of individuals through-out the timescale of prolonged diapause makes it possible to distinguish between a high incidence of prolonged diapause in heavy individuals from a scenario where small individuals enter prolonged diapause at the same rate but simply do not survive. It will be interesting to know if, at the onset of diapause, there is a molecular signature differentiating a single-year diapause from a prolonged diapause.

The prevailing environment may also impact the incidence of prolonged diapause. **Soil moisture** levels affect the incidence of prolonged diapause in larvae of the wheat blossom midge *S. mosellana*: Soil moisture of 40–60% yields a 6% incidence of prolonged diapause while a soil moisture level of 20% results in a prolonged diapause incidence of 43.8% (Cheng et al. 2017). Similarly, a portion of Colorado potato beetles overwintering in light (dry) soils in Ukraine enters prolonged diapause, but those overwintering in wetter clay soils do not (Ushatinskaya 1976b). In Ukraine, prolonged diapause is more prevalent for females of the Colorado potato beetle (63%) than for males (37%) (Ushatinskaya 1976b).

Geography is also a determinant of prolonged diapause. The incidence of pro-longed pupal diapause in the pine processionary moth *T. pityocampa* increases with elevation in an alpine valley in the South Tyrol region of northern Italy (Salman et al. 2016). Across their geographic range in southern Europe and northern Africa, *T.pityocampa* and *T. wilkinsoni* are more likely to be univoltine in localities with intermediate low winter temperatures but undergo prolonged pupal diapause when winter temperatures are either lower or higher (Salman et al. 2019a). These results, based on long-term observations of the incidence of prolonged diapause at 36 sites in 7 countries, suggest a bet-hedging strategy to possibly mitigate the uncertainty of climate at the edges of the species range. In processionary moths, prolonged diapause comes at a cost: Diapause mortality increases with an increase in the incidence of prolonged diapause. Altitude, latitude, prevailing temperatures, drought, soil type, age of host plants, and duration of the host plant's presence have all been implicated as potential causative factors determining the incidence of prolonged diapause in certain species (Saulich 2010).

The variable incidence of prolonged diapause in a population, and even among siblings derived from the same clutch of eggs, implies individual variability in responsiveness to diapause-terminating cues. As discussed later, the temperature is the dominant environmental factor impacting diapause termination, and periods of **chilling** are sometimes required. The observed variability among individuals in

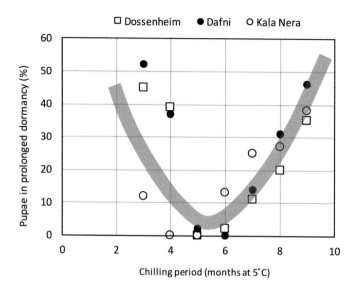

Figure 3.3 The relationship between chilling duration and the incidence of prolonged diapause in pupae of the cherry fruit fly *Rhagoletis cerasi* collected from three highland localities in Greece. From Moraiti and Papadopoulos (2017), with permission from John Wiley & Sons.

response to diapause-terminating factors is the most likely cause for differences in diapause duration. Although the source of this environmental variability has received little attention, there is good evidence in some systems that prolonged diapause is linked to individual variation in the temperature response. Prolonged diapause in larvae of the chestnut weevil *C. sikkimensis* can be broken by repeated chilling and warming cycles (Higaki et al. 2010, Higaki and Toyama 2012). In this species, warm periods simulating summer intensify the diapause. Interestingly, chilling for up to 120 days at 5°C is effective in breaking diapause, but not chilling periods exceeding 120 days. Beyond that time, larvae require a substantial period of higher temperature before they again respond to chilling. This response pattern presumably prevents untimely emergence during summer, intensifies the diapause, and allows the insect to overwinter a second time. A similar response appears to operate in pupae of the European cherry fly *R. cerasi* (Moraiti and Papadopoulos 2017): A two-to-four-month period of chilling at 5°C enables a portion of pupae to break diapause and emerge as adults, but extending the chilling period beyond six months boosts the incidence of prolonged diapause (Figure 3.3). Thus, in this species, both insufficient chilling and extended chilling force pupae to remain in diapause (Moraiti et al. 2013). In a dry, subtropical region of south-central Mexico, a humid period early in the pupal stage boosts the incidence of prolonged diapause in *R. pomonella* (Rull et al. 2019).

Diapause itself can take a toll on other life-history traits, as discussed in Chapter 4, but it is evident that prolonged diapause exacts an even greater **cost** than a single-season diapause in some species. When adults of *R. cerasi* emerging from a single season of pupal diapause are compared to those emerging from prolonged diapause, adult lifespan is not affected, but fecundity is reduced and females delay the onset of oviposition (Moraiti et al. 2012). Prolonged pupal diapause in the pine processionary moths *T. pityocampa* and *T. wilkinsoni* also correlates with higher pupal mortality (Salman et al. 2019a). But, somewhat surprisingly, in the chestnut weevil *C. elephas* (Menu and Debouzie 1993) and Colorado potato beetle *L. decemlineata* (Ushatinskaya 1976b), neither adult longevity nor fecundity appears to be significantly impacted by the second year of diapause. Other studies with *L. decemlineata*, however, suggest that prolonged diapause can exact a toll on fecundity: Females from a Vermont population produce fewer eggs, and survival is poorer in progeny produced by females that have spent two, rather than one, years in diapause (Margus and Lindström 2020). No cost is noted for males of *L. decemlineata* that have been in diapause for two years. An additional year in diapause results in shorter eyestalks for males of the weevil *E. leucopis*. This protrusion affects male fitness during male–male conflicts, thus males with shorter eyestalks are at a competitive disadvantage (Matsuo 2006). Mortality also increases in *E. leuopis* as the length of prolonged diapause is extended.

Prolonged diapause can have profound implications for **pest management** in agricultural and forest systems. The northern corn rootworm *D. barberi*, a major pest of corn in the northern United States, was historically controlled by diligently rotating fields between corn and soybeans. Since soybeans are not attacked by the corn rootworm, populations of the pest, with its one-year life cycle, cannot be sustained when the planting of corn is rotated with soybeans. But, with time, corn rootworms with a two-year life cycle began to appear, disrupting the cultural control measures attained by alternating the plantings of these two crops. The basis for this new two-year life cycle was the intense selection favoring prolonged diapause within populations of this pest species (Krysan et al. 1986, Levine et al. 1992). Thus, the capacity for prolonged diapause, previously unknown in this species, emerged as a mechanism allowing rootworms to synchronize development with the new pattern of crop rotation. Presumably, a low incidence of prolonged embryonic diapause was present in earlier populations but only became the dominant form of diapause under the intense selective pressure of crop rotation. Persistent low levels of prolonged diapause in adults of the Colorado potato beetle *L. decemlineata* in upstate New York suggest a similar potential for this pest species to quickly evolve resistance to crop rotation (Tauber and Tauber 2002). Prolonged diapause means that control measures may need to be extended over multiple years. The pine processionary moth *T. pityocampa* can diapause for up to seven years as pupae, allowing outbreak numbers of this forest pest to be sustained annually as a result of adults emerging from cohorts that are in prolonged pupal diapause. Not knowing the size of the population in prolonged diapause makes it challenging to predict future outbreaks (Salman et al. 2019b). Annual applications of the biocontrol agent *Bacillus thuringiensis kurstaki* fail to suppress the population until after the eighth year, at which time the stock of pupae in prolonged diapause is depleted (Salman et al. 2016). Thus, the evolution of prolonged diapause can easily subvert attempts to suppress pest populations and cannot be overlooked in developing effective strategies for pest management.

4   The Cost of Diapause and Some Diapause Alternatives

Diapause offers huge dividends in terms of survival, but it is not cost-free. Costs may be incurred by the individual experiencing diapause or may show up later as reduced fitness in the progeny of individuals that have been through diapause. And, sometimes no apparent cost is evident. This chapter examines costs linked to investing in diapause and also explores alternatives to diapause that may be employed to survive adverse seasons. Whether the alternatives are more or less costly than diapause has yet to be determined.

4.1   The Cost of Diapause

Consequences of diapause may include high mortality, depletion of energy reserves, reduction in size, low fecundity, delay in oviposition, shortened post-diapause longevity, and loss of generations. One of the problems in interpreting this body of literature is that, from field results, it is not always easy to distinguish whether costs are imposed as a direct consequence of diapause or an indirect consequence emanating from environmental stress (*e.g.,* cold and desiccation) concurrently experienced during diapause. From an ecological perspective it may not matter too much whether the costs of diapause are direct or indirect, but I think it is useful to be aware of the distinction. Experiments that control the level of environmental stress while manipulating the duration of diapause are particularly valuable for assessing the cost of diapause itself, but here too, it is not always possible to distinguish the cost of diapause from the cumulative costs of a stressful environment experienced over the timescale of diapause.

4.1.1   Missed Opportunities

One obvious consequence of diapause is that **fewer generations** can be completed each year when a portion of the year is devoted to diapause. Insertion of a 3–4 month period of summer diapause into the life cycle of a southern Iberian population of *Pieris brassicae* means that those populations lose three generations per year, compared to more northern populations that do not enter summer diapause (Spieth and Schwarzer 2001). The literature is replete with numerous such examples showing reduction in the number of generations, and hence a loss of reproductive output, resulting from inclusion of diapause in the life cycle.

4.1.2   Diapause Mortality

Although we assume diapause is better than no diapause for bridging unfavorable seasons, diapause does not guarantee high survival. Mortality can be quite high in spite of diapause. Summer dormancy in larvae of the hemlock woolly adelgid *Adelges tsugae* exacts a high toll on survival (Sussky and Elkinton 2015). It is not clear if the summer dormancy in this species is a diapause or a simple quiescence, but what is clear is that the period of summer arrest is extremely stressful, as reflected in survival rates that can be <1%, survival far lower than noted for the overwintering diapause (30%). This unusually high summer mortality, coupled with rather high overwintering mortality, may account for the slow spread of this invasive species in North America and may reflect the challenge of a new arrival adapting to its new environment.

Low survival is noted as well in a number of species with an overwintering diapause, even in species with a robust diapause. In the Arctic aphid *Acyrthosiphon calvulus*, only 54% of diapausing eggs survive winter along the western coast of Spitsbergen, Norway, in spite of having supercooling points down to −40°C (Gillespie et al. 2007). Mortality of overwintering eggs of the sycamore aphid *Drepanosiphum platanoidis* in the United Kingdom frequently exceeds 80% (Wade and Leather 2002), as it does in Switzerland for diapausing adults of the apple blossom weevil, *Anthonomus pomorum* (Toepfer et al. 2000). Mortality in diapausing aggregations of ladybird beetles can be extremely high: Adults of *Coccinella septempunctata,* a species that migrates and diapauses in huge aggregations in mountainous regions of southern Turkey, succumb in high numbers, up to 93% mortality in some years, mainly caused by entomopathogenic fungi (Guven et al. 2015). Field survival of overwintering populations of the bark beetle in Europe varies depending upon the microclimate at different altitudes and specific microhabitats (Koštál et al. 2014). Overwintering mortality for diapausing larvae of satyrine butterflies in Sweden can be as high as 51% for *Lasiommata maera* and 38% for *Lopinga achine* (Gotthard et al. 1999). While 76% of nondiapausing larvae of the parasitoid *Doryctobracon areolatus* successfully emerge as adults, only 40% adult emergence is noted for those that have gone through diapause (Cruz-Bustos et al. 2020).

In late summer, most adults of the hover fly *Episyrphus balteatus* from Central Europe migrate to the Mediterranean region, although a few remain behind and enter diapause (Hondelmann and Poehling 2007). Few of these adults survive. By March, all males are dead and mortality is 87% for females. Thus, diapause does not consistently result in high overwintering survival.

4.1.3   Reduction in Fitness

One of the most striking and recurring themes concerning cost is the reduced fitness that accompanies diapause, a cost frequently reflected in survival time and reproductive output. Costs may differ between **capital breeders** and **income breeders**. Capital breeders rely on energy reserves stored in previous life stages for reproduction. In contrast, income breeders base reproduction on current food intake. Capital breeders can be expected to suffer greater losses to future reproduction than income breeders when diapause is long and energy reserves are severely depleted (Varpe and Ejsmond 2018).

By simulating natural conditions, Bradshaw et al. (1998) calculated that overwintering, diapausing larvae of the mosquito *Wyeomyia smithii* suffer a 60% loss in fitness compared to larvae in an optimal summer environment, losses that occur through reduced survivorship, fecundity, fertility, and adult longevity. The earlier overwintering larvae of *W. smithii* are oviposited, the lower their odds of surviving winter (Bergland et al. 2005), an observation underscoring the significance of seasonal timing for success of diapause.

Adult lifespan may be shortened by the experience of diapause in a prior life stage, but that is not always the case, and diapause-altered longevity is not the only impact on post-diapause fitness. Adults of the beet webworm, *Loxostege sticticalis*, actually live longer if they experience larval diapause, but in spite of increased longevity, females that have been through diapause delay oviposition, oviposit over a shorter interval, and lay fewer eggs (Jiang et al. 2010). A rich supply of carbohydrates fed to newly emerged adults of *L. sticticalis* can mitigate some of these negative reproductive consequences (Xie et al. 2012). Other attributes also differ in webworm adults that have been through larval diapause. Such adults display higher flight capacity, represented by longer flight duration, greater distance covered, and higher flight velocity, possibly an adaptation enabling adults to better escape their depauperate winter environment. Adults of the bruchid beetle *Kytorhinus sharpianus* that have experienced larval diapause also live longer as adults but oviposit later and lay fewer eggs (Ishihara and Shimada 1995).

In the aphid parasitoid *Praon volucre*, the experience of larval diapause shortens adult life if adult feeding is restricted (Colinet et al. 2010). The experience of adult diapause also shortens adult life in the boll weevil *Anthonomous grandis grandis*, and the number of eggs produced is much reduced, but curiously, the time required for development of larval progeny is extended if the mother overwinters in diapause (Greenberg et al. 2007).

4.1.4   Consequences of Size Reduction

Though metabolic depression certainly helps to conserve energy reserves, a long diapause still exacts a cost reflected in a loss in energy stores, commonly monitored as loss in weight or depletion of a specific metabolic reserve, most notably lipids (Hahn and Denlinger 2011). The **smaller size** sometimes seen in insects that have been through diapause frequently correlates with lower reproductive output. Larval diapause in the adzuki bean borer, *Ostrinia scapulalis*, impacts female reproductive output, a consequence of the smaller size of females that emerge from diapause as well as their truncated longevity (Win and Ishikawa 2015). By contrast, males of the adzuki bean borer that have been through diapause display no impairment of reproductive capacity and transfer a similar number of eupyrene and apyrene sperm as their nondiapause counterparts. In the parasitoid *Doryctobracon areolatus*, females that have been through larval diapause are smaller than those that do not enter diapause, flight ability is impaired, and fecundity is lower (Cruz-Bustos et al. 2020). Adults of the pierid butterfly *Pieris napi* that have been in pupal diapause are smaller than their nondiapause counterparts and have lower dispersal capacity, but lifetime fecundity is higher, suggesting a trade-off between investment in egg production and the expensive activity of flight muscle development essential for dispersal (Karlsson and Johansson 2008). Interestingly, adults of *P. napi* that emerge from diapausing pupae are more sexually mature than adults emerging from nondiapausing pupae. This is reflected in earlier egg production by females and earlier synthesis of sex pheromone and mating by males (Mellström et al. 2010).

Size distinctions are not as apparent in insects that diapause in the adult stage. In fact, diapausing adults may be larger. Adults of the overwintering generation of the scorpionfly, *Panorpa vulgaris*, are larger than those of the summer, nondiapausing generation, and the larger size of males means the sperm are also considerably longer, a trait thought to contribute to superiority in sperm competition (Vermeulen et al. 2009).

Depletion of reserves may disproportionately affect specific body parts. An example cited in Section 3.3 points to an indirect effect of larval diapause duration on eyestalk length in adult males of the weevil *Exechesops leucopis* (Matsuo 2006). In this species, the male eyestalk is a sexually selected trait: Eyestalk size affects the outcome of male-male combat. When larval diapause is prolonged by an additional year, large males are still able to produce long eyestalks, but smaller males do not retain sufficient reserves and thus produce shorter eyestalks, placing them at a competitive disadvantage for securing mates.

4.1.5   Relying on the Environment to Reduce Costs

Environmental conditions may directly impact the cost of diapause. A **warm winter**, perhaps counterintuitively, can be energetically more expensive than a **cold winter**, simply because insects draw down their metabolic reserves more quickly at high temperatures. The negative impact of a mild winter is nicely demonstrated in the freeze-tolerant goldenrod gall fly, *Eurosta solidaginis* (Irwin and Lee 2000). Mortality is higher and fecundity lower in adults emerging from larvae held for 3 months at 12°C when compared to those held unfrozen at 0°C or in a frozen state at −22°C. Although freezing causes some mortality (but not as much as noted with individuals at 12°C) fecundity is not reduced among those individuals that survive. The level of environmental stress can be expected to differ from year to year, implying that the overwintering cost may also vary from one year to the next (Marshall and Sinclair 2015).

Environmental costs affecting post-diapause performance may be incurred even before the insect enters diapause, as exemplified by solitary bees in the genus *Osmia*

(Bosch et al. 2010, Sgolastra et al. 2011, 2012). An early start of the pre-diapause period for these solitary bees in the warm months of late summer and early autumn results in rapid depletion of lipid reserves and increased winter mortality. Such results underscore the importance of optimal timing for diapause entry. Too late is obviously bad, but too early can have equally dire consequences. In spite of pre-diapause duration exacting a cost by depleting energy reserves, *O. lignaria* has the potential to feed as an adult after diapause has been completed, a response that helps mitigate the negative effects of pre-diapause duration on post-winter performance (Sgolastra et al. 2016). Species lacking the ability to feed after diapause termination, the capital breeders, are more likely to be impacted by pre-diapause and overwintering conditions that deplete nutritional reserves.

4.1.6   Impact on Progeny Development

Certain other effects of diapause are not obvious costs but are manifested in progeny of individuals that have been through diapause. Adults of the swallowtail butterfly *Papilio xuthus* are smaller if they emerge from diapausing pupae, and interestingly, the larval progeny of such females exhibits a faster growth rate and shorter developmental time than progeny produced by females that do not undergo diapause (Komata and Sota 2017). Alterations in **progeny growth rates** are also noted in a few other species. Adult diapause in the cabbage beetle, *Colaphellus bowringi* is quite variable in nature, thus setting the stage for evaluating the impact of diapause duration on post-diapause events (Wang et al. 2006a, Wei et al. 2010). In this species eggs laid by parents that have been through diapause appear to develop faster than eggs from nondiapausing parents. Unlike the example of *C. bowringi*, progeny of *Helicoverpa zea* that experienced pupal diapause develop more slowly during embryonic and larval stages than progeny produced by females that did not undergo pupal diapause (Akkawi and Scott 1984).

The onion fly, *Delia antiqua*, shows differences in adult **eclosion rhythmicity**, based on pupal developmental status: Adults emerging from nondiapausing pupae are more sensitive to light as a Zeitgeber for eclosion than are those originating from pupae that experienced diapause, but the opposite is true for responsiveness to thermoperiod (Watari 2005). Thermoperiodic cues are likely more accessible, and hence more relevant, to diapausing pupae that are buried more deeply in the soil.

The **stage of diapause** may have consequences evident only in later stages, as exemplified for the butterfly *Pararge aegeria* (van Dyck and Wiklund 2002, Gotthard and Berger 2010). In southern Sweden, a portion of the population overwinters in larval diapause and another portion in pupal diapause. Females that overwinter in larval diapause are considerably heavier as adults (90 mg) than those that overwinter in pupal diapause (75 mg), and heavier females have higher fecundity. But, pupal diapause allows earlier adult eclosion, thus the disadvantage of lower weight may be offset by an earlier start in the spring, suggesting interesting trade-offs in life-history traits.

4.1.7   Post-diapause Benefits

Some potentially positive attributes linked to diapause persist after diapause has been terminated. For example, cold tolerance associated with adult diapause in the *Drosophila virilis* group is retained after diapause ends, a feature not seen in females that develop without diapause (Salminen et al. 2015).

4.1.8   Testing the Cost of Diapause Duration

Several experiments have manipulated diapause to directly test the effects of diapause duration on survival and fecundity. In the flesh fly *Sarcophaga crassipalpis*, mortality is 26% in flies that overwinter in pupal diapause but only 7% in flies that develop immediately to adulthood (Denlinger 1981). Females that overwinter in pupal diapause mate at a lower frequency than females that have not been through diapause, they lay fewer eggs and fewer of those eggs are fertile (Figure 4.1), thus experiencing pupal diapause increases adult female mortality and decreases both **egg production and fertility**, a cost magnified by negative consequences at several levels. The longer the diapause, the greater the reproductive consequences (Figure 4.1). Interestingly, there are no such costs for males. Males that have gone through pupal diapause survive well, mate successfully with females that have not gone through diapause, and those females produce the same number of viable eggs as females mated with males that have not experienced diapause (Denlinger 1981). This difference in cost of diapause between males and females leads to an interesting sexual difference in the

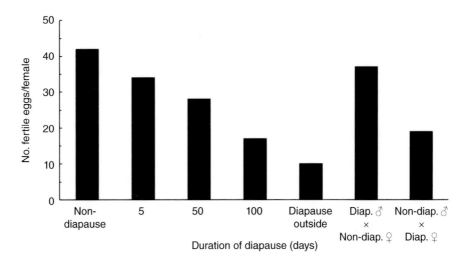

Figure 4.1 Female fertility in the flesh fly, *Sarcophaga crassipalpis*, is dependent upon the duration of pupal diapause she experiences. Male diapause history is not relevant. Diapause duration based on time at 20°C, and flies that had been in diapause for 100 days were used for crosses.
Adapted from Denlinger (1981).

timing of diapause entry. Females enter diapause later than males. Most likely, the high cost of a long diapause in females is the selective force behind the female's late entry into diapause. Male pupae, lacking such consequences, enter diapause early, while females "gamble" by delaying diapause entry, thereby reducing the cost of diapause by shortening its duration.

In the bruchid, *Acanthoscelides pallidipennis*, the experience of larval diapause in either parent exacts a toll reflected in delayed oviposition (Sadakiyo and Ishihara 2012). By experimentally manipulating duration of larval diapause in another bruchid, *Kytorhinus sharpianus*, Ishihara and Shimada (1995) demonstrated that longer diapause duration results in a longer preoviposition period in adult females and also increases longevity but has no impact on the overall number of eggs produced. As in the above example of *S. crassipalpis*, males of *K. sharpianus* are not affected by the experience of diapause. Similar manipulations with the parasitoid, *Asobara tabida*, demonstrate that increased diapause duration results in higher mortality, decreased egg production, and lower fat reserves and dry weight in females that have experienced larval diapause (Ellers and van Alphen 2002). The correlation between lower survival, reduced egg production, and smaller fat reserves in this parasitoid points to depletion of energy reserves as the likely cost of a lengthy diapause.

Population differences in diapause duration offer yet another method for evaluating costs of diapause. For example, increasing diapause duration for one population (collected on the host plant *Orixa* japonica) of the Kanzawa spider mite, *Tetranychus kanzawai*, negatively impacts overwintering survival but has no effect on either the number of offspring produced by the survivors or on post-diapause longevity, while diapause duration in another population (from the host plant *Hydrangea hirta*) has little impact on survival, but diapause reduces reproductive output by nearly 50% and longevity decreases (Ito 2004). The correlation between diapause duration and reduced fecundity in *T. kanzawai* is genetically based, indicating that natural selection acts simultaneously on these two life-history traits (Ito 2007).

4.1.9   Stage-Specific Costs

There may be unique costs associated with specific stages of diapause. For example, adult diapauses, especially in species that maintain mobility, usually do not exhibit the same degree of metabolic suppression as diapauses in larval or pupal stages (Section 7.3). An adult diapause, however, offers the benefit of early and immediate exploitation of a host in early spring. A field study of the nut weevil *Curculio nucum* nicely demonstrates this point (Bel-Venner et al. 2009). This weevil has a 2-year life cycle, with the first winter spent as a larva and the second as an adult. It is only by emerging as an adult in early spring that the weevil can perforate a mature nut to lay an egg. Later in the season, nuts are too hard to penetrate. But, there is a cost: Overwintering adults show a more profound loss of lipids than overwintering larvae, and it is these lipid reserves that the female depends on for egg production.

4.2 No Costs?

In some cases, there appear to be no or few repercussions from having experienced diapause. Female survival and fecundity are not noticeably impacted by a long diapause in the Colorado potato beetle *Leptinotarsa decemlineata* (Jansson et al. 1989), although male survival is significantly reduced (Piiroinen et al. 2011). In the southern green stink bug, *Nezara viridula*, a long adult diapause reduces winter survival, but for those that survive, diapause appears to alter neither reproductive output nor longevity (Musolin et al. 2007). A long diapause can sometimes offer post-diapause reproductive benefits. In cases where diapause results in larger body size and more metabolic reserves, adults derived from diapausing individuals have even greater fecundity, as in the cabbage beetle, *Colapellus bowringi* (Wang et al. 2006a). In the corn stalk borer, *Sesamia nonagrioides*, diapausing larvae attain a larger size by feeding a bit during diapause and undergoing supernumerary molts. This results in larger adults capable of producing more eggs (Fantinou et al. 2004).

4.3 Alternatives to Diapause

Diapause offers an attractive survival package for bridging adversity, but there are alternatives. Not all insects diapause, yet many nondiapausing species are highly successful and have invaded the far reaches of Earth. In brief, I suggest a few alternatives that enable insects to survive without diapause.

4.3.1 Quiescence

Quiescence is one viable alternative, and it has the advantages of being a state that can be accessed at any developmental stage and can be quickly entered and exited. On the negative side, it does not offer a preparatory phase that can be used to sequester energy reserves nor does it allow time for behavioral responses that may enable a diapause-destined insect to find a site suitable for long-term occupancy. It may also fall short by denying the full repertoire of defense responses that are common to a diapause program. Yet, in some environments, quiescence may be a perfectly suitable overwintering strategy. Among other examples, the egg parasitoid *Trichogramma funiculatum*, rather than entering diapause, simply responds to low temperatures of the Australian winter by slowing its rate of larval development and increasing adult longevity (Rundle and Hoffmann 2003). In spite of the ubiquity of quiescence, the physiological dimensions of quiescence and its limitations regarding duration and extent of the adversity it can sustain have not been adequately examined.

4.3.2 Cold Hardening without Diapause

The cold of winter is a conspicuous environmental challenge mitigated by diapause, but cold hardening is not restricted to diapause (Section 7.13.6). Robust mechanisms protecting against cold injury can also be found in nondiapausing species as well as in nondiapausing stages of insects that do have a diapause (Denlinger 1991). Although diapause usually enhances cold hardiness, a respectable level of cold hardiness can be attained in nondiapausing stages or species by long-term acclimation to low temperatures or within minutes or hours through **rapid cold hardening** (Lee et al. 1987a, Lee and Denlinger 2010). Like quiescence, rapid cold hardening allows quick tracking of temperature change and deployment of protective mechanisms. Cold hardening, including rapid cold hardening in nondiapausing individuals, is a complex response that may include elevation of polyols, sugars, amino acids, heat shock proteins, alterations of the cell membrane, and activation of p38 mitogen-activated protein kinase (MAPK) (Lee and Denlinger 2010). In other words, cold-hardening responses in nondiapausing insects evoke many of the same mechanisms noted during cold hardening of diapausing insects. The difference is the extent of protection generated. The level of cold hardening during diapause commonly exceeds that attained by nondiapausing individuals (Denlinger 1991).

4.3.3   Extended Life Cycles without Diapause

A frequently touted advantage of diapause is the idea of bet-hedging, evoking diapause in at least a portion of the population to increase chances that some individuals will survive a calamity. But, there are mechanisms other than diapause that can achieve this same outcome. Prolonging development by spreading the risk across different life stages could achieve the same end.

For the cricket *Modicogryllus siamensis* the alternative to continuous nondiapause development under long days at 25°C is not diapause but a short-day **prolongation of nymphal development** (from approximately 50 to 130 days from egg hatch to adult eclosion) (Figure 4.2), a trait accompanied by an increased number of nymphal molts

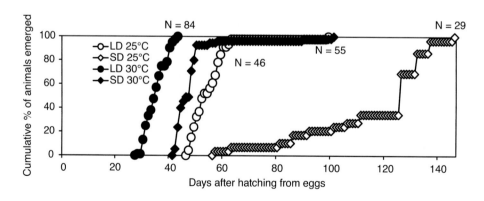

Figure 4.2 Effect of photoperiod and temperature on nymphal development time (from egg hatch to adult emergence) in the cricket *Modicogryllus siamensis* under long-day (LD 16:8) or short-day conditions (SD 12:12) at 25°C and 30°C.
From Miki et al. (2020), with permission from National Academy of Sciences, USA.

(from 7–8 to 9) and increased adult weight (180 mg vs. 320 mg)(Miki et al. 2020). The distinction is evident as well at 30°C, but the short-day delay is not as pronounced at the higher temperature. Development never really halts as it does in most diapauses, but it slows down considerably. Granted, a number of diapausing insects also continue to develop slowly during diapause, so the distinction between those cases and this response for the cricket is subjective and more a matter of degree. Short-day crickets are consistently heavier than the quickly developing long-day crickets, suggesting greater fecundity as a possible compensation for their longer development time.

Nondiapausing species of flesh flies (Sarcophagidae) present another interesting example of an alternative to the bet-hedging strategy of diapause (Denlinger et al. 1988a). Species of flesh flies from Africa, Europe, Asia, and North America diapause as pupae, and in all of these flies larval development is rapid and pupariation occurs after a rather brief period of wandering. Following adult emergence, females deposit one, two, and sometimes three clutches of larvae (these flies deposit first-instar larvae rather than eggs) and die within a month. In contrast, we have been unable to demonstrate pupal diapause, or diapause in any stage, in three neotropical species of flesh flies from Panama and Costa Rica. The species that lack diapause show a distinctly different suite of life-history traits: Instead of wandering for 2–3 days after feeding the neotropical fly larvae wander for up to 16 days, adults live up to 70 days and deposit smaller, but more, clutches (up to 5) of larvae. The neotropical flies thus spread larviposition over a much longer period, enhanced by both a longer and more variable wandering larval stage and longer adult life (Denlinger et al. 1988). This suggests that, among fly species with the capacity for diapause, risk is invested primarily in the diapausing pupal stage and other life stages are brief. The slower and more variable developmental rate and increased adult longevity in flies lacking diapause capacity desynchronizes development and permits the nondiapausing species to distribute an environmental risk over different stage of the life cycle, thus offering an alternative to diapause as a bet-hedging strategy.

An impressively **long adult life**, approaching 300 days, in several species of tropical butterflies from a Uganda forest (Molleman et al. 2007) and the subtropical tephritid *Bactocera opiliae* (Malod et al. 2020) may, like the sarcophagids, represent a diapause alternative, although the possibility of adult reproductive diapause in these studies cannot be excluded. Interestingly, all of these examples are from species inhabiting subtropical or tropical regions, suggesting that this may be an option restricted to warm areas of Earth lacking the obstacle of a formidable winter.

4.3.4   Seeking a Favorable Environment

Another viable alternative to diapause is simply to exploit an environment that remains favorable. Many household pests, including cockroaches and bed bugs, lack the capacity for diapause but have been enormously successful outside the tropics by affiliating with humans and occupying "tropical habitats" such as houses or other heated buildings. Surprisingly, the overwintering strategy for a number of important human and animal pests, such as the house fly *Musca domestica*, remain undefined, although the fly's presence in animal confinements during winter suggests it escapes the brunt of the cold season by retreating to warm habitats.

4.3.5   Remaining Active

A number of insects and spiders forgo diapause and remain active during winter, a strategy that allows them to maximize growth opportunities at a time of the year when competitors are few. Winter-active wolf spiders, *Schizocosa ocreata* and *S. stridulans*, in Kentucky remain active at the verge of their lower lethal temperatures and thus risk freezing during winter, but presumably the advantages of continuing to feed and grow during winter merit the risk inherent in living at the brink (Whitney et al. 2014).

Although food resources are frequently dramatically restricted during winter, some species appear to bridge this period by surviving on the limited resources available. The aphid parasitoid *Aphidius rhopalosiphi* thrives during the summer in Brittany on its favored aphid hosts, but this braconid parasitoid can take advantage of an array of hosts and thus can utilize the few aphid species that remain active in winter (Andrade et al. 2013). Similarly, a few volunteer maize plants and irrigated plots of maize enable the leafhopper *Dalbulus maidis* to survive the off-season in Brazil (Oliveira et al. 2013).

4.3.6   An Alternative for Life Cycle Synchronization

Synchronization of the life cycle, another key adaptive feature achieved through diapause, can also be accomplished by other means. Although neotropical flesh flies lack the capacity for diapause, they are still highly seasonal (Denlinger et al. 1991), suggesting that factors other than diapause can dictate seasonal patterns of abundance. *Opisoplatia orientalis*, an ovoviviparous cockroach found in subtropical Japan, lacks the capacity for diapause, but it exploits seasonal change in photoperiod to synchronize timing of adult eclosion and mating in the spring (Zhu and Tanaka 2004). Daylength during the fourth nymphal instar is particularly important in this species. Cohorts of this cockroach that hatch in the summer experience short daylengths during the fourth instar and in response undergo seven nymphal instars, while the autumn cohort overwinters in early instars and by the time they reach the fourth-instar spring has arrived and daylengths are long. This results in adults being produced after only six nymphal instars for the fall cohort compared to seven instars in the summer cohort and yields synchronous emergence of adults from these two cohorts. Adults produced after seven nymphal instars are larger and produce more eggs than adults emerging after six instars, so there is indeed some cost associated with this life-history tactic.

Thus, as pervasive as diapause is among insects, there are other ways to circumvent adverse seasons. Quiescence, prolonging developmental stages without entering diapause, enhancing defenses such as cold hardening in the absence of diapause, remaining active in a protected environment, feeding on the rare resources available during adverse times, and finding alternative means for synchronizing development are all viable tactics exploited by a few species as substitutes for diapause. Although there are alternatives for escaping the perils of seasonal adversity, the preponderance of diapause in insect life cycles suggests diapause is the best option for most species.

5 Interpreting Seasonal Cues to Program Diapause Entry

How do insects know when to engage the diapause program? This decision relies on an accurate assessment of seasonal environmental cues and translation of this information into a decision to either continue development or to bring it to a halt. Deciding the optimal time to enter and exit diapause is central to the success of this life-history strategy. Reading the right signals allows the insect to avoid inimical seasons, choose a period when food is available, and to synchronize population emergence at the end of diapause. Entering diapause too early results in excessive energetic costs but waiting too long is a gamble that may prove lethal. Thus, the insect needs to reliably predict when the adverse season is approaching and not wait until the adversity is on hand. For species with an obligate diapause the decision to enter diapause at a particular stage is encoded within their genetic makeup, thus they do not rely on seasonal cues for diapause entry, but they too frequently rely on environmental cues for reinitiating development once diapause has ended. But, for the majority of species residing in temperate latitudes, diapause is facultative as discussed in Chapter 1, meaning that they decide when to enter diapause based on some external environmental signal.

The signal most widely exploited to monitor the seasons is daylength, an immutable, precise indicator of seasonal change. The term **photoperiodism** refers to reliance on seasonal changes in daylength to regulate seasonal patterns of development, such as diapause and migration. Daylength is a classic **token stimulus**: It is not the feature that needs to be avoided, but it offers an unerring indicator foretelling the advent of the unfavorable season. Low temperature, the obstacle most temperate species need to avoid, is, by contrast, highly variable on a daily basis, and early autumn bouts of low or high temperature result in a signal that pales in comparison to the reliability of photoperiod. Yet, temperature, host plant quality and rainfall patterns frequently provide supplemental cues or may substitute for photoperiod in certain situations.

5.1   The Key Role of Photoperiod

Several excellent resources provide in-depth coverage of daily (circadian) and seasonal (photoperiodic) timekeeping mechanisms in insects (Saunders 2002), discuss molecular mechanisms regulating such rhythms (Hall 2003, Hardin 2005), and offer an overview of photoperiodism in plants, vertebrates, and invertebrates (Nelson et al. 2010).

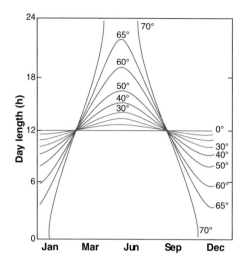

Figure 5.1 Geographic and seasonal variation in daylength (sunrise to sunset) at different latitudes (°N) in the Northern Hemisphere.
From Bradshaw and Hozapfel (2010a), with permission from Annual Reviews.

It is not my goal to extensively duplicate such information here but to briefly summarize major points relevant to the photoperiodic programming of insect diapause.

As shown in Figure 5.1, the seasonal pattern of daylength varies with latitude, yielding no annual variation on the equator and extreme variation in polar regions, ranging from constant daylight in summer to complete darkness in winter. But, at intermediate latitudes daylength progressively decreases after the summer solstice, reaching its shortest duration at the winter solstice, and thereafter increases. The annual cycle of photoperiod in the Southern Hemisphere is, of course, a mirror image of that in the Northern Hemisphere. Though there are latitudinal differences in daylength, the systematic seasonal changes in daylength at any given latitude offer an absolutely inerrant seasonal indicator allowing insects to accurately monitor the changing seasons.

Absolute daylength, defined as the interval between sunrise and sunset, is indicated in Figure 5.1, but crepuscular periods should be included in daylength calculations because most insects also interpret at least a portion of twilight as a component of the light period. Thus, a more accurate calculation of daylength in any given area would also include twilight (Beck 1968). **Civil twilight** includes the time required for the upper limit of the sun to drop 6° below the horizon, and nautical twilight includes the time for the sun to reach 12° below the horizon. Depending on the specific sensitivity of the species, either civil or nautical twilight should be included in a calculation of relevant daylength, but such calculations can be further complicated because some species distinguish between dawn and dusk light intensity, thus the two daily periods of twilight are not always weighed equally. **Astronomical twilight**, defined by the sun reaching a point 18° below the horizon, is likely of no significance for photoperiodism due to the extremely low light intensity during that interval. The duration of twilight in equatorial regions is extremely brief, while in polar regions the summer sun drops only slightly below the horizon at night, yielding high light intensity even after the sun has set.

Light plays a signaling role, meaning that the critical feature is its presence or absence. **Intensity** frequently is not important, in spite of the huge variation that occurs throughout the day. The light sensitivity threshold is normally quite low, but generally above the intensity of a full moon (0.1 lux). And, as mentioned above, the light intensity threshold for some species is much lower at dawn than at dusk (Bradshaw 1974, Saunders 1975a, Takeda and Masaki 1979). The actual threshold of light impinging on the photoreceptors may be far less than recorded from the body surface or from the surface of the substrate: Light penetrating the core of an apple is detected and used for programming diapause in the oriental fruit moth *Grapholita molesta* (=*Laspeyresia molesta*) (Dickson 1949), embryos developing within the abdomen of the flesh fly *Sarcophaga crassipalpis* perceive external daylength (Denlinger 1971), pupae of the silk moth *Antheraea pernyi* detect light penetrating their dense cocoon (Williams et al. 1965), and parasitoids developing within the body of their hosts respond to light signals external to their hosts (Maslennikova 1958). Approximately 20% of ambient light penetrates brood cells within cavities used by the leafcutting bee *Megachile rotundata*, an intensity that enables photoperiodic cues to complement thermoperiod as a cue for synchronizing spring emergence (Bennett et al. 2018). Such results underscore the fact that low intensities of light may adequately provide the needed photoperiodic signals. The sensitivity threshold, however, is not immutable: High light intensity during the photophase may raise the sensitivity threshold. For example, 0.5 lux is interpreted as dark by the fall webworm *Hyphantria cunea* after exposure to 9,000 lux but as light after exposure to 200 lux (Takeda and Masaki 1979). Though most diapause experiments use rectangular light regimes (lights turned on and off abruptly), that sort of on/off switch is not realistic and may generate results that do not reflect natural conditions. Under a rectangular design, adults of *Drosophila melanogaster* exhibit limited photoperiodicity, but a semi-natural photoregime that incorporates intensity changes boosts the photoperiodic response (Nagy et al. 2018).

Not all wavelengths of the visual spectrum are of equal importance. **Blue light** provides the most meaningful information for most species, while red wavelengths are less commonly interpreted as "light." This can be tested by supplementing diapause-inductive short days with an additional light period consisting of different wavelengths. For example, in the flesh fly *Sarcophaga crassipalpis* a 12 hours photophase induces diapause while a 15 hours photophase results in nondiapause, but by adding 3 hours of light of different wavelengths to the 12 hours photophase, it is possible to evaluate what portion of the spectrum is interpreted as "light." As shown in Table 5.1, when a 12 hours diapause-inductive photophase is supplemented with 3 hours of light with wavelengths exceeding 550 nm, pupae still enter diapause, indicating that light of those wavelengths is not seen. By contrast, wavelengths less than 540 nm are

Table 5.1 Action spectrum for induction of pupal diapause in the flesh fly *Sarcophaga crassipalpis*.

Wavelength (nm)	N	Diapause (%)
390–440	347	6.1 + 4.8
490–535	921	1.2 + 0.3
500–540	436	1.4 + 1.0
550–595	89	97.7 + 1.6
550–620	362	90.0 + 13.3
550–650	545	98.0 + 0.7
580–660	413	95.2 + 2.8
650–750	550	95.2 + 3.9

Exposing photosensitive embryos to a diapause-inductive 12 hours photophase extended by 3 hours of light with restricted wavelengths reveals wavelengths involved in the photoperiodic response. Wavelengths greater than 540 nm are not perceived, yielding a short day response (diapause), while shorter wavelengths are recognized, resulting in a long day response (nondiapause).

Source: Adapted from Gnagey and Denlinger (1984).

interpreted as light, thus yielding a long day response of no diapause (Gnagey and Denlinger 1984). This is a classic blue-light mediated response, as also noted, for example, in the midge *Chaoborus americanus* (Bradshaw 1974), aphid *Megoura viciae* (Lees 1981), and spider mite *Tetranychus urticae* (van Zon et al. 1981). But, in some species sensitivity is fairly broad and may also incorporate red portions of the spectrum, as noted in the Colorado potato beetle *Leptinotarsa decemlineata* (de Wilde and Bonga 1958), pink bollworm *Pectinophora gossypiella* (Pittendrigh et al. 1970), and parasitoid *Nasonia vitripennis* (Saunders 1975). And, sensitivity may differ according to time of day, as nicely demonstrated for pupal diapause induction in the flesh fly *Sarcophaga similis* (Goto and Numata 2009): Early in the scotophase flies are sensitive to light of wavelengths 470 nm or shorter, but late in the scotophase they are sensitive to a much broader range of wavelengths, ranging from 395 to 660 nm. Action spectra can be used to predict characteristics of the pigment used in photo-reception, but there are pitfalls in such interpretations, especially because proteins associated with pigment molecules may alter absorption properties and shift peaks of sensitivity (Cheesman et al. 1967).

Light (**photophase**) and dark (**scotophase**) segments of the daily cycle are not of equal value in photoperiodic measurement. The scotophase is usually more important. Again, this can be illustrated in the flesh fly *S. crassipalpis* (Gnagey and Denlinger 1984). In this species, exposure to 13.5 hours of light/day or less induces pupal diapause, but this could also mean that the fly requires at least 10.5 hours of darkness. Which is more important? To answer that question, a long-day photophase (15 hours) can be combined with a short-day scotophase (12 hours), and a short-day photophase (12 hours) combined with a long-day scotophase (9 hours). As shown in Table 5.2, it is clearly scotophase length that dictates the diapause response in this species. The central importance of the scotophase is well documented for numerous insects (Beck 1980, Saunders 2002), including more recent examples from moths *Lobesia botrana*

Table 5.2 Pupal diapause incidence in the flesh fly *Sarcophaga crassipalpis* when photosensitive embryos are exposed to various combinations of "long" or "short" photophases or scotophases yielding cycles of 21, 24, and 27 hours.

Embryonic photoperiod	N	Diapause (%)
15L:9D	309	0
12L:12D	645	98.2
12L:9D	393	0
15L:12D	479	94.3

Results demonstrate that scotophase, not photophase, length governs diapause expression.
Source: Adapted from Gnagey and Denlinger (1984).

(Roditakis and Karandinos 2001), *Pseudopidorus fasciata* (Hua et al. 2005a), *Scrobipalpa ocellatella* (Ahmadi et al. 2018), and *Ostrinia furnacalis* (Yang et al. 2014). There are, however, examples in which photophase is more important than scotophase, for example, the moth *Thyrassia penangae* (He et al. 2009). In this species, all light:dark cycles from LD 12:12 to 12:72 generate a short-day diapause response, while cycles from LD 14:4 to 14:72 result in a long-day nondiapause response, clearly indicating that this moth is measuring short days, not long nights.

Critical daylength (CDL), also referred to as **critical photoperiod**, defines the inflection point in a curve plotting diapause incidence as a function of daylength. For example, the critical daylength for programming pupal diapause in populations of the flesh fly *Sarcophaga bullata* from Illinois (40° 15′ N) and Missouri (38° 30′ N) is 13.5 hours at 25°C (Figure 5.2) (Denlinger 1972b). At longer daylengths, very few enter diapause and at shorter daylengths, nearly all enter diapause. The transition between generating a diapause and nondiapause phenotype is remarkably rapid in this species, occurring within a narrow 15 minutes window, a response that underscores the amazing precision of insect timekeeping mechanisms. Transitions are sometimes not quite so rapid, for example, in the stink bug *Nezara viridula* the CDL is much broader, generating a response curve with the transition occurring over the time course of an hour (Musolin and Numata 2003a). Although there are a few exceptions, at the level of the individual, the photoperiodic response is all-or-none. Either the insect fully enters diapause or it does not. At the population level, not all individuals respond the same way, but at the individual level it is a binary switch, the insect either enters diapause or it does not.

From an ecological perspective, daylengths in Figure 5.2 that are shorter than 9 hours 20 minutes or longer than 15 hours 3 minutes are not biologically relevant for the Illinois population of *S. bullata* because such daylengths do not naturally exist at 40° N (Denlinger 1972a), and even the shortest daylengths that do occur naturally are not particularly relevant because the flies are already in diapause by that time. The tapering off of the photoperiodic response curve seen for the Illinois population of *S. bullata* at biologically irrelevant short daylengths is a common response, but one that is not fully understood. Such responses, however, offer clues that may be helpful for interpreting clock mechanisms (Saunders 2020a).

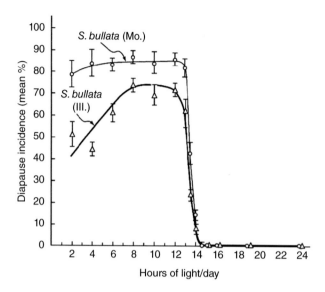

Figure 5.2 Photoperiod response curves for populations of the flesh fly *Sarcophaga bullata* from Illinois (40° 15′ N) and Missouri (38° 30′ N) at 25°C. Critical daylength for the programming of pupal diapause in both populations of this long-day species is 13.5 hours of light/day. From Denlinger (1972b).

Critical photoperiod curves have been defined for hundreds of insect species, many of which have been cataloged by Danilevskii (1965), Beck (1980), Tauber et al. (1986), Danks (1987), and Saunders (2002). The most common pattern (Figures 5.2 and 5.3A) is for diapause to be programmed by short daylengths. Although this is sometimes referred to as a long-day photoperiod curve because it promotes development at long daylengths (Beck 1980), I prefer to emphasize the diapause response and thus refer to this as a short-day response, nomenclature more in keeping with common usage in the animal and plant literature. This is by far the most commonly described type of photoperiodic response, but we must recognize the bias based on the fact that the preponderance of experimenters are working on temperate latitude species. Insects relying on this type of response foretell the advent of winter by detecting the short daylengths of late summer and respond by entering an overwintering diapause.

A mirror image of the short-day response curve, the long-day response curve, promotes diapause entry only at long daylengths, a response commonly seen for induction of summer diapause or aestivation (Figure 5.3B). Examples beyond those provided by Danilevskii (1965) include response curves for summer diapause in the stink bug *Picromeru bidens* (Musolin and Saulich 2000) and the white butterfly *Pieris brassicae* (Spieth et al. 2011). *P. brassicae* has both a winter and summer diapause; short days trigger winter diapause, while long days trigger summer diapause, a response similar to that shown in Figure 2.2 for summer and winter diapause in *Mamestra brassicae* (Grüner and Masaki 1994).

Although less common, photoperiodic response curves may reveal only a narrow range of daylengths that are diapause-inductive (Figure 5.3C), as seen for pupal

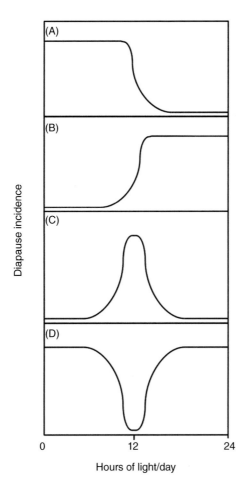

Diapause incidence

(A)

(B)

(C)

(D)

0   12   24

Hours of light/day

Figure 5.3 Schematic representation of photoperiod response curves for (A) an insect that enters diapause only at short daylengths, (B) an insect that enters diapause only at long daylengths, (C) an insect that enters diapause only in response to a narrow range of daylengths, and (D) an insect that enters diapause in all but a narrow range of daylengths. Hour designations are arbitrary.

diapause in the moth *Scrobipalpa ocellatella* (Ahmadi et al. 2018) and for adult diapause in the plant bug *Lygus hesperus* (Beards and Strong 1966), or at the opposite extreme, a narrow range of daylengths that do not program diapause (Figure 5.3D), as seen for programming adult summer diapause in the cabbage butterfly *Pieris melete* (Xiao et al. 2008a) and the coccinellid beetle *Epilachna admirabilis* (Imai 2004). In these cases, diapause is either the dominant response across a range of photoperiods or can be induced during only a narrow seasonal window. Diapause or a lack thereof in insects with such response curves can easily be overlooked unless responses are carefully monitored over a wide range of daylengths. And, in some cases, the shape of the response curve within a single species may differ depending on temperature. For example, at temperatures below 21.5°C, the response curve for the stalk borer

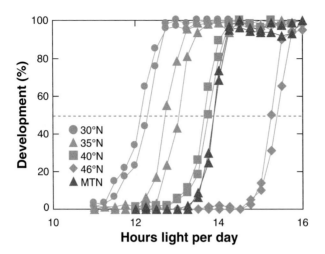

Figure 5.4 Geographic variation in photoperiodic response for larval diapause in populations of the pitcher plant mosquito *Wyeomyia smithii* from different localities in eastern North America. MTN (red) is a mountain population collected at 900–1,000 m, 35° N. Note the MTN shift in response from that noted in a low altitude population from the same latitude. Critical daylength is indicated by dashed line. (A black and white version of this figure will appear in some formats. For the color version, please refer to the plate section.)
From Bradshaw and Hozapfel (2010a), with permission from Annual Reviews.

Sesamia nonagrioides is similar to that shown in Figure 5.3A, but at 25°C, no diapause is seen at very short daylengths and the curve at that temperature more closely resembles the curve depicted in Figure 5.3C (Eizaguirre et al. 1994). Thus, the shape of the curve is not an immutable feature of diapause for any particular species.

What is also clear from Figure 5.1 is that a short daylength at one latitude may not be interpreted as such at another latitude, thus this environmental information is site-specific and finely tuned to the local seasonal environment (Hut et al. 2013). A rule of thumb, proposed by Danilevskii (1965) and supported by many more recent observations, is that critical daylength increases by approximately 1 hour for every 5° increase in latitude. A refinement of this relationship, based on a large dataset (57 studies) proposes a linear increase of 48 minutes per 5° increase in latitude (Joschinski and Bonte 2021). The example in Figure 5.4 for populations of the pitcher plant mosquito *Wyeomyia smithii* across a range of latitudes between 30 and 46° N nicely demonstrates the latitudinal shifts commonly seen for distinct geographic populations (Bradshaw et al. 2003b, Bradshaw and Holzapfel 2010a).

The CDL for programming adult diapause in the Colorado potato beetle *Leptinotarsa decemlineata* varies systematically across Europe from extremes of 15.2 hours in Padua, Italy (45° 48′ N) to 16.8 h in Petroskoi, Russia (61° 49′ N) (Lehmann et al. 2015b). The latitudinal shift in CDL is especially steep for populations of the tiger swallowtail *Papilio glaucus* in eastern North America: CDL increases by 0.4 hours with each increasing degree in latitude (Ryan et al. 2018). In Japan, CDLs for pupal diapause in *Sarcophaga similis* range from approximately 12.5 hours in the south (32.7° N) to 14 hours in the north (43.2° N) (Yamaguchi and Goto 2019). An especially notable study using 41 populations of *Drosophila littoralis* ranging from the Black Sea (41.6° N) to northern Finland (69.0° N) reports CDLs ranging from 11.6 to 20.3 hours for adult diapause induction (Lankinen 1986). Within Finland, the CDL for adult diapause in *Drosophila montana* ranges from 16.6 hours in the south (60° 59′ N) to 18.8 hours in the north (67° 06′ N) (Tyukmaeva et al. 2011), and a broader dataset based on four latitudinal clines of *D. montana* from both North America and Europe shows a consistent 1 hour increase in CDL for each 5° increase in latitude (Tyukmaeva et al. 2020). But, the CDL also is responsive to local adaptation. For example, populations of *D. montana* from a harsh environment in the Kamchatka Peninsula in Asia (56° N) have a growing season that is quite short (120 days) and a CDL quite similar to populations from Finland and Alaska (60–61° N) that have a similarly short growing seasons (117–122 days).

Numerous additional examples of systematic geographic shifts in CDL as a function of latitude are provided by previous publications, especially in the work of Danilevskii (1965). As latitude increases, summer daylengths are longer, winter arrives earlier and the climate is usually more severe, thus prompting most insects to enter diapause at longer daylengths than are used to program diapause in more southern populations of the same species. And, as Alexander von Humboldt astutely noted over 200 years ago, from the perspective of climate change, a few kilometers of elevation are equivalent to thousands of kilometers of latitude at sea level. This point is nicely demonstrated for CDL in the mosquito *Ochlerotatus triseriatus* (Holzapfel and Bradshaw 1981): A 1° change in latitude is equivalent to a 254 m altitude change. The altitude effect (nearly a 45-minute shift) is also evident in a comparison of high and low altitude populations of *W. smithii* collected at 35° N, as noted in Figure 5.4. Altitudinal adaptations of the life cycle are also well documented for the crickets *Allonemobius fasciatus* (Tanaka and Brookes 1983) and *Phlidoptera griseoaptera* (Černecká et al. 2021), as well as numerous other examples (Hodkinson 2005). For most insect populations the photoperiodic response is finely tuned to their specific geographic site. Although fidelity to photoperiod as a seasonal signal is usually a highly desirable trait, it does present an obstacle in a changing climate, as discussed in Chapter 11.

Occasional CDL observations defy expectations for latitude. For example, the CDL of the green stink bug *Nezara viridula* in Japan varies between 13.5 and 12.25 hours within latitudes ranging from 26 to 34° N, but there is no obvious relationship with latitude (Musolin et al. 2011), possibly because this species is highly migratory and can also be carried long distances by typhoons, resulting in a destabilized photoperiodic response. The beet webworm, *Loxostege sticticalis*, is remarkable in having a nearly invariant CDL of 14 hours across its broad geographic distribution in Asia, Europe, and North America (Kutcherov et al. 2015), a result suggesting that this highly mobile species may adapt to contrasting environments using alternative adjustments of life-history traits.

5.2   Do Insects Monitor Changes in Daylength?

Two major scenarios can be envisioned for monitoring daylength. The insect could simply measure length of the day and "decide" whether the day should be interpreted as being long or short. In this case, it doesn't matter if daylength is increasing, decreasing, or stationary. For example, all long days greater than the CDL would be considered of equal value and are interpreted as nondiapause-inducing, while all daylengths shorter than the CDL are also of equal value and are interpreted as diapause-inducing. What is important is simply the length of the day in relation to the CDL. Alternatively, the insect could monitor change in daylength, determining whether days are progressively getting shorter or longer. Although the former is more common, there are good cases documenting a response to change in daylength in a few species.

In cases where the insect is sensitive to daylength during a short interval, measuring changes in daylength would be problematic. For example, in the flesh fly *Sarcophaga crassipalpis* the period of sensitivity to daylength is restricted to a 4-day window (Denlinger 1971), making detection of changes in daylength unlikely, and indeed these flies and many others appear to rely on a simple binary decision interpreting the daylength as being either short or long. Numerous examples cited in previous reviews (Danilevskii 1965, Beck 1968, Tauber et al. 1986, Danks 1987) support responses that rely on measurement of absolute daylength rather than changing daylength, and a more recent example includes induction of larval diapause in the Indianmeal moth *Plodia interpunctella* (Kikukawa et al. 2008). In this species, a gradual increase or decrease of photoperiod (2 minutes/day) yields an effect no different than a step-wise change.

Some of the earliest and most convincing evidence showing that some insects do indeed respond to change in daylength rather than absolute daylength is provided by experiments on adult diapause in the lacewing *Chrysopa carnea* (Tauber and Tauber 1970b, 1973). In this species progressive shortening of daylength both increases diapause incidence and diapause duration, assuring that diapause persists at least through the winter solstice. The incidence of embryonic diapause in the Australian plague locust *Chortoicetes terminifera* progressively increases with a gradual decline in daylength, suggesting a response that is graded rather than an abrupt response to a specific threshold (Deveson and Woodman 2014). In the Mediterranean tiger moth *Cymbalophora pudica* a change in daylength during larval life causes a dramatic effect on diapause duration (Koštál and Hodek 1997). Short constant photoregimes result in a long diapause (88 days) while long constant photoregimes elicit a short diapause (52 days), but if larvae are switched to longer or shorter photoregimes diapause can be extended up to 138 days or reduced to 10 days.

But, as we see for many diapause responses, there are cases that do not fall neatly into one of these two categories but instead show a qualitative response that differs with length of the day and developmental stage that receives the photoperiodic cues (Tauber et al. 1986). For example, the corn earworm *Helicoverpa zea* enters pupal diapause in response to short daylengths, but the response is considerably greater when mothers and eggs receive long days and only the larvae receive short days (Wellso and Adkisson 1966). Exposure to only short days throughout development is suboptimal for the programming of diapause in this species.

5.3   When Are Photoperiodic Cues Received?

The **photosensitive period** refers to the developmental window used to detect the photoperiodic cues involved in programming diapause. Daylengths outside this developmental window are irrelevant. This sensitive period can be as short as a single day, for example, the phantom midge *Chaoborus americanus* (Bradshaw 1969), or, more frequently, may extend over many days and across several developmental stages. The sensitive period usually occurs well in advance of the onset of diapause, thus enabling the insect to engage in preparatory steps that enhance the diapause response (Chapter 6), but in some cases the sensitive period extends until shortly before diapause entry.

Examples in Figure 5.5 represent a range in lengths and timing of the photosensitive period. In the silkworm *Bombyx mori*, the photoperiod received by the mother determines whether her female progeny will lay diapausing eggs (Kogure 1933). Transgenerational effects such as this are discussed more fully in Section 5.13. In the flesh fly *Sarcophaga crassipalpis*, a 4-day window (last 2 days of embryonic development and first 2 days of larval life) is the sensitive period for programming pupal diapause (Denlinger 1971). As in *S. crassipalpis* highest sensitivity for pupal diapause programming in *S. similis* also occurs during development of the embryo within the uterus of the female (Tanaka et al. 2008). In the tobacco hornworm, *Manduca sexta*, the photosensitive period is especially long, extending from embryonic development through to the final larval instar, shortly before the onset of pupal diapause (Denlinger and Bradfield 1981). A wide range in timing and duration of the photosensitive period is noted, a feature that likely reflects adaptive peculiarities that are species-specific.

5.4   Central Role for the Brain in Photoreception and Storage of the Diapause Program

The brain is central to both reception of photoperiodic cues and translation of this information into a meaningful developmental program that prompts both the onset and termination of diapause. The photoperiodic signals are either perceived directly by photoreceptors in the brain (**extraretinal photoreception**) or channeled to the brain through the **compound eyes** of adults or **stemmata** of larvae. In some systems both conduits are operative. Ocelli appear not to be important for photoperiodism. A newly found photoreceptor, present in *Drosophila melanogaster*, could also be a conduit for photoperiodic signals (Helfrich-Förster et al. 2002): Beneath the posterior margin of

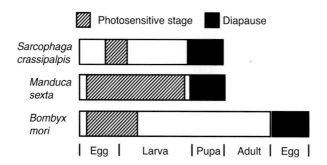

Figure 5.5 Relationship between the photosensitive period and the stage of diapause in three insect examples.
From Denlinger (1985).

the compound eye lies a pair of neurons, **extraretinal eyelets**, that contain rhodopsin and have axons projecting into the pacemaker center of the brain. These neurons participate in regulating circadian rhythms, but whether they are also involved in photoperiodism is unknown.

In a classic experiment with aphids, A.D. Lees used thin rays of light directed at either the compound eyes or the cuticle overlying the brain (Lees 1964). Aphids fail to respond to light directed on the eyes, but are responsive to light impinging on the brain, thus nicely demonstrating that light used for the aphid's photoperiodic response is channeled directly to the brain. Extraretinal reception is fairly common, and some insects enhance channeling of light into the brain through special structures, as seen in translucent windows over the brain in photosensitive larval stages of the cabbage butterfly *Pieris brassicae* (Claret 1966) and in pupae of several giant silk moths, including *Antheraea peryni* (Williams et al. 1965), as shown in Figure 5.6. Blackening visual centers with an opaque paint or destruction of such centers using thermocautery fail to eliminate the photoperiodic response in most species, for example, Colorado potato beetle *Leptinotarsa decemlineata* (de Wilde et al. 1959), further underscoring the role of extraretinal photoreception.

But, reception of photoperiodic cues through the compounds eyes may be nearly as prevalent as extraretinal photoreception. The involved receptors are restricted to a small portion of the compound eye. By selectively removing ommatidia from various regions of the eye in the bean bug *Riptortus clavatus*, Morita and Numata (1997) demonstrated that ommatidia located only in the central region of the eye are involved. Compound eyes are also the conduit for photoperiodic information in the carabid beetle *Pterostichus nigrita* (Ferenz 1975). Beetles retain photosensitivity when one of the compound eyes is destroyed by thermocautery, but the response is lost if both eyes are destroyed. Phosphorescent paints are another tool that can be used for determining regions of light sensitivity. These paints absorb light energy and emit phosphorescence in the dark. When such paints are applied to compound eyes of the bug *Poecilocoris lewisi* the photoperiodic response is greatly diminished, while the diapause response remains intact if the central region of the head is painted

Figure 5.6 Pupa of the giant Chinese silk moth *Antheraea pernyi*, showing the translucent window over the brain that channels photoperiodic cues directly to the brain.
Photograph courtesy of Yanqun Liu (Shenyang Agricultural University).

(Miyawaki et al. 2003). Compound eyes are the site of entrainment of circadian rhythms in the cricket *Gryllus bimaculatus*, and experiments based on severance of one of the optic nerves demonstrates cross talk between the two lobes by coupling through medulla bilateral neurons, a communication channel possibly regulated by serotonin and pigment dispersing factor (Koga et al. 2005).

Though adults of *P. nigrita* channel light through their compound eyes, another beetle, *L. decemlineata*, relies on extraretinal photoreception (de Wilde et al. 1959). The site of photoreception can even differ at different stages of development within a single species. Larvae of the carabid beetle *Leptocarabus kumagaii* switch the site of photoreception at metamorphosis: Larval diapause is programmed by light impinging on the stemmata, but light used to program adult diapause is received through the compound eyes (Shintani et al. 2009). At metamorphosis the stemmata migrate into the brain, remain rudimentary within the optic lobe, and cease to function in photoperiodism. Larvae of another beetle, the yellow-spotted longicorn beetle *Psacothea hilaris*, do not channel light through their stemmata but instead rely exclusively on extraretinal photoreception (Shintani and Numata 2010). Among Diptera, the brain of the adult calliphorid *Calliphora vicina* hosts the photoreceptor (Saunders and Cymborowski 1996), but the compound eyes serve that function in a closely related calliphorid, *Protophormia terraenovae* (Shiga and Numata 1997). There appear to be no good phylogenetic correlations defining which species rely on retinal versus extra-retinal photoreceptors (Numata et al. 1997, Shiga and Numata 2007, Goto et al. 2010),

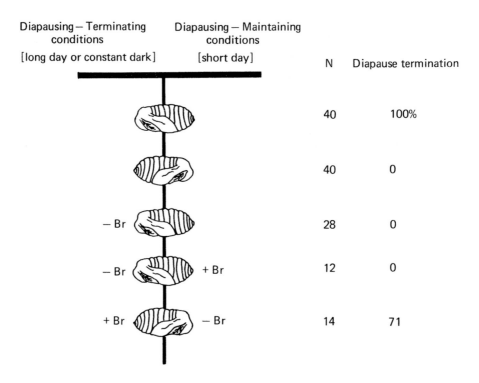

Diapausing – Terminating conditions [long day or constant dark]	Diapausing – Maintaining conditions [short day]	N	Diapause termination
		40	100%
		40	0
– Br		28	0
– Br	+ Br	12	0
+ Br	– Br	14	71

Figure 5.7 Schematic representation of the divided photoperiod chamber used to identify the pupal brain of *Antheraea pernyi* as the photoreceptor site that coordinates the diapause response. Adapted from Williams and Adkisson (1964), as presented in Denlinger (1985), with permission from Elsevier.

although some trends are emerging. For example, extraretinal photoreception seems to be particularly common among Lepidoptera, whereas reliance upon both compound eyes and direct brain photoreception is evident among Hemiptera, Coleoptera, and Diptera.

Surgical transplants provide some of the most powerful evidence revealing the role of the brain as the ultimate recipient of photoperiodic information. The silk moth *Antheraea pernyi* breaks pupal diapause in response to long daylength or constant darkness. Using a divided photoperiod chamber Williams and Adkisson (1964) exposed heads and abdomens of diapausing pupae to contrasting photoperiods (Figure 5.7). Among intact pupae, only those whose heads are exposed to long daylengths terminate diapause. Pupae whose brains are extirpated fail to develop. Brains transplanted into abdomens exposed to short daylengths fail to stimulate development, but brains transplanted into abdomens exposed to diapause-breaking conditions stimulate development. Sensitivity can thus be transferred from the head to the abdomen with a brain transplant. A similar response is seen in photosensitive larvae of *Pieris brassicae* (Claret 1966, Seuge and Veith 1976): Brains transplanted into abdomens of larvae or pupae convey the capacity for photoreception to the abdomen. Experiments with pupal diapause of the flesh fly *Sarcophaga crassipalpis*

(Giebultowicz and Denlinger 1986) and adult diapause in the linden bug *Pyrrhocoris apterus* (Hodková 1992) also provide evidence that the diapause program resides within the brain and can be transferred from one individual to another by transplantation of the programmed brain.

Even more striking are experiments that program the brain with short daylengths *in vitro* and then monitor the diapause response in recipients receiving those brains. Such brains elicit diapause when implanted into debrained pupae of the tobacco hornworm *Manduca sexta* (Bowen et al. 1984). Likewise, larval brain-subesophageal ganglion complexes of the commercial silkworm *Bombyx mori* can be photoperiodically programmed for diapause *in vitro* and prompt a diapause response when re-implanted into larvae (Hasegawa and Shimizu 1987). Such experiments provide some of the most compelling evidence for extraretinal photoreception and underscore the fact that the brain dictates the diapause program.

In addition to functioning as the receptor of photoperiodic information, the brain is also the repository of the diapause program. Transplanted brains from the tobacco hornworm *M. sexta* retain their commitment for pupal diapause while brains from nondiapause-programmed pupae remain committed to continuous development (Safranek and Williams 1980). The brain of *M. sexta* also clearly determines how long diapause should last. In this species the duration of diapause is variable, depending upon how many short days it receives (Denlinger and Bradfield 1981). If larvae receive many short days, indicative of early autumn, the diapause is short, but if they receive only a few short days late in larval development, indicative of late summer, pupae remain in diapause much longer, a distinction that enables pupae to fine-tune their diapause duration based on when they enter diapause. This program for diapause duration can be transferred readily to other pupae by brain transplantation (Figure 5.8), thus implying that the brain is not only the receptor of environmental information but also the site where such information is stored.

Identifying precisely where this information is received and stored within the brain is more challenging (Goto et al. 2010). For insects that utilize extraretinal photoreception, the neurons involved should contain photoreceptor pigments, as well as clock genes that monitor time, and they should be closely aligned with cells that evoke hormonal responses known to regulate diapause. In the silk moth *Antheraea peryni*, photoreceptors are located within the dorsolateral region of the brain, as demonstrated by implanting select portions of the brain (Williams 1969). This same region contains lateral neurosecretory cells that produce prothoracicotropic hormone (PTTH), the neuropeptide that regulates pupal diapause (Sauman and Reppert 1996). Ablation of a specific set of lateral neurosecretory cells (type-Ia$_1$) from brains of *Manduca sexta* larvae prevents entry into pupal diapause, thus implicating these cells in photoperiodism (Shiga et al. 2003b). These cells, comparable to cells that produce the circadian clock protein PERIOD in *A. peryni*, have processes in close proximity to processes from PTTH cells. Thus cells involved in photoreception and the effector region that produces PTTH are located in close proximity within the brain, an ideal location for communication.

The **dorsal protocerebrum** is implicated in the photoperiodic response in diverse species (Shiga and Numata 2007). In the vetch aphid, *Megoura viciae*, antibodies

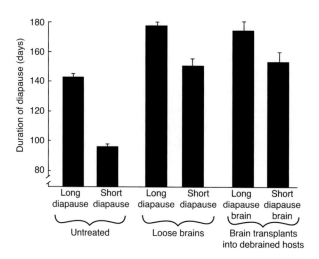

Figure 5.8 Transfer of a program for long or short diapause duration in pupae of the tobacco hornworm, *Manduca sexta,* by brain transplantation. Pupae were programmed for short diapause by being held at short days throughout larval development. Long diapause was programmed by transferring larvae from long day to short day midway through larval development.
From Denlinger and Bradfield (1981).

directed against opsins and phototransduction proteins consistently label an anterior ventral neuropil region within the protocerebrum (Gao et al. 1999), a region consistent with previous experiments using localized illumination and microlesions to identify the photoreceptor region. In the monarch butterfly *Danaus plexipus* the central circadian clock involved in the migration response is localized to four cells in the dorsolateral protocerebrum (Sauman et al. 2005). Prime candidates as regulatory sites for diapause are cells homologous to eight clusters of brain cells in *Drosophila melanogaster* that have been implicated in circadian rhythmicity: ventral lateral neurons (s-LN$_v$, l-LN$_v$), dorsal lateral neurons (LN$_d$), dorsal neurons (DN$_3$, DN$_{1p}$, DN$_{1a}$, DN$_2$), and lateral posterior neurons (LPN) (Helfrich-Förster 2003, 2006). These cell clusters are interconnected and coordinate daily rhythmic cycles of behavior. Although there is growing evidence that the circadian clock also is involved in the diapause response, as discussed below, whether these cells function the same way in coordinating the diapause response remains unknown.

Circadian clock cells that synthesize the protein **pigment-dispersing factor** (PDF) innervate neurosecretory cells in the dorsolateral protocerebrum of the blow fly *Protophormia terraenovae* (Shiga and Numata 2009). PDF is an important output of many circadian clock cells, thus suggesting a site in the protocerebrum that may be key to regulating the diapause program. By contrast, mushroom bodies do not appear to be involved. Chemical ablation of mushroom bodies using hydroxyurea destroys olfactory learning in the fly *P. terraenovae* but not the photoperiodic response of diapause (Ikeda et al. 2005). In adults of the bean bug *Riptortus pedestris* an anterior region of the medulla contains PDF-immunoreactive cell bodies that, when removed, eliminate the photoperiodic response (Ikeno et al. 2014).

R. pedestris also offers an interesting link between specific large neurons in the **pars intercerebalis** and the core clock gene *per* (Hasebe and Shiga 2021). Frequent bouts of spontaneous firing are noted in these neurons under long-day conditions that result in reproduction, while these neurons are silent under short days that lead to diapause. Knockdown of *per* with RNAi diminishes the photoperiodic response of these neurons. The change in firing frequency is suggested to regulate release of several neuropeptides, including insulin-like peptides and diuretic hormone 44. This example thus provides a powerful connection within the central brain between the photoperiodic signal, expression of *per*, activity of key neurons in the pars intercerebralis, neuropeptide release and the downstream enactment of the diapause program.

Thus, in all of these cases, the complete brain is not needed, but instead a limited number of cells within the brain are essential for receiving and storing the diapause program. But, programming of diapause involves not only reception of the photoperiodic signal but also downstream components of the pathway. Disruption of both input and output components of the timekeeping mechanism can result in an altered diapause response, thus when the diapause response is disrupted it is not always easy to be certain which step has been derailed.

5.5   Photoreceptor Pigments

For light to exert its biological effect it must interact with a pigment, a reaction that marks the first step in translating the light signal into a meaningful photoperiodic response. Identifying such pigments and knowing where they are located offer a door to the pathway leading to diapause. Two major classes of pigments are involved (Goto et al. 2010, Saunders 2012): a blue-light-sensitive flavoprotein **cryptochrome** (CRY), and an orange-red light-sensitive **carotenoid/Vitamin A-based photopigment conjugated with opsin**.

CRY is directly linked to the circadian pacemaker of the fruit fly *Drosophila melanogaster*, where it acts to daily reset the clock at the onset of the photophase (Stanewsky et al. 1998, Hall 2003, Peschel and Helfrich-Förster 2011). Two forms of CRY, CRY1 and CRY2, are present in most insects (Yuan et al. 2007), but only CRY1 is known from *D. melanogaster*, a feature that underscores the need for caution in generalizing results observed from this model insect. In *D. melanogaster*, CRY1 binds directly to the clock gene TIMELESS (TIM) in the presence of light and degrades both TIM and itself, thus it plays a direct role in transduction of the light signal. CRY2, more recently discovered in the monarch butterfly *Danaus plexippus* (Zhu et al. 2005, Zhang et al. 2017), as well as several Hymenoptera and lower Diptera, differs from CRY1 in that it is light insensitive, thus it is not involved directly in reception of the light message. CRY2 acts as a repressor of two of the clock genes, CLOCK and CYCLE, and appears to replace the role of TIM in some species. Thus, CRY2 plays an important role in clock function but not at the level of light recognition. Several species lack CRY1, for example, the parasitoid *Nasonia vitri-pennis*, the honey bee *Apis mellifera* and the red flour beetle *Tribolium casaneum* (Yuan et al. 2007, Bertossa et al. 2014, Benetta et al. 2019), but thus far, a substitute for the transducing photoreceptor CRY1 has not been found in these species.

The carotenoid/Vitamin A/opsin link was first noted in association with diapause photoreception by rearing insects and mites on carotenoid-free diets. In several species, the diapause response is lost when dietary carotenoids are absent, for example, adults of the predatory mite *Amblyseius potentillae* (Veerman et al. 1985) and spider mite *Tetranychus urticae* (Bosse and Veerman 1996), embryos of *Bombyx mori* (Shimizu and Kato 1984), and pupae of *Pieris brassicae* (Claret and Volkoff 1992). Adding carotenoids back into the diet restores the capacity for diapause. Carotenoids, in turn, exert their effect through association with the visual pigment opsin. The compound eyes, implicated in regulation of nymphal diapause in the cricket *Modicogryllus siamensis* (Sakamoto and Tomioka 2007), contain three distinct opsins responsive to ultraviolet, blue and green light, respectively (Tamaki et al. 2013). All three appear to be involved in the diapause response as shown by knock-down experiments using RNAi (Tamaki et al. 2013). RNAi effects for ultraviolet- and blue-sensitive opsins are more pronounced than that for green-sensitive opsin under long-day conditions, whereas a response is noted only for ultraviolet-sensitive opsin under short days, results suggesting that different opsins play distinct roles in the photoperiodic response. The fact that ultraviolet- and blue-sensitive opsins are local-ized in the dorsal rim of the compound eye suggests this as a candidate region for photoreception involved in the cricket's diapause response. **Bocereopsin**, a novel opsin found in the brain of the silk moth *Bombyx mori*, responds to long wavelengths and is another candidate photoreceptor possibly implicated in diapause (Shimizu et al. 2001).

Opsins may also participate in the diapause response through mechanisms other than photoreception. Correlations exist between opsin expression and diapause fea-tures such as latitudinal differences in diapause incidence in the butterfly *Limenitis arthemis* (Frentiu et al. 2015), responsiveness to daylength (short vs. long days and day vs. night) in the pea aphid *Acyrthosiphon pisum* (Collantes-Alegre et al. 2018), incidence of diapause in the crustacean *Daphnia magna* (Roulin et al. 2016), and the transition from diapause initiation to diapause maintenance in the Colorado potato beetle *Leptinotarsa decemlineata* (Yocum et al. 2009b). The nature of these results suggests some yet undefined roles for opsins beyond that of photoreception. Experimental findings in the monarch butterfly are particularly convincing that the Vitamin A pathway in this species operates downstream of the clock in this species, possibly through **retinoic acid (RA) signaling** (Iiams et al. 2019). RNA-seq results show brain differences in diurnal expression that are season-dependent, and among these are changes in components of the Vitamin A pathway, including genes involved in carotene uptake (*santa-maria 1* and *2*) and conversion of Vitamin A to retinal (*rdh13, ninaB1*). When this pathway is disrupted the photoperiodic response is lost without affecting circadian rhythms. Inactivating the clock using loss-of-function mutants for the circadian activators CLOCK and BMAL1 and the circadian repressor

CRY2 abolishes rhythmic expression of genes in the Vitamin A pathway and consequently the photoperiodic response is lost. A possible link with RA signaling is especially intriguing because RA is involved in seasonal regulation of transcription and brain remodeling in mammalian photoperiodism.

In spite of the important role for carotenoids in many photoperiodic responses, most insects lack the ability to synthesize carotenoids and instead rely on plant sources to obtain this critical pigment. Interestingly, the twospotted spider mite, *Tetranychus urticae*, and a few related species have acquired the carotenoid biosynthetic gene by horizonal transfer and can therefore make their own carotenoids, which they rely upon for pigmentation as well as diapause induction (Bryon et al. 2017b). Several genes in the pathway to carotenoid synthesis are differentially expressed in association with diapause in *T. urticae* (Bryon et al. 2013).

Insects may not rely exclusively on one or the other type of pigment. The fact that both CRY1 and opsin-based photoreceptor systems cooperate in behavioral responses evoked by UV light in *D. melanogaster* (Baik et al. 2017) suggests that similar coordinated roles for the two pigment systems are possible in other photoperiodic responses. Light signaling that regulates the evening activity cycle in *D. melanogaster* relies on input from both CRY1 (extraretinal receptors) as well as light impinging on the compound eyes (Kistenpfennig et al. 2018), but the two systems exert counter effects: CRY1 limits the ability of the evening peak to track dusk, while input from compound eyes increases the ability to track dusk. Although these results do not directly relate to diapause, this observation suggests the sort of complexity that could be relevant for the differences in dawn and dusk photosensitivity mentioned above.

It seems logical that the same photoreceptor pigments used to entrain circadian rhythms would be used for photoperiodism, but this may not be the case. In the spider mite *Tetranychus urticae*, photoperiodic time measurement relies on orange-red light (580 nm and above), characteristic of a Vitamin A-based opsin, but circadian activity, monitored by the Nanda–Hamner protocol, appears to rely on a blue-light sensitive pigment, most likely cryptochrome (Veerman and Veenendaal 2003).

As the experimental base of species increases it becomes increasingly more challenging to arrive at conclusions that universally apply. Here, we have seen that different sites of photoreception are well documented (extraretinal and compound eyes), distinct regions of the brain are involved and may differ with species, and at least two pigment categories are common. But, based on phylogeny we cannot reliably predict which of these limited options operate in species that have not been studied. Yet, the possibilities are restricted to a narrow range of options. The ecological relevance of light signals varies among species depending on microhabitat, cloud cover, time of day, and seasons (Danks 2005), features that likely reflect some of the species differences noted.

5.6   Involvement of Circadian Clocks

The daily rhythms of insect life, the conspicuous patterns of feeding, locomotion, sleeping, mating, egg laying, physiological activities, as well as events such as adult eclosion that occur only once during an insect's life, are all driven by precise circadian clocks. **Circadian rhythms** have a periodicity of approximately 24 hours and persist (**free-run**) even if the organism is shifted from a daily light-dark cycle to complete darkness. Onset of the light period (**photophase**) or darkness (**scotophase**) serves as a *Zeitgeber* (time giver) that fine-tunes accuracy of the cycle, maintaining the 24 hours periodicity. Free-running periodicity under complete darkness may be slightly longer or shorter than 24 hours, but the period quickly resets to a 24 hours cycle when the insect is returned to a 24 hours light-dark cycle. The fact that oscillations persist in the absence of light cues is strong evidence that the rhythm is endogenous and not simply a direct response to prevailing light-dark cycles. Circadian cycles are also **tempera-ture compensated**, even in ectothermic organisms, so that whether temperatures are high or low, the clock cycles at the same rate, further underscoring the endogenous nature of the clock. The circadian clock accurately tells the time of day and retains a memory of time even in the absence of a *Zeitgeber*.

Such timekeeping features are also desirable for a **photoperiodic clock** that monitors seasonal changes used to program diapause, but the two clocks have different tasks. Circadian clocks primarily identify a specific time during a 24 hours light-dark cycle for an activity to occur, while the photoperiodic clock primarily measures length of the day (or, more commonly, length of the night). The number of short (or long) days are counted and used to initiate a response (diapause) at a later stage of development. Photoperiodism thus involves two distinct components: a timer that measures length of the day (or night) and a counter that calculates the number of short days (or nights) accumulated and uses this information to prompt the diapause or nondiapause response. Does this difference foretell distinctions in the clock mechanisms as well? Genetic cross experiments indicate that the circadian adult eclosion rhythm and the photoperiodic response of diapause induction are genetically distinct, but likely closely linked, as in *Drosophila littoralis* (Lankinen and Forsman 2006). Geographic populations of the pitcher plant mosquito *Wyeomyia smithii* provide good evidence that the photoperiodic timer and counter have evolved separately from the circadian clock (Bradshaw et al. 2003a,b, Emerson et al. 2009a). Similar distinctions between circadian rhythmicity and photoperiodism are reported in *D. melanogaster* (Saunders et al. 1989), the spider mite *Tetranychus urticae* (Veerman and Veenendaal 2003), and blow fly *Calliphora vicina* (Saunders and Cymborowski 2003), among others.

Several reviews debate the relationship between circadian rhythms and photo-periodism in depth (Saunders 2002, 2020a,b, Danks 2005, Bradshaw and Holzapfel 2010c, Koštál 2011, Saunders and Bertossa 2011, Goto 2013, Meuti and Denlinger 2013), thus the account provided here briefly summarizes the major points highlight-ing the relationship between these two types of clocks. Clearly, certain common clock genes are involved in regulating both circadian rhythmicity as well as photoperiodism, but the confounding feature is that a number of experiments define distinct differences in these two timing mechanisms. How is this reconciled? The alternatives are either (a) a fully integrated mechanism in which the circadian clock provides the mechanistic basis for the photoperiodic timer or (b) independent mechanisms in which the circadian clock and photoperiodic timer are distinct physiological mechanisms, with one or more genes in the circadian clock being co-opted to act pleiotropically as a time reference point (Bradshaw and Holzapfel 2010c). Unfortunately, it is not easy to determine experimentally which of these two options is correct. I share the view put forth by Saunders (2010b) and Koštál (2011), that the two types of clocks are neither identical nor completely separate but rather cooperate in generating the photoperiodic response, albeit through interactive mechanisms that remain poorly understood.

5.6.1   Classical Approaches for Evaluating Circadian Rhythmicity

The rhythmic daily movement of bean seedling leaves, even under complete darkness, prompted the German plant physiologist, Erwin Bünning (1936), to hypothesize that plants and animals rely on an endogenous oscillator to measure daylength for photoperiodic responses. The **Bünning model** assumed a daily 12 hours light-requiring phase (subjective day, photophil) and a 12 hours dark-requiring phase (subjective night, scotophil). Restriction of light to the photophil results in a short-day response, but extending light into the scotophil generates a long-day response. This model was later refined as the **external coincidence model** (Pittendrigh 1966, Saunders 2016), a descriptor that emphasizes the fact that an external light source must coincide with a specific phase of the subjective night to elicit a photoperiodic response.

The first tests of the Bünning hypothesis used **skeleton photoperiods (night-interruption experiments)**. This approach uses a short photoperiod followed by brief pulses of light administered at different times during the night. If light falls on a light-sensitive phase of the subjective night the light pulse will elicit a long-day response. Very brief light pulses are sufficient to elicit a long-day response if received at the correct time, a time that coincides with a specific phase of the circadian oscillation. Experiments by Adkisson (1964, 1966) on the pink bollworm *Pectinophora gossypiella* scanned a 14 hours scotophase with a 1 hour light pulse and found two points during the scotophase when diapause is prevented: Point A, occurring a few hours after lights-off and Point B, occurring a few hours before lights-on (Figure 5.9). Light pulses at other times, including the middle of the night, had little effect. Similar bimodal responses of light pulses on the programming of diapause have been noted repeatedly, for example, for the flesh fly *Sarcophaga argyrostoma* (Saunders 1975b, 2020a) and the rice stem borer *Chilo suppressalis* (Xiao et al. 2010). In *C. suppressalis*, diapause incidence is reduced by a 1 hour light pulse administered 3 hours after lights-off (Point A) or 3 hours before lights-on (Point B) in a 12:12 LD cycle. In the moth *Thyrassia penangae* point A occurs 3 hours after lights-off and point B at 5 hours before lights-on at LD 9:15, 25°C (He et al. 2009), although timing of the sensitive points varies according to temperature as well as the photoregime being tested, for example, grape berry moth *Lobesia botrana* (Roditakis and Karandinos 2001). Pulses as short as 5 minutes and even as brief as a photoflash have proven effective in disrupting the programming of diapause when administered at the right time (Saunders 2002), but longer pulses may be needed, as in the Indianmeal moth *Plodia interpunctella*, which requires a 2-hour light pulse to effectively disrupt the

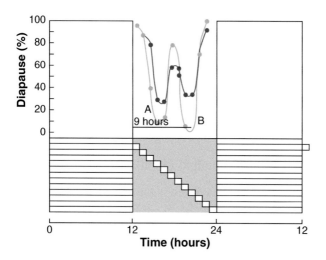

Figure 5.9 Results from two replicates of a night interruption experiment in which the long night of a diapause-inducing cycle is systematically interrupted by a 1-hour pulse. In most insects, a low incidence of diapause is observed at two phases: point A, early in the night and point B, late in the night. Data for the pink bollworm *Pectinophora gossypiella*.
From Adkisson (1964) and reproduced by Saunders (2020a), with permission from Annual Reviews.

larval diapause response (Kikukawa et al. 2009). The fact that light pulses administered at different times elicit distinct effects suggests some sort of endogenous circadian oscillation of photosensitivity, but some species showing robust night interruption results fail to display circadian control of the diapause response when other tests of rhythmicity (e.g., Nanda–Hamner or Bünsow experimental protocols discussed below) are used, for example, *T. penangae* (He et al. 2009), suggesting that points of sensitivity may not be iron-clad evidence of circadian control of photoperiodic responses.

The external coincidence model, depicted in Figure 5.10, displays critical features of this model developed from experiments on pupal diapause induction in the flesh fly *Sarcophaga similis* (Yamaguchi and Goto 2019), as well as other species. Oscillations of the circadian clock are noted under both long and short days, but it is only during the long nights associated with a short daylength that a **photoinducible phase** (φ_i) falls within the scotophase, leading to diapause induction. Night interruption experiments in which the light pulse is administered early in the scotophase causes a circadian phase delay, essentially resetting the circadian clock and forcing φ_i into the next photophase, hence preventing diapause. By contrast, a light pulse received late in the scotophase does not reset the clock and thus φ_i is susceptible to disruption by light. As implied in Figure 5.11(B5), the early light pulse that causes a phase delay of the circadian clock usually has a more profound impact on the diapause decision than the later light pulse occurring during the photoinducible phase (φ_i) of the scotophase (Saunders 2002, Yamaguchi and Goto 2019). Interestingly, the location

(A): Photoperiodic response curve

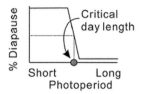

(B): External coincidence model

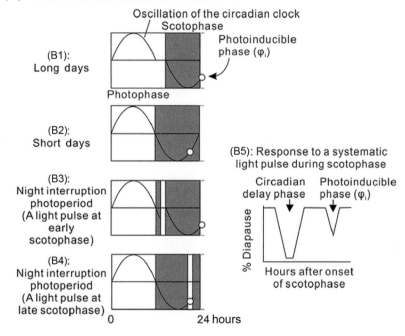

Figure 5.10 Photoperiodic response curve (A) and conceptual diagram of the external coincidence model (B) for photoperiodic induction of pupal diapause in the flesh fly *Sarcophaga similis*. The model is based on an oscillation of the circadian clock, which sets its phase at dawn and dusk, and on the position of the photoinducible phase (φ_i) in the later half of the scotophase. During the long days of summer, φ_i is exposed to light and a long-day response is induced (B1). During short days of autumn φ_i falls in the dark phase, eliciting a short-day response (B2). When a light pulse is applied during early scotophase, the phase of the clock is delayed and thus φ_i is pushed into the main photophase and exposed to light, yielding a long-day effect (B3). A light pulse in late scotophase directly illuminates φ_i and induces a long-day response (B4). When the scotophase is systematically scanned with a light pulse, two light-sensitive phases averting diapause are observed; the first is the circadian delay phase promoting the phase delay of the clock, the second is the consequence of φ_i (B5).
From Yamaguchi and Goto (2019), with permission from SAGE Publications.

of φ_i within the scotophase varies systematically with latitude, thus suggesting that selection for shifts in the timing of φ_i may account for clinal changes in critical daylength.

A second approach to testing the Bünning hypothesis couples a light phase with an extended night period to create unnatural light-dark cycle lengths (T) ranging from 36 to 72 hours, or more. In this **Bünsow protocol**, extended nights are interrupted with 1- or 2-hour light pulses (Bünsow 1953). As exemplified in the parasitoid *Nasonia vitripennis* sensitivity to a 2-hour light pulse occurs with a periodicity of approximately 24 hours (Saunders 1970), as shown in Figure 5.11. In contrast to the Bünsow protocol, the **Nanda–Hamner** or **resonance protocol** combines cycles of a short fixed photophase with a wide range of scotophases of unnatural lengths, typically with durations up to $T = 72$ hours (Nanda and Hamner 1958), as shown in Figure 5.12. Or, in some cases length of the scotophase remains constant and photophase duration is progressively changed to attain T's of different lengths. Positive responses, as seen for *Sarcophaga argyrostoma* and *Calliphora vicina* (Saunders 2020a), show peak diapause responses occurring when the total cycle length is equal to 24, 48, and 72 hours, providing strong evidence for involvement of a circadian clock in the programming of diapause. Note that the aphid *Megoura viciae* fails to display the same patterns of peaks and troughs seen for the two fly species (Figure 5.13), as discussed below.

In contrast to the external coincidence model, the **internal coincidence model** assumes two or more internal oscillations that change their phase relationship during the annual seasonal cycle of daylength (Pittendrigh and Minis 1964, Saunders 2016). In this model, light entrains separate dawn and dusk oscillators, and according to the changing phases generated by change in daylength, diapause is either induced or averted. The attraction of this model is that it can accommodate multiple oscillators that may interact with each other in different ways. Such a model is more consistent with the growing evidence that multiple oscillators, especially distinct dawn and dusk oscillators, likely cooperate in the programming of diapause, as seen in *N. vitripennis* (Saunders 1974) and other species.

Though many insect species display rhythmicity when subjected to skeleton photoperiods, the Bünsow protocol or the Nanda–Hamner protocol, some do not. The alternative is an **hourglass timer**, a scenario that does not rely on circadian rhythmicity but on the simple daily accumulation of a chemical during the day or night that triggers a diapause response when a critical threshold is reached (Veerman 2001, Saunders 2010b). Such a mechanism is frequently cited as the default mechanism when it is not possible to show evidence for circadian involvement in photoperiodic responses, but circadian and hourglass models are not mutually exclusive (Saunders 2005). In his classic experiments with the vetch aphid, *Megoura viciae*, Lees (1966, 1973) proposed an hourglass timer to measure night length and induce diapause. But, the challenge of proving a negative, in this case that circadian rhythms are not involved, is underscored by further experiments with *M. viciae*. Using more extended periods of darkness than used by Lees, Vaz Nunes and Hardie (1993) found an oscillation in diapause incidence when periods equaled multiples of 24 hours,

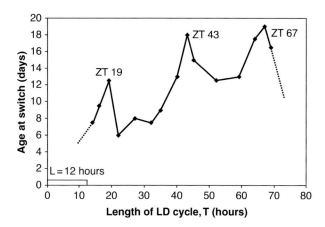

Figure 5.11 The Bünsow protocol in this example from the parasitoid *Nasonia vitripennis* couples an initial 12 hours photophase with a 60-hour dark period interrupted by a 2-hour light pulse administered at different times. Results show the age of females at the time they switch to production of diapausing progeny. Sensitivity occurs with a periodicity of approximately 24 hours, at *Zeitgeber* time (ZT) of 19, 43, and 67 hours in the extended night.
Schematic from Saunders (1970, 2010a), with permission from Oxford University Press.

suggesting some sort of circadian-based mechanism is involved here as well. An extensive series of experiments probing larval diapause in the moth *Pseudopidorus fasciata* reveals no evidence for circadian involvement in the photoperiodic clock used to program diapause, leading to the conclusion that in this species the photoperiodic clock is best described as a long-night measuring hourglass (Wei et al. 2001). Likewise, no evidence for circadian rhythmicity is noted for programming larval diapause in the Indian meal moth *P. interpunctella* (Kikukawa and Ohde 2007, Kikukawa et al. 2009), a result again suggesting an hourglass mechanism that measures night length. Other species that fail to display a positive response to Nanda–Hamner or Bünsow protocols include larvae of the pine caterpillar *Dendrolimus punctatus* (Huang et al. 2005), Asian corn borer *Ostrinia furnacalis* (Yang et al. 2014), the moth *Thyrassia penangae* (He et al. 2009), and adults of *Drosophila montana* (Kauranen et al. 2013), suggesting that if an oscillator is present it is heavily dampened in some species. In both the mosquito *Wyeomyia smithii* (Bradshaw et al. 2003a,b) and moth *T. penangae* (He et al. 2009) an hourglass timer best describes the response, but in these cases it is length of the day rather than night that is measured.

   Drosophila ezoana, a northern species with a reproductive diapause, also appears to use an hourglass to measure night length (Vaze and Helfrich-Förster 2016), but again the challenge is distinguishing an hourglass from a highly dampened circadian clock. Vaze and Helfrich-Förster propose the intriguing idea that an hourglass timer may be more appropriate for northern species such as *D. ezoana* that have weak circadian rhythms and live in an environment with a rapidly changing photoperiod. The idea is that weak circadian rhythms or hourglass timers may be more plastic and

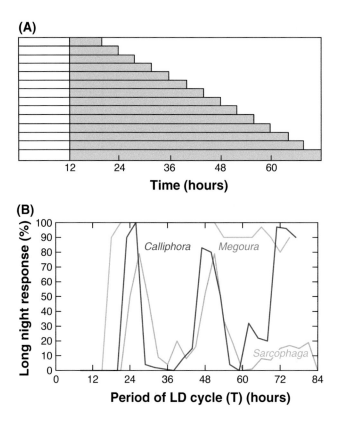

Figure 5.12 The Nanda–Hamner or resonance protocol combines cycles of a short fixed photophase with a range of scotophases of unnatural lengths, typically with durations up to $T = 72$ hours. The examples (A) couple a 12 hours photophase with different durations of darkness. Diapause incidences are recorded for (B) the flesh fly *Sarcophaga argyrostoma,* blow fly *Calliphora vicinia,* and the aphid *Megoura viciae.* Note peak responses at intervals of approximately 24 hours for the flies but the absence of such peaks and troughs for the aphid. From Saunders (2020a), with permission from Annual Reviews.

thus more easily synchronized with changing photoperiods. Further support for a loss of rhythmicity in darkness at high latitudes is provided by another fly, *Chymomyza costata* (Bertolini et al. 2019). While dark rhythmicity is retained in two related species of *Chymomyza* from low latitudes, this feature is lost in the high-latitude species.

Numerous time-related steps are involved in early phases of programming diapause: receiving the light-dark cycles, deciding whether they should be interpreted as long day or short day, and counting the number of short or long days received. Are all of these steps using the same clock? Perhaps not. Interestingly, the spider mite *Tetranychus urticae* demonstrates circadian rhythmicity in photoperiodic responses using the Nanda–Hamner protocol, but not under certain skeleton photoperiods (Vaz Nunes and Veerman 1982), leading to the idea that *T. urticae* relies on an hourglass to distinguish long and short days but a circadian-based counter to tally the number of long nights (Veerman and Veenendaal 2003). Experiments with the pitcher plant

(A)

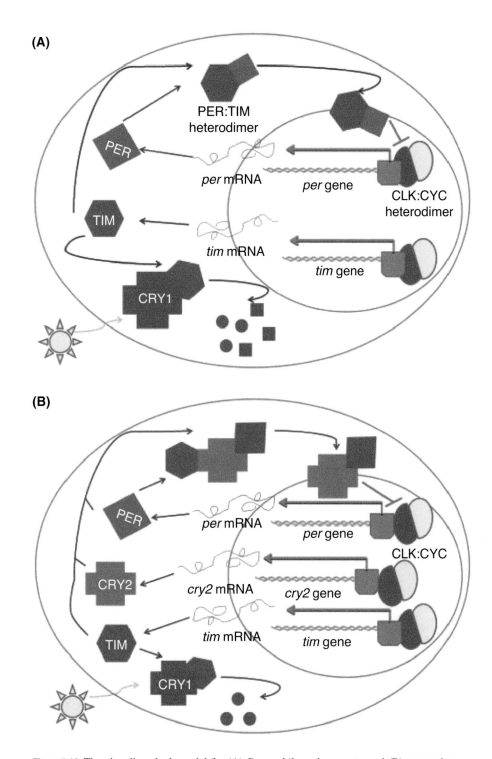

(B)

Figure 5.13 The circadian clock model for (A) *Drosophila melanogaster* and (B) most other insects. In the nucleus of *D. melanogaster* circadian clock cells, CLOCK (CLK) and CYCLE

mosquito *Wyeomyia smithii* also suggest different time-monitoring mechanisms operate in different processes (Emerson et al. 2008). Their experiments suggest an hourglass timer is involved in the photoperiodic counter mechanism, but Saunders (2013) argues that a circadian basis for the counter could also be inferred from their experiments.

Clearly there is strong experimental evidence that a circadian oscillatory mechanism is involved in most examples of photoperiodism. In cases where cycling is not evident, Saunders (2005, 2016) suggests that cycling is simply dampened in response to external conditions such as temperature. But, in light of the diverse evolution of diapause there is no reason to assume that all photoperiodic responses will conform to the same mechanism for monitoring the light-dark cycle nor for orchestrating related events such as the photoperiodic counter. Features of clock models invoked for photoperiodic time measurement (Vaz Nunes and Saunders 1999, Saunders et al. 2004, Koštál 2011, Saunders 2011, 2016, 2020b, Numata et al. 2015) indicate some commonality but also sufficient diversity to preclude the idea that a single model will be universally applicable.

The challenge with both the circadian and hourglass models is that they rely on theoretical oscillations or accumulation of unknown substances. Although these models offer valued frameworks for developing theories for how the photoperiodic clock works, as long as the physiological and molecular bases for the observations remain unknown, the models will, of necessity, not be fully validated. Since conclusions from these experiments are largely based on the final output (does the insect enter diapause or not?), the theoretical models offer no way of knowing where in the pathway the disruption occurs. Is it at the level of light reception, at the level of distinguishing long days from short days, at the level of the counter mechanism that determines how many diapause-inductive days have been received, at an integration point of multiple oscillators, or at some downstream step in the path toward initiating the diapause program?

(CYC) proteins form a heterodimer that acts as a transcription activator by binding to the E-box promoter region of the *period* (*per*) and *timeless* (*tim*) genes. *per* and *tim* mRNA are translated in the cytoplasm of the cell. PER and TIM proteins then form a heterodimer and translocate back into the nucleus where PER inhibits the action of CLK:CYC, thereby suppressing the transcription of *per* and *tim*. CRYPTOCHROME1 (CRY1) protein degrades TIM and itself in the presence of light. This results in increasing levels of *per* and *tim* mRNA throughout the day when CLK:CYC activity is uninhibited, and decreasing levels of *per* and *tim* mRNA during the night. (B) Circadian clocks in most other insects differs in that they possess a light-insensitive CRY2 protein that acts as the major negative transcriptional regulator of the core circadian clock. In this case, PER appears to assist CRY2 in nuclear translocation, whereas TIM helps to stabilize PER and CRY2. (A black and white version of this figure will appear in some formats. For the color version, please refer to the plate section.)
From Meuti and Denlinger (2013).

5.6.2   Molecular Approaches Evaluating the Role of the Circadian Clock in Diapause

Defining the molecular mechanisms regulating the circadian clock, a story that culminated in awarding the 2017 Nobel Prize in Physiology or Medicine to Jeffrey Hall, Michael Rosbash, and Michael Young, has yielded extremely useful tools for probing insect diapause by identifying key players likely to be involved in a time-keeping mechanism. Based on a model developed for *Drosophila melanogaster* (Hardin 2005), the proteins **CLOCK** (CLK) and **CYCLE** (CYC) act as positive transcriptional regulators that bind to an E-box promoter on the genes *period* (*per*) and *timeless* (*tim*) within the cell nucleus (Figure 5.13A). They recruit RNA polymerase, thereby increasing abundance of *per* and *tim* RNAs. The two RNAs then move to the cytoplasm where they are translated into proteins PER and TIM, that form a heterodimer which is then translocated back into the nucleus where the PER component of the heterodimer suppresses transcription of itself and *tim* by inhibiting action of CLK and CYC. The clock is informed by light acting through **CYTOCHROME1** (CRY1), which binds to TIM in the presence of light and degrades both TIM and itself.

This elegant model was assumed to be widely applicable to other insects as well, but the discovery of an additional cryptochrome, **CRYPTOCHROME2** (CRY2), in the monarch butterfly *Danaus plexippus* (Zhu et al. 2005) suggests that not all insects comply to the *Drosophila* playbook. As mentioned above (Section 5.5), we now know that CRY2 is present in most insects except *Drosophila* and perhaps other higher Diptera. Unlike CRY1, CRY2 is not sensitive to light but instead acts as a potent repressor of CLK and CYC (Figure 5.13B). CRY2, not PER, is likely the most important negative regulator of circadian clocks in most, if not all, non-drosophilids. Discoveries with non-model insects continue to add new twists. CRY1 and CRY2, present in compound eyes of the cricket *Gryllus bimaculatus*, participate in photic entrainment of circadian rhythms by forming an oscillatory loop that operates independently of PER and TIM, a loop that is regulated upstream by **C-FOS** (Kutaragi et al. 2018). C-FOS in crickets appears capable of resetting the clock, a feature that may also operate in other species. Recent work with mice proposes that mechanisms not dependent on the canonical clock genes participate in regulating circadian oscillations (Ray et al. 2020), a result suggesting still-unknown clock mechanisms await discovery.

The clock genes, especially *per* and *tim*, are now known from many species, making it possible to probe function of the clock components in relation to diapause. Null mutants, and more recently RNAi and CRISPR-Cas9, provide powerful tools for assessing the function of genes contributing to diapause. In *D. melanogaster*, flies that are null for *per* have arrhythmic locomotory behavior but are still capable of entering diapause (Saunders et al. 1989, Saunders 1990). The critical daylength shifts slightly in *per*-null flies, but *per* does not appear to be involved in the decision to enter diapause in this species.

Two natural variants of *tim* are present in European populations of *D. melanogaster* (Tauber et al. 2007): One population exclusively produces a short form of the mRNA (*s-tim*), while the other produces both short and long forms (*ls-tim*). The *ls-tim* variant correlates with higher levels of diapause, a feature that would appear to make it more adaptive at higher latitudes, but the opposite is observed: The *s-tim* variant is more common at higher latitudes. The S-TIM protein appears to bind more tightly to CRY1 than does LS-TIM, making the shorter form more readily degraded (Sandrelli et al. 2007), but again, the significance of this for diapause in natural conditions remains unclear. The LS-TIM variant appeared rather recently, approximately 300–3,000 years ago in southeastern Italy (Zonato et al. 2018) and appears to be spreading, as noted by the increased frequency of this allele in seasonal European habitats (Kyriacou et al. 2008). Flies that are *tim*-null retain the ability to enter diapause but lose their photoperiodic response, thus allowing a portion of the population to enter diapause regardless of daylength (Tauber et al. 2007). This suggests that *tim*, but not *per*, may be involved in measuring daylength or modulating other features involved in induction of diapause in *D. melanogaster*. There is strong evidence for an interaction between TIM and **EYES ABSENT** (EYA) in the diapause decision (Abrieux et al. 2020). TIM-null mutants yield low diapause under simulated winter conditions, while TIM overexpressed lines yield some diapause even under long daylengths. EYA, well known for its role in development of the eye and other organs, peaks in *D. melanogaster* at night in short days and accumulates at high levels in the cold, and TIM is thought to contribute to the response by stabilizing EYA. A role for EYA in insect diapause is consistent with what is already known from studies of photoperiodism in sheep (Hut 2011).

   In spite of the wonderful genetic tools available in *D. melanogaster*, this species is not ideal for probing the circadian clock's potential role in regulating diapause. Species with a more robust diapause response may prove more insightful. But, the potential importance of *tim* noted in *D. melanogaster* is evident in other species as well. The drosophilid *Chymomoyza costata* relies on photoperiod to regulate larval diapause, and in this species *tim* is a potential link between a circadian oscillator and diapause. A non-photoperiodic diapausing mutant strain of the fly (Riihimaa and Kimura 1988, Koštál and Shimada 2001) has a deletion in the 5′ sequence of the *tim* gene (Pavelka et al. 2003). This deletion contains sites for transcription activation and regulatory motifs including the E-box promoter and results in inhibition of *tim* transcription, leading to low TIM protein levels in nondiapause mutants (Stehlik et al. 2008), a result that again points to *tim* as a critical player in the photoperiodic response. The *C. costata* experiments show robust cycling of *tim* in short-day larvae, but in long-day larvae cycles are considerably less pronounced. Brains of wild-type fly larvae contain two TIM-positive neurons in each hemisphere, but no such staining is present in mutants that lack the diapause response (Stehlik et al. 2008). These results consistently point to the clock gene *tim* as a critical player in this fly's photoperiodic response. Mutation in another gene, *facet*, also eliminates diapause, suggesting a possible link between *tim* and *facet* in the diapause response of *C. costata* (Shimada 2005). The importance of *tim* is also evident in the pitcher plant mosquito *Wyeomyia smithii* (Mathias et al. 2005): In this species *tim* expression varies not only by time of day, but also by the larval instar in which diapause occurs and by latitude. The latitude effect, similar to observations in *D. melanogaster*, highlights the potential for *tim* to affect photoperiodic responses across a range of species.

A nondiapausing strain of the flesh fly *Sarcophaga bullata* displays an arrhythmic pattern of adult eclosion (Goto et al. 2006), suggesting a malfunction of the clock. But, curiously, both *per* and *tim* are more highly expressed in the mutants. A comparison of *per* sequences isolated from wild-type, high and low diapause strains of *S. bullata* shows that incidences of pupal diapause are higher in strains having shorter PER C-terminal regions (Han and Denlinger 2009a), suggesting that this region is important for the photoperiodic response, but what proteins interact with this region of PER remain undefined, and thus it is still unclear how this region exerts an effect on diapause.

Several studies show different expression patterns for clock genes under long- and short-day conditions, a feature that could be informative for the photoperiodic clock, but the results do not reflect a consistent pattern across species. Although these sorts of distinctions may prove to be informative, at this point the results are simply correlative. In adults of the flesh fly *Sarcophaga crassipalpis* the peak in *per* expression shifts with timing of scotophase onset (Goto and Denlinger 2002). A shift in *tim* is less pronounced but the amplitude of the *tim* peak is severely dampened under long days. Photosensitive larvae of *S. crassipalpis* show peak *per* and *tim* expression in the brain during the light-dark transition, but also a conspicuous difference between long- and short-day photoregimes: Peak expression of both mRNAs occurs during the light phase under long days but during the dark phase under short days (Koštál et al. 2009). Although not directly related to diapause, observations in adults of the silk moth *Bombyx mori* show short and long day differences in both expression patterns and cycle amplitude: Peak expression of both *per* and *tim* is advanced under short days relative to long days, and levels of *tim* mRNA are considerably lower under short days (Iwai et al. 2006). Cycling of *per* and *tim* is more pronounced under long days than under short days in females of the cabbage beetle *Colaphellus bowringi*, and consistently both genes are more highly expressed under the long days that avert diapause (Zhu et al. 2019). Diurnal rhythms of *per* expression in the head are weak in the linden bug *Pyrrhocoris apterus* under both long and short days, but *per* expression is nearly 10x higher in short-day bugs (Hodková et al. 2003). Whole body expression patterns in *P. apterus* are similar to results found in the head (Doležel et al. 2005), but in this case the major source of expression is the fat body rather than the brain (Doležel et al. 2008). Although this does not negate a role for clock genes in the photoperiodic response, it does raise the point that peripheral clocks complicate the interpretation of results based on expression patterns derived from whole bodies. At this point, we have little information on the contribution of peripheral circadian clocks to the programming of diapause (Section 5.6.3).

Distinct expression patterns of core circadian clock genes are noted in a number of additional species. Short days are linked to higher expression of several clock genes in the pea aphid *Acyrthosiphon pisum* (Barberà et al. 2017). Differences in amplitude and timing of expression peaks of *per*, *tim*, *cyc*, and *cry1* are evident in comparisons of heads from nondiapausing and diapausing larvae of the moth *Sesamia nonagrioides*;

oscillation peaks occur 5 hours after scotophase onset under long days but are advanced to 2 hours after scotophase onset in short days (Kontogiannatos et al. 2016, 2017, Kourti and Kontogiannatos 2018). Head expression levels of *per, tim,* and *takeout* are all higher prior to the onset of pupal diapause in the onion maggot *Delia antiqua* (Ren et al. 2018). In adults of the *Drosophila virilis* group, *per* is consistently more highly expressed in diapausing females (Salminen et al. 2015). Females of *Locusta migratoria* programmed to lay diapausing eggs have lower levels of *takeout*, a gene downstream of the circadian clock, than long-day females programmed to lay nondiapausing eggs (Jarwar et al. 2019). Clearly, intriguing distinctions of these sorts could be relaying meaningful information for the programming or maintenance of diapause, but if and how this information is used remains unknown.

Not all insects show expression patterns that differ under short and long days. In the bean bug *Riptortus pedestris*, which has an impressive response to knockdown of the clock genes, mRNA levels of *per, cyc, vrille,* and *cry* are nearly indistinguishable under short- and long-day photoregimes (Ikeno et al. 2008). Likewise, no distinctions in expression are noted between short and long days for diapause-destined larvae of the peach fruit moth *Carposina sasakii* (Cao et al. 2020). The Antarctic midge, *Belgica antarctica,* overwinters in a larval diapause (Spacht et al. 2020), but none of the clock genes cycle even during the short austral summer when the insect is in its active phase (Kobelkova et al. 2015).

RNAi experiments have proven useful for linking clock genes to the diapause response in a number of insect species, and in several cases circadian activity has been monitored simultaneously with the photoperiodic response. When the bean bug *R. pedestris* is reared under diapause-inducing short days or diapause-averting long days, *per* is expressed at nearly identical levels (Ikeno et al. 2008), but when the negative clock regulators *per* or *cry2* are knocked down using RNAi both the daily rhythm of cuticle deposition (a circadian response) and ovarian arrest (a photoperiodic response) are disrupted under short-day conditions (Ikeno et al. 2010, 2011a,b,c). Such females produce a single dark layer of cuticle rather than the alternating dark and light layers seen in control females and develop their ovaries rather than entering reproductive arrest. RNAi knockdown of the positive clock regulators *cyc* or *Clk* generates the opposite effect: Females produce a single layer of light cuticle and fail to develop ovaries when reared under long-day conditions (Ikeno et al. 2010, 2013, Goto 2013). A similar response is noted in males of *R. pedestris* (Ikeno et al. 2011a,c). RNAi directed against *Clk* in bean bug females also reduces expression of the JH-inducible transcription factor *Krüppel homologue 1* (Dong et al. 2021), further supporting a link between the clock response and the downstream effector controlling ovarian development. Knockdown of *cyc* in the bug *Plautia stali* results in malfunction of both the circadian rhythm of locomotion as well as the photoperiodic response that dictates entry into adult diapause (Tamai et al. 2019). In the cricket *Modicogryllus siamensis* nymphal development time, a photoperiodic event, and the circadian rhythm of locomotion are both disrupted by *per* knockdown (Sakamoto et al. 2009). Such experiments that demonstrate an effect on both the circadian rhythm and photoperiodism are especially powerful evidence that the two types of clocks have a common basis.

A growing list of species shows that impairment of one or more of the clock genes results in an altered diapause response. Knockdown of *tim* in larvae of the drosophilid *Chymomyza costata* prevents diapause under the short daylengths that normally lead to larval diapause (Pavelka et al. 2003). In the linden bug *Pyrrhocoris apterus* knockdown of *cyc* and *Clk* during the photosensitive nymphal stage results in a diapause-like state in adults (Kotwica-Rolinska et al. 2017), a result consistent with Ikeno's bean bug observations. The transition from diapause to reproduction in *P. apterus* requires long daylengths, but knockdown of *Clk, cyc,* or *cry2* prevents this transition from occurring (Urbanová et al. 2016). In adults of the cabbage beetle, *Colaphellus bowringi*, RNAi directed against *per* and *tim* fail to alter the female's decision to enter diapause, but such beetles do not accumulate the fat reserves normally associated with diapause (Zhu et al. 2019). Short days prompt females of the parasitoid *Nasonia vitripennis* to produce eggs that will result in larvae entering diapause, but when RNAi is directed against *per* diapause in the progeny is averted (Mukai and Goto 2016). Interestingly, low temperature is an alternative way to induce diapause in this species, and when females with RNAi-suppressed *per* expression are exposed to low temperature they can still produce diapausing progeny, a result suggesting that *per* directly affects the photoperiodic response but not the downstream capacity to produce diapause-destined eggs. In the Chinese oak silk moth *Antheraea pernyi* diapause termination is prompted by exposure to long days, but RNAi directed against *cyc, Clk,* or *per* derails the photoperiodic response and permits moths to emerge under short days as well (Mohamed et al. 2014). Loss-of-function mutants for the circadian activators *Clk* and *cyc-like* (*Bmal1*) elicit a typical long-day boost in egg laying in short-day reared females of the monarch *Danaeus plexippus,* while mutants for the circadian repressor, *cry2,* generate the opposite response, reduced egg output typical of diapause in long-day reared butterflies (Iiams et al. 2019). Quite consistently, knockdown experiments provide compelling evidence that a functional clock is essential for diapause. But, there are some species such as *P. apterus* in which RNAi directed against the clock genes *per, cry, clock, cyc,* and **Par domain protein 1** (*pdp1*) fails to influence diapause, yet downstream genes in the gut are affected (Bajgar et al. 2013a,b).

Experiments with the northern house mosquito *Culex pipiens* include a wide range of approaches evaluating contributions of clock genes to the diapause decision, as well as genes downstream of the clock. Expression levels of *clock, per, tim,* and *vrille* are all dampened in pupae of the mosquito *Culex pipiens* that are being programmed for adult diapause (Hickner et al. 2015), but no differences in expression are noted in adults, the actual stage of diapause (Kang et al. 2016). Those conclusions are based on transcriptomic analyses of whole bodies of females at a single time point, but altered expression of these genes is noted when brain expression is monitored at different time points. Distinct responses are noted when these genes are knocked down with RNAi (Meuti et al. 2015). Knocking down *per, tim* and *cry2* with RNAi causes females programmed for adult diapause by short days to avert diapause, while knockdown of a downstream circadian-associated gene *pigment dispersing factor* (*pdf*) elicits the opposite effect, generating a diapause-like state in long-day females. A secondary transcriptional feedback loop involving the transcriptional repressor **Vrille** (VRI) and the transcriptional activator PDP1 regulates rhythmic expression of *Clk* in *D. mela-nogaster,* and RNAi knockdown of either of these transcriptional factors influences the diapause phenotype in *Cx. pipiens* (Chang and Meuti 2020). Knockdown of *vri* produces variable results on egg follicle development but never affects fat content. Knocking down *Pdp1,* however, does not affect follicle development but prevents short-day females from accumulating the fat reserves expected in diapause, a result suggesting that these two transcriptional factors affect different features of the dia-pause phenotype. Though *vri* mRNA oscillates under both long and short days, *Pdp1* expression oscillates only under long days and is constitutively upregulated in short-day diapausing females. Two genes downstream of the clock, *takeout* (*to*) and *Nocturnin* (*Noc*), have expression patterns nearly identical to that of *Pdp1*, suggesting they may be regulated by PDP1 or that all are regulated by the same transcription factor. A connection between *Pdp1, to* and *Noc* is further suggested by the observation that *Pdp1* knockdown results in elevated expression of *to* and *Noc*. *Noc* possibly plays a role in increasing fatty acid synthesis, and *to* is best known for its role in diapause-related functions such as boosting sugar-feeding behavior, enhancing survival, and inhibiting JH signaling. Interestingly, *to* is also upregulated by the transcription factor **FoxO**, a leading candidate for regulation of many of the features of diapause in *Cx. pipiens* (Sim et al. 2015). Though definitive links in this pathway are not firmly established, experiments such as these are beginning to make the necessary connec-tions between the canonical clock genes and the downstream action that generates the diapause phenotype, including involvement of FoxO (Section 9.3), microRNAs (Section 9.11), and the JH signaling pathway (Section 9.1.3.2). *Cx. molestus,* a sister species that lacks diapause, harbors a nine-nucleotide deletion in the gene encoding Helicase domino (*dom*) (Epstein et al. 2021). Since *dom* is known from *Drosophila* to affect expression of *per* and *tim* this deletion could be responsible for the absence of diapause in *Cx. molestus.*

Surgical ablation offers another tool for assessing roles of clock-related genes. By localizing cells that show immunoreactivity for clock proteins or proteins associated with the clock it is possible to ablate or otherwise destroy such cells and monitor the response. Photoperiodic control of pupal diapause in the tobacco hornworm *Manduca sexta* is lost after ablation of *per*-expressing neurons (Shiga et al. 2003b), and following ablation of PDF-containing somata in the blow fly *Protophormia terraeno-vae* the ability to discriminate photoperiods is lost, and flies enter reproductive diapause under either a short- or long-day photoperiod (Shiga and Numata 2009). These responses in *M. sexta* and *P. terraenovae* are consistent with RNAi results in *Cx. pipiens* (Meuti et al. 2015) implicating *pdf* as a downstream player in the photoperiodic response leading to diapause. Interestingly, a PDF receptor is present in the prothoracic gland of *Bombyx mori*, and PDF is capable of stimulating the gland's production of ecdysone (Iga et al. 2014), thus offering a potential link between the circadian clock mechanism and hormone action. In the bean bug *Riptortus pedestris,* ablation of the brain region containing PDF-immunoreactive neurons also disrupts the photoperiodic response, allowing ovarian development to ensue even under short days, but in this case, RNAi directed against *pdf* is ineffective, perhaps suggesting *pdf* transcription is not an essential component of the response in *R. pedestris* (Ikeno et al. 2014). These differences among species on whether PDF is essential for the photoperiodic response underscore how little we know about the detailed pathway from photoreception to diapause.

One caveat that must be considered is the plague of pleiotropy (Emerson et al. 2009b, Bradshaw and Holzapfel 2010b, Meuti and Denlinger 2013, Denlinger et al. 2017). Diapause is a complex trait and its initiation requires perception of photoperiod, storage of this information, and eventually translation of this information into the endocrine signals that elicit the diapause program. Most clock-related experiments examine the effect of photoperiod by examining the final downstream response, diapause. As Emerson et al. (2009b) correctly point out, the central circadian clock, or individual genes that comprise the clock, may be interacting pleiotropically at any one of these steps. This is especially noteworthy since circadian clocks regulate numerous behavioral and physiological processes that may be implicated in diapause. One elegant example of this complication is the interaction of CLK and CYC with juvenile hormone (JH) in the gut of the linden bug *P. apterus* (Bajgar et al. 2013a,b). These two clock proteins interact downstream of JH to elevate expression of the clock gene *Pdp*1 and suppress expression of *cry2* under long days. In contrast, *cry2* expression is elevated and *Pdp*1 is suppressed under short days. These clock genes regulate the diapause status of the gut in *P. apterus*, but they do so independently of the brain's daily oscillations. It is thus challenging to determine where and whether circadian clocks as a whole entity (modular pleiotropy) or specific clock genes (gene pleiotropy) are exerting their effect on the diapause program.

To empirically determine whether an individual clock gene or the clock as a functional module affects diapause, Emerson et al. (2009b) make a strong case for examining the effects of multiple, single gene knockouts. Excellent experiments mentioned above (Ikeno et al. 2010, 2011a,b, Goto 2013) nicely exemplify this type of approach by separately examining the effects of several genes on both a circadian event (cuticle deposition) as well as diapause. In their experiments with the bean bug, the induction of diapause in long-day reared bugs following RNAi directed against *cyc* could be reversed with application of JH (Ikeno et al. 2010). This implies that the effect of *cyc* knockdown occurs upstream of JH release from the corpora allata, but it is still not possible to know with certainty whether the central circadian clock is involved in photoreception, measurement of daylength, or neurosecretory signaling to the corpora allata. Experiments showing consistent results from knocking down multiple positive and negative circadian clock regulators help strengthen the argument that the entire circadian clock is involved in diapause initiation, and the forward genetics approach advocated by Bradshaw and Holzapfel (2007) offers promise for identifying an inclusive set of genes involved in photoperiodic diapause responses.

Insects that currently offer the best models for probing clock properties, *Drosophila melanogaster* and the honey bee *Apis mellifera*, unfortunately lack a robust photoperiodic response (Beer and Helfrich-Förster 2020), while species with well-documented diapause responses are missing the full set of genetic tools, including both loss and gain-of-function mutations, that are urgently needed to further advance this exciting field of discovery.

5.6.3   Are Peripheral Clocks Involved?

Circadian clocks are not just operating in the brain. Cells throughout the body possess circadian clocks, some of which are entrained centrally by input from the brain, but others are entrained directly by light and are thus autonomous (Giebultowicz 2001). The fact that peripheral clocks are so widespread suggests that they too could be conduits for photoperiodic information, but currently there is no evidence for direct control of diapause by peripheral clocks. Yet, the very idea that multiple autonomous clocks are undergoing oscillations within the insect body suggests the possibility that they may be involved in executing or coordinating certain aspects of the diapause phenotype.

There is at least one interesting example of a peripheral clock contributing to the overall diapause response. Remarkably, a clock operating within **antennae** of the monarch butterfly *Danaus plexippus* is critical for navigation processes involved in the butterfly's migratory flight to its overwintering site (Froy et al. 2003, Reppert et al. 2016). Monarchs use the sun as a compass for flight orientation and adjust their orientation relative to the sun's azimuthal position to maintain a constant light bearing, and it is the role of the antennal circadian clock to provide the timing component needed to compensate for the sun's movement across the sky during the course of a day. Similar circadian clock control of flight orientation could be operating in other migratory insects as well.

5.7   The Photoperiodic Counter

Distinguishing a short day from a long day is one of the essential components of photoperiodism, but the insect must also be able to sum the number of such days it has received. As noted in Section 5.3, the photosensitive period encompasses a specific developmental interval during which the insect monitors daylength and uses this information to either progress with development or enter diapause. The diapause response is thus dependent on the insect receiving a specific number of days with the "correct" daylength. Simply stated, the insect has a counter capable of reckoning the number of short (or long) days it has received (Saunders 1981). The number of days needed to render this developmental decision is referred to as the **required day number** (RDN). Too few such days means no diapause, while days equaling or exceeding that number translates into diapause. The photoperiodic counter, at least in the pitcher plant mosquito *Wyeomyia smithii*, appears to count daily light:dark cycles using an hourglass or external coincidence mechanism rather than some endogenous circadian system (Emerson et al. 2008). The photoperiodic counter in

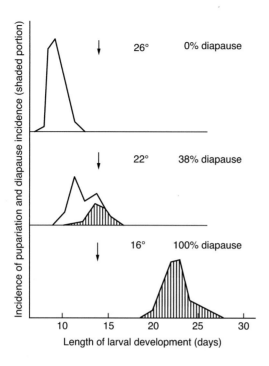

Figure 5.14 Effect of temperature on the time of pupariation and incidence of pupal diapause in the flesh fly *Sarcophaga argyrostoma*. Arrow indicates 14 days, the number of short days required for diapause induction.
Adapted from Saunders (1971), as shown in Denlinger (1985), with permission from Elsevier.

Pyrrhocoris apterus stops in continuous darkness but information previously accumulated is retained (Hodková 2015).

Depending on the species, long days, short days or both may be accumulated, and the diapause-promoting effect of short days can be countered by long days, reversing the accumulation of short days (Saunders 2002). The fact that daylength and temperature can also impact the rate of development implies that these environmental factors will sometimes determine how quickly an insect will reach its RDN. In an attractive model proposed for the flesh fly *Sarcophaga argyrostoma*, the RDN for pupal diapause induction is 14 short days received during larval development regardless of temperature (Saunders 1971). Whether larvae reach the RDN depends on their development rate. Larvae of this species are sensitive to short daylengths throughout larval development, but as shown in Figure 5.14, development time at 26°C is so rapid that pupariation and the end of photosensitivity is attained before the larvae can accumulate 14 short days, hence they fail to enter diapause. At the opposite extreme, 16°C, all larvae experience more than 14 days and all enter diapause. At an intermediate temperature, 22°C, a portion of the population pupariates on either side of 14 days, resulting in a mixed diapause response, but clearly the incidence of diapause is higher in those that pupariate later. Under this model, the RDN is **temperature compensated** (i.e., remains a fixed duration regardless of temperature), while the photosensitive stage is not. A similar scenario nicely accounts for induction of adult diapause in *Drosophila montana* (Salminen and Hoikkala 2013). The RDN in this species remains fixed at 3 days regardless of temperature, thus more adults enter diapause when exposed to lower temperature or other factors that retard development such as inadequate food supplies.

Attractive as the above story is, there is a caveat. *S. argyrostoma*, as well as other species of *Sarcophaga,* first become photosensitive during embryonic development while they are still within the uterus of the female (Denlinger 1971, 1972b), and when that period of sensitivity is included, a brief 4-day interval (the last 2 days of embryonic development and the first 2 days of larval development, Figure 5.5) suffices to induce pupal diapause. Thus, the scenario outlined above, requiring 14 days, only applies if the earlier photosensitive stage is skipped. Short days during larval life can compensate for missing the full photosensitive period, but a much longer exposure period is required. A similar scenario is evident in the linden bug *P. apterus*. The photosensitive period in this species extends from late nymphal stages into adulthood, but 3 short days experienced during nymphal development is sufficient to induce diapause while 9–10 short days are required if the programming is delayed until onset of the adult stage (Hodková 2015). Even within the sensitive period, not all days are equivalent. Larvae of the butterfly *Pieris brassicae* are broadly photosensitive during the fourth and fifth larval instars, but maximum sensitivity for the programming of pupal diapause occurs one day after the molt to the fifth instar (Spieth 1995). Similarly, sensitivity for the programming of larval diapause in the moth *Pseudopidorus fasciata* persists over the first 7 days of larval life, but highest sensitivity is on day 4 (Hua et al. 2005a).

RDN values vary widely among species, for example, 14 days for pupal diapause in *S. argyrostoma* (Saunders 1971) but only 4 days for larval diapause in *Chymomyza constata* (Koštál et al. 2000b), and as noted above the number can vary depending on which components of the photosensitive stage are included. Geography can also impact the RDN, as seen in the mosquito *Aedes albopictus*: An RDN of 4 days is noted in a population from 45° N but 9 days are required for a population from 14° N (Beach 1978).

The attractive idea that the RDN is temperature-compensated does not hold for all species, and the idea has not been widely tested, although it is frequently assumed. A case in point is the Asian corn borer *Ostrinia furnacalis* (Yang et al. 2014). In this species RDNs for reversing the short-day programming of larval diapause with long days are 13.5, 6.5, and 5 days at 22, 25, and 28°C, respectively. By contrast the short day RDNs for programming diapause are 19.3, 18.5, and 14 days at those same temperatures, values showing up to threefold differences at 25 and 28°C. These results imply that more days are needed to program diapause than to reverse the program. A similar result is noted for larval diapause induction in the moth *Pseudopidorus fasciata* (Hua et al. 2005b): RDNs for short days between 20 and 28°C range from 8.5 to 11.5 days, but much briefer RDNs (3–5 days) are required to reverse the effect of long-day cycles. Thus, RDNs may vary widely depending on rearing temperature, portions of the photosensitive stage examined, as well as species and population geography.

Models attempting to explain the photoperiodic counter imply that some biochemical factor gradually accumulates, reaching a threshold that triggers the diapause response (Gibbs 1975). Identifying this factor, of course, is a more challenging question, but a question of considerable importance if we hope to understand how insects translate photoperiodic cues into a meaningful developmental program. A viable candidate for such a role should be closely aligned to or be a component of the clock cell neurons that distinguish short and long days and be located within neurons that project to cells involved in triggering the release of hormones involved in diapause. The factor must also have the capacity to be stored and to release its information at the correct developmental time, usually at a much later point in time.

Recent experiments with the silk moth *Antheraea peryni* suggest that **arylalkylamine N-acetyltransferase** (aaNAT) has many of the properties one would expect to see in a candidate counter molecule (Mohamed et al. 2014, Wang et al. 2015a), although the work is based on diapause termination, not entry. Diapausing pupae of *A. pernyi* break diapause in response to long daylength after a period of cold storage, and in this system, melatonin is implicated in the response. In turn, the enzyme aaNAT is essential for generation of melatonin, and cycling of the clock genes *clk* and *cyc* direct the cycling of *nat*. Although melatonin has thus far not been widely considered of importance in regulation of insect circadian systems, a role for melatonin is consistent with the observations in *A. pernyi*, and rhythmic oscillation of aaNAT activity seems to direct rhythmic oscillation of melatonin. Cycling of the *nat* transcript and melatonin synthesis is an output of the circadian clock cells in *A. pernyi*, and interconnections between these cells and the cells that regulate release of prothoracicotropic hormone (PTTH) suggest a mechanism and pathway for conveying photoperiodic information to the desired endocrine switch. If melatonin directs the action of the PTTH-producing neurons one would expect to find melatonin receptors present on these neurons, and this appears to be the case. This information suggests that melatonin or the *nat* transcript progressively increases in response to long days and reaches a threshold that stimulates release of PTTH, a scenario that fulfills the essential role of a photoperiodic counter. The *A. pernyi* system offers the advantage that diapause is terminated quickly and directly in response to long days and thus one does not have to account for a counting mechanism that has the sort of long delay between photoreception and endocrine action that is more common in insect diapause. As discussed more fully in Section 8.4, roles for *per, pdf, cyc,* and *nat* have also been proposed for the timing of photoperiodic termination of larval diapause in the European corn borer *Ostrinia nubilalis* (Kozak et al. 2019). Cycling of melatonin and *nat* in response to daylength in the pea aphid *Acyrthosiphon pisum* (Barberà et al. 2013, 2018) represents another case where this pathway contributes to the photoperiodic response, possibly again as a counter. Scenarios similar to this could be operating in other insects, but there are a number of different clock outputs, many of which could be used as counters.

5.8   Storage of Photoperiodic Information

The fact that photoperiodic signals usually are received well in advance of the onset of diapause implies that the insect not only needs a timer to distinguish short and long days (the photoperiodic timer) and a mechanism to count the number of inductive days that have accumulated (the photoperiodic counter), but it must also store that information within the brain to evoke the appropriate endocrine signals at some time in the future. What is stored and where is it stored?

Brain transplant experiments discussed above (Section 5.4) clearly identify the brain as the organ that stores the information, but little beyond that can be stated with certainty. Several scenarios can be imagined. Perhaps a transcript or its product accumulates and triggers a response at a later date when an appropriate receptor is produced. This would be similar to the response noted above for the silk moth *A. pernyi*, in which the *nat* transcript gradually accumulates under long days, leading to photoperiodic termination of diapause in the pupae (Mohamed et al. 2014). In this case, the diapause termination response immediately follows long-day exposure and thus does not require long retention of the signal, but in most cases there is a long interval between reception of the environmental signal and enactment of the diapause response. A transcript or its product could perhaps simply persist for a long time, enabling the response at the appropriate developmental stage. Alternatively, one could imagine an inhibitor that is gradually degraded and allows diapause to proceed only after a critical point is reached. Epigenetic marking of chromatin and small noncoding RNA molecules have features suggesting they could possibly emerge as regulators of the long-lasting effects noted in the diapause response.

Parallels can also perhaps be derived from our understanding of **memory processes** (Kandel 2001, Kandel et al. 2014). Although memory storage largely remains a mystery, there are two ways proposed to save a memory, both involving stable, self-reproducing states. One involves a transcriptional stable state, in which a cell is shifted to a state where a transcription factor in a positive feedback loop initiates its own continuous expression. Presence of the transcription factor represents the memory and any other cell can "query" for its presence to determine the state of the memory. An alternative way to achieve this is the synaptic route, having a small group of neurons simultaneously computing the same response, thereby creating a self-stabilizing neuronal network that can be "queried" at the appropriate time. These putative memory sites would need to be integrated with or act upstream of the neurosecretory cells that govern the hormonal response. Some of the pharmacological disruptors of diapause, discussed in Section 12.3.2, could be targeting such sites of memory.

5.9   Circannual Rhythms

Although the term circannual rhythms is sometimes used simply to refer to an insect's annual cycle of development, I restrict use of the term to refer to the rather rare endogenous rhythms having a periodicity of approximately 1 year. A few long-lived species, most notably the varied carpet beetle *Anthrenus verbasci*, achieve seasonal synchrony in this manner (Blake 1958, Nisimura and Numata 2001, 2003). Larval diapause is broken in the spring and pupation is elicited in response to changes in photoperiod but not to a particular critical value. Evidence that the rhythm is endogenous is suggested by the fact that, under complete darkness or unchanging photoperiod, beetles display a free-running pupation rhythm of approximately 40 weeks, a duration considerably shorter than a year. Across a range of temperatures (17–25°C) the interval between peaks remains constant, indicating the rhythm is **temperature compensated**. Under natural changes in photoperiod the periodicity is exactly 52 weeks, indicating that the beetles rely on seasonal change in daylength as the *Zeitgeber* for synchronization, much like other insects rely on the daily light-dark cycle to synchronize circadian rhythms. Results using the Nanda–Hamner protocol provide evidence that the circadian system is involved in this response (Miyazaki et al. 2009), and phase shifts that reset the annual rhythm can be achieved by subjecting larvae to low temperature pulses (Miyazaki et al. 2016).

Relatively few species appear to rely on circannual rhythms. A review by Miyazaki et al. (2014) suggests it is mainly seen in long-lived insects, especially those living at high latitudes. *A. verbasci*, an inhabitant of bird nests, requires two or more years to complete its life cycle. Another long-lived carpet beetle, *A. sarnicus*, displays a similar long-term rhythm, as do some northern ant species (Kipyatkov 1995), but less is known about the details of these rhythms. Although the adaptive advantage of a circannual clock is not clear, for a long-lived insect living in an environment that may be nutrient-poor and unstable, it may add a layer of precision that facilitates coordination of seasonal development among populations that show considerable inherent variation in developmental timing.

5.10 Role of Temperature

Temperature usually plays a somewhat secondary role in programming diapause, presumably because it lacks the reliability and seasonal precision of photoperiod. Though low temperature sometimes triggers winter diapause independently of photoperiod, it is more common for low temperature to simply boost the diapause response if the appropriate short day signal is received, for example, the subtropical swallowtail *Euryades corethrus* (Caporale et al. 2017), and the swallowtail *Sericinus montelus* (Wang et al. 2009), among numerous other examples. Without the short day signal, diapause may not be induced even if temperatures are low. For example, in the flesh fly *Sarcophaga crassipalpis*, long days completely avert diapause at any temperature, but at short days and 25°C, 85% of the pupae enter diapause. Lowering the rearing temperature to 18°C raises the incidence of diapause to nearly 100% (Denlinger 1972b). Input from both photoperiod and temperature are essential in some cases, for example, pupal diapause in subtropical *Helicoverpa armigera* cannot be induced by short daylength or low temperature alone: Both cues are needed (Shimizu and

Fujisaki 2006). By contrast, photoperiod by itself can induce diapause in the closely related temperate species *H. assulta*.

In *S. crassipalpis* the incidence of diapause increases at low temperature without **altering the critical photoperiod**, as depicted schematically in Figure 5.15A, but in other species, such as the butterfly *Pieris brassicae* and moth *Acronycta rumicis*, lowering the temperature also lengthens the critical photoperiod (Danilevskii 1965), thus indicating that the insect will enter diapause at a longer daylength (earlier in the season) if temperatures are low (Figure 5.15B). In larvae of the Asian corn borer *Ostrinia furnacalis* the CDL remains quite stable at 13.5 hours across a temperature range of 25–30°C, but increases to 14 hours when temperatures are lowered to 22°C (Yang et al. 2014). For the programming of larval diapause in the bruchid *Bruchidius dorsalis* the CDL is 12 hours at 24°C but increases to 12.5 hours at 20°C (Kurota and Shimada 2001, 2003a). In the moth *Thyrassia penangae* the CDL is 12.3 hours at 28°C but increases to 13.5 hours at 25°C (He et al. 2009). Likewise the CDL of the zygaenid moth *Pseudopidorus fasciata* is constant at 25 and 28°C but increases 0.5 hour when temperature drops to 22°C (Hua et al. 2005a). In the northern drosophilid *Drosophila montana*, CDL remains rather stable at 16–19°C but increases by approximately 1 hour at 13°C (Tyukmaeva et al. 2020). The CDL for the hover fly *Episyrphus balteatus* is 11.9 hours at 20°C, but increases to 12.7 hours at 15°C (Hondelmann and Poehling 2007). Rather consistently, CDL values remain constant over a certain temperature range but may increase at lower temperatures, thereby shortening the active period, or conversely decrease at higher temperatures, postponing diapause and possibly allowing production of an additional generation in the autumn.

Unlike winter diapause, the incidence of summer diapause frequently increases at high temperatures (Masaki 1980). High temperatures in the 33–39°C range prompt summer pupal diapause in the cotton bollworm *Helicoverpa armigera* (Liu et al. 2006), as well as in its sister species, the tobacco budworm *H. assulta* (Liu et al. 2016d). For induction of summer diapause in pupae of the onion fly *Delia antiqua*, long daylengths coupled with temperatures around 24°C are optimal; lower temperatures are less effective (Ishikawa et al. 2000). This, however, is not always the case, for example, summer diapause in the leaf beetle *Phaedon brassicae* is programmed by long daylength but the photoperiodic response is not enhanced, and in fact is slightly diminished by high temperature (Wang et al. 2007a), and in the cabbage beetle *Colaphellus bowringi* low temperatures are critical for programming both summer and winter diapause (Xue et al. 2002).

In some species with a winter diapause low temperature can override the photoperiodic response, generating diapause under long daylengths that would prohibit diapause at higher temperatures. For example, the brown marmorated stink bug does not normally enter adult diapause at long daylengths, but if the rearing temperature is lowered to 20°C nearly half the population will enter diapause, even under an extremely long daylength (L:D 18:6) (Musolin et al. 2019). Similarly, long days normally avert diapause in the leaf beetle *Phaedon brassicae*, but nearly half the beetles enter adult diapause under long days if reared at 12°C (Wang et al. 2007a). The seed-sucking bug *Dybowskyia reticulata* responds to short days by entering diapause,

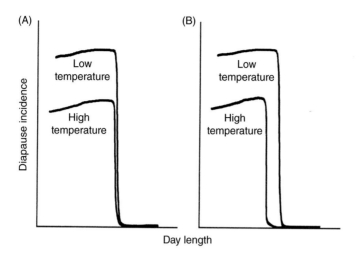

Figure 5.15 Low temperature can impact the photoperiodic response curve in two ways. (A) In some cases, the critical daylength remains the same but the incidence of diapause at short daylengths is higher. (B) In other instances, the critical daylength is also shifted toward longer daylengths at low temperature.
From Denlinger (2001).

but it will enter diapause under long days at temperatures of 25°C or lower (Numata and Nakamura 2002). At temperatures below 20°C, adults of the cabbage beetle *Colaphellus bowringi* enter diapause regardless of the photoregime (Xue et al. 2002). The bean blister beetle *Epicauta gorhami,* diapauses as a pseudopupa, a larval stage unique to hypermetamorphosis in meloid beetles, and in this species photoperiodic control of diapause is restricted to a narrow temperature range (around 25°C); at 20–22.5°C, all enter diapause regardless of photoperiod, and at temperatures above 27.5°C, none enter diapause under any photoregime (Shintani et al. 2011). Short days prompt pupal diapause across a wide temperature range in the cotton bollworm *Helicoverpa zea*, but diapause can also be elicited under long days at temperatures of 17°C and lower (Clemmensen and Hahn 2015).

As discussed in Section 5.7, the low temperature enhancement of diapause is, at least sometimes, the consequence of slowing the rate of development (Figure 5.15). *S. argyrostoma* appears to count the number of short days received, and the higher the number of short days received, the greater the incidence of diapause (Saunders 1971). The slower development rate at low temperatures thus increases the number of short days to which the larvae are exposed and thereby boosts the incidence of diapause.

Constant temperatures, the conditions most frequently used in laboratory experiments, are, of course, not ecologically relevant. Night temperatures are usually considerably lower than those during the day. Coupling low temperature (**cryophase**) with the scotophase and high temperature (**thermophase**) with the photophase usually results in a higher diapause incidence than observed if the insect is exposed to the mean of the two extremes or if the cryophase is coupled with the photophase instead

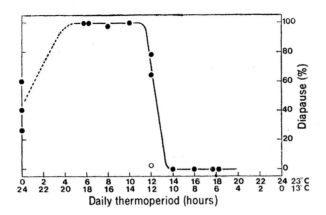

Figure 5.16 Thermoperiod can substitute for photoperiod, as demonstrated for induction of larval diapause in the parasitoid *Nasonia vitripennis*. Under continuous darkness, a daily temperature cycle (23/13°C) is highly effective for diapause induction.
From Saunders (1973), with permission from AAAS.

of the scotophase. Linking temperature and photoperiod can impact numerous developmental responses, as reviewed by Beck (1991). Thermoperiodic responses are also clock controlled and linked to physiological processes governed by circadian rhythms (Beck 1991, Saunders 2002). In the pitcher plant mosquito *Wyeomyia smithii* a phase delay in onset of the cryophase is considerably more effective for diapause induction than a phase advance (Bradshaw 1980), a response that makes ecological sense because the nighttime decline in temperature normally lags the onset of darkness. Such phase responses have seldom been explored but are possibly common.

Thermoperiods not only enhance photoperiodic responses but sometimes can substitute for photoperiod as a meaningful signal for diapause induction if the insect is held under constant darkness or constant light, as seen in the European corn borer *Ostrinia nubilalis* (Beck 1987), the southwestern corn borer *Diatraea grandiosella* (Chippendale et al. 1976), the Indianmeal moth *Plodia interpunctella* (Masaki and Kikukawa 1981), the stalkborer *Sesamia nonagrioides* (Eizaguirre et al. 1994, Fantinou and Kagkou 2000, Fantinou et al. 2002), the drosophilid fly *Chymomyza constata* (Koštál et al. 2016b), the cabbage beetle *Colaphellus bowringi* (Wang et al. 2004, 2007b), the parasitoid *Nasonia vitripennis* (Saunders 1973), among others. Thermoperiodic response curves are usually quite similar to photoperiodic response curves (Beck 1991), as seen for *N. vitripennis* (Figure 5.16), but transitions may not be quite as sharp and may be shifted a bit as in *N. vitripennis*, which has a critical thermoperiod of approximately 13 hours (Figure 5.16) but a longer 15.5 hours critical photoperiod (Saunders 1973). As with photoperiodic response curves, which rely mainly on night length, thermoperiodic induction of diapause depends mainly on cryophase duration. A certain threshold temperature is essential for recognition of a cryophase. For *O. nubilalis*, this threshold is 17.5°C (Beck 1982), and the optimum diapause response is obtained by coupling a cryophase temperature of 10°C with a thermophase temperature of 20°C (Beck 1985). In the cabbage beetle *Colaphellus bowringi* the threshold for the cryophase is 19°C and the thermophase threshold is 31°C; the amplitude (difference between the cryophase and thermophase) is particularly important (Wang et al. 2004). Thermoperiods, however, cannot be universally substituted for photoperiods, for example, *Platynota idaeusalis* (Rock 1983) and *Sarcophaga argyrostoma* (Saunders 1984).

In certain geographic regions and habitats, temperature cues may completely supplant photoperiod as the environmental signal for diapause induction. As discussed below (Section 5.14), such is the case for tropical flesh flies (family Sarcophagidae) living near the equator. Similarly, diapause in populations of the *Drosophila* parasitoid, *Leptopiliana boulardi*, rely exclusively on low temperature (17°C and lower) for larval diapause induction across a wide geographic range (4° S to 44° N); no contribution from photoperiod is detected (Carton and Claret 1982). The diapause response of a temperate-region meloid beetle, *Mylabris phalerata*, appears to depend strictly on temperature, with no evidence of photoperiodic input (Zhu et al. 2006), and at least some populations of *Drosophila melanogaster* enter their dormant period in response to temperature rather than photoperiod (Emerson et al. 2009c). Both pupal and adult diapauses in low- and high-altitude populations of dung flies (Sepsidae) also appears to rely on low temperature cues for the induction of diapause (Blanckenhorn 1998a). Low temperature, in full darkness, is capable of evoking larval diapause in the egg parasitoid *Trichogramma cordubensis* (Ventura Garcia et al. 2002), but the range is amazingly narrow: 30 days at 10°C will induce diapause but not an equivalent or longer time at 7 or 12°C. Summer diapause in pupae of the onion maggot *Delia antiqua* relies strictly on temperatures (below 24°C) received during an extremely short developmental window between puparium formation and pupation; photoperiod appears to be irrelevant (Ishikawa et al. 2000). Temperatures at or below 17°C prompt embryonic diapause in the spider mite *Eotetranychus smithi*, a response that appears to be independent of photoperiod (Takano et al. 2017). Temperature cycles may generally be more important for entraining the endogenous clock of immature stages than they are for entraining the clock of adult insects, as noted in a switch from temperature to light sensitivity in some insects as they develop into adults (Beer and Helfrich-Förster 2020).

One of the fascinating questions posed by thermoperiodic responses is how they relate to photoperiodism. Clearly, thermoperiod is somehow feeding into the same neural control mechanism as photoperiod to produce the diapause outcome, but it is doing so using an alternative pathway that doesn't involve circuitry used by the light receptors. Temperature cycles and pulses are fully capable of entraining circadian rhythms (Saunders 2002), thus suggesting a temperature conduit that circumvents the light receptors but connects to the circadian timekeeping mechanism.

The interaction of EYES ABSENT (EYA) and TIM, mentioned above in Section 5.6.2, provides insight into one possible mechanism connecting temperature and photoperiod at the molecular level in *D. melanogaster* (Abrieux et al. 2020). Thermosensitive splicing of *tim* leads to production of a temperature-sensitive isoform of TIM that stabilizes EYA, facilitating its accumulation at night and tripping the switch leading to diapause. Unlike *tim*, *eya* does not appear to be clock regulated but instead is a cold-induced gene whose protein product, EYA, is stabilized by a TIM isoform to elicit the seasonal response. Whether similar mechanisms operate in other species remains unknown. How temperature cues are integrated into the photoperiodic response remains one of the fascinating questions yet to be explored.

5.11   Host Influences on Diapause

As the seasons progress, plants that insects feed upon undergo dramatic changes in composition as they initiate growth in the spring, reach maturity, and then senesce in the autumn. Through the growing season, water and nitrogen content usually declines, while leaf toughness, phenolic compounds, or other defense agents increase. Likewise, insects that harbor parasitoids undergo seasonal changes as they progress through development and make their own preparations for diapause. The intimate relationships between insects and their hosts demand seasonal synchrony to maximize the relationship and to enable a plant-feeding insect or a parasitoid to effectively capitalize on the host resources. Though photoperiod and temperature provide the major seasonal cues, input relaying changing conditions of the host often provide secondary cues modifying the diapause decision.

5.11.1   Plant Feeders

One of the earliest and most extensively studied systems reporting host plant influences on diapause is the Colorado potato beetle, *Leptinotarsa decemlineata*. Early experiments with adult diapause in this species demonstrate an enhanced diapause response when beetles feed on older leaves (de Wilde et al. 1969, Tauber et al. 1988) or if food is withheld for 10 hours/day, even under continuous light or long days (de Wilde et al. 1959). Adults feeding on older leaves appear to compensate for poor leaf quality by eating more (nearly 50% more) than when fed young leaves, but the final weight differs little between the two groups, suggesting the additional food consumption is compensation for poorer food quality (Noronha and Cloutier 2006). Nitrogen, but not carbon, levels are considerably higher in short-day reared potato plants, a feature that nicely correlates with diapause induction in *L. decemlineata* and highlights the interplay between photoperiodism, plant physiology and the diapause response (Izzo et al. 2014). *L. decemlineata* exemplifies diverse aspects of host plant interactions affecting diapause and demonstrates that food can impact diapause both qualitatively and quantitatively.

Bouts of **starvation** are also elicitors of larval diapause in the cerambycids *Monochamus alternatus* (Togashi 2014) and *Psacothea hilaris* (Munyiri et al. 2004) and pupal diapause in the cotton bollworm *Helicoverpa armigera* (Liu et al. 2010). Timing of the onset of starvation is important. In *P. hilaris*, only 11% of the larvae enter diapause if denied food on the first day of the fifth instar, but all enter diapause if food is withheld on days 4–8 of the sixth instar (Munyiri et al. 2004). A threshold weight of approximately 600 mg is required for diapause entry; larvae weighing less fail to enter diapause. In addition, diapause duration is greater in beetles that are starved. Starvation also impacts diapause duration in the swallowtail *Atrophaneura alcinous*: Food deprivation every other day for the first 10 days of larval life results in a longer pupal diapause than observed in pupae that are not food limited (Kozuki and Takeda 2004). But, starvation does not universally affect diapause. No starvation effects are noted for larval diapause of the fly *Chymomyza costata* (Koštál et al. 2016b).

As the growing season progresses **host plant quality deteriorates**, a change that frequently is translated into a seasonal cue that boosts the diapause response, as noted above for *L. decemlineata* and in numerous other species. In the cabbage beetle, *Colaphellus bowringi*, the incidence of adult diapause is low when larvae are fed young radish leaves, but diapause increases when larvae are fed more mature leaves (Wang et al. 2006b). Similarly, the leaf beetle *Phratora vulgatissima* has a lower adult diapause incidence when fed young willow leaves rather than mature leaves (Dalin and Nylin 2012). The willow leaf beetle *Plagiodera versicolora* shows a similar response: Beetles reared under controlled laboratory conditions, with the age of willow leaves as the only variable, respond to the seasonal decline in leaf quality by increasing both development time and the incidence of adult diapause (Ishihara and Ohgushi 2006). Larvae of the swallowtail *Byasa alcinous*, even larvae from the same mother, that feed on tough, older leaves of its host plant, a woody vine *Aristolochia kaempferi*, enter pupal diapause at higher levels than larvae feeding on young tender leaves (Takagi and Miyashita 2008). Larval development proceeds more slowly on the tough leaves, but the swallowtails attain the same pupal weight as those reared on young leaves, suggesting that larvae compensate for poor quality food by feeding longer. Adults of the monarch butterfly *Danaus plexippus* reared on potted old milkweed plants enter diapause more readily than those reared on potted young plants (Goehring and Oberhauser 2002), and rearing larvae on a tropical milkweed, not native to North America, appears to prevent the diapause/migration response, leading to the establishment of permanent residency in North America (Majewska and Altizer 2019). Larval diapause incidence increases in *Chilo* stemborers when reared on aged maize stems that characteristically have low protein content (Scheltes 1978), a response that may be the dominant environmental cue for diapause in these tropical insects.

And, among species that feed on a range of plant species, it is clear that different hosts sometimes differently affect the diapause decision. When the leafroller *Choristoneura rosaceana* feeds on chokeberry the incidence of larval diapause is low and most larvae go on to produce a second generation; those reared on red maple or black ash exhibit a much higher incidence of diapause, and an intermediate response is noted on paper birch (Hunter and McNeil 1997). Feeding *C. rosaceana* low- and high-quality artificial diets demonstrates that nutritional quality of the diet has a direct effect on diapause induction, with a low-quality diet favoring diapause. The incidence of adult diapause in the cabbage beetle *Colaphelllus bowringi* is low when reared on Chinese cabbage variety Shanghaiqin, but much higher when reared on Chinese cabbage variety Suzhouqin, stem mustard or radish (Wang et al. 2006b), but the effect is restricted to a narrow temperature range and is not observed at temperatures of 20°C or lower. The cotton bollworm *H. armigera* feeds on a wide variety of plants, but as seen in Figure 5.17 the host plant utilized affects the incidence of pupal diapause: Feeding on cotton results in a high diapause incidence (nearly 80%), while, at the other extreme, the incidence of diapause when larvae are fed tobacco drops to less than 50% at 25°C (Liu et al. 2010). The distinction does not appear to be the consequence of different feeding times or differences in pupal weight because these values are quite similar for cotton and tobacco, although for each host plant, those entering diapause are consistently heavier than their nondiapausing counterparts and contain higher levels of lipids and glycogen (Liu et al. 2007). In *H. armigera* lowering the rearing temperature to 20°C eliminates the host plant effect; at the lower temperature all enter diapause regardless of the host plant.

Body size of the rice leaf beetle, *Trigonotylus caelestialium*, is greatly affected by the species of grass used as a food source and by host-specific seasonal changes in food quality, effects thought to be responsible for the increased incidence of embryonic diapause as summer progresses (Shintani 2009a). When females of *T. caelestialium* are fed a seasonally deteriorating host (orange foxtail) they lay diapausing eggs, but when transferred to a host that is not deteriorating (wheat) females switch to laying nondiapause eggs (Shintani and Nagamine 2020). Nymphs of the spider mite *Tetranychus kanzawai* readily feed on a range of host plants, including *Orixa japonica* and *Phaseolus vulgaris*, but the incidence of adult diapause is consistently higher on *O. japonica* (Ito 2010). The first summer generation of the cabbage bug *Eurydema rugosum* reared on seeds of brown mustard enter adult diapause, but those reared on the leaves do not, yet in the second generation, those reared on leaves also enter diapause (Ikeda-Kikue and Numata 2001). Diet quality also influences the proportion of diapause in adults of *Drosophila melanogaster*: Ovarian arrest occurs less frequently in field cages provided a mixture of corn meal and molasses than for flies offered fruit (Erickson et al. 2020). Offering live yeast also diminishes the incidence of diapause, a result suggesting that feeding yeast or the deactivated yeast present in the cornmeal and molasses mixture directly impacts the diapause decision.

The host plant used by the leaf beetle *Phratora vulgatissima* influences not only diapause incidence but also critical daylength (Dalin and Nylin 2012). The beetles readily feed on a wide variety of willow species, but critical daylength increases by nearly an hour when beetles feed on a non-preferred willow host, a response leading to earlier entry into diapause. The host plant may impact not only the decision to enter diapause but also success and duration of diapause, as exemplified in the cotton bollworm *H. armigera*. Larvae reared on cotton survive pupal diapause considerably better than those reared on corn, and diapause duration is longer for cotton-raised individuals (Ge et al. 2005). When larvae of the leaf beetle *Chrysomela lapponica* feed on a host plant (*Salix borealis*) that previously suffered heavy damage from herbivory, larval feeding duration is prolonged and overwintering mortality of the diapausing adults is high, a result suggesting that herbivory can elicit long-term effects on host quality that may impact success of diapause (Zvereva 2002).

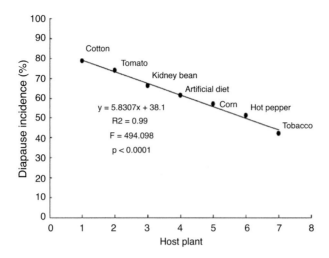

Figure 5.17 The relationship between pupal diapause incidence in the cotton bollworm
Helicoverpa armigera and different host plants. All were reared at 25°C and a short daylength
(LD 8:16).
From Liu et al. (2010), with permission from Elsevier.

Nutritional quality of a plant can also be altered by fungal infestations.
A mutualistic relationship exists between the grapevine moth *Lobesia botrana* and a
phytopathogenic fungus *Botrytis cinerea*: Moth larvae offered fungus-infested food
survive better, complete larval development more quickly, and display better synchrony of adult emergence after pupal diapause as well as higher post-diapause
fecundity (Mondy and Corio-Costet 2004), a result suggesting that the fungus somehow improves the nutritional quality of the larval diet, which in turn may affect timing
of diapause termination and adult eclosion.

The complicating factor in evaluating the effect of host plants on diapause is
knowing whether the effect is a direct consequence of host composition or an indirect
effect resulting from qualities of a less-than-optimal host that may influence growth
rate, efficiency of conversion, storage of energy reserves, or other factors impinging
on the diapause decision.

5.11.2   Parasitoids and Predators

Just as quality and quantity of host plants change through the season for plant-feeding
insects, the availability and quality of hosts used by parasitoids and the prey fed upon
by predatory insects can be expected to change seasonally and thus may offer
important cues for coordinating diapause. Numerous examples suggest the use of
host- or prey-based signals in determining the diapause fate of parasitoids and
predators. The alternative, also employed commonly, is for both hosts and parasit-
oid/predators to independently rely on the same environmental cues for diapause

induction. The net effect for either option is to closely synchronize the timing of development and diapause in parasitoids and predators with the host life cycle.

Food deprivation, known to increase the incidence of diapause in plant feeders, also generates a parallel response for some parasitoids. Denying parasitoids access to suitable hosts can increase the incidence of diapause, but in this case the effect is reflected in the progeny's diapause incidence rather than in the female herself, an effect first noted for the parasitoid *Nasonia vitripennis* (Schneiderman and Horwitz 1958). No access to a host for a few days after eclosion of the female wasp results in increased production of diapausing larvae by that female. A similar, but less profound, impact on progeny diapause is also noted if the female is deprived access to a host in the middle of her reproductive life or if she is given access to a host only on alternate days (Saunders 1965).

Parasitoids of **univoltine hosts** usually have only one generation per year and consequently enter an obligate diapause, whereas parasitoids of **polyvoltine hosts**, which have many generations per year, usually rely on external signals to program diapause (Tauber et al. 1986). Aphid parasitoids, which have been studied most extensively, are polyvoltine, as are their aphid hosts, and their diapause response depends on both abiotic (temperature, photoperiod) and biotic (host insect and/or host plant) signals (Polgar and Hardie 2000).

As aphid prey species are depleted the incidence of summer diapause increases in progeny of the aphid parasitoids *Aphidius avenae* and *A. rhopalosiphi.* To disentangle the effect of prey depletion from increased competition among parasitoids, Tougeron et al. (2018) reared parasitoids alone with different host densities (5–140 aphids) or reared parasitoids together with competing females (2–20 parasitoids) and 50 aphids. Low aphid densities do not induce diapause in the parasitoid's progeny, but the incidence of progeny diapause increases with increasing levels of **competition** experienced by females, a result that nicely underscores the importance of direct maternal competition in the programming of diapause. High competition is likely a proxy for measuring the risk of future host shortage and superparasitism. How the female monitors such competition remains a fascinating but unanswered question.

Diapause in progeny of the aphid parasitoid *A. ervi* is influenced by the morph of the pea aphid *Acyrthosiphon pisum* (Tougeron et al. 2019). The pea aphid produces both an **asexual viviparous morph** (spring and summer) and a **sexual oviparous morph** (autumn) that lays eggs that will overwinter in diapause. *A. ervi* can apparently distinguish these two morphs and deposits more diapause-destined progeny in the oviparous morph. This effect, however, is only evident for Canadian populations at 20°C, not at 17°C, and is not evident in French populations at either temperature. Higher fat, sugar and polyol levels are found in oviparous morphs, but exactly what cue is used to distinguish the two morphs remains unknown. Diapause status of the host is proposed as a cue for diapause in *Encarsia scapeata*, a parasitoid of the whitefly *Bemisia tabaci*: The parasitoid enters diapause in diapausing hosts but fails to do so if the host is not in diapause (Gerling et al. 2009). Similar responses are noted in the parasitoid *Pauesia unilachni*, hosted by the conifer lachnid *Schizolachnus pineti* (Polgar et al. 1995) and in *Praon volucre* and *Aphidius matricaariae*, hosted by the black bean aphid *Aphis fabae* (Polgar et al. 1991). Senescing plants used as food by the host aphid can sometimes also enhance the parasitoid's diapause response, as noted in certain parasitoids of *Uroleucon jaceae* (Polgar and Hardie 2000), but whether the effect is a direct response to declining quality of the plant or is indirectly mediated by the host aphid is difficult to discern.

Some parasitoids, such as *Nasonia vitripennis*, have the option of parasitizing a range of host fly species as well as pupae that are either destined for diapause or for nondiapause. How does the host species and its diapause program influence the diapause decision of the parasitoid? When *N. vitripennis* parasitizes pupae of the flesh fly *Sarcophaga barbata*, the parasitoid's life span is short, fecundity is reduced, and it switches to production of diapausing larvae after 8–11 days, but when puparia of either *Calliphora erythrocephala* or *Phormia terrae-novae* are parasitized, the parasitoid lives longer, fecundity is increased, and the switch to producing diapausing larvae is delayed (Saunders et al. 1970). A comparison of diapausing and nondiapausing host pupae of *Sarcophaga bullata* demonstrates that diapausing hosts are less suitable for *N. vitripennis* (fewer eggs laid, fewer adults emerge, more males produced), but diapause status of the host does not affect the parasitoid's diapause decision (Rivers and Denlinger 1995a). Failure of diapausing hosts to respond to envenomation by redirecting metabolism, especially fat metabolism, as seen in envenomated nondiapausing pupae of *S. bullata* (Danneels et al. 2013) may account for the lower production of adult parasitoids on diapausing hosts (Rivers and Denlinger 1994). Attacking a diapausing host, however, appears to convey a benefit for overwintering. Nondiapausing parasitoid larvae feeding on a diapausing fly host acquire high levels of the cryoprotectants glycerol and alanine from their hosts, an acquisition that translates into surviving at subzero temperatures $4–9\times$ longer than wasps reared on nondiapausing pupae that contain lower levels of cryoprotectants (Rivers et al. 2000). This feature can be exploited by augmenting the host diet with cryoprotectants to enhance cold hardiness in *N. vitripennis* (Li et al. 2014, 2015), an attribute that may be especially attractive for generating cold tolerant parasitoids for biological control.

Interestingly, the act of parasitization appears to sometimes enhance cold hardiness of the host. Larvae of the stem borer *Chilo suppressalis* host a number of endoparasitoids, and a comparison of parasitized and nonparasitized field-collected hosts shows higher glycerol levels and lower supercooling points in parasitized hosts (Quan et al. 2013), features consistent with greater cold hardiness induced in the host by parasitization.

The diapause fate of the parasitoid *Microctonus hyperodae* appears to be independent of the diapause status of its host, the Argentine stem weevil *Listronotus bonariensis* (Goldson et al. 1993). The parasitoid responds to prevailing daylength rather than being influenced by the adult host's developmental status. A similar independence from the host is noted for *M. vittatae* (Wylie 1980) and for *Oobius agrili*, an egg parasitoid of the emerald ash borer *Agrilus planipennis* (Petrice et al. 2019).

The predatory coccinellid *Hippodamia convergens* enters adult diapause if **denied prey**, and quality of prey is important as well: Over half the beetles enter diapause when offered a suboptimal prey, frozen eggs of the flour moth *Ephestia kuehniella*, while few enter diapause if offered living greenbugs *Schizaphis graminum*, a preferred prey species (Michaud and Qureshi 2005). When the coccinellid *Harmonia axyridis* is given an abundance of aphids (10–50/day) females continue to reproduce under both long and short days, but reducing prey abundance to one aphid/day or every 2 days boosts the diapause incidence in short-day females (Ovchinnikova et al. 2016).

The spider mite *Tetranychus urticae* appears to use **predation-related odors** as a cue for diapause entry (Kroon et al. 2008). Odors from the predatory mite *Typhlodromus pyri* or perhaps odors emanating from the events of predation cause spider mites to disperse from the leaf surface, crawl into protected crevices and enter diapause. Although such reports are rare for insects, similar responses are seen in water fleas (*Daphnia magna*), where diapause is elicited by the presence of a fish kairomone (Hairston 1987, Slusarczyk and Rybicka 2011).

5.11.3   Moisture Levels

A decline in water content is associated with plant senescence, a feature likely to be used by insects to monitor seasonal changes in host plant quality, but water content of the food may be, by itself, a factor modifying diapause. In the case of the flesh fly *Sarcophaga crassipalpis*, the response to water can be separated from plant senescence because this is a species that feeds on carrion. Adding 10% water to a diet of ground beef liver boosts pupal diapause incidence by approximately 10% (Denlinger 1972b). Tropical flesh flies show the same response: Adding 10% water to the diet of *Poecilometopa spilogaster* increases diapause incidence by 10%, while removal of 10% water from the meat diet drops the diapause incidence by 18% (Denlinger 1979). Water content by itself does not appear to elicit diapause in flesh flies, but it can modify responses to photoperiod or temperature. A contrasting effect of moisture content is noted for a closely related fly that has a larval diapause, *Lucilia sericata*: In this species dry meat produces a higher incidence of diapause (Mellanby 1938). The basis for these modifying effects of water remain unexplored but the results for each species are consistent and capable of modifying the incidence of diapause.

5.12   Signals May Be Interpreted Differently by Sexes

It is not uncommon for males and females to exhibit **different thresholds** for diapause induction, a response suggesting the cost of diapause may differ between the sexes. Males frequently have a lower threshold for diapause entry, implying that they enter diapause earlier in the season and consequently undergo a longer diapause. This also means that nondiapausing late season populations may be female-biased and overwintering populations are male-biased. This is the case for the flesh fly *Sarcophaga crassipalpis* (Denlinger 1972b) and related flies in this genus. The sex ratio approaches 1:1 when the diapause incidence is 100%, but at low diapause incidences, males are more highly represented in the diapause component, for example, 87% males in diapause when diapause incidence is <10%. As discussed in Section 4.1.1, early entry into pupal diapause for these flies carries little cost for males (Figure 4.1), but the **cost of diapause** for females, as noted in impaired reproductive output, is high and rises as the duration of diapause increases (Denlinger 1981). No such skewing of the sex ratio for diapausing pupae is noted in the tropical flesh fly *Poecilometopa spilogaster* (Denlinger 1979), a result suggesting that the short diapause of tropical flies does not exact an unequal toll on the sexes.

Males of the butterfly *Pieris napi* also enter pupal diapause at a longer daylength (i.e., earlier) than females (Wiklund et al. 1992). In this case there does not appear to be a fitness cost associated with diapause as seen in flesh flies, suggesting there may be alternative explanations for this sex difference. Perhaps the abundance of males that have built up by the end of summer makes it more advantageous for males to enter diapause and delay mating until the spring when competition may not be as intense. This seems especially advantageous for *P. napi* because it exhibits protandry (males emerge earlier than females), and males emerging late in the summer, when males are abundant, are unlikely to be as successful as males emerging early the following spring (Nylin 2013). Males of the yellow dung fly *Scathophaga stercoraria* also enter diapause earlier in the season than females (Blanckenhorn 1998b). Males of the monarch butterfly *Danaus plexippus* are more abundant at overwintering sites (Frey and Leong 1993), but whether this is due to higher female mortality during migration or to females opting to continue reproducing rather than migrating remains unclear.

Although the basis for sexual differences is often not known, such differences are rather common. In *Sarcophaga similis* the critical daylength for males is longer than for females (Tagaya et al. 2010), a response that advances the male's entry into diapause. Although such a distinction is evident for males from southern Japan, that is, they enter diapause earlier than females, no such difference is seen in populations from northern Japan (Yamaguchi and Goto 2019). Males of *S. similis* also appear to be less sensitive to a light pulse administered during the photoinducible phase (φ_i), noted above in Figure 5.11. The incidence of diapause in males of *Helicoverpa armigera* is consistently higher than in females, across populations and under different photoregimes (Shimizu and Fujisaki 2002, 2006), and the overwintering duration is shorter for females than males (Ge et al. 2005).

More males of the parasitoid *Nasonia vitripennis* are produced on less suitable (diapausing) fly hosts than on the more attractive nondiapausing hosts, a trait accompanied by production of fewer eggs (Rivers and Denlinger 1995). In spite of the widespread use of *N. vitripennis* for studies of sex allocation the influence of the host's diapause status has not been extensively examined.

One example that appears to run counter to the propensity for males to enter diapause earlier is the stem borer *Chilo suppressalis* (Lu et al. 2013). In this case, females tend to enter larval diapause earlier than their male counterparts. The basis for this distinction remains unclear. And, as one might expect, there are also good examples of species showing no differences between the sexes in the tendency to enter diapause, for example, the brown marmorated stink bug *Halyomorpha halys* (Musolin et al. 2019).

5.13   Letting Mother Decide

A particularly fascinating route for programming diapause involves letting mother decide. In these cases, the mother receives the photoperiodic (and sometimes thermal) cues and somehow transfers this information to her progeny. Transgenerational effects such as this, when the phenotype of the mother or the environment she experiences causes phenotypic effects in her offspring, are referred to as **maternal effects**. These responses are independent of genes the mother contributes to her offspring as well as environmental effects acting directly on the offspring (Mousseau and Fox 1998). Distinct genetic contributions from the mother and father do obviously influence diapause, a topic discussed in Chapter 10, but the discussion here focuses on effects elicited by the mother's environment rather than her genetic contribution. The best early documentation of maternal effects regulating diapause includes Japanese work on embryonic diapause of the silk moth *Bombyx mori* (see Section 9.1.1.1) and regulation of larval diapause in the parasitoid *Nasonia vitripennis* (Schneiderman and Horwitz 1958, Saunders 1965). In both examples, it is clearly the mother's environmental history that dictates the diapause fate of her progeny.

Maternal programming of diapause is particularly valuable for **embryonic diapause**, a situation in which the individual has limited access or time to evaluate seasonal conditions on its own. But, maternal effects are not limited to embryonic diapause. Mothers may influence the diapause fate of their progeny at larval and pupal stages as well, sometimes even adult progeny, and occasionally the effect can extend over multiple generations, for example, by a grandmother effect in several species of the parasitoid *Trichogramma* (Voinovich et al. 2013) and an effect persisting for up to five generations in *T. principium* (Reznik and Samartsev 2015). Maternal effects influencing diapause are especially well documented for Diptera and Hymenoptera, but are also known for a number of additional insect orders (Orthoptera, Homoptera, Hemiptera, Lepidoptera, Coleoptera), and among the acarines. Among the Aedini tribe of mosquitoes maternal control of embryonic diapause appears to be the ancestral state (Bova et al. 2019). Transgenerational effects are not unique to insects, and are also nicely documented in crustaceans, for example, diapause in *Daphnia* (Arbaciauskas 2004); plants, for example, flowering time in *Arabidopsis* (Chen et al. 2014); and vertebrates, for example, embryonic diapause in the annual killifish (Romney and Podrabsky 2017). A rather comprehensive list of insect species showing transgenerational effects is provided by Danks (1987), and Table 5.3 includes a few representative species examined more recently.

The mother's **thermal history**, as well as her **photoperiodic history**, may be influential. In the parasitoid *T. telengai* temperature to which pupae are exposed impacts the diapause fate of the adult female's larval progeny (Reznik and Voinovich 2016). Low temperature (17°C) during a 2-day window late in pupal life prompts diapause in progeny of the adult female, while exposure to high temperature (30°C) prevents diapause in the progeny. This brief developmental window coincides with the interval of photosensitivity, suggesting that temperature and photoperiod possibly elicit their responses through a similar mechanism. And, as mentioned above, even the grandmother's thermal and photoperiodic history are important in *T. telengai* (Voinovich and Reznik 2020): Low temperature and short-day rearing of the grand-mother's generation increase diapause incidence two generations later! A contrasting temperature response for summer diapause is noted in females of the hemlock woolly adelgid *Adelges tsugae*: Raising mothers at low temperature (12–14°C) results in larvae that fail to enter summer diapause, while diapause in the progeny is observed if adult females are maintained at a higher temperature of 17°C (Salom et al. 2001). Adult photoperiod, not temperature, coupled with embryonic temperature are the primary determinants of embryonic diapause in *Locusta migratoria* (Wang et al. 2021). The mother's diet can also impact the diapause fate of her progeny. Females of the rice leaf beetle *Trigonotylus caelestialium* deposit diapausing eggs when reared on a deteriorating host plant but produce nondiapausing eggs when reared on a high-quality host (Shintani 2017).

Though maternal effects can dictate the capacity for diapause in the offspring, environmental conditions to which offspring are exposed also commonly play a critical role. In the blow fly *Calliphora vicina* short daylength received by the mother capacitates her larvae to enter diapause, but whether or not they do so depends on the temperature the larvae receive: only a few larvae enter diapause when reared at 16°C, approximately 50% do so at 12–13°C, and nearly all enter diapause at 7–9°C (Vinogradova and Reznik 2013). The maternal effect continues to operate even in a far northern population of *C. vicina* from Lofoten Islands, within the Arctic Circle in northern Norway, but in this case larval temperature must be quite low, 4–10°C, and at that temperature some larvae enter diapause even if the mothers receive long daylengths (Vinogradova and Reznik 2015). In the band-legged ground cricket *Dianemobius nigrofasciatus* temperatures experienced by both the mother and her progeny affect the incidence of embryonic diapause: High temperatures during either stage reduce the incidence of embryonic diapause, but the reduction is consistently more pronounced if the mother is exposed to long days (Fukumoto et al. 2006).

Diapause duration, as well as diapause incidence, may be influenced by the mother's photoperiod and thermal history. In the blow fly *Lucilia sericata* short days and high temperatures (25 vs. 17.5°C) during the parental generation results in a longer larval diapause for the progeny, a response that likely prevents untimely termination of larval diapause in warm autumns (Tachibana and Numata 2004). A maternal effect on diapause duration is also noted for *Calliphora vicina* (Vinogradova 1974, McWatters and Saunders 1998). In this species, diapause dur-ation is longer when the mother is exposed to short days, but, unlike *L. sericata*, exposing *C. vicina* females to a higher temperature (20 vs. 15°C) shortens diapause duration in her progeny. Duration of embryonic diapause in the walking stick *Ramulus irregulariterdentatus* is dictated by the mother: Long days result in a longer diapause (Yamaguchi and Nakamura 2015). In the mite *Halotydeus destructor* summer embry-onic diapause duration of the offspring is longer if parents are reared under hot, dry conditions (Cheng et al. 2019).

Table 5.3 Examples of maternal effects influencing the diapause fate of the female's progeny

Taxa	Diapause stage	Reference
Acarines		
Halotydeus destructor	Embryo	Cheng et al. (2019)
Orthoptera		
Allonemobius socius	Embryo	Huestis and Marshall (2006)
Ephippiger ephippiger	Embryo	Hockham et al. (2001)
Dianemobius nigrofasciatus	Embryo	Goto et al. (2008)
Chortoicetes terminifera	Embryo	Deveson and Woodman (2014)
Locusta migratoria	Embryo	Jarwar et al. (2019), Wang et al. (2021)
Homoptera		
Adelges tsugae	Larva	Salom et al. (2001)
Hemiptera		
Trigonotylus caelestialium	Embryo	Shintani et al. (2009b)
Lepidoptera		
Bombyx mori	Embryo	Sato et al. (2014)
Coleoptera		
Colaphellus bowringi	Adult	Yang et al. (2007)
Diptera		
Aedes albopictus	Embryo	Lacour et al. (2014)
A. japonicas japonicas	Embryo	Bova et al. (2019)
Lucilia sericata	Larva	Tachibana and Numata (2004)
Calliphora vicina	Larva	McWatters and Saunders (1998)
Sarcophaga bullata	Pupa	Henrich and Denlinger (1982a)
Scathophaga stercoraria	Pupa	Scharf et al. (2010)
Hymenoptera		
Trichogramma telengai	Larva	Voinovich and Reznik (2017)
T. principium	Larva	Reznik and Samartsev (2015)
T. embryophagum	Larva	Reznik et al. (2011)
T. chilonis	Larva	Ghosh and Ballal (2018)
T. cacoeciae	Larva	Pizzol and Pintureau (2008)
Nasonia vitripennis	Larva	Saunders (1965)
Aphidius ervi	Larva	Tougeron et al. (2020)

Diapause duration of the parental generation exerts an impact on the incidence of diapause in the progeny of some species, as demonstrated in the cabbage beetle *Colaphellus bowringi* (Yang et al. 2007). This is a species having both a summer and winter diapause. Females that have been through winter diapause produce a higher incidence of progeny that will enter adult diapause than those that have been through summer diapause, and in both cases the progeny diapause incidence is higher the greater the duration of the parental diapause. Diapause duration of both the mother and father impact progeny diapause incidence, but the effect from the mother is more pronounced. In these beetles, a delay in mating also boosts progeny diapause incidence, and females that mate only once produce more diapausing progeny than those that mate multiple times, an effect that is more pronounced during the first 10 days after adult eclosion than for older adults. Progeny diapause in *C. bowringi* can also be impacted by food the parents consume (Lai et al. 2008): The incidence of diapause is highest when parental beetles are fed radish (71.6%), followed by Chinese cabbage (64.6%) and stem mustard (31.1%). The results observed for *C. bowringi* are a bit unusual in that several aspects of the parents' diapause, reproductive and feeding history influence the next generation. Such diverse effects possibly operate in other species as well but have simply been overlooked.

5.13.1   Transfer of Information from Mother to Progeny

How is maternal information transferred? It is interesting that egg size and lipid content of diapausing embryos are frequently greater than in their nondiapausing counterparts, for example, eggs of *Aedes albopictus* (Reynolds et al. 2012), but it is most likely that the larger size is a function of the diapause program rather than a cause. Large eggs, laden with more lipid, also predict the pupal diapause fate of the yellow dung fly *Scathophaga stercoraria* (Scharf et al. 2010).

In the migratory locust, *Locusta migratoria*, mothers raised under short days produce egg pods with diapausing embryos while those reared under long days produce nondiapausing embryos. A transcriptomic comparison of the CNS from the two types of mothers reveals more than 600 differentially expressed genes, with expression levels both up (360 transcripts) and down (250 transcripts) (Jarwar et al. 2019). One gene of potential interest is a circadian-regulated gene *takeout*. RNAi directed against *takeout* has no effect on long-day locusts, but suppressed expression in short-day locusts slightly raises the progeny diapause incidence (from 71 to 94%). Numerous other expression differences are associated with FoxO and related signaling pathways in *L. migratoria*, but most have not been functionally evaluated.

In the one case where the basis for the maternal effect is known, the commercial silk moth *Bombyx mori*, the decision to diapause is dictated by the release of diapause hormone (DH) within the female (Yamashita 1996), as discussed further in Section 9.1.1.1. DH acts on the ovaries to alter carbohydrate metabolism and boost levels of sorbitol in the yolk of the egg. It was initially assumed that the high levels of sorbitol were strictly contributing to cold hardiness, but Horie et al. (2000) nicely demonstrated that it is the presence of sorbitol in the yolk that causes the developmental arrest. Eggs artificially depleted of sorbitol proceed with development immediately, while addition of sorbitol to a medium surrounding embryos programmed for nondiapause brings about a developmental halt. The mother's sensitivity to photoperiod and temperature for the programming of diapause in her progeny begins when she is an embryo.

Recent progress has been made in identifying how the temperature response in *B. mori* is generated. A **thermosensitive transient receptor potential** (TRP) channel in the embryo (BmTrpA1) is activated at temperatures above 21°C as an early component of the pathway leading to induction of diapause in the female's progeny (Sato et al. 2014). The TRP channel appears to act as a molecular switch generating the long-term response in the following generation, a conclusion supported by RNAi results as well as responses to chemical activators of BmTrpA1. Diapause in the Kosetsu strain of

B.mori relies strictly on temperature for the programming of diapause (Tsuchiya et al. 2021), and as discussed more fully in Section 9.1.1.1, this strain has proven useful for probing fine details of the temperature response, revealing a temperature-dependent signaling pathway (Figure 9.4) that relies on both **corazonin** and **GABA** for release of DH. Though a few links in the regulatory scheme remain undefined, these results connecting temperature to the signaling pathway leading to DH synthesis and release offer significant strides in demonstrating how temperature can exact this transgenerational effect. But, there is little evidence that DH elicits a transgenerational effect in species other than *B. mori*, thus we are left with a phenomenon that, for most species, remains largely unexplained at the physiological level.

5.13.2   A Maternal Effect Blocking Diapause in the Next Generation

An interesting maternal effect operates in the flesh fly *Sarcophaga bullata* to shut down the capacity for diapause in the following generation if the mother herself has been through pupal diapause (Henrich and Denlinger 1982a). The beauty of this is that it enables the first generation of spring-emerging flies to override the short daylengths prevailing in early May and develop without diapause, even though the photoperiodic cues at that time should dictate diapause (Figure 5.18). It is not actually the experience of diapause that causes the effect, but the short-day photoperiodic cues received by the mother prior to diapause entry. This can be demonstrated by programming the flies for diapause by short days but then artificially preventing diapause. The diapause history of the father has no effect on this response, thus it is strictly a maternal effect. Only by rearing a generation under long-day conditions can the capacity for pupal diapause be restored in the progeny. Experiments based on ovarian transplants suggest the information is transferred from the mother to her germline sometime between the end of

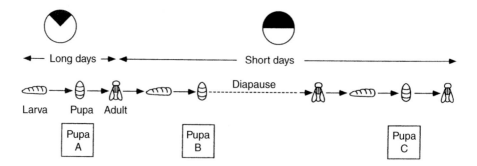

Figure 5.18 The influence of daylength on the pupa's developmental fate in the flesh fly *Sarcophaga bullata*. Pupa A does not enter diapause because it was exposed to long daylength as an embryo and larva. Pupa B can respond to short daylength by entering diapause because its mother was reared at long daylengths. Pupa C *cannot* respond to short daylength by entering diapause because its mother was reared at short daylength. It is the influence of the mother's photoperiodic history that is referred to as the maternal effect.
From Denlinger (1998).

larval life and the third day after adult eclosion, prior to fertilization and the onset of embryogenesis (Rockey et al. 1989). A nervous connection between the mother's brain and ovary is not essential for expression of the maternal effect, as demonstrated by transection of the nerves innervating the ovary, but a blood-borne factor from the CNS of the female may be involved. None of the common insect hormones are capable of mimicking the effect, but a slight reversal of the maternal effect can be elicited with picrotoxin, an agent that blocks the action of GABA, as well as by the neurotransmitter octopamine and the cholinergic agonist pilocarpine (Webb and Denlinger 1998). These agents are active only within a rather narrow developmental window (e.g., day 1 of adult life for picrotoxin), a response suggesting that the maternal effect can be intercepted at a fairly specific time. Several differences in protein abundance, including elevated levels of three actin-binding proteins in ovaries of short-day females, are also linked to the maternal effect (Li et al. 2012), but contributions of such proteins to regulation or execution of the maternal effect remain unknown.

A similar maternal effect preventing diapause operates in *Trichogramma* parasitoids (Reznik and Samartsev 2015). The incidence of diapause in progeny of females that have undergone larval diapause is extremely low, even under strong diapause-inducing conditions. As proposed for *S. bullata*, this maternal effect may be an adaptation preventing diapause in early spring when the prevailing environmental conditions are similar to conditions in the autumn. In *T. telengai* the capacity to enter diapause is restored in two generations, but five generations are required to restore the diapause capacity in *T. principium*. Unlike *S. bullata*, the maternal effect in *T. telengai* is caused by the mother's experience of diapause, not the environmental cues that lead to diapause (Voinovich and Reznik 2017). Although the photoperiod exposure time needed for the mother to change from producing nondiapausing to diapausing progeny varies with species, the switch can sometimes be remarkably swift, as noted in *T. embryophagum*, in which a single day spent under a different photoperiod is sufficient to trigger a switch in the type of progeny produced (Reznik et al. 2011).

These maternal effects are akin to prevention of diapause in the first post-diapause generation of aphids (Hardie 2010, Le Trionnaire et al. 2012, Matsuda et al. 2017), as discussed in Section 6.9. The growing toolbox available for probing diapause in the pea aphid offers considerable promise for examining this system in depth.

5.13.3   Impact of an Aging Mother

Age of the mother can influence the incidence of diapause in her progeny. If the mother is maintained under the same environmental conditions throughout, the diapause incidence of her progeny may change, either increasing or decreasing, as the mother ages. Pupal diapause incidence in successive broods gradually increases in progeny of the flesh fly *S. bullata*, increasing from 37% in the first clutch to >70% in the third clutch (Rockey and Denlinger 1986). These results are for flies that had a long-day history and thus are fully capable of responding to short days by entering pupal diapause. For flies expressing the maternal effect discussed above (i.e., the mothers had a short-day history and had gone through pupal diapause) the first clutch, as expected, produces no diapausing progeny (i.e., they displayed the maternal effect), but a few pupae from the second clutch enter diapause and even more (24%) from the third clutch do so. Thus, the maternal effect gradually erodes as the mother ages, thus both types of flies show an increase in diapause incidence correlating with increased age of the mother.

Similar increases in progeny diapause are noted in other Diptera, for example, blow flies *Lucilia caesar* (Ring 1967) and *Calliphora vicina* (Nesin et al. 1995); Orthoptera, for example, bushcricket *Ephippiger ephippiger* (Hockham et al. 2001), as well as in Hymenoptera, for example, *Nasonia vitripennis* (Saunders 1965) and *Doryctobracon areolatus* (Cruz-Bustos et al. 2020). *N. vitripennis* differs a bit from the Diptera examples in that the switch to producing diapausing offspring is more abrupt and dramatic, going from 0 to 100% diapause production in females between days 10 and 15 of adult life.

In some species, progeny diapause incidence decreases with the mother's age. In the parasitoid *Trichogramma embryophagum* the maximum diapause response is noted in progeny of young mothers (15% in females 1–2 days after adult eclosion) but drops to 0–5% by days 9–11 before rebounding slightly in older females (Reznik et al. 2002). Aging produces numerous physiological and biochemical changes, and it is not at all clear what specific aspect of aging elicits changes in the diapause fate of the progeny.

5.14   Reading Seasonal Cues near the Equator

Diapause is common in the tropics, as discussed in Chapter 2, and developmental arrests that bridge the dry season are especially well documented. Restricted plant growth during the dry season limits feeding options and would appear to be the driving force for diapause in many plant-feeding insects. But, dry seasons are not uniformly bad and may indeed be the favored season for some species. For example, the seed bug, *Jadera aeola,* relies on seeds of Sapindaceae, and hence the bug's peak activity is late in the dry season in Panama, when the seeds are mass produced (Tanaka et al. 1987c). In this same environment, the fungus beetle *Stenotarsus rotundus* feeds on puff ball fungi that appear briefly after the onset of the rainy season; the beetle then enters adult diapause for the remaining months of the rainy season as well as the following dry season, and becomes active again only when the next rainy season begins (Wolda and Denlinger 1984, Tanaka et al. 1987a). For flesh flies (family Sarcophagidae) in Kenya, pupal diapause coincides with the coldest portion of the year, an interval occurring during an interlude between the long and short rainy seasons (Denlinger 1974). Thus, depending on the species, different seasons may be favored for diapause, and thus environmental signals can be expected to differ across species and lack the consistency observed in temperate latitudes, but in far too few cases do we actually know the environmental signals used to program diapause in tropical species.

Photoperiod, the dominant environmental signal used at higher latitudes, offers a rather subtle signal near the equator, and I am unaware of any evidence for a photoperiodically regulated diapause operating within 5° of the equator. One interesting aspect of the natural light-dark cycle at the equator is that the times of sunrise and sunset (**solar time**) vary throughout the year, even though day and night lengths remain fixed at 12 hours each. This variation, a consequence of the elliptic shape of the Earth's orbit and the tilt of the Earth's axis relative to its orbit, generate an annual 30 minute variation in absolute times of sunrise and sunset. A songbird, the African stonechat, from Nakuru, Kenya (0° 14′ S), exploits this subtle seasonal change in solar time to coordinate its molt cycle (Goymann et al. 2012), and several tropical trees appear to rely on solar time to synchronize flowering (Calle et al. 2010). I am not aware of any insects that monitor solar time, but it has not been rigorously tested. Most evidence for tropical insects living near the equator points to the use of environmental signals other than photoperiod for programming diapause.

As reviewed previously (Denlinger 1986), photoperiod is used as a diapause regulator for many tropical insects that live 7 to 10° beyond the equator, even though the seasonal differences seem small by temperate latitude standards. The Asian tiger mosquito *Aedes albopictus*, a species widely distributed, from the tropics to northern latitudes, uses short daylengths to program embryonic diapause in northern populations, but populations from Thailand (9° N), eastern Malaysia (5° N), western Malaysia (4° N) are not sensitive to photoperiod (Hawley et al. 1987). The response is variable near the Tropic of Cancer: A population from Hanoi (21° N) responds to photoperiod (Tsunoda et al. 2015), but a population from Hong Kong (22° N) does not (Hawley et al. 1987). What is not completely clear, though, is whether tropical populations of *A. albopictus* lack diapause completely or if they simply rely on other environmental cues to program a diapause.

In the Cairns region of Australia (16° S), short daylengths during larval development are critical for induction of diapause in adults of the nymphalid *Hypolimnas bolina* (Pieloor and Seymour 2001). Short days in Cairns normally coincide with a decline in seasonal rainfall and larval resources, thus this environmental signal provides a reliable cue for initiating diapause. Temperature appears not to be important. But, daylength is not the only essential factor. Larvae must also feed sufficiently to attain a pupal weight of 0.72 g or more in order for adult diapause to be manifested. Even if this weight is attained and the correct photoperiodic cues are received, not all females enter diapause. A small portion (18%) fail to enter diapause even under optimal diapause-inducing conditions.

In Nairobi, Kenya, 1° S of the equator, only a 7 minutes difference distinguishes the longest day from the shortest day of the year. Yet, flies from Nairobi still enter pupal diapause during July and August, the coolest season (Denlinger 1974, 1979). In this environment the flies are not influenced by photoperiod, but low daytime (not night) **temperatures** during larval development shunt pupae toward diapause. It is important that the low temperature occurs during daytime, conditions that best predict the July–August period when daytime temperatures are low. Lowest nighttime temperatures occur in January–February, thus low daytime temperature provides the most reliable cue for programming a July–August diapause. It is, however, still unclear why diapause would be employed at that time because the temperatures, though cool, are well above developmental thresholds and remain suitable for continuous development, and food (carrion) would not appear to be seasonally limited. Quite likely this diapause is not used to circumvent an unsuitable physical environment but to perhaps periodically escape a biotic constraint such as parasitism or predation. Or, the diapause could be functioning primarily as a seasonal synchronizer. Regardless of the ultimate forces selecting for diapause, reliance on cool daytime temperatures enables the flies to enter diapause at a precise time during the African year. And, like the example of *H. bolina*, cited above, only a portion of the sarcophagid population enters diapause. This less-than-full commitment to diapause appears to be especially prevalent among tropical species and may reflect the rather unpredictable timing of tropical seasons.

Though diapause in the dry season is common, there are few examples demonstrating that **lack of rainfall** or decrease in relative humidity are used as cues for diapause induction. By contrast, rainfall is widely used by tropical insects to prompt resumption of development at the end of diapause (Section 8.8). One example in which diapause is elicited by lack of water is for embryos of *Stictophaula armata*, a tettigonid living in the evergreen forests of northern Thailand (18° 47′ N). Under humid conditions favorable for development, eggs hatch within 60 days, but at 60% R.H. no hatching occurs (Ingrisch 1996). Eggs stored at 60% R.H. hatch only if they are stored for 2–12 months and then allowed to absorb water, thus the initial lack of water in their environment prompts the diapause response.

Lack of rainfall likely exerts its most profound effect on diapause indirectly, through influences on **host plants**. As plants age, moisture content drops, levels of protein and lipid drop, and sugar levels increase. Such changes appear to be especially important promoters of diapause for tropical plant-feeding insects, including the maize stem borer *Busseola fusca* in Nigeria (Usua 1970), *Chilo* stem borers on the Kenyan coast (Scheltes 1978), and the pink bollworm *Pectinophora gossypiella* in Trinidad (Squire 1939), but it is not always clear what features of the aging plant are being monitored. When only the age of the maize host plant is altered under constant conditions in the laboratory, the incidence of diapause in *B. fusca* fed young maize (6–9 weeks old) is 24%, compared to 91% diapause in larvae fed old maize (12–15 weeks). A similar increase in the incidence of diapause with plant age is noted for larvae of *B. fusca* feeding on sorghum in Kenya (Okuda 1991). A decline in water content within the maize plant is suspected to be the driving force for increases in larval diapause in *Chilo partellus* and *C. orichalcociliella* along the Kenyan coast (Scheltes 1978). Maize that remains well hydrated in irrigated fields does not harbor diapausing larvae.

Attempts to induce diapause by manipulating environmental conditions have met with modest success thus far for members of the *Anopheles* complex of malaria-transmitting vectors. Some members of this complex, such as *An. arabiensis* likely do not enter diapause (Simard et al. 2000), but others such as *An. coluzzi* appear to employ diapause to bridge the dry season (Dao et al. 2014). But, exactly what environmental conditions prompt entry into diapause remains unclear. By mimicking cool conditions and short daylengths that naturally occur in Mali (13.66° N) late in the wet season and early in the dry season, some adults survive more than 100 days (Krajacich et al. 2020), but this is still considerably shorter than field survival that can extend beyond 200 days.

Environmental signals prompting induction of tropical diapause thus remain rather poorly known, with only a few examples convincingly defining the signals used. As discussed in Section 8.8, many more studies have examined the role of the environment in terminating such diapauses.

5.15   Variability in the Environment

Most experiments on environmental regulation of diapause have focused on, at best, one or two environmental cues, but a **variety of environmental factors** can simultaneously exert modifying influences on the dominant photoperiodic response. Temperature, diet quality and quantity, moisture, sex and population density all can modify the response, but seldom have their interactions been thoroughly examined in a single species. In the flesh fly *Sarcophaga crassipalpis,* the effect of short daylengths and low temperature on pupal diapause induction is augmented by adding 10% water to the larval medium, and the threshold for diapause induction is lower for males than females (Denlinger 1972b). Larval crowding in *S. argyrostoma* results in undersized larvae that pupariate early and escape the programming of pupal diapause (Saunders 1997). Larval diapause in the fly *Chymomyza costata* is dependent not only on photoperiod and temperature but also on population density (Koštál et al. 2016). Overcrowding in *C. costata* lengthens duration of larval life and thereby influences the diapause response. In the silver-spotted skipper *Epargyreus clarus* the influence of short daylengths on the programming of diapause is enhanced not only by low temperature but also by low quality of its host food, but the diapause-promoting effect of short days can be countered by exposure to high temperatures and high quality food (Abarca 2019). Temperature increases from 16 to 20°C under short days result in greater lean and fat body mass, a consequence of concurrent increases in food consumption and longer development times for diapause-destined pupae of the corn earworm *Helicoverpa zea* (Clemmensen and Hahn 2015). Results such as these nicely exemplify the need to integrate multiple seasonal cues to fully understand how the diapause phenotype is generated. Similar interactions are likely common and highlight the interplay expected between diverse environmental factors.

Not only are multiple factors involved, but, unlike most laboratory experimental setups, single factors seldom remain constant in the real world. Insects in natural conditions are subjected to a highly **variable environment**, especially for such factors as temperature. Although it is easier to examine adaptations to a single constant environmental factor, it is important to recognize that insects confront a variable environment that, as pointed out by Dillon and Lozier (2019), can perhaps best be evaluated using integrative approaches that combine fine-scale reductionist studies with large-scale biogeographic patterns.

5.16   Aborting the Diapause Program

As noted above, programming diapause is seldom a one-step process. A series of events must occur in the correct sequence, and there are few options for short-circuiting the system by bypassing any of the steps. For illustration, Figure 5.19 shows the sequence of events required to generate pupal diapause in flesh flies (genus *Sarcophaga*). The flies must have the genetic capacity for diapause, mothers must lack a history of having been in diapause, embryos must be exposed to short days, larvae must receive short days, larval temperature must be cool, and temperatures following pupariation must be cool (Denlinger 1985). Failure to meet any of these requirements will either completely avert diapause or greatly reduce the diapause incidence. There are only limited possibilities for short-circuiting the system, and doing so requires input of an even greater stimulus further down the programming pathway. For example, if embryos receive long days rather than short days, a low incidence of diapause can still be induced by rearing larvae at short days, but such larvae must then be reared at much lower temperatures.

While possibilities for inducing diapause by skipping steps in the sequence are quite limited, the diapause program can **readily be averted** at many different times, right up to the very onset of diapause. Thus, the programming of diapause can be viewed as a sequence of developmental criteria, all of which must be fulfilled for diapause to be exhibited. Failure to meet any of these rigid criteria irreversibly erases the program and channels the insect toward continuous development. The diapause program thus has an easy reset switch allowing the insect to avert diapause at the last minute if conditions change. One can imagine, for example, that high post-pupariation temperatures for flesh fly pupae would signal an unfavorable pupariation site, and rather than entering diapause, pupae make a last minute developmental decision to avert diapause and avoid the high metabolic cost of a diapause that could not be sustained. The risk of certain death during a high temperature diapause can be traded for the possible lower risk of not being able to complete an additional generation. A number of environmental, chemical, and physical manipulations are capable of aborting the diapause program in diverse species (Section 12.3), underscoring vulnerability of the environmental programming phase of diapause, a feature likely to have considerable adaptive value.

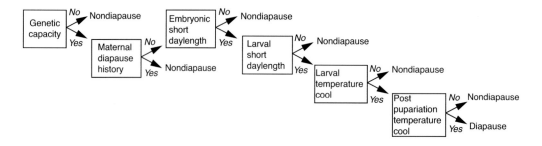

Figure 5.19 Sequence of events required for the programming of pupal diapause in flesh flies (*Sarcophaga*).
From Denlinger (1985).

6   Preparing for Diapause

Diapause is a complex phenotype, with distinctions starting before diapause begins and persisting long after diapause has ended. These differences include behavioral, structural, physiological, molecular, and fitness attributes that collectively lend credence to the idea that diapause can impact the entire life cycle and is thus more appropriately viewed as an **alternative life cycle**. In this chapter on preparing for diapause and in the following chapter on the diapause state, my goal is to highlight the diverse features of diapause, demonstrating the complexity of this response. Some attributes are common, others less so. Unlike a period of quiescence that quickly starts and stops, diapause is programmed well in advance of its inception, enabling the insect to engage in preparative steps for entry into a successful diapause. Such preparative steps can be considered as components of the diapause syndrome.

It is important to distinguish whether differences noted prior to diapause are attributes of the diapause program or are consequences of direct environmental effects that may impact the growing insect differently in different seasons, *e.g.*, differences in temperature or the changing quality of host plants. Factors such as low temperature or poor quality host plants can have direct impacts on growth rates and size, so it may not be immediately obvious if differences between diapause and nondiapause are consequences of such differences or if they can be attributed to the diapause program. It is not readily possible to distinguish between these two options in the field, but rigorous laboratory experiments can reveal which option is in play. The term **anticipatory** or **cued plasticity** refers to phenotypic plasticity induced by signals foretelling a future developmental event such as diapause, while the term **responsive** or **direct plasticity** refers to plasticity evoked as an immediate response to the environment (West-Eberhard 2003, Whitman and Agrawal 2009). The characteristics described below represent anticipatory plasticity and set the stage for the upcoming diapause state.

6.1   Prolongation or Acceleration of Pre-diapause Stages

The rate of pre-diapause development is frequently altered by the diapause program, resulting in either prolongation or acceleration of development. This is not a response restricted to insects but is also seen in diapauses of insect relatives. In the twospotted spider mite *Tetranychus urticae*, a species in which only adult females overwinter, there is a striking prolongation of development in diapause-destined deuteronymphs

(Suzuki and Takeda 2009). A similar delay is seen in *Daphnia*, a freshwater crustacean (Chen et al. 2018). Embryos of *Daphnia* enter diapause when they reach a cell number of approximately 3,500; at that point mitotic activity ceases, but prior to that time, embryos destined for diapause develop more slowly (mitotic activity is $2.5\times$ slower) than their nondiapausing counterparts.

Among insects, a distinction is noted for embryos as well as larvae. Diapause in the Asian tiger mosquito *Aedes albopictus* occurs as a pharate first-instar larva (at the end of embryonic development when the larva is still encased within the chorion), and in this species embryonic development prior to the onset of diapause proceeds more slowly for embryos that are diapause-destined (Lacour et al. 2014). In some cases, slower development means a longer feeding period, which can result in larger individuals that consequently stockpile more energy reserves. For example, larvae of the mosquito *Culex pipiens*, when reared under conditions that lead to adult diapause, extend larval life by one day, become larger, and consequently store more reserves than individuals that do not enter adult diapause (Zhang and Denlinger 2011a). Likewise, larvae of the green lacewing *Chrysopa formosa* destined for diapause feed three to four days longer than their nondiapause-destined counterparts and attain a greater body mass (Li et al. 2018b). Larvae of the cerambycid beetle *Psacothea hilaris* reared under long-day, nondiapause-inducing conditions pupate at the end of the fifth larval instar, while those reared under diapause-inducing conditions frequently undergo two additional larval molts and enter a seventh-instar diapause (Munyiri and Ishikawa 2004). Acquisition of extra weight is critical in this species: Nondiapausing larvae weighing as little as 180 mg survive to adulthood, but going through diapause and surviving to adulthood requires a starting weight of at least 600 mg (Munyiri et al. 2004). Longer developmental time and hence larger size attained by the diapausing generation may also be a "luxury" of developing later in the season when bird predation rates are lower (Teder et al. 2010). Extended feeding periods and concomitant weight gain for diapause-destined individuals are common, but certainly not universal.

In some cases, extension of the pre-diapause interval does not protract the feeding period but instead occurs after the insect is done feeding. Flesh fly larvae destined for pupal diapause feed for the same duration as their nondiapause counterparts, but the post-feeding, wandering phase of the final larval (third) instar is greatly extended (Denlinger 1972b). While nondiapause-destined larvae of *Sarcophaga bullata* pupariate one to two days after leaving their food (Figure 6.1A), those destined for diapause may wander as long as two weeks before pupariating (Figure 6.1B). Within the diapausing cohort, the incidence of pupal diapause increases with the length of the delay (Figure 6.1C). A possible adaptive advantage of this longer wandering period may be that it offers the diapause-destined fly more time to secure a pupariation site suitable for a long overwintering hiatus. A similar, but less dramatic, prolongation of the post-feeding, wandering larval stage is seen in some Lepidoptera. Larvae of the swallowtail butterfly *Papilio machaon* programmed for pupal diapause wander nearly twice as long before pupating as their nondiapause-destined counterparts, but in this case the prolongation is measured in hours rather than days (Wiklund et al. 2017). Prolongation of the wandering

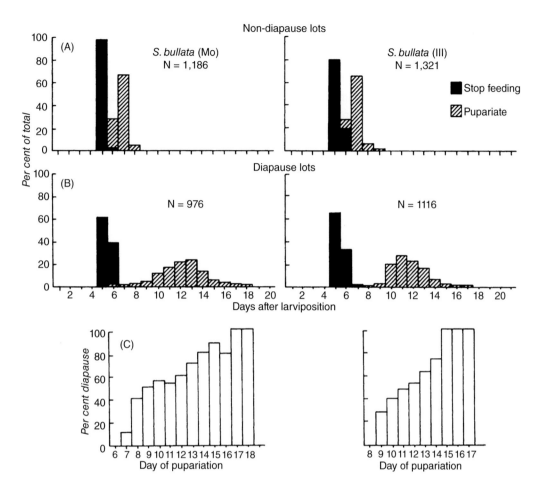

Figure 6.1 Time of pupariation in two populations (Illinois and Missouri) of (A) nondiapause-destined, long-day reared larvae of the flesh fly *Sarcophaga bullata* and (B) diapause-destined larvae reared under short-day conditions. (C) Incidence of pupal diapause within the short-day group, based on the day of pupariation. All were reared at 26°C.
From Denlinger (1972b).

larval phase in the swallowtail *Iphiclides podalirius* enables diapause-destined larvae to move away from their host plants to pupate in the leaf litter; by contrast those not destined for diapause pupate directly on their host plants (Esperk et al. 2013). Delays of this sort enable the larvae to perhaps not only find a "better" overwintering site, but more time wandering also translates into more distance from their food source. Since many predators and parasitoids rely on host odors to find their prey, there are likely adaptive advantages in moving further away from their larval food source.

   In the burnet moth *Zygaena trifolii* diapause most frequently occurs in the third larval instar, and the previous instar is prolonged. But, in less than six generations of laboratory selection, diapause in later larval instars can dominate, and interestingly, this shift in instars is consistently accompanied by prolongation of the previous larval instar (Wipking and Kurtz 2000), underscoring the linkage between diapause and pre-diapause development.

There are also numerous examples of development being more rapid in individuals destined for diapause. Larvae of *Drosophila montana* reared under short-day conditions simulating autumn complete egg-to-adult development more rapidly than do their counterparts reared under long days, and the resulting adults are considerably smaller (Salminen et al. 2012). The period of photosensitivity regulating the rate of larval development in *D. montana* occurs early, during egg-larva development, and is distinct from the sensitive period used for programming adult diapause, an interval that occurs only after adult eclosion. This early use of short daylengths to speed up development enables this far-northern species to reach the adult stage before it's too late! In the Colorado potato beetle, *Leptinotarsa decemlineata*, diapause-destined, short-day pupae and their nondiapause-destined, long-day counterparts attain similar weights, but, beginning in the third instar, short-day larvae accelerate development during both the feeding and post-feeding phases of larval life and pupate more than six days earlier than larvae reared under long days at the same temperature of 25°C (Doležal et al. 2007). A comparison of developmental rates in two northern populations (53 and 63° N) of the damselfly *Lestes sponsa* shows more rapid larval development in both populations when reared under a northern photoregime, a response resulting in smaller adult size (Sniegula and Johansson 2010). The developmental acceleration seen at high latitudes is likely an adaptation to the temporal constraints of summers at such latitudes.

A similar, but less pronounced, acceleration of pre-diapause development is noted in nymphs of the pirate bugs *Orius minutus* and *O. sauteri*: Those destined for adult diapause as a consequence of short-day exposure develop more rapidly as nymphs (Musolin and Ito 2008). And, in the linden bug *Pyrrhocoris apterus*, a combination of short days and low temperatures prompts acceleration of nymphal development time, allowing the bug to reach adulthood more quickly (Lopatina et al. 2007). Acceleration of pre-diapause development is particularly prevalent among the true bugs. Other examples include the brown marmorated stink bug *Halyomorpha halys*, which completes nymphal development in 530 degree days under short days but requires 590 degree days under long days (Musolin et al. 2019) and the mirid plant bug *Dicyphus errans*, a species that develops most rapidly under a 12:12 LD cycle but requires an additional 6 days at 14:10 LD (Pazyuk et al. 2018).

Short-day-induced, pre-diapause acceleration of development in the ladybird beetle *Harmonia axyridis* results in smaller adults with fewer energy reserves and impaired winter survival (Reznik et al. 2015), a trade-off that enables the beetles to complete development before winter but at the cost of being small. These beetles, however, can compensate somewhat for their small size by feeding as adults if environmental conditions permit. Interestingly, the short-day responses governing developmental acceleration and diapause induction in *H. axyridis* are not identical (Reznik and Vaghina 2011): Daylengths between 16 and 14 hours elicit developmental acceleration, while daylengths between 14 and 12 hours are required to bring about the adult's reproductive arrest.

Species having both bivoltine and univoltine populations reveal an intriguing interplay linking diapause and pre-diapause development time. To switch from being univoltine to bivoltine implies that development must either be accelerated in the bivoltine population and/or they may have to settle for a smaller size. In North Wales the lycaenid butterfly *Aricia agestis* shows both patterns of voltinism in nearly the same geographic area. Several trade-offs are apparent (Burke et al. 2005): To achieve a bivoltine pattern, the rate of larval development is accelerated, adult weight is reduced, and the critical photoperiod for diapause induction is shifted to a shorter daylength by more than 1.5 hours to accommodate completion of an additional summer generation. Similar trade-offs are noted for another lycaenid butterfly *Lycaena hippothoe* (Fischer and Fiedler 2002): The bivoltine population found in Hungary has both a faster development time and is also smaller than diapausing larvae from univoltine populations collected in western Germany and the Alps.

Within a single species, daylength may affect growth rate differently depending on the season. The satyrine butterfly *Lasiommata maera* overwinters as a larva midway through larval development, thus the same individual feeds both in the autumn and again in the spring (Gotthard et al. 1999). The declining autumn daylength accelerates growth rate while the opposite effect is elicited in the spring. Short daylength in a closely related butterfly *Lopina achine* also accelerates development in the autumn, but daylength in the spring appears not to impact growth rate. In these butterflies, and presumably many other species, it is the growth rate that is being altered by the diapause program. Getting bigger is not necessarily just a function of feeding longer.

Two possible scenarios could explain the relationship between developmental time and diapause: Does slow development foster the decision to enter diapause or is slow development a consequence of the diapause decision? Few experiments have directly tested these two scenarios, but experiments with the butterfly *Pieris napi* suggest that the slower growth rate in larvae destined for pupal diapause is largely determined by the decision to enter diapause, although a few individuals may enter diapause because they are growing at a slower rate (Friberg et al. 2012). It can be difficult to disentangle relationships between developmental rate, photoperiod, and the diapause response (Kutcherov et al. 2018), but in most cases the pre-diapause developmental rate is thought to be a component of the diapause program, regardless of whether this means retardation or acceleration of development prior to diapause. This linkage is further underscored by results with flesh flies *Sarcophaga bullata* (Henrich and Denlinger 1982b) and *S. similis* (Goto 2009), showing that selection for larval duration directly impacts pupal diapause incidence. But, again, there are exceptions. Another flesh fly *Boettcherisca* (*Sarcophaga*) *peregrina* distributed across Asia and Oceania fails to diapause at short daylengths and 20°C in certain tropical regions, yet short daylengths still result in prolongation of the post-feeding wandering phase of the third instar (Moribayashi et al. 2008). Selection for faster pre-adult development time in *Drosophila montana* increases variance in the adult diapause response and lengthens the critical daylength (Kauranen et al. 2019a), again demonstrating connections between what would appear to be distinct diapause-related traits.

There is considerable phenotypic variation in development time, allowing selection to act rather quickly. This can be seen in the cinnabar moth *Tyria jacobaeae*, a European species introduced to North America for biological control of ragweed. After being introduced to the Willamette Valley (83 m) in Oregon in 1960, it expanded its range into the Cascade Mountains (1,572 m), while concomitantly shortening the duration of larval development (from 30 to 24 days) to accommodate the shorter growing season at high altitudes (McEvoy et al. 2012).

6.2   Acquisition of Energy Reserves

Stockpiling additional energy reserves prior to diapause is a common and logical adaptation to prepare for the long period of fasting implicit in diapause (Hahn and Denlinger 2007, 2011). And, in some cases, reserves stored prior to diapause must not only fuel diapause but also subsequent metamorphosis, reproduction, and flight activity in the adult. The demand is particularly acute for species in which adults lack functional mouthparts and are totally dependent on reserves sequestered during larval life, *e.g.*, the giant silk moth *Hyalophora cecropia* and fall webworm *Hyphantria cunea*.

Good examples of diapausing individuals storing **extra reserves** can be seen in all diapause stages. In embryonic diapause of the Asian tiger mosquito *Aedes albopictus*, the mother sequesters approximately 30% more lipid reserves into diapause-destined eggs than into those not programmed for diapause (Reynolds et al. 2012). Diapausing larvae of the burnet moth *Zygaena trifolii* contain twice as much fat as their non-diapausing counterparts (Wipking et al. 1995), and the dry weight of diapausing larvae of the parasitoid *Praon volucre* is nearly 30% greater than that of nondiapausing larvae, most of which is a disparity in lipid content (Colinet et al. 2010). Diapausing larvae of the southwestern corn borer *Diatraea grandiosella* (Brown and Chippendale 1978) and diapausing adults of the Colorado potato beetle *Leptinotarsa decemlineata* (de Kort and Koopmanschap 1994, Piiroinen et al. 2011) and the linden bug *Pyrrhocoris apterus* (Sula et al. 1995) produce vastly more storage hexamerins if they are destined for diapause. The pierid butterfly *Pieris napi* stockpiles larger quantities of the storage lipid triacylglycerol if it is programmed for pupal diapause (Lehmann et al. 2016a), and adult females of the mosquito *Culex pipiens* accumulate twice as much fat if they are programmed for diapause (Mitchell and Briegel 1989, Zhou and Miesfeld 2009). Diapausing adults of the chrysomelid beetle *Aulacophora nigripennis* weigh nearly 50% more than their nondiapausing counterparts and contain nearly four times the amount of lipid, but by the end of diapause lipid content drops from 4 to 1 mg/adult, a value equivalent to the amount present in nondiapausing beetles (Watanabe and Tanaka 1998). Adults of the cabbage beetle *Colaphellus bowringi* programmed by short daylength for diapause accumulate nearly four times the amount of lipid noted in nondiapausing beetles (Zhu et al. 2019).

Sex differences are sometimes noted. In the ladybird beetle *Harmonia axyridis*, diapause-destined females accumulate both carbohydrates and lipids, while males accumulate mainly carbohydrates (Gao et al. 2019). Diapause-destined females of the hover fly *Episyrphus balteatus* more than double their fat reserves at select short daylengths, but males fail to do so; in spite of this disparity, males survive the winter equally well (Hondelmann and Poehling 2007).

Weight is a good indicator of the additional energy accumulation associated with diapause. Diapausing larvae of the corn-stalk borer *Sesamia nonagrioides* weigh nearly twice as much as larvae that do not enter diapause (Gadenne et al. 1997), diapausing pupae of the apple leaf miner *Phyllonorycter ringoniella* weigh approximately 1.6× more than nondiapausing pupae (Li et al. 2002a), diapausing pupae of the chestnut leaf miner *Cameraria ohridella* are 1.3× the weight of nondiapausing pupae (Weyda et al. 2015), diapausing pupae of the potato tuber moth *Phthorimaea operculella* weigh 1.2× as much as nondiapausing pupae (Hemmati et al. 2017), diapausing pupae of the fall webworm *Hyphantria cunea* weigh 1.3× more than their nondiapausing counterparts (Zhao et al. 2021), and pupae and adults developing from diapausing larvae of the copper butterfly *Lycaena tityrus* weigh approximately 1.25× more than their counterparts that do not enter diapause (Fischer and Fiedler 2001).

The greater length of forewings in diapausing adults of the tropical butterfly *Hypolimnas bolina* also implies a larger size of diapausing individuals (Kemp and Jones 2001). Based on multiyear sampling of 12 small moth species from Estonia, body size in nearly all adults from the overwintering generation is larger than in the early summer generation, in spite of the poorer food quality available in late summer (Teder et al. 2010).

Greater body mass within the diapausing cohort has a demonstrated positive effect on winter survival in some species, *e.g.*, diapausing larvae of satyrine butterflies *Lasiommatata maera* and *Lopinga achine* (Gotthard et al. 1999), adults of the harlequin ladybird *Harmonia axyridis* (Knapp and Rericha 2020), and is likely true for many other species as well. The greater body size associated with adult diapause in the parasitoid *Habrobracon hebetor* also results in a larger body size for the female's progeny (Reznik et al. 2020). The importance of pre-diapause feeding opportunities and the associated weight gain, especially accumulation of lipids, can be seen for adults of the nymphalids *Aglais urticae* and *Inachis io* by observing the consequences of restricting the duration of feeding prior to diapause entry (Pullin 1987). Nearly all diapausing adults allowed to feed for 30 days after adult emergence survive at least 160 days at 4°C, those granted a 7-day feeding opportunity survive less well, and feeding limited to 3 days results in 100% mortality for *A. urticae* and 30% mortality for *I. io* within the same 160-day period. These nymphalids emerge as adults with nearly no lipid reserves, but if allowed to feed, increase lipid levels in their abdomens to 10 mg (*A. urticae*) or 55 mg (*I. io*) within two weeks. Without the opportunity to build these lipid reserves, diapausing adults are unlikely to survive the winter.

Among insects displaying a mixture of diapausing and nondiapausing individuals, diapause incidence frequently correlates positively with weight, which in turn affects overwinter survival, as noted in pupae of the cotton bollworm, *Helicoverpa armigera* (Liu et al. 2010). The **host plant** can indirectly influence the insect's fate; pupae of *H. armigera* reared as larvae on optimal host plants, cotton or tobacco, attain higher

pupal weights, amass more lipid stores, and have lower winter mortality than diapausing pupae reared on a suboptimal host plant, kidney beans (Liu et al. 2009). How
much and what quality of food is offered to the larvae is also a determining factor in
the diapause decision of the solitary bee *Megachile rotundata* (Fischman et al. 2017).
Field surveys and laboratory manipulations of the bee's diet demonstrate that an
abundance of food increases the incidence of diapause. Supplementing the diet with
honey bee royal jelly, a food *M. rotundata* would not normally receive, also enhances
the diapause response, even though larvae fed royal jelly are smaller in size. How
royal jelly exerts its effect on diapause is not currently understood. When diapausing
and nondiapausing adults of the scorpionfly *Panorpa vulgaris* are offered unlimited
food, diapausing adults become larger, and interestingly, males produce larger sperm
(3.5 mm vs. 3.2 mm) than their nondiapausing counterparts (Vermeulen et al. 2009), a
result that could have implications for sperm competition.

Diapausing pupae of *H. zea* are both larger and have a higher fat mass than their
nondiapause counterparts, a feature attributed to a longer period of larval development
and greater food consumption by larvae programmed for pupal diapause
(Clemmensen and Hahn 2015). Interestingly, lean mass of both diapausing and
nondiapausing pupae of *H. zea* increases linearly with temperature, as does fat mass
in diapausing pupae, but in nondiapausing pupae fat accumulation levels off at high
temperature, indicating a profound distinction in **temperature-size relationships**
prompted by the diapause program (Figure 6.2). Temperature, host plant, and family
lineage all interact in the Glanville fritillary *Melitaea cinxia* to determine how much
fat is stored in preparation for larval diapause (Verspagen et al. 2020). Since the
mother selects the site for oviposition, she too ultimately participates in the amount of
fat stored by her pre-diapause larval progeny. Experiments such as these nicely
demonstrate that the high accumulation of energy stores associated with diapause is
not simply a function of food availability but reflects other nuances of the diapause
program as well.

What is perhaps more surprising is that not all species selectively accumulate more
reserves in preparation for diapause! A few species show no differences in fresh or dry
weight between diapausing and nondiapausing individuals, *e.g.*, larvae of the blow fly
Calliphora vicina (Saunders 1997) and pupae of the tobacco hornworm *Manduca
sexta* (Siegert 1986). And, in a few species, diapause-destined individuals are actually
smaller than their nondiapausing counterparts, *e.g.*, larvae of the budworm
Choristoneura fumiferana (Harvey 1961), pupae of the swallowtails *Papilio polyxenes* (Blau 1981) and *Iphiclides podalirius* (Esperk et al. 2013), and adults of
Drosophila montana (Salminen et al. 2012). Some of these cases may be the consequence of poor nutrition due to a late-season decline in host plant quality, but the
examples of *C. vicina*, *M. sexta*, *I. podalirius*, and *D. montana* are clearly not the
result of food deficiencies.

Food is not necessarily off-limits during diapause, especially for species that
diapause as larvae or adults (Bradshaw 1973, Tauber et al. 1986, Danks 1987, Hahn
and Denlinger 2007), but for embryos or pupae feeding is not an option. When
feeding does occur during diapause, the threshold for accepting food may be higher,

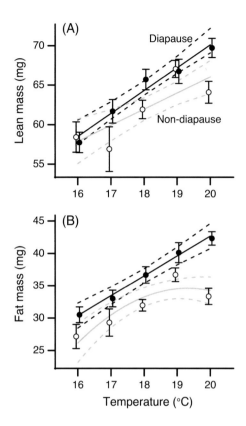

Figure 6.2 Temperature-size responses for (A) lean mass and (B) fat mass in diapausing (closed circle) and nondiapausing (open circle) pupae of the corn earworm *Helicoverpa zea*. Diapausing pupae have an overall greater lean mass and linear increase in fat mass, while nondiapause individuals show a quadratic relationship that levels off at higher temperature.
From Clemmensen and Hahn (2015), with permission from Springer.

as noted in the food-sensing tarsal response of diapausing adult face flies *Musca autumnalis* (Stoffolano 1968). The rate of food consumption is commonly far lower than in nondiapausing individuals, *e.g.*, in adults of the black blow fly *Phormia regina* (Stoffolano 1975) and in numerous species of adult *Drosophila* (Matsunaga et al. 1995). Larvae of the damselfly *Lestes eurinus* that enter winter undersized can compensate for their poor nutritional state by feeding during winter diapause, a feature not observed in well-fed individuals (de Block et al. 2007). Feeding during larval diapause in the dermestid beetle *Trogoderma granarium* extends the time larvae can survive in diapause (Burges 1960), and diapausing larvae of the drosphilid *Chymomyza costata* denied food survive less than two months (Koštál 2006). Feeding during diapause is not an option for adults lacking functional mouthparts. Caddisflies do not feed as adults, so a species such as *Mesophylax aspersus*, which enters summer diapause soon after adult eclosion, must carry over reserves for survival and egg production from larval life, resulting in strong selection for large body size in females but not males (Salavert et al. 2011). Female mortality in these caddisflies is especially high during diapause, as indicated by a change in the sex ratio from 1:1 in early diapause to a male-biased 4:1 ratio near the end of diapause. Only the largest females survive, whereas males, not under the same selective pressure, survive well. Being a large female not only increases the odds of surviving diapause but also results in the production of more progeny.

How long an insect remains in diapause impacts depletion of energy reserves. Different European populations of the blow fly *Calliphora vicina* stockpile approximately the same amount of fat during larval diapause (20% of dry weight), and all populations consume their fat at nearly the same rate during diapause. Consequently, populations from southern Spain that have a shorter diapause than their counterparts from Finland emerge as adults with more residual fat than their northern counterparts (Saunders 2000), a feature likely to have repercussions for reproductive success.

Diets just prior to diapause may differ from the insect's diet at earlier stages. A **dietary shift** is apparent in the ladybird beetle *Coccinella septempuntata* in central Italy (Ricci et al. 2005). When adults first emerge in the agricultural valley they feed almost exclusively on aphids, but when wheat and broadbeans are harvested in late July most adults migrate to highland meadows where, prior to diapause, pollen and fungal spores comprise the bulk of their diet. Acquisition of additional or unique resources prior to diapause may also be reflected in behavioral distinctions. Diapause-destined adults of the mosquito *Culex pipiens* not only avoid taking blood meals from their avian hosts but they sugar feed more frequently than their nondiapausing counterparts and extend their period of sugar feeding over a greater portion of the night (Bowen 1992). An adequate diet of nectar and pollen is essential for the overwintering success of bumble bee queens (Amsalem et al. 2015, Costa et al. 2020). This diet is not only needed for sequestering glycogen and lipid stores but also impacts later gene expression patterns during diapause. Queens of the bumble bee *Bombus impatiens* provided a sugar-rich diet prior to diapause entry display enhanced expression of numerous cytochrome P450 monooxygenase genes in the fat body during diapause, suggesting that such diapausing queens may have an enhanced defense response as a consequence of their pre-diapause diet (Costa et al. 2020).

Triacylglycerides (TAGs) are the dominant form of stored energy, but elevated levels of glycogen and hexamerins are also commonly noted (Hahn and Denlinger 2007, 2011; Sinclair and Marshall 2018). Fat offers the conspicuous advantage of high caloric content, coupled with a low hydration state and high yield of metabolic water, features that are further discussed in Chapter 7. Transcriptomic studies frequently reveal a boost in TAG-associated gene expression during preparation for diapause, *e.g.*, transcripts indicative of increased TAG synthesis are more highly expressed in ovaries from females of *Bombyx mori* destined to produce diapausing eggs (Chen et al. 2017).

To some extent, what is stored may reflect the **food available** as the insect prepares for diapause. When diapause-destined females of the mosquito *Culex pipiens* are fed different carbohydrate sources, different reserves accumulate (Chang et al. 2016). Females fed sucrose primarily accumulate glycogen, while those feeding on glucose accumulate both glycogen and lipids. Females offered both sucrose and glucose preferentially use sucrose for glycogen synthesis and glucose exclusively for biosynthesis of TAGs, features that may reflect selective feeding on nectars with a desired carbohydrate composition.

RNA-seq comparisons between diapause-destined individuals and those not destined for diapause commonly reflect distinctions in **gene profiles** associated with nutrient storage. During the days prior to onset of adult diapause in the cabbage beetle *Colaphellus bowringi*, genes associated with lipid (fatty acid biosynthesis) and carbohydrate (glycolysis) metabolic pathways are more highly expressed (Liu et al. 2017, Tan et al. 2017a). Fat metabolism is a dynamic process throughout adult diapause in the mosquito *Culex pipiens*, but of special note here is the early upregulation of fatty acid synthase (*fas-1*) as newly eclosed females prepare to enter diapause (Sim and Denlinger 2009b). Suppression of this gene and others involved in lipid storage results in failure to sequester the lipids needed for overwintering.

6.3   Migration

Selecting a suitable site for diapause frequently involves departure from the feeding site. This may involve **local migration** to a nearby protected hedgerow, woodland, or sheltered structure, or **long-distance migration** that may be several miles or thousands of miles from the summer habitat. Homeowners in many parts of North America are acutely familiar with migratory behavior each autumn as huge numbers of Halloween ladybird beetles *Harmonia axyridis*, a beetle that overwinters on white cliffs in its native China, migrate toward prominent light-colored houses as substitute overwintering sites in their newly colonized, North American habitat (Koch 2003). Similarly, the brown marmorated stink bug *Halyomorpha hayls* retreats to crevices of trees and crawls under loose bark in natural environments, but it also frequently invades homes each autumn in search of an overwintering site (Lee et al. 2014). Diapause-destined adults of the mosquito *Culex pipiens* seek dark, damp underground environments such as caves or basements for overwintering (Vinogradova 2000). In this environment, the temperature is more moderate and high humidity levels protect against dehydration. Adults of the bluegrass weevil, *Listronotus maculicollis* move off golf course fairways in the autumn and take refuge in high cut grass and mixed tree litter at a treeline that may be 60 m from the fairways (Diaz and Peck 2007).

These local movements to more protected sites offer tangible benefits, reflected in higher overwinter survival. Some diapausing adults of the bean leaf beetle *Cerotoma trifurata* remain in soybean fields during the winter, but many migrate to nearby woodland, where temperatures in the leaf litter seldom drop below $-5°C$ (Lam and Pedigo 2000a). The buffering of the woodland site offers considerable protection as reflected in 77–89% mortality during three winters among beetles that stayed in soybean fields versus mortality of 49–82% in woodlands (Lam and Pedigo 2000b).

In addition to short-distance, autumn movements of diapause-programmed insects a few species undertake long-distance migrations that are as impressive as those of migratory birds. The iconic insect migrant, the monarch butterfly *Danaus plexippus*, journeys from eastern northern North America and southern Canada to its overwintering site in Mexico, a distance exceeding 3,000 miles (Figure 6.3). A smaller population west of the Rocky Mountains migrates a shorter distance to Southern California. The most famous overwintering site for the western population is Pacific Grove on the Monterey Peninsula. Although sites in Mexico were "discovered" by western scientists only in the mid-1970s (Urquhart and Urquhart 1976, Brower et al. 1977), both the Mexican and Southern California sites were well known to indigenous people for ages and feature prominently in their folklore. Overwintering destinations in Mexico are within the tropics, but the fir-forested mountain sites are at high elevations (9,800–11,000 feet above sea level), subjecting diapausing butterflies to low temperatures as well as snow and ice storms. Thus, diapausing monarchs must be capable of resisting many of the same environmental insults that confront insects overwintering at temperate latitudes.

Interestingly, monarchs invaded Australia and New Zealand around 1871, likely island-hopping across the Pacific with the aid of cyclonic winds, and are now firmly established in this new habitat (Zalucki and Clarke 2004). In Australia, many display migratory behavior, flying from west and southwestern highlands to regions near Sydney and Adelaide in the autumn, though distances traveled are not nearly as great as seen in North America (Dingle et al. 1999, James and James 2019). But, not all do so. Some monarchs in Australia, in spite of having what would appear to be an intact migratory mechanism, show a loss of migration in this new environment and instead breed year-round in the same locale (Freedman et al. 2018), as do certain populations of monarchs in the Caribbean (Dockx 2007), other regions of Central and South America (Zhan et al. 2014), as well as southern coastal North America (Majewska and Altizer 2019). Permanent residency in North America appears to be favored by plantings of a non-native tropical milkweed *Asclepias curassavica*.

Like birds, autumn migration of monarchs to warmer climes is achieved by a single individual, but, unlike birds, the same individual does not complete the entire northward journey in the spring. Overwintering adults from Mexico typically return to the Gulf Coast in search of milkweed for depositing the first spring generation of eggs. Two or three successive nondiapausing generations are required for the monarch to reclaim its previous high-latitude habitat the following summer.

Short days of late summer in the monarch's northern habitat, combined with dropping temperatures and diminished quality of their host plants, prompt entry into adult diapause, evidenced by undeveloped ovaries in females, small mass of the male ejaculatory ducts, and onset of the long migration (Goehring and Oberhauser 2002). Only monarchs that mature in late summer display the directional, oriented flight characteristic of diapause (Zhu et al. 2009), a southward migration facilitated by a sophisticated magnetic compass. Recent success in rearing monarchs in the laboratory, along with access to new genomic resources (Zhan et al. 2011), have allowed monarchs to emerge as an exciting model for probing the molecular and neural basis of long-distance migration (Reppert et al. 2010, 2016).

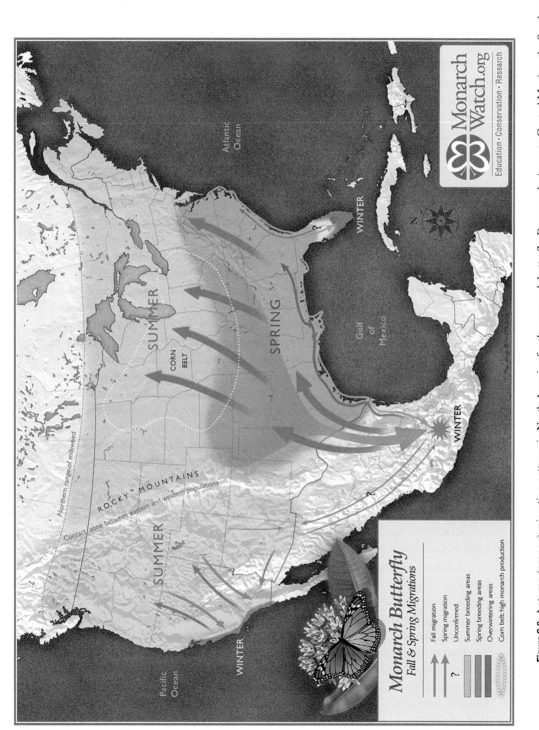

Figure 6.3 Autumn (orange) migration patterns in North America for the monarch butterfly *Danaus plexippus* to Central Mexico, the Southern California coast, and southern Florida, and the spring (green) northward return. (A black and white version of this figure will appear in some formats. For the color version, please refer to the plate section.)

Courtesy of Monarch Watch (monarchwatch.org).

One other North American species rivals the monarch as a long-distance, diapausing migrant: the large milkweed bug *Oncopeltus fasciatus*. Its smaller size has made it less noticeable than the charismatic monarch, although both species feed exclusively on milkweed. Like monarchs, the milkweed bug responds to short daylengths of late summer by becoming reproductively inactive and departs its summer habitat in northern USA and southern Canada to overwinter as far south as Mexico (Dingle 1974). Like the monarch, overwintering milkweed bugs first reproduce near their overwintering sites in the spring, and successive generations then reinvade the most northern sites the following summer. Even though the bugs overwinter in subtropical and tropical regions, their diapause shows the same high tolerance of cold and heat stress that characterize diapause in species overwintering at higher latitudes (Dean et al. 2016). Mating ceases, reproduction is halted, and the flight response is enhanced in those initiating the long migratory flight associated with diapause.

More modest migratory flights are documented for other species. Most ladybird beetles (Coccinellidae) migrate at least short distances to find an overwintering site, but some, such as *Hippodamia convergens* migrate over a somewhat longer distance from their summer habitats to overwintering sites at or near mountain summits (Rankin and Rankin 1980, Hodek 2012). Adults of *H. convergens* depart agricultural areas of California's Central Valley in the autumn for overwintering sites in the Sierra Nevada, first aggregating at intermediate altitudes and then moving on to higher altitudes near the mountain summits (Hagen 1962). Likewise, the hemipteran sunn pest *Eurygaster integriceps* departs wheat fields in the lowlands of Ankara, Turkey and heads for the mountains of the Beynam Forest in late summer (Guz et al. 2014). Many cutworms, including species in the genera *Agrotis* and *Euxoa*, emerge as adults in their breeding ground and immediately depart for alpine areas to enter a summer diapause (Oku 1983). In North America, cutworm moths leave the Great Plains and fly more than 100 miles to the Rocky Mountains where they remain reproductively inactive in diapause. Similarly in Japan, the cutworm moths depart soon after adult eclosion for adjacent alpine areas, a scenario also seen in Australia where newly emerged adults leave the tablelands of New South Wales and enter summer diapause in the Australian Alps. In all of these cases the onset of ovarian development coincides with the return of the moths to their low altitude breeding ground. For cutworms, as well as *H. convergens* and *E. integriceps*, distances between overwintering sites and summer habitats are substantial but not as great as for the monarchs, and the same individuals make the return trip following diapause.

Though migration is frequently associated with diapause, as in the monarch butterfly and ladybird beetles, the two events are not always linked. The large migratory dragonfly, the green darner *Anax junius*, exercises the option of either entering a nymphal diapause in the north or migrating without diapause (Hallworth et al. 2018). Green darners that emerge as adults early in the summer in northern USA and Canada do not diapause but migrate south for the winter and give rise to a nondiapausing generation in southern USA and Mexico, which, in turn, produce a nondiapausing generation that returns north, a distance that may exceed 400 miles. But, a portion of the northern cohort of green darners, those that develop more slowly, remain in the north and overwinter as diapausing nymphs. Likewise, hoverflies migrate south in autumn from northern Germany to the Mediterranean, but some stay behind in unnatural hibernation sites such as greenhouses (Hondelmann and Poehling 2007).

The experience of diapause sometimes affects migration patterns following diapause. For example, in Germany, the multivoltine white cabbage butterfly, *Pieris brassicae*, overwinters in pupal diapause and then migrates northward. By contrast, adults emerging from nondiapausing pupae head south (Spieth and Cordes 2012). Although the adaptive basis for this migration is unknown, presumably the spring northward migration directs butterflies to abundant summer host plants, while adults developing from nondiapausing pupae are directed to a habitat that will allow the following generation to successfully overwinter.

Just because an insect migrates does not mean it will enter diapause in its new location. Some species simply migrate to new locations on a regular basis and continue to develop in their overwintering site. Such is the case for the red admiral butterfly *Vanessa atalanta* (Stefanescu 2001). In autumn, populations depart northern and central Europe for the Mediterranean region, where they initiate a period of intense breeding and produce a new generation of adults that migrates northward to recolonize their northern range the following spring. The painted lady butterfly *V. cardui* migrates across the Sahara to winter-breeding grounds in sub-Saharan Africa and then returns to summer-breeding grounds in Europe (Hu et al. 2021). Rather than escaping unfavorable conditions in one geographic area, part of the driving force for such migrations may simply be to avoid high levels of parasitism (Stefanescu et al. 2012). In Spain, levels of parasitism in the painted lady butterfly *V. cardui* are four times higher in late summer than in spring, and in Morocco levels of parasitism are five times higher at the end of winter than in the autumn, suggesting that in both regions parasitoid levels build up in populations that breed in the same general area over successive generations. This possibly represents a selective force favoring seasonal migration to enemy-free space.

The number of migrating insect species may very well be underestimated. Recent long-term radar studies suggest an annual seasonal migration of 3.5 trillion high-flying insects over southern UK, a flux of approximately 3,200 tons of insect biomass (Hu et al. 2016). It has also been recognized recently that two noctuid moths from central Europe, the red underwing *Catocala nupta* and the large yellow underwing *Noctua pronuba*, likely engage in long-distance nocturnal migrations to the Mediterranean region in the autumn (Dreyer et al. 2018). Though migrations of large-bodied monarchs, green darners, and some hoverflies are well known, results such as this suggest that many more insects may engage in seasonal migrations. This, of course, does not necessarily mean such migrations are all linked to diapause, but quite likely many are diapause related.

6.4   Choosing a Suitable Diapause Site

To survive the long duration of diapause, it is imperative that insects select sites with microenvironments well buffered from upcoming environmental onslaughts and protected from predators, parasites, and pathogens. While nondiapausing individuals, especially immobile embryonic or pupal stages, also need protected sites their risk is far less because the duration of their immobile stages is considerably shorter than that of diapausing insects. Many overwintering stages burrow underground or seek refuge under detritus on the soil surface, under loose bark, in crevices of trees, in rocky crevices, or other microhabitats that offer protection from the brunt of winter.

When given options, diapause-destined adults of the apple blossom weevil *Anthonomus pomorum* select dry leaf litter as their top choice for overwintering, followed by rough bark, rather than smooth bark, grass, or soil, a selection that is also reflected in lower mortality at their favored sites (Toepfer et al. 2000). While the nondiapausing summer generation of the leaf beetle *Chrysomela lapponica* feeds and pupates on leaves of their willow hosts, adults destined for an overwinter diapause burrow into the soil (Zvereva 2002). Nondiapausing pupae of the swallowtail *Iphiclides podalirius* are found on their host plants, but larvae destined for diapause pupate in the leaf litter, a less conspicuous location (Esperk et al. 2013). Larvae of the lacewing *Chrysopa perla* crawl into small cavities in the substrate before spinning a cocoon, and those destined for diapause hide themselves more so than their counterparts that develop without diapause (Canard 2005).

Overwintering underground is particularly common among diapausing insects, and like other ectotherms that retreat underground in winter (Huey et al. 2021), insects face consequences associated with the depth selected for overwintering. Deep sites are usually warmer, reducing the risk of cold injury, but warm sites concurrently boost energy expenditure as a result of increased metabolic rates. Shallow sites can be expected to increase the risk of cold injury, but cooler conditions will aid in lowering energy costs. Thus, one would predict a trade-off between maximizing protection while minimizing energy expenditure. Other factors such as ease of reaching the soil surface in the spring also likely come into play. Diapause-destined individuals may dig deeper into the soil than their nondiapausing counterparts, *e.g.*, both diapause and nondiapause-destined individuals of the Colorado potato beetle *Leptinotarsa decemlineata* complete metamorphosis underground, but those entering adult diapause burrow more deeply (Minder 1976). A heavy clay soil prevents deep penetration and results in higher winter mortality than seen among diapausing beetles that are able to dig deeply into sandy soils. Soil temperature also influences how deeply the Colorado potato beetle burrows: At 11.5°C, pre-diapause adults burrow to a depth of 30 cm in laboratory soil columns, but at a higher temperature, 17°C, they burrow to 45 cm, a depth that further reduces the risk of cold injury (Noronha and Cloutier 2006). Somewhat surprisingly, under simulated winter conditions in the laboratory, a few diapausing adults of *L. decemlineata* occasionally resurface during winter and then again return to the soil (Piiroinen et al. 2011).

Diverse protected sites may be sought as winter approaches. Diapause-destined adults of the northern house mosquito *Culex pipiens* head for caves and protected human constructions such as basements, culverts, and cisterns (Vinogradova 2000, Rozsypal et al. 2021). Diapausing females of the predatory mite *Euseius finlandicus* can be found in peach orchards under loose bark, inside empty scales of coccids, or in deep crevices surrounding old scars, canker wounds, and splintered ends of dying twigs (Broufas 2002). Diapausing eggs of the red-legged earth mite *Halotydeus destructor* are protected within cadavers of adult female mites (Ridsdill-Smith et al. 2005). Many parasitoids remain encased within the exocuticles of hosts they have just consumed: Diapausing larvae of *Nasonia vitripennis* are well protected within the puparium of their fly host (Rivers and Denlinger 1995), larvae of *Praon volucre* gain extra protection by remaining inside mummies of their host aphids (Colinet et al. 2010), and diapausing pupae of *Psyllaephagus pistaciae* are protected inside the bodies of their mummified psyllid hosts (Mehrnejad and Copland 2005). Galls offer well-protected sites for gall insects living at both high latitudes, *e.g.*, *Eurosta solidaginis* (Irwin and Lee 2003) as well as the tropics, *e.g.*, *Eurytoma sp.* (Barbosa and Fernandes 2019). Adults of *D. melanogaster* overwinter under manure piles, in garden sheds, at the base of fruit trees, or in holes within tree trunks (Nagy et al. 2018). Most Lepidoptera stem borers remain encased in the protective stems of their host plants (Chippendale and Reddy 1972, Scheltes 1976). When larvae of the sugarbeet root maggot *Tetanops myopaeformis* complete the feeding phase of the third instar, they burrow 5–35 cm underground where they spend the winter in larval diapause, but rather than remaining deep in the soil until adult eclosion, they move upward, to within 10 cm of the surface, to pupariate (Chirumamilla et al. 2008). Presumably, it is much easier for larvae to move up through the soil than it is for newly eclosed adults.

The precise site selected for overwintering may vary with locality. In the Central European lowlands, most diapausing adults of the European spruce bark beetle *Ips typographus* remain under bark of their host spruce trees during winter, but in more severe localities of northern Europe and at higher elevations most adults leave the brood tree and seek overwintering habitats under snow in moss, leaf litter, and under bark of fallen trees (Schebeck et al. 2017).

In some cases selecting a cooler site may be beneficial, as nicely demonstrated for diapausing larvae of the goldenrod gall fly *Eurosta solidaginis* (Irwin and Lee 2003). Galls located below the snow line are subjected to higher winter temperatures than those above the snow, thus diapausing larvae in galls beneath the snow burn through their energy reserves more quickly, leading to lower overwinter survival and lower fecundity as adults. But, in certain winters, the additional buffering provided by snow cover may very well be a positive feature. In more severe localities, such as montane habitats where snowpack is especially variable in autumn and spring, depletion of the snow cover can dramatically increase overwintering mortality, as seen in the montane Sierra leaf beetle *Chrysomela aeneicollis* (Boychuk et al. 2015). Eggs of western spruce budworms *Choristoneura* are deposited on foliage of conifer trees, and newly hatched first-instar larvae head for the bole of the tree and, without feeding, molt to the second instar, construct a hibernaculum, and enter diapause. High mortality is noted during this dispersal phase, but larvae benefit by moving to a cooler summer environ-ment that retards loss of energy reserves during the warm summer months and also provides a more highly buffered shelter for winter (Nealis and Regniere 2016). Diapausing larvae remaining on branches rather than on the central bole have higher mortality, thus the advantages of finding the best site for diapause appear to override the high cost of traveling to the bole. Whether adults of the bark beetle *Pityogenes chalcographus* overwinter at ground level or 3 m above ground in the same spruce forest impacts survival (Koštál et al. 2014), again underscoring the profound effect of microclimate on diapause survival. In interior Alaska, diapausing adults of the aspen leaf miner *Phyllocnistis populiella* survive better in leaf litter beneath white spruce than under the leaf miner's aspen host (Wagner et al. 2012). Spruce sheds snow more effectively than the deciduous host tree, resulting in litter temperatures that are 7–9°C lower than under aspen. The moths are thus precariously teetering on the verge of their lower lethal temperature, but the dry environment of the deeper leaf litter under spruce may offer the advantage of protection against inoculative freezing. Diapausing adults of the mosquito *Culex pipiens* seek protected shelters for overwintering, but not all shelters are of equal value. In a study performed in the Netherlands, adults overwin-tering in an unheated shed survived well (70% survival after four months), whereas all females that overwintered in unheated rooms within heated houses died within that time frame (Koenraadt et al. 2019). The unheated rooms in houses are warmer, offer a more stable temperature, and are less humid than the unheated sheds, but the colder, more humid sheds are better for overwintering. For diapausing adults of the invasive Asian ladybird *Harmonia axyridis*, warmer is better (Labrie et al. 2008, Knapp and Rericha 2020): Nearly all beetles that overwinter outside die, while those that take refuge inside survive quite well.

Selection of a suitable site for overwintering impacts not only the insect's vulner-ability to temperature conditions but also exposure to parasites and pathogens, as nicely demonstrated for the ladybird beetle *Coccinella septempunctata* (Ceryngier 2000). At a low altitude site (800 m) in Poland, braconid parasitization exceeds 70%, while parasitization at high altitude sites (1,360 and 1,600 m) is much lower, 14–28%. But, it's a trade-off. Fungal infection is considerably higher at the two mountain sites than at the lower altitude.

Large saturniid pupae are particularly vulnerable to woodpecker and mouse preda-tion during winter. The risk of woodpecker predation can be mitigated in *Hyalophora cecropia* by selecting a protected site for cocoon placement. When the larva ceases feeding on wild cherry, it departs the tree and seeks a site to spin its cocoon among the stems of low shrubs that are frequently covered by fallen leaves, out of woodpecker view (Scarbrough et al. 1977). By contrast, *Callosamia promethea*, a smaller sat-urniid, remains on its host tree, moves to the tip of a tree branch, and uses its silk to secure a leaf to the tree before spinning its silken cocoon within the leaf, a location safe from white-footed mice and woodpeckers (Waldbauer 1996). Yet, a third strategy is evident in another saturnid, *Antheraea polyphemus*. This species also spins its cocoon at the tip of a branch on the host tree and incorporates a leaf, but the cocoon is loosely attached to the branch and in early autumn falls to the ground, scattered among fallen leaves on the forest floor, making it challenging for mice and wood-peckers to find (Waldbauer 1996). Diapausing adults of the butterfly species *Inachis io* and *Aglais urticae*, commonly found in attics of unheated buildings in Sweden, are vulnerable to high rates of rodent predation (>50%) within the first two weeks after entering their overwintering sites, but those roosting in sites less accessible to rodents survive quite well (Wiklund et al. 2008).

Sites selected for overwintering may also impact duration of diapause. Diapausing adults of the tarnished plant bug *Lygus lineolaris* can be found in non-nutritious plant debris as well as in association with henbit, a food source. Those overwintering on henbit terminate diapause nearly a month earlier than those overwintering in plant debris (Snodgrass et al. 2012). Selecting a hot spot for overwintering sometimes conveys an advantage. Although most adults of the alpine leaf beetle *Oreina cacaliae* overwinter in soil next to their host plants, many also leave the area of the host and fly to overwintering sites that receive extensive sun exposure (Kalberer et al. 2005). These warm locations allow the beetles to break diapause two months earlier than conspecifics that remain in diapause near their snow-covered host plants, enabling the beetles that overwinter in hot spots to be the first to exploit new host plants in the spring.

6.5   Enhancing the Hibernaculum

Once the insect has found a suitable site for diapause, it may further enhance the site by lining the refuge with thick layers of silk, constructing a cocoon, or adding waterproofing agents. Danks (1987, 2004b) nicely summarizes distinct differences in the quality of cocoons constructed by some diapause-destined insects. Typically, cocoons used by diapausing individuals are thicker and more tightly woven than those produced by their counterparts that do not enter diapause, and in some cases cocoons of diapausing insects have distinct shapes, *e.g.*, cocoons made by nondiapausing larvae of the peach fruit moth *Carposina sasakii* are spindle-shaped while cocoon made by diapausing larvae are compressed (Zhang et al. 2016b). Robust winter cocoons can facilitate acquisition of solar heat, provide waterproofing, as well as protect against abrasion of the insect cuticle, protect against ultraviolet light exposure, impose a challenging barrier for penetration of natural enemies, and, through special properties of silk, help ward off bacterial and fungal attacks. Enhancing desiccation resistance in a dry winter environment is frequently important, but a critical role for cocoons in keeping the insect free of water in waterlogged soil is equally important for some species such as the soybean pod borer *Leguminivora glycinivorella* (Sakagami et al. 1985). Gregarious larvae of the Glanville fritillary butterfly *Melitaea cinxia* spin a communal nest; the nest feature that best correlates with overwinter survival is silk density (Duplouy et al. 2018). The anti-fungal agent spodomicin is also incorporated into the silk nest of *M. cinxia*.

Waterproofing may be bolstered by addition of resin and wax to the hibernaculum, as noted in the neotropical bee *Epicharis zonata* (Roubik and Michener 1980), or by the layering of extra hydrocarbons, as noted inside puparia of diapausing flesh flies (Yoder et al. 1992a). The puparium in higher Diptera is the tanned, hardened third-instar larval cuticle, a structure that provides a well-protected shelter for the pupa within. A rich wax coating not only retards water loss but also provides a barrier against penetration of external ice crystals (Crosthwaite et al. 2011). The puparium of a diapausing flesh fly *Sarcophaga crassipalpis* appears to offer more than just a lipid water barrier; some unknown additional feature of the puparium protects against inoculative freezing (Kelty and Lee 2000). Modifications of this sort offer microenvironments with humidities and temperatures that differ considerably from those of the environment at large and offer additional protection from flooding, inoculative freezing, desiccation, mechanical injury, as well as protection from predation, parasitization, and pathogen attack.

Immobile diapausing stages are especially vulnerable to predation by vertebrates or other invertebrates. Plus, these immobile stages are unable to groom themselves to fend off parasites or microbes. A well-constructed cocoon provides protection, although predation rates on some insects, such as the winter moth *Operophtera brumata*, can sometimes be as high as 100% (Frank 1967). Cocoons with greater mass, which presumably translates into greater thickness, offer more protection against predators of the slug caterpillars *Acharia stimulea* and *Euclea delphinii* (Murphy and Lill 2010). Although predation on these two species is rather high (22–29%), heavier cocoons are more successful in avoiding predation than their lighter counterparts when both types of cocoons are placed in leaf litter beneath a range of host plants.

Diapausing embryos of the mite *Halotydeus destructor* are encased in a chorion that is much thicker and darker than that of their nondiapausing counterparts, a feature associated with elevated heat and desiccation resistance (Cheng et al. 2018, 2019). Among some other invertebrates, even more extreme protective coatings may be employed for protection. Embryos of the brine shrimp *Artemia franciscana* enter diapause at the gastrulae stage and are encased in a hard, chitinous shell, termed as cyst (Ma et al. 2013), a structure particularly impervious to environmental insult (MacRae 2016).

6.6   Formation of Aggregations

Local as well as long-distance migrations frequently culminate in the formation of large aggregations of diapausing individuals. Such is the case for monarchs that reach their overwintering sites in Mexico and Southern California. Over 14 million monarchs, clustered on 2,375 firs, pines, and cypresses, are estimated to occupy approximately 3.7 acres at one site in Mexico (Brower et al. 1977). Clusters of up to 40 million individuals of the coccinellid *Hippodamia convergens* can be found near the snowline in canyons of the Sierra Nevada Mountains of California during the winter (Hagen 1962), and aggregations of thousands of adults of an invasive coccinellid *Harmonia axyridis* invade homes and other sheltered sites in North America each autumn (Labrie et al. 2008). Diapausing aggregations are well known for numerous additional species of adult coccinellids living in both temperate (Hodek 2012) and tropical latitudes (Poulton 1936). Diapause aggregations of stink bugs can be as small as a few dozen bugs to thousands of individuals (Musolin and Saulich 2018). Aggregations sometimes include more than one species. Though diapausing adults of *H. undecimnotata* are the dominant species at aggregation sites in southern Europe, other ladybird species including aphidophagous *Coccinella septempunctata* and mycophagous *Tytthaspis sedecimpunctata* and *Psyllobora vigintiduopunctata* are also found comingled within the aggregations (Susset et al. 2017a), a situation that raises interesting questions about the dynamics of competition and cooperation (Boulay et al. 2019).

Aggregations of diapausing adults are well documented for species of several Coleoptera, Lepidoptera, and Hemiptera from both the New and Old World tropics (Denlinger 1986). In the tropical region of Northern Australia, small pockets of monsoon forest are surrounded by much drier zones of open forest and grassland. During the dry season the moist forest becomes an important refuge for many animal species, including several species of Hemiptera, Lepidoptera, and Diptera that can be found in large aggregations (Monteith 1982). Diapause has not been confirmed in many of these species, but its occurrence is suspected since the adults remain inactive and refrain from feeding. More recent studies have confirmed diapause in some of these Australian species, including the butterflies *Euploea core* and *E. sylvester*, species that aggregate during the tropical dry season in large numbers along shaded gullies (Canzano et al. 2003, 2006), locations that offer high humidity and protection from the prevailing winds (Kitching and Zalucki 1981). The tropical rain forest on Barro Colorado Island, Panama hosts an aggregation of more than 70,000 individuals of the handsome fungus beetle, *Stenotarsus rotundus* (Wolda and Denlinger 1984) as well as aggregations exceeding 100,000 individuals of the rhopalid bug *Jadera obscura* (Denlinger 1986). Aggregations of this sort are best known for species that are distasteful to predators and display warning coloration.

Diapause aggregations are not restricted to adults, although that is most commonly reported. Larvae of the Glanville fritillary butterfly *Melitaea cinxia* feed gregariously on their host plants and then in autumn spin a large communal web in which the diapausing larvae spend the winter (Verspagen et al. 2020).

There are advantages to forming an aggregation that goes beyond protection from predators. Aggregations of the beetle *S. rotundus* can be up to eight layers deep. These high densities modify the insect's microenvironment by reducing evaporative water loss and lowering incident radiation. Group size affects the beetle's metabolic rate (Tanaka et al. 1988): The rate of oxygen consumption declines as group size increases (Figure 6.4), a feature that also enhances water conservation (Yoder et al. 1992b). Adults of the shield bug *Parastrachia japonensis*, clustered in aggregations during diapause, reduce their oxygen consumption rates by half in comparison to single individuals (Tojo et al. 2005). Aggregations thus generate a unique microhabitat that both reduces metabolic expenditure and promotes conservation of valuable water resources. Temperature and humidity, however, may not always be ideal at

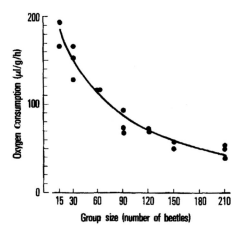

Figure 6.4 Oxygen consumption rates in different size aggregations of the fungus beetle *Stenotarsus rotundus* at 25°C.
From Tanaka et al. (1988), with permission from John Wiley & Sons.

aggregation sites. Temperatures at three overwintering sites used by the ladybird beetle *Hippodamia undecimnotata* in France frequently vary beyond the −5 to +5°C preferred range, and high humidity promotes growth of ectoparasitic fungi, resulting in overwintering mortality ranging from 18 to 28% at the three sites (Susset et al. 2017b). Since mating frequently occurs when the diapause season is ending and adults are preparing to leave the aggregation (*e.g.*, Tanaka et al. 1987b, Susset et al. 2018), aggregations also function as a mechanism for bringing the sexes together prior to dispersal. Communal nests constructed in the autumn by diapausing larvae of the browntail moth *Euproctis chrysorrhoea* are used not only for overwintering but also as a site of retreat in the spring when larvae start to feed at the time of budbreak (Frago et al. 2010).

It is notable that many diapause aggregation sites are used year after year by the same species. The same individuals normally do not survive for a second bout of overwintering, thus they are members of one of the following generations that come to the same site the following year, a cycle that is broken only when the landscape changes (Hodek 2012). The very same palm tree in the lowland rain forest of Barro Colorado Island, Panama has hosted huge aggregations of the beetle *S. rotundus* for at least 20 years (Denlinger 1994). The same dry season aggregation sites have also been used for decades, and likely much longer, by the edible stink bug *Encosternum delegorguei* in Malawi, Zimbabwe, and South Africa (Dzerefos et al. 2015). How are these sites found repeatedly? Do they have distinct features that allow repeated discovery? Or, do the overwintering individuals leave behind chemical cues that mark the territory? It's a fascinating question, but the answer remains unknown. Attempts to address this question for *S. rotundus* have been made by painting alternative trees with beetle extracts, but so far the beetles remain loyal to a single tree, yet there is nothing particularly striking about the location or profile of the tree!

Some aggregating species move *en masse* to different locations during diapause. The tropical seed bug *Jadera obscura* moves from site to site as an aggregation during adult diapause in Barro Colorado Island, Panama (Tanaka and Wolda 1987). The monarch butterfly *Danaus plexippus* has a limited capacity to move *en masse* to a new site if conditions at the original site deteriorate (Brower et al. 2011), although this does not preclude mass mortality at their favored sites if a catastrophic winter storm strikes (Brower et al. 2004).

Aggregations also have a dark side! Diapausing adults of the boxelder bug *Leptocorisa trivittatus* have been observed feeding on dead conspecifics, a strategy that enhances overwintering survival among the remaining boxelder bugs (Brown and Norris 2004). The brown marmorated stink bug *Halyomorpha halys* goes beyond necrophagy and engages in **cannibalism** (Papa and Negri 2020). At a study site in northern Italy, overwintering adults of *H. halys* were observed feeding on conspecifics that were still alive, and roughly 25–70% of the dead insects found at the study sites showed perforation holes and saliva residue on the ventral surface of the metathorax, clear signs of cannibalism.

6.7   Color Distinctions

Insects destined for diapause may assume distinct coloration or color patterns. Though sometimes misleading, color many times serves as an accurate proxy for the diapause phenotype. These color distinctions of diapause commonly allow the insect to better blend with its winter or dry season environment, but in other cases may simply reflect features of diapause physiology or rearing conditions not specifically related to the diapause program.

Distinctions can be observed in embryos as well as in diapauses at later stages of development. Nondiapausing eggs of *Bombyx mori* are pale yellowish-white, while diapausing eggs of this species are quite dark, a consequence of high quantities of ommochromes and their precursor 3-hydroxy-kynurenine in both the chorion as well as the underlying serosa (Zhang et al. 2017a). These pigments are totally lacking in the chorion of nondiapausing eggs. Color distinctions are also evident in adults of the Diazo race of *B. mori*: Autumn morphs that give rise to diapausing eggs have dark brown bands on the outer surface of their anterior wings, a feature lacking in summer morphs that give rise to nondiapausing eggs (Tsurumaki et al. 1999). In the mite *Halotydeus destructor* the chorion surrounding a diapausing embryo is much darker than the chorion of a nondiapausing embryo (Cheng et al. 2018).

Color polyphenism is especially dramatic among certain tropical butterflies and moths. Wet-season (nondiapause) adult morphs of the West African nymphalid *Precis octavia* are bright orange with black markings, while dry-season (reproductive diapause) morphs have a complex pattern of black markings with blue spots and very little orange (Owen 1971). Dry-season morphs that occupy open habitats tend to be light-colored; inhabitants of shady, forested areas tend to be dark (Young 1982). In the Australian tropics, adults of the nymphalid butterfly *Hypolimnas bolina* are duller and

Figure 6.5 Diapause is associated with a color change in the stink bug *Nezara viridula.* (A) Green adults are not in diapause and (B) brown adults are in diapause. (A black and white version of this figure will appear in some formats. For the color version, please refer to the plate section.) Photographs courtesy of Dmitry Musolin (St. Petersburg State University).

darker during their dry season diapause than nondiapausing adults that are active during the wet season, a color distinction that nicely matches the dry and wet season habitats (Kemp and Jones 2001). In these tropical examples, changes in color patterns effectively track seasonal shifts in background coloration.

In temperate habitats, color distinctions are also frequently apparent between diapausing and nondiapausing individuals. The nondiapausing summer phenotype of the lacewing *Chrysoperla* (=*Chrysopa*) *carnea* is green and matches the summer-foliage, while diapausing adults that develop in autumn are brown and blend extremely well with the lacewing's winter habitat (MacLeod 1967, Tauber et al. 1970). Summer, nondiapausing adults of the southern green stink bug *Nezara viridula* are green (Figure 6.5A), while russet-colored adults (Figure 6.5B) predominate as the overwintering diapause form (Musolin 2012). Similarly, abdomens of the western tarnished plant bug *Lygus hesperus* are dark green in summer adults, but yellow in diapausing bugs that overwinter (Brent 2012).

Color polyphenism is especially evident in the butterfly families Papilionidae, Pieridae, and Nymphalidae. Diapausing pupae of the papilionid *Papilio polyxenes* are exclusively brown, while nondiapausing cohorts can be either green or brown (Sims 2007). Diapausing pupae of *P. xuthus* are orange, while nondiapausing pupae are brown (Yamanaka et al. 2006). Winter-collected, diapausing pupae of the Japanese papilionid *Byasa alcinous* are dark brown while summer forms are bright yellow (Yamamoto et al. 2011). Diapausing adults of the comma butterfly *Polygonia c-album* are considerably darker than summer adults that do not enter diapause (Söderlind and Nylin 2011).

(A)

(B)

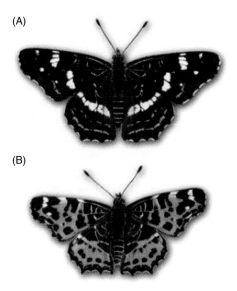

Figure 6.6 Seasonal polyphenism in the butterfly *Araschnia levana*. (A) Adults from a summer generation that have not experienced pupal diapause emerge with black and white coloration. (B) Those emerging in spring after overwintering in pupal diapause are predominantly orange. (A black and white version of this figure will appear in some formats. For the color version, please refer to the plate section.)
From Morehouse et al. (2013), with permission from John Wiley & Sons.

Likewise, diapausing larvae of the aphid parasitoid *Praon volucre* are much darker than their nondiapausing counterparts (Colinet et al. 2010). Diapausing larvae of the aphid parasitoid *Aphidius gifuensis* are a golden yellow color while nondiapausing larvae of the same stage (fourth instar) remain white (Zhang et al. 2018a). A similar color difference is noted in the cerambycid *Psacothea hilaris*: Larvae entering diapause acquire a deep yellow color that contrasts with the lighter coloration of nondiapausing larvae (Munyiri et al. 2004). Diapausing pupae of the chestnut leaf miner *Cameraria ohridella* are yellow, while nondiapausing pupae are dark brown (Weyda et al. 2015). Color distinctions are not restricted to a certain diapause stage but can be found in diapauses of all stages.

Color differences are sometimes not evident during diapause but are noted in the subsequent stage. One striking example is the Euroasian nymphalid *Araschinia levana* (Koch and Bückmann 1987, Morehouse et al. 2013). Adults that emerge in summer without having experienced diapause are black and white (Figure 6.6A), while those that emerge in spring after overwintering in pupal diapause are predominately orange (Figure 6.6B). Although a number of physiological parameters distinguish these two morphs (Morehouse et al. 2013), it remains unclear what selective forces are driving this seasonal polyphenism. In Costa Rica, adults of the saturniid *Rothschildia lebeau* display a range of color morphs from light rust to dark chocolate (Janzen 1984). Light morphs normally emerge from diapausing pupae that have bridged the dry season. As the rainy season progresses, more dark forms appear. In this case, color pattern is not a direct consequence of diapause. Dark morphs can be experimentally produced by keeping diapausing pupae moist and cool, but under natural conditions the heat and dryness that prevail during diapause elicit production of light-colored morphs.

Although color distinctions frequently coincide with diapause and are sometimes firmly linked to the photoperiodic cues that determine diapause, environmental factors determining color may be independent of the photoperiodic response. Even in a species such as the southern green stink bug *Nezara viridula*, a species in which color is commonly used as a proxy for diapause (Musolin and Numata 2003a), color is not always a reliable diapause indicator, *e.g.*, at the northern edge of its distribution near Osaka, Japan (Musolin and Numata 2004). Low temperature, humidity, surface texture, light intensity, and food availability are frequently more important than photoperiod. Rearing larvae of the painted lady butterfly *Vanessa cardui* at 16°C generates a dark pupal cuticle, while rearing at 32°C yields a white pupa, independent of the photoregime (Yamanaka et al. 2009). The net result is that a diapausing individual pupating under the short days and low temperatures of autumn would be brown, rather than white, but the manifestation of diapause and the dark coloration are independently determined. Color distinctions in *V. indica* carry over to adults as well: Winter adults are distinctly darker than their counterparts that develop in the summer, and again, colors and patterns on adult wings can be manipulated in these nymphalids by temperature (Ohtaki 2008). Nondiapausing pupae of the swallowtail butterfly *Papilio xuthus* can be either green or brown, depending on surface texture and light intensity: A smooth pupation site yields green pupae and a rough pupation site yields brown pupae, although bright light can suppress brown coloration (Hiraga 2006). Similarly, pupal color (green or brown) in the tropical nymphalid *Mycalesis mineus* depends on whether pupation occurs on a green leaf or a brown substrate (Mayekar and Kodandaramaiah 2021). In at least one species of swallowtail *Battus polydamas archidamas*, a desert species living in Chile, food availability determines pupal coloration (Pinones-Tapia et al. 2017). Larvae offered host leaves *ad libitum* turn into green pupae, while larvae on a restricted diet, fed only on alternate days, become brown pupae. This color distinction is reflected in lower predation of green pupae on green cacti and a slightly lower risk for brown pupae on dark rocks. Dark coloration of the spotted wing drosophilid *Drosophila suzukii* appears to be generated by low larval rearing temperatures prevailing in autumn (Toxopeus et al. 2016, Wallingford and Loeb 2016). Young adults of the six-toothed bark beetle *Pityogenes chalcographus* are pale in early autumn, but most become darker at the onset of winter diapause; the dark forms are heavier, store more lipids, and, in some localities, survive better than the less abundant pale beetles (Koštál et al. 2014).

Dark cuticles in winter morphs would appear to offer several adaptive advantages. A dark color is more likely to be cryptic, allowing the overwintering insect to blend more easily into tones of its winter environment. A dark cuticle also allows the winter insect to capture more heat, thus functioning as an accelerator of development by maximizing heat absorption, as nicely demonstrated in subarctic populations of the leaf beetle *Chrysomela lapponica* (Zverev et al. 2018). Dark cuticles may also offer other benefits such as desiccation resistance and enhanced immune functions (Parkash et al. 2009, Evison et al. 2017, but see Karl et al. 2010).

The southwestern corn borer *Diatraea grandiosella* loses its spotted cuticular pigmentation during larval diapause, becoming a yellowish-white "immaculate" larva, a distinction that serves as an effective marker for diapause (Chippendale and Reddy 1972). This is not a gradual loss of pigmentation but a loss following a stationary larval molt undergone by diapause-destined larvae. Similarly, pigmented spots on the larval cuticle of the East African stem borers *Chilo partellus* and *C. orichalcoiliella* disappear at the onset of diapause, yielding a consistently light-colored larva that spends the dry season in diapause (Scheltes 1978). A color change of this sort is unlikely to play an adaptive role in cryptic coloration because larvae are hidden within stalks of their host plant, and in the case of *D. grandiosella*, the portion of corn stalk harboring the larvae is below ground, well out of visual range for a predator.

Color distinctions noted in these stem borers underscore the point that not all such distinctions are necessarily adaptive. Some color differences are by-products of distinct features of the diapause program (Nylin 2013), and features such as the transition from pink-red to purple-red hindwings at diapause termination in the red locust *Nomadacris septemfasciata* may simply be a consequence of aging (Franc and Luong-Skovmand 2009). The bright orange color specific to diapausing females of the twospotted spider mite *Tetranychus urticae* seems unlikely to offer cryptic protection. The color is due to an abundance of the keto-carotenoid astaxanthin, an agent that likely helps protect against damage from reactive oxygen species (Kawaguchi et al. 2016, Bryon et al. 2017a). Other bright colors, as in diapausing aggregations of the handsome fungus beetle *Stenotarus rotundus* may be aposematic (Denlinger 1994), reminding predators that they are distasteful.

To our eyes, the dark brown winter colors appear to be cryptic, but do they also appear that way to potential predators? This has been tested experimentally in relatively few cases, but where it has been done, the results do indeed suggest that winter colors and patterns provide protection from predators that rely on visual cues. Light, nondiapausing summer morphs of the comma butterfly *Polygonia c-album* are preyed upon by great tits (*Parus major*) at higher rates than dark, diapausing winter forms when the butterflies are resting on tree trunks, a natural overwintering site (Wiklund and Tullberg 2004), but interestingly the lighter forms also face higher mortality on stinging nettles, their summer host plant, suggesting that additional features may be involved in dictating color distinctions. Great tits were also used to monitor prey preference for two color forms of the striated stink bug (Johansen et al. 2010). Diapausing adults of this bug are pale brown with black striations, but at termination of diapause in the spring the same individuals undergo a color change to red and black striations (Figure 6.7). When both types of bugs are placed on dry vegetation, the red and black forms are discovered much more quickly, experimentally supporting the idea that coloration of the diapausing bugs is cryptic not only to humans but also to potential avian predators.

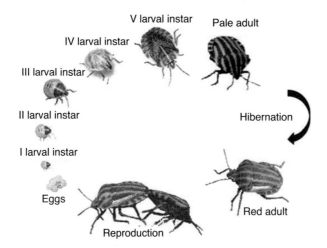

V larval instar   Pale adult

IV larval instar

III larval instar

II larval instar   Hibernation

I larval instar

Eggs   Red adult

Reproduction

Figure 6.7 Life cycle of the striated stink bug *Graphosoma lineatum* showing the transition from a pale and black striated adult that enters diapause in the autumn to a red and black adult that breaks out of diapause in the spring. (A black and white version of this figure will appear in some formats. For the color version, please refer to the plate section.)
From Johansen et al. (2010), with permission from John Wiley & Sons.

6.8   Structural Distinctions

Like color, structural distinctions sometimes readily and reliably distinguish pheno-types programmed for diapause. In some species, wing size correlates with diapause. The first summer generation of the water strider *Gerris lacustris* consists primarily of short-winged (**brachypterous**) forms that have reduced flight muscles and reproduce early, while adults produced in late summer are long-winged (**macropterous**), have well-developed flight muscles, and initiate reproduction a bit later, an apparent trade-off that sacrifices some reproductive output for the ability to fly to suitable overwintering sites (Pfenning et al. 2008). The opposite scenario is seen in the white-spotted tussock moth *Orgyia thyellina*: Long days of summer produce flight-capable, long-winged adults that lay nondiapausing eggs, while the short days of autumn yield flightless, short-winged adults that lay diapausing eggs (Uehara et al. 2011). Adults of the linden bug *Pyrrhocoris apterus* reared under short-day conditions enter reproductive diapause, and nearly all are short-winged, while those reared at the same temperature under long days initiate reproduction immediately and are long-winged, a distinction that has little to do with function because both forms are flightless (Socha 2007).

But structural distinctions sometimes fail as diapause markers. A case in point is the mite *Halotydeus destructor*, a species that diapauses as an embryo. The chorion encasing the embryo is usually thick and dark in diapausing individuals, but a few embryos that are clearly in diapause lack the thick, dark chorion, a condition aptly referred to as a **cryptic diapause** (Cheng et al. 2018). Interestingly, the cryptic diapause also tends to be of shorter duration, thus suggesting a suite of distinctions for cryptic diapause that go beyond obvious structural differences (Cheng et al. 2019).

6.9   Switch of Reproductive Modes in Aphids

One of the most striking switches preceding diapause is seen in aphids. In response to photoperiodic cues in late summer, aphids such as *Megoura viciae* (Lees 1966), *Rhopalosiphum padi* (Austin et al. 1996), *Acyrthosiphon pisum* (Shingleton et al. 2003), and others (Hardie 2010) switch their reproductive mode from parthenogenesis to sexual reproduction as a prelude to production of cold-resistant diapausing eggs. Aphid life cycles are extremely variable; many species migrate between different hosts, switch from wingless to winged forms, and switch from being parthenogenetic in the summer to reproducing sexually in the autumn (Hardie 2010). The genetic resources currently available for the pea aphid, *A. pisum*, make it possible to probe the fascinating life cycle of this species in considerable detail (*e.g.*, Le Trionnaire et al. 2008, 2009, 2012; Gallot et al. 2010). Overwintering eggs of the pea aphid hatch into asexual morphs that use clonal propagation to reproduce during the long days of spring and summer (Figure 6.8). Even if subjected to short daylengths in the spring the

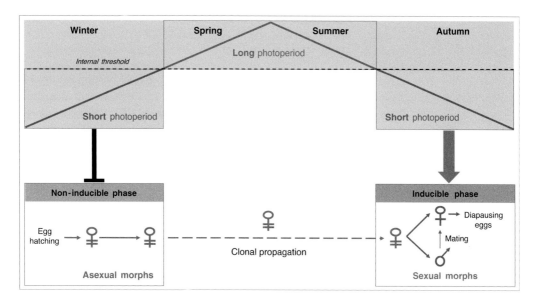

Figure 6.8 Schematic representation of the seasonal cycle of the pea aphid *Acyrthosiphon pisum.* Aphids that emerge from overwintering eggs that have been in diapause are unable to respond to short days (non-inducible phase) by producing sexual morphs but instead reproduce by clonal propagation. Only after a fixed period of approximately 90 days have elapsed are the aphids able to respond to short days by producing sexual morphs (inducible phase) capable of laying diapausing eggs.
Drawing courtesy of Gaël Le Trionnaire (University of Rennes).

aphids fail to reproduce sexually. This inhibitory response preventing the production of sexual morphs, termed the "**foundress effect,**" lasts approximately 90 days. Both long days and short days appear to be counted in this calculation (Matsuda et al. 2017), and only after this specific interval, referred to variously as the **interval or seasonal timer**, has elapsed are pea aphids capable of responding to short days of early autumn. The short autumn days elicit the production of sexual morphs that lay fertilized eggs destined for an overwintering obligate diapause. The late summer consequence of these complex life cycle transitions thus involves a switch from asexual to sexual reproduction (**cyclic parthenogenesis**), a prerequisite for the production of diapausing eggs. Interestingly, not all clones of the pea aphid incorporate a sexual phase into their life cycle. Some continue to reproduce asexually throughout the year (**obligate parthenogenesis**). The two clonal types sometimes coexist within the same field, but those that are persistently parthenogenetic are more common in southern Europe where winters are less severe (Jaquiery et al. 2014).

7   The Diapause State

This chapter examines the actual state of diapause. What are the features that characterize this developmental hiatus and what are the physiological and molecular underpinnings supporting its maintenance? As evident from the previous chapter, attributes of diapause frequently precede its onset and may indeed impact life-history traits noted after diapause has been terminated, but the focus here will be on diapause itself, the attributes that characterize this state and facilitate survival when development is arrested. The huge and diverse transcriptional changes noted during diapause foretell the pervasive impact that the diapause program can be expected to exert on multiple functional systems. The types of responses are fairly universal, consistent across different stages of diapause, but the downstream phenotypes they reflect can be attained through different molecular mechanisms, a feature indicative of the multiple evolutionary origins of diapause.

Hallmark attributes of diapause include arrested development (or severely retarding the rate of developmental progression), a global metabolic shift in metabolism, metabolic suppression, conservation of energy and water resources, and bolstering defense responses. Sometimes there are obvious structural distinctions characterizing diapause as well. Success in diapause also implies that the insect has mechanisms in place to maintain the arrest until favorable conditions return.

7.1   Developmental Arrest

The most obvious feature of diapause is cessation of development or at least a dramatic developmental slowdown. Development is, in many cases, halted, and while the nondiapausing insects' counterparts continue to mature, those in diapause remain locked in developmental arrest. Developmentally, nondiapausing insects pass through the same ontogenetic stage as their diapausing counterparts, but rather than remaining in that stage they quickly progress to a more advanced developmental stage. For example, pupae of the flesh fly *Sarcophaga argyrostoma*, that fail to diapause pass through a stage identical to diapause (the pupal head has everted and imaginal discs remain undifferentiated, referred to as the phanerocephalic stage), but they remain in that stage for less than a day, while those programmed for diapause remain in this undifferentiated phanerocephalic pupal stage for many months. Throughout diapause, the fly brain, as well as other organs, remain indistinguishable from organs seen in nondiapausing pupae 40 hours after the formation of the puparium (Fraenkel and Hsiao 1968). The combined ganglion present in larvae has not yet separated into ganglia of the head and thorax, fat body is dissociated into single cells but the cells have not disintegrated further, and cells of the heart-pericardial complex remain large as in mature larvae. Likewise, brain development in both nondiapausing and diapausing pupae of the butterfly *Pieris napi* proceeds synchronously until three days after pupation, at which time the brain of diapausing pupae ceases to develop and remains that way for the duration of diapause (Lehmann et al. 2017a). By this time, the subesophageal ganglion, which is clearly separated from the brain in larvae, has migrated forward and is fused with the brain. By day 3, regions of the brain involved in vision are well developed, but neuropils linked to olfaction are not. Although developmental processes are halted in the brain, the brain remains electrically active throughout diapause, as evidenced by persistent spontaneous electrical activity (Schoonhoven 1963).

In adults, the terminally differentiated stage, a developmental halt is also manifested, but in this case it is observed as a failure to proceed with reproductive development and maturation. Females fail to mature eggs, sperm maturation may cease in males, and accessory glands in both sexes remain undeveloped. The contrast in the appearance of reproductive tracts between diapause and nondiapause is nicely illustrated in the linden bug *Pyrrhochoris apterus* for males (Figure 7.1A and B, adapted from Urbanová et al. 2016) and females (Figure 7.1C and D, adapted from Smykal et al. 2014). As in these examples from *P. apterus*, differences in the status of reproductive organs between diapause and nondiapause are usually quite conspicuous.

The normal progression of senescence is stopped or retarded during adult diapause. In *Drosophila melanogaster*, mortality rates in post-diapause flies are similar to rates in newly eclosed adult flies that never entered diapause, suggesting that senescence during diapause is minimal (Tatar and Yin 2001).

While developmental arrest is frequently a clear halt that persists until diapause is terminated, in a few species some progression of development occurs during diapause, albeit at a much slower pace, as discussed in Chapter 1 for the northern house mosquito *Culex pipiens* (Readio et al. 1999), the pea aphid *Acyrthosiphon pisum* (Shingleton et al. 2003), and other aphids in the genus *Cinara* (Durak et al. 2020). Testes from males of the tropical endomychid beetle *Stenotarsus rotundus* remain small throughout much of diapause but double in size from February to March, prior to resumption of reproductive activity when the rains arrive in late April (Tanaka et al. 1987a). By contrast, the onset of ovarian development and flight muscle maturation in both sexes occurs much later, shortly before the rains arrive. Even within a single individual, developmental fates of tissues may differ, as in pupae of the tobacco hornworm *Manduca sexta*. Most tissue development is halted, but testicular stem cells continue to undergo mitotic division (Friedlander and Reynolds 1992). During larval diapause of the drosophilid *Chymomza costata* morphogenesis of most organs is halted, but the central nervous system and prothoracic wing discs increase in size (Koštál et al. 2000a). Diapause-destined adults of the comma butterfly *Polygonia c-album* start adult life with smaller olfactory neuropils than their nondiapause

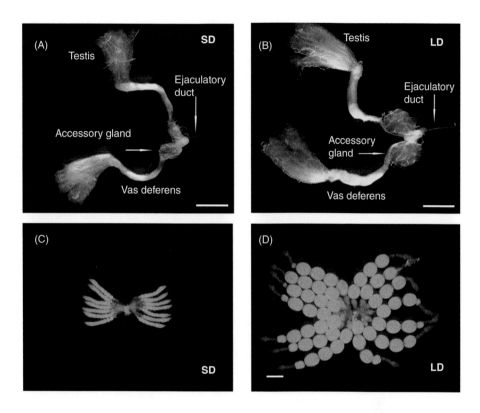

Figure 7.1 Male reproductive tract from (A) diapausing (short day, SD) and (B) nondiapausing (long day, LD) individuals of the linden bug *Pyrrhocoris apterus*. Note the smaller size of the accessory glands in diapausing males. Ovarian dissections from (C) diapausing and (D) nondiapausing females. Scale bar, 1 mm.
Photo courtesy of David Doležel (Czech Academy of Sciences).

counterparts, but both the antennal lobe (involved in sensory input) and mushroom body (involved in learning) slowly increase in volume as diapause progresses, reaching peak volume in mid-diapause (Eriksson et al. 2020). This delay in development perhaps serves as an energy-saving mechanism reducing costly maintenance of nervous tissue until the olfactory system is needed for mating and host-seeking after diapause ends. And, whether or not gross developmental changes are noted during diapause, when diapause is examined at deeper physiological or molecular levels, it is apparent that covert changes are occurring. Transcript profiles monitored throughout pupal diapause in the apple maggot *Rhagoletis pomonella* reveal that diapause, in spite of being a developmental stasis, is not static but reflects gradual shifts in gene expression, especially at the beginning and end of diapause (Dowle et al. 2020), as discussed further under diapause dynamics in Section 7.16. Such observations are indicative of the fact that diapause does not always neatly conform to simple definitions, and arrested development does not mean that development has completely halted.

Some insects even molt during diapause, a feature that is especially prevalent in certain pyralid and noctuid larvae, as well as crickets. Examples include the southwestern corn borer *Diatraea grandiosella* (Yin and Chippendale 1973), the Mediterranean corn borer *Sesamia nonagrioides* (Pérez-Hedo et al. 2010), the stem borers *Chilo suppressalis* (Yagi and Fukaya 1974) and *Busseola fusca* (Kfir 1991), and the cricket *Modicogryllus siamensis* (Taniguchi and Tomioka 2003). Remarkably, some of the stem borers such as *S. nonagrioides* molt five to six extra times during diapause. These molts during diapause are referred to as **stationary molts** rather than **progressive molts** because they do not reflect growth in size or weight, or at best only slight weight gain (*e.g.*, *S. nonagrioides,* as noted by Gadenne et al. 1997) or loss (*e.g.*, *D. grandiosella*, as noted by Yin and Chippendale 1973). This seems like an expensive option for a diapausing insect, but there are possible benefits in periodically shedding an old cuticle that may be laden with fungal or pathogenic microbes.

7.2   Cell Cycle Arrest

Implicit in arrest of development is a halt in the cell cycle. Cell cycle progression is essential to allow growth and proliferation of new tissues, a situation one would not expect to see during diapause. Four distinct phases of the cell cycle are recognized: G1, S, G2, and M (Figure 7.2). The **S phase** is a period of DNA replication, the **M phase** represents mitotic division, **G1** is the gap phase between mitosis and DNA replication, and **G2** is the gap phase between DNA replication and mitosis. The **G0** phase refers to cells that have exited the cell cycle, but quiescent cells in G0 are capable of re-entering the cycle (Cook et al. 2000, Wang and Levin 2009). Driving the cell cycle is a sequence of timed events that involve protein signaling cascades, proteolysis, and differential gene expression (Sherr 1996). Even in adults that have ceased growing, the cell cycle continues to be active in producing new gametes. Suppression of the cell cycle translates into developmental arrest in immature stages and arrested reproduction in adults. One of the most conspicuous features of diapause is cell cycle arrest (Tammariello 2001), but it is interesting that not all diapausing

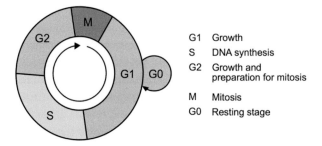

Figure 7.2 Progression through the cell cycle, including the four major phases indicated. During a resting phase, G0, the cell leaves and then returns to the G1 phase. Progression is regulated by the activity of complexes composed of cyclins and cyclin-dependent kinases.

insects arrest the cell cycle at the same phase, and since there are numerous check-points in the cell cycle it is also not surprising that different checkpoints are exploited to halt the cell cycle. Thus, the net phenotypic observation is the same: The cell cycle is halted, but how it is halted appears to differ among species.

The cell cycle during embryonic diapause in the silk moth *Bombyx mori* is arrested at G2 (Nakagaki et al. 1991), a result verified by flow cytometry and expression patterns of the cyclin-dependent kinase inhibitors Cdc2 and Cdc25 (Iwasaki et al. 1997).

Brain cells in diapausing pupae of the flesh fly *Sarcophaga crassipalpis* are arrested in G0/G1 (Tammariello and Denlinger 1998), an arrest that nicely correlates with downregulation of the transcript ***proliferating cell nuclear antigen*** (*pcna*) (Flannagan et al. 1998). Downregulation of *pcna* is of special interest because the protein it encodes is involved in the G1/S phase transition. Expression levels of several other genes involved in the G1/S transition, such as *cyclin E, p21*, and *p53*, do not appear to be altered by diapause in *S. crassipalpis*, thus it is possible that diapause entry relies extensively on a shutdown in the expression of *pcna*. But, other players are also likely involved: A phosphoproteomic analysis of the brain of *S. crassipalpis* shows a 3.6-fold increase in phosphorylation of Cyclin-dependent kinase 1 (Cdk1), suggesting that this cell cycle regulator may be inactivated during flesh fly diapause as well (Pavlides et al. 2011).

The optic lobe anlagen, cells that will develop into the compound eye, are arrested at G2 during pupal diapause in the tobacco hornworm *Manduca sexta* (Champlin and Truman 1998). Incorporation of bromodeoxyuridine (BUdR) and the combined use of the mitotic inhibitor colcemid and wash-out experiments reveal that cells of the optic lobe anlagen must first go through mitosis before entering the S phase after diapause termination, evidence that the cells during diapause are arrested at G2.

The cell cycle has been examined in two species that diapause as larvae, the drosophild *Chymomyza costata* (Koštál et al. 2009) and the jewel wasp *Nasonia vitripennis* (Shimizu et al. 2018a). In both species the majority of brain cells are arrested at G0/G1, while a small but significant portion of cells (13–20%) are arrested in G2. This is unlike the pupal diapause of *S. crassipalpis* in which nearly all (98%) arrested cells are in G0/G1. As in *S. crassipalpis*, the transcript *pcna* is significantly downregulated in *C. costata*, although the downregulation is not quite as pronounced as it is in *S. crassipalpis*. A broader transcriptomic approach indicates that numerous cell cycle regulators, including *cyclin E, cyclin B*, and *Cdks 1, 2*, and *7*, are also downregulated during diapause in *C. costata* (Poupardin et al. 2015). These factors, components of the control system driving progression through the cell cycle, act at two critical checkpoints: the G1/S and G2/M transitions. Several cell cycle regulators, including *pcna,* are also downregulated during diapause in the brains of *N. vitripennis*, but a link to diapause is less clear in this case because a similar downregulation is also seen in brains of nondiapausing larvae at the same age (Shimizu et al. 2018a). At the termination of larval diapause in the European corn borer *Ostrinia nubilalis, pcna* is upregulated, again suggesting an important role for this gene in maintaining cell cycle arrest during diapause (Wadsworth and Dopman 2015). The cell cycle checkpoint protein RAD9 contributes to cell cycle arrest during embryonic diapause of the brine shrimp *Artemia sinica* (Huang et al. 2019), but there is no similar information on the action of this protein in insects or other invertebrates.

Large-scale transcriptomic studies of other species, including the apple maggot *Rhagoletis pomonella* (Ragland et al. 2011) and cotton bollworm *Helicoverpa armigera* (Bao and Xu 2011), both of which have a pupal diapause, and the mosquito *Aedes albopictus*, which has a maternally induced embryonic diapause (Huang et al. 2015), provide further evidence for perturbations in expression of cell cycle regulators as a feature of diapause. Interestingly, downregulation of *pcna* is common to all three of these examples. In the oil-collecting bee *Tetrapedia diversipes*, a solitary bee that diapauses as a larva during the tropical dry season, two transcripts involved in repression of cell proliferation, *Rho-related BTB domain-containing protein 2* and *SAM/SH3 domain-containing protein 1*, are upregulated during diapause (Santos et al. 2018). Transcripts encoding numerous components of the cell cycle increase in abundance at larval diapause termination in the alfalfa leafcutting bee *Megachile rotundata*, e.g., cyclin-e, cyclin-k, cyclin-dependent kinase-2, and s-phase kinase-associated protein 2 (Yocum et al. 2015). These results with *M. rotundata* suggest that the cells are preparing to enter the S phase of the cell cycle upon resumption of development.

Although cell cycle arrest emerges as a common theme of diapause, the stage of arrest (G0/G1, G2, or both), as well as the specific gene regulators, appear to vary among species. Numerous **cell cycle checkpoints** regulate the cell cycle, and we can expect an expanded list of checkpoints used in diapauses of different species. Will reliance on two stages of cell cycle arrest within a single species, as seen in our two examples with a larval diapause, emerge as a trend for larval diapause? If so, why? Although it is interesting that arrest at two stages is observed in both examples of larval diapause, data from more larval diapauses is needed before suggesting this to be a common feature of larval diapause. Cell cycle arrest is likely a nearly universal component of dormancy in diverse animal taxa, e.g., a G1 cell cycle arrest is also noted during diapause in the annual killifish *Nothobranchius furzeri* (Dolfi et al. 2019). Among species that do not bring development to a complete halt but simply develop much more slowly, e.g., embryos of the *Cinara* genus of aphids (Durak et al. 2020), mitotic cell division continues, albeit at a slow rate, thus implying that a cell cycle arrest may not be a trait shared by all diapausing species.

7.3   Metabolic Depression

Low metabolic rate is a key component of diapause. It allows the diapausing insect to carefully parse its energy reserves and survive for months, in some cases years, without feeding. The fact that diapausing insects are considerably less active than their nondiapausing counterparts and have halted development implies they have less energy demands than insects not in diapause. And, for species that diapause during winter, the prevailing low temperatures also help to drive down energy costs.

Metabolic depression, however, is not dependent on low temperature and is seen during diapause at high temperatures as well. Unlike mammalian hibernators, insects, as ectotherms, have no need to maintain elevated body temperatures for survival during their overwintering dormancy and can thus lower their metabolic rates to extremely low levels in the winter environment.

Recordings of metabolic rates in the literature are based on direct measurements of oxygen consumption or the release of carbon dioxide, either of which provides a good indicator of the metabolic rate. A simultaneous record of both oxygen consumption and carbon dioxide release is, of course, essential for the calculation of the **Respiratory Quotient** (RQ), a valuable indicator revealing the source of energy being used, as discussed in the later section on energy utilization (Section 7.8).

Depression of metabolism during diapause is best documented by comparing oxygen consumption and/or carbon dioxide release rates of diapausing and nondiapausing individuals at the same stage of development and at the same temperature. For species with an obligate diapause, it is not possible to observe both diapausing and nondiapausing individuals at the same developmental stage, thus the next best option is to compare metabolic rates of diapausing individuals with individuals before diapause entry and after diapause has been terminated. The lowest diapause rates of oxygen consumption are noted in diapausing pupae (10–20% of the lowest rate in nondiapausing individuals) and embryos, stages that remain immobile during diapause. In contrast, the depression observed in diapausing adults and larvae may be much less, often in the range of a 40–50% reduction compared with nondiapausing individuals at the same temperature, and sometimes as little as a 15% reduction, as noted in the flight-capable monarch butterfly *Danaus plexipus*, during its overwintering diapause (Chaplin and Wells 1982). The drop in metabolic rate during diapause in adults of the linden bug *Pyrrhocoris apterus* is approximately 25% of the rate prior to diapause entry (Koštál et al. 2008), but rates vary throughout the day in *P. apterus*, with rates during the photophase being nearly twice that recorded during the scotophase (Kalushkov et al. 2001). The rate of oxygen consumption in diapausing larvae of tropical stalk borers in the genus *Chilo* is approximately 20% of the rate observed in larvae before diapause (Scheltes 1978). Adults and larvae in diapause frequently retain some mobility, a feature linked to the higher overall diapause metabolic rates in those stages, or occasional bursts of high metabolic rates associated with flight, *e.g.*, *Culex pipiens* (Rozsypal et al. 2021). But, adults that become immobile during diapause may show rates of metabolic depression on par with diapausing pupae, *e.g.*, the metabolic rate in diapausing adults of the Colorado potato beetle *Leptinotarsa decemlineata*, is reduced as much as 90% (Lehmann et al. 2020).

As discussed for aggregating insects in Section 6.5, size of diapause **aggregations** can influence metabolic rate: The rate decreases as aggregation size increases. The extent of metabolic depression may also vary with **latitude**, as shown for populations of the Colorado potato beetle *Leptinotarsa decemlineata*: Across a range of temperatures, diapausing adult beetles from higher latitudes in Russia display lower metabolic rates than those from a lower latitude in Italy (Lehmann et al. 2015a). And, metabolic rates do not remain constant throughout diapause as demonstrated in adults of the

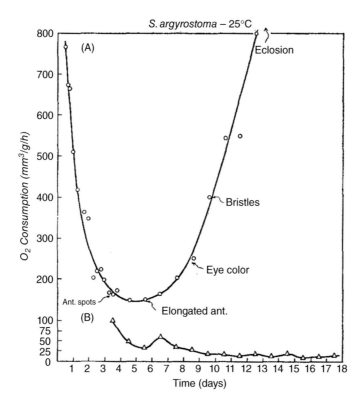

Figure 7.3 Comparison of the oxygen consumption rate in (A) a nondiapausing individual of the flesh fly *Sarcophaga argyrostoma* from pupariation to adult eclosion, showing the U-shaped curve characteristic of metamorphosis and (B) a group of 10 diapausing pupae during the same interval, all at 25°C.
From Denlinger et al. (1972).

mountain pine beetle *Dendroctonus ponderosae* (Lester and Irwin 2012) and pupae of *Sarcophaga crassipalpis* (Denlinger et al. 1972), a point emphasized in Section 7.16, where the dynamics of diapause is discussed. Commonly, metabolic depression is most pronounced in mid-diapause. A good listing of metabolic rates for species in different diapause stages is provided by Danks (1987), thus I offer only a few examples here.

The example of the flesh fly *Sarcophaga argyrostoma* displayed in Figure 7.3 shows a U-shaped curve for oxygen consumption rate in nondiapausing individuals from the time of pupariation until adult eclosion (Denlinger et al. 1972), a curve typical of holometabolous insects as they progress through metamorphosis. Diapause is entered at the developmental stage represented by the nadir of the curve for nondiapausing flies, but the diapause metabolic rate does not simply remain at the low point of the curve (approximately 150 µl/g/h) but drops considerably further to a low of 10–20 µl/g/h, a rate approximately 10% of the lowest rate seen in nondiapausing pupae (Denlinger et al. 1972). Only when diapause is broken does the oxygen consumption rate return to the nadir of the curve noted for nondiapausing pupae and then continue to climb as the fly approaches adult eclosion, following the right-hand U-curve seen in Figure 7.3. This reflects a **biphasic response**, first a rise to the nadir of the U-curve and then a progressive increase in the oxygen consumption rate until adult ecolsion (Sláma and Denlinger 1992). This biphasic response is also nicely recorded at diapause termination in pupae of the apple maggot *Rhagoletis pomonella* (Ragland et al. 2009). When diapause is first broken in *R. pomonella* the rate of carbon dioxide release initially rises to a plateau, presumably equivalent to the nadir of the U-shaped curve for a nondiapausing fly.

Most laboratory measurements of metabolic rate during diapause are calculated at relatively high temperatures, but, as expected, there is an obvious decline in metabolic rate with temperature. For example, mean rate of oxygen consumption in diapausing pupae of the flesh fly *Sarcophaga crassipalpis* is 18.4 µl/g/h at 25°C but drops to 8.5 µl/g/h at 18°C (Denlinger et al. 1972). Metabolic rates for those subjected to natural overwintering temperatures are considerably lower and may be barely detectable at temperatures below 10°C. Recordings made under constant temperatures in the laboratory may also be a bit misleading. As demonstrated in overwintering sixth-instar larvae of the butterfly *Erynnis propertius*, variable temperatures that occur naturally, especially in the autumn, exact a higher metabolic cost than predicted from calculations based on stable temperatures (Williams et al. 2012).

Although warm **tropical environments** do not facilitate lowering of the metabolic rate to the same extent as winter, there is little difference in the level of metabolic depression during diapause in insects from these contrasting environments. Tropical flesh flies, *Sarcophaga inzi* and *Poecilometopa spilogaster*, attain approximately the same level of metabolic depression (13–15% of minimum nondiapause levels) during pupal diapause as achieved by their temperate latitude relatives (Denlinger 1979). And, the metabolic rate noted in diapausing adults of the tropical beetle *Stenotarsus rotundus* (22 µl/g/h for an aggregation of 300 adults, Tanaka et al. 1988) is lower in comparison to rates for other diapausing adults from temperate latitudes recorded at the same temperature (25°C). Metabolic depression is somewhat less pronounced for the tropical butterflies *Euploea core* and *E. sylvester* (28% lower than reproductively active butterflies) than in some other diapausing adults, a feature that may be associated with the shallow diapause of these butterflies and their ability to rapidly respond to new host plant odors for diapause termination (Canzano et al. 2006).

Metabolic depression during diapause, especially in large saturniid moth pupae, captured the interest of Carroll Williams' laboratory in the early days of diapause research. Diapause in pupae of *Hyalophora cecropia*, was reported to be largely insensitive to cyanide and carbon monoxide (*e.g.*, Schneiderman and Williams 1954), suggesting some fundamental difference in components or function of the electron transport chain that utilizes oxygen for the generation of ATP. Using detection methods available at that time, it was evident that a number of the cytochromes were less abundant during diapause. But, cyanide insensitivity is no longer thought to indicate a fundamentally different system in operation during diapause but simply that low levels of functional cytochromes are adequate to meet the low energy demands of diapause.

Activity of **cytochrome c oxidase**, the terminal enzyme in the final step of the mitochondrial electron transport chain, is an excellent marker for metabolic activity, and a number of diapause studies, *e.g.*, embryonic diapause in the ground cricket *Allonemobius socius* (Reynolds and Hand 2009a), larval diapause in the bamboo borer *Omphisa fuscidentalis* (Singtripop et al. 2007), pupal diapause in the sweet potato hornworm *Agrius convolvuli* (Uno et al. 2004) and cotton bollworm *Helicoverpa armigera* (Yang et al. 2010, Xu et al. 2012), and adult diapause in the linden bug *Pyrrhocoris apterus* (Koštál et al. 2008), show low levels of enzyme activity and/or the transcript encoding cytochrome c oxidase subunit I (*cox1*) or other *cox* subunits during diapause and elevation at diapause termination. Low enzyme activity and expression levels of *cox1* during diapause are thus consistent with the expectations of metabolic depression.

Denying access to critical substrates needed by the mitochondria for energy-yielding processes is one effective means to suppress metabolism. Activities of numerous mitochondrial enzymes, including citrate synthase, NAD-isocitrate dehydrogenase, and glutamate dehydrogenase, are reduced during diapause, sometimes by as much as 50%, as noted in the goldenrod gall fly *Eurosta solidaginis* (Joanisse and Storey 1994, McMullen and Storey 2008a), and dramatic reductions in number of mitochondria (Levin et al. 2003, McMullen and Storey 2008a) contribute significantly to suppression of oxygen-based metabolism during diapause in some species. During embryonic diapause of the brine shrimp *Artemia franciscana*, an insect relative, trehalose is the sole fuel source, and during diapause mitochondrial function is downregulated primarily by restricting substrate supply through inhibition of key enzymes of carbohydrate metabolism (Hand et al. 2018). Similarly, fat reserves that fuel energy demands during diapause in the mosquito *Culex pipiens* are less available because genes encoding enzymes of β-oxidation are severely downregulated at this time (Sim and Denlinger 2009a, Zhou and Miesfeld 2009).

Big datasets examining transcriptional changes associated with diapause routinely identify downregulation of genes involved in energy generation as a major feature of diapause, *e.g.*, pupal diapause in the flesh fly *Sarcophaga crassipalpis* (Ragland et al. 2010) and apple maggot *Rhagoletis pomonella* (Ragland et al. 2011), embryonic diapause in the Asian tiger mosquito *Aedes albopictus* (Reynolds et al. 2012, Poelchau et al. 2013a), and adult diapause of the northern house mosquito *Culex pipiens* (Sim and Denlinger 2009). Surprisingly, some transcripts that one might expect to be downregulated, such as those encoding cytochrome c oxidase subunits I and III (COI and COIII), are actually upregulated at the onset of diapause in *C. pipiens* (Robich et al. 2007), possibly a consequence of the intense flight activity and nectar-feeding evident in this early phase of diapause. Only in late diapause is expression of the transcript encoding COI downregulated in this mosquito.

Cellular processes that are especially **ATP-expensive** are expected to be a key target leading to metabolic suppression (Storey and Storey, 2004, 2007), and the greatest single energy-consuming process in cells is **active transport of ions** across the cell membrane. The ATP-dependent sodium-potassium ion pump (Na$^+$K$^+$-ATPase), the pump controlling movement of ions across the cell membrane, is thus a likely candidate for inhibition during diapause. Suppression of this activity could effectively aid in suppressing diapause metabolism. This indeed appears to be the case for diapausing larvae of the goldenrod gall fly *Eurosta solidaginis* (McMullen and Storey 2008b): Na^+K^+-ATPase activity is high in field-collected larvae in October (0.56 nmol min/mg), drops by nearly 85% in November, and rises again in April. This seasonal suppression of activity appears to be linked to enzyme phosphorylation, an action that suppresses enzyme activity. Phosphorylation of the enzyme appears to be a seasonally mediated response linked to diapause rather than a direct response to environmental factors such as cold shock, freezing, or hypoxia.

Though metabolism is greatly suppressed during diapause, physical injury can provoke an elevation in metabolic rate, presumably associated with wound healing. This sort of **injury metabolism**, first reported in diapausing pupae of *Hyalophora cecropia* (Harvey and Williams 1961), results in rapid elevation of metabolic activity, and the effect may persist for more than a month, as seen in diapausing pupae of the flesh fly *Sarcophaga argyrostoma* (Denlinger et al. 1972). In neither *H. cecropia* nor *S. argyrostoma* does this elevation in metabolic rate reach levels seen in nondiapausing pupae, and in neither case does the injury result in diapause termination. But, in diapausing pupae of the saturniid *Antheraea polyphemus,* such injury can break diapause (McDaniel and Berry 1967).

As in other animal studies that examine metabolic depression (Storey and Storey 2004), insect diapause transcriptomic datasets provide ample examples of differential gene expression related to the generation of protein products that adjust metabolism. The signatures of diapause reflect less mitochondrial activity and a shift from oxidative phosphorylation to glycolysis and gluconeogenesis, as discussed further in Section 7.8.

7.4   Cycles of Oxygen Consumption

Curiously, in flesh flies (*Sarcophaga*) and a few other diapausing pupae, the rate of oxygen consumption is not constant. Days showing barely detectable oxygen consumption are interspersed with periods of much higher metabolic rates. Cycles are obscured in Figure 7.3 because those measurements are based on groups of pupae, but when individuals are monitored separately, a cyclic **infradian** pattern emerges (Figure 7.4). Diapausing pupae of *S. crassipalpis* have cycles with a periodicity of 4–5 days at 25°C and 10 days at 18°C (Denlinger et al. 1972). And, interestingly, frequency of the peaks changes systematically throughout diapause (Figure 7.5): Cycles are close together in early diapause, become further apart in mid-diapause and then again become closer together as the end of diapause approaches, a pattern reminiscent of the periodic arousal cycles noted in hibernating ground squirrels (Pengelley and Fisher 1961).

The rise in oxygen consumption, which lasts approximately 34 hours at 25°C, is rather abrupt and then declines more slowly (Sláma and Denlinger 1992). The peak of the oxygen consumption cycle is approximately 50 μl/g/h, a level nearly 7× greater

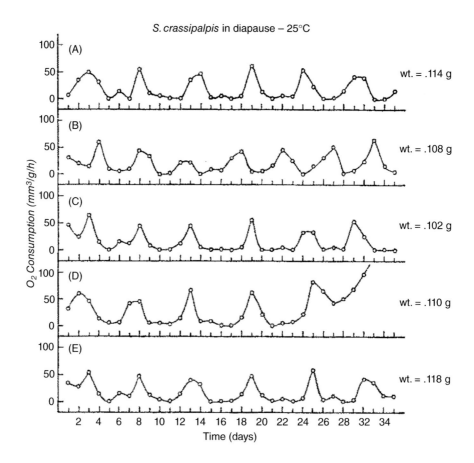

Figure 7.4 Infradian cycles of oxygen consumption in five (A–E) individual diapausing pupae of the flesh fly *Sarcophaga crassipalpis* at 25°C.
From Denlinger et al. (1972).

than during the trough, but a level well below the lowest point of the U-shaped curve for nondiapausing pupae (150 µl/g/h) as seen in Figure 7.4. A boost in protein synthesis is noted during the peaks (Joplin and Denlinger 1989), suggesting that numerous metabolic processes are turned on and off periodically during diapause.

Juvenile hormone (JH) appears to drive the cycles: The JH titer gradually increases during the cycle trough and application of JH elicits a premature rise in metabolic rate, suggesting JH triggers onset of the oxygen consumption peak (Denlinger et al. 1984). Application of JH to larvae destined for pupal diapause completely obliterates the cycles, and the rate of oxygen consumption is sustained at a high rate (equivalent to the peak rate in cycling flies) throughout pupal diapause.

The reason for these cycles is not clear. Might it simply be more efficient to periodically turn on a burst of metabolism than to maintain a constant low rate? Limiting periods of intense oxygen consumption may also help reduces oxidative damage to the insect during the long period of diapause.

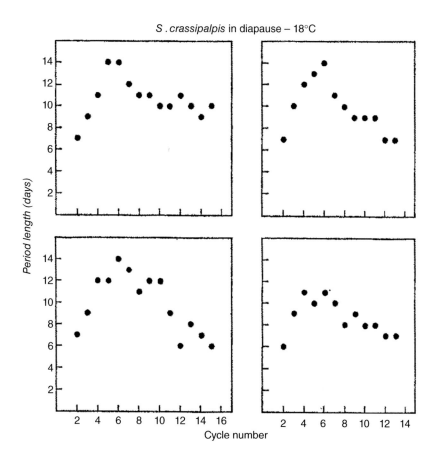

Figure 7.5 Change in periodicity of oxygen consumption cycles throughout the course of pupal diapause in four individuals of the flesh fly *Sarcophaga crassipalpis*.
From Denlinger et al. (1972).

This cyclic pattern suggests that metabolic end products may accumulate during the trough of the cycle, triggering a period of high metabolic activity that oxidizes the end products and concurrently generates the high energy compounds needed to sustain the fly until the next metabolic peak (Denlinger et al. 1972), a hypothesis supported by recent experimental observations (Chen et al. 2021). Though diapausing fly pupae rely primarily on anaerobic glycolysis during the cycle trough, they shift to aerobic respiration through the TCA cycle during the oxygen consumption peaks. Like mammalian hibernators during arousal, diapausing pupae clear anaerobic by-products during the peak and regenerate metabolic intermediates that have accumulated since the last burst of oxygen consumption. A decrease in ROS levels elicits the metabolic burst, and it does so by changing the activity of two critical metabolic enzymes, PDH (pyruvate dehydrogenase) and CPT1 (carnitine palmitoyltransferase 1). ROS signaling, in turn, is modulated through the action of HIF (hypoxia-inducible transcription factor) and phosphorylation of AMPK (AMP activated protein kinase). Interestingly,

ROS signaling has also been implicated as a regulator of periodic arousal in hibernating 13-lined ground squirrels (Brown et al. 2011) and Daurian ground squirrels (Wei et al. 2018), further underscoring the similarity between arousal during hibernation and insect diapause. As mentioned above, JH also appears to contribute to regulation of the cycles in flesh flies (Denlinger et al. 1984), but the link between JH and ROS signaling has not yet been established in this system.

Metabolic cycles are not universal among diapausing insects. In fact, they may be rather uncommon, but they have also been noted in diapausing pupae of several tropical species of Sarcophagidae (Denlinger 1979), as well as in diapausing pupae of the cabbage butterfly *Pieris brassicae* (Crozier 1979). Why some diapausing insects utilize a cyclic pattern and others have metabolic rates that remain stable during diapause remains unknown.

7.5   Discontinuous Gas Exchange (DGE)

Low demand for oxygen during diapause allows some insects the luxury of keeping the spiracular valves that guard entrance to their tracheal system closed much of the time, a strategy that reduces water loss from the trachea. How frequently the **spiracular valves** open is dependent on oxygen demand, thus the spiracles remain closed for longer intervals at lower temperatures. For example, diapausing pupae of the giant silk moth *Hyalophora cecropia* open their spiracles for a few minutes once every 7 hours at 25°C but only once every three days at 10°C (Levy and Schneiderman 1966). Water is a valued commodity for a diapausing insect, especially for those that lack the ability to drink, and water loss is measurably reduced when spiracles are closed. But, additional advantages may be accrued by keeping the spiracles closed. The resulting decrease of oxygen within the tracheal system helps reduce oxidative damage during lengthy periods of diapause (Hetz and Bradley 2005), and experiments with diapausing pupae of the silk moth *Samia cynthia* suggest that avoiding oxidative damage may indeed be the primary driver of discontinuous gas exchange (DGE), while water retention, though important, may play a lesser role (Terblanche et al. 2008). Very little oxygen is required during diapause: The demand can be met by conductance through a single spiracle, as reported for diapausing pupae of the saturniid moths *Attacus atlas* (Wobschall and Hetz 2004) and *S. cynthia* (Moerbitz and Hetz 2010). The pupal diapause pattern of DGE noted in the butterfly *Anthocharis cardamines* gradually transitions to a state of continuous gas exchange at diapause termination, with the spiracles thereafter remaining continuously open (Stålhandske et al. 2015).

Patterns of DGE are most evident and best studied in pupal diapause of large saturniid moths, but there is also evidence for DGEs of much shorter periodicity (approximately 10 minutes) during adult diapause in the Colorado potato beetle *Leptinotarsa decemlineata* (Vanatoa et al. 2006, Lehmann et al. 2020) or longer (21 minutes periodicity) during prolonged diapause in *L. decemlineata* (Ploomi et al. 2018). DGE is not observed in active, nondiapausing individuals of these moths and beetles.

Though some details of DGE remain somewhat controversial and may vary among species (Lighton 1996), the basic pattern reflects the opening and closing of the spiracles and air movement based on changes in partial pressure of gases within the tracheal system. When the spiracular valves open, partial pressure of oxygen and carbon dioxide within the tracheal system approximates that of the atmosphere. When the valves close, partial pressure of oxygen within the system progressively drops as the insect consumes oxygen. Although carbon dioxide is also released into the trachea, quite a bit of the carbon dioxide is released into the hemolymph, thus resulting in an overall decline of partial pressure within the trachea. When the partial pressure of oxygen reaches approximately 4 kPA, the spiracles begin to flutter, allowing fresh air to move into the trachea, elevating tracheal pressure to near that of the atmosphere. Meanwhile, carbon dioxide continues to accumulate within the trachea and triggers opening of the spiracular valves when levels reach a threshold of approximately 5 kPa. The subsequent rapid release of carbon dioxide (a carbon dioxide burst) completes the DGE cycle. The spiracles again close and a new cycle begins. Throughout the cycle the rate of oxygen consumption remains rather constant in diapausing pupae, but the release of carbon dioxide to the atmosphere is periodic, hence the term discontinuous gas exchange in these diapausing pupae refers mainly to the periodic carbon dioxide bursts when the spiracles are open. The flutter phase of the spiracles observed by many of the early workers (*e.g.*, Burkett and Schneiderman 1974) is not always evident. Instead, more recent detailed observations of the spiracles in diapausing pupae of *Attacus atlas* suggest that the spiracles are usually either open or closed and do not display an intermediate fluttering phase prior to opening unless the partial pressure of oxygen oscillates near the point that triggers opening of the spiracles (Förster and Hetz 2010). Even in pupae with all but one of their spiracles blocked with wax, the DGC cycles persist (Moerbitz and Hetz 2010). Movement of air within the tracheal system is also facilitated by subtle undulations of the abdomen and internal organs (Sláma 1988, Wasserthal 1996).

Although DGE is nicely documented in large-bodied diapausing pupae of Lepidoptera, especially the Saturnidae, it is also well known in many nondiapausing insects, including ants, grasshoppers, and beetles, especially those inhabiting dry environments, so it is not a feature unique to diapause. How widespread DGE may be among diapausing insects is not known. Not all diapausing insects display DGE. While the Colorado potato beetle *Leptinotarsa decemlineata* respires using DGE during adult diapause, another closely related, diapausing adult chrysomelid living in the same geographic area, the alder leaf beetle *Agelastica alni*, shows no evidence of DGE and instead shows a pattern of continuous gas exchange (Lehmann et al. 2020). Why some do and others do not is not clear.

The DGE cycles reported for saturniid pupae are superficially similar to the metabolic cycles noted for pupae of *Sarcophaga* discussed above (Section 7.4), but they are clearly distinct and represent two very different processes. The DGE cycles largely represent periodic discharge of carbon dioxide while oxygen consumption remains fairly constant, whereas the fly oxygen consumption cycles reflect periodic increases in the overall metabolic rate. Metabolism is being turned on and off in the diapausing fly pupae while there are no such cycles of metabolism in the saturniid pupae, only cycles purging the built-up carbon dioxide from within the tracheal system.

7.6   Heartbeat

As one might expect, heartbeat rate drops concomitantly with suppression of metabolism during diapause. In diapausing pupae of the flesh fly *Sarcophaga crassipalpis* the heart may stop beating for up to 30 minutes (Sláma and Denlinger 2013). These prolonged periods of cardiac rest are interrupted by short bouts (few seconds to 2 minutes) of fast heartbeats (40–60 beats/minute). All pulsations of the heart are exclusively **anterograde** (forwardly directed) in flesh flies during diapause. During oxygen consumption peaks (Figure 7.4), rest periods are shortened and heartbeat frequency increases more than twofold (Figure 7.6). At diapause termination, **retrograde** (posteriorly directed) pulsations are also initiated within the abdomen, directing hemolymph in a posterior direction, and heartbeat frequency increases (up to 300 anterograde and 125 retrograde beats/minute). The appearance of retrograde pulsations correlates with the higher metabolic demands of development, but molecular triggers initiating this switch to higher complexity have not yet been identified. The exclusive use of anterograde pulsations seen in diapausing flesh fly pupae is not true for all diapauses. Both anterograde and retrograde pulsations are noted during diapause in some other species, including pupae of the tobacco hornworm *Manduca sexta* (Dulcis et al. 2001). Heartbeat rates observed during diapause in the large pupae of *M. sexta* (Sláma and Miller 2001) are also considerably lower (11–22 anterograde and 6–11 retrograde pulses/minute) than observed in flesh flies. The heartbeat persists in diapausing adults of the Colorado potato beetle *L. decemlineata* at temperatures as low as 0°C, and in this species heartbeat frequencies are consistently higher under humid conditions across a range of temperatures (Vanatoa et al. 2006).

Extracardiac pulsations of hemolymph pressure, generated by neurogenically controlled rhythmic contractions of intersegmental abdominal muscles, are a feature common to most developmental stages (Sláma 2000), but such pulsations are notably absent during pupal diapause in flesh flies (Sláma and Denlinger 2013). Diapausing fly pupae lack functional intersegmental musculature, the musculature largely responsible for extracardiac pulsations, thus it is not surprising that such pulsations are noted only after adult development has been initiated. But, low-frequency pulsations of the midgut (1/minute) are seen during diapause. These pulsations in flesh fly pupae coincide with the bouts of anterograde heartbeats and help facilitate hemolymph circulation.

7.7   Structural Modifications

In many cases tissues of diapausing insects differ little from tissues of their nondiapausing counterparts at the same stage of ontogenetic development, but there are

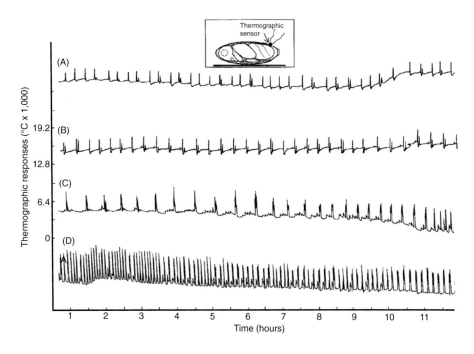

Figure 7.6 Thermographic records of discontinuous cycles of anterograde heartbeats in diapausing pupae of the flesh fly *Sarcophaga crassipalpis,* recorded during a four-day infradian cycle of oxygen consumption. (A–C) are for days during the trough of the cycle and (D) is a record from the day of peak oxygen consumption. Recorded at 23°C.
From Sláma and Denlinger (2013).

notable exceptions. Flight muscles, which are notoriously expensive to maintain, degenerate in some, but not all, adults during diapause. The most detailed examination of flight muscles is based on classic diapause studies from the de Wilde laboratory on the Colorado potato beetle *Leptinotarsa decemlineata* (de Wilde and de Boer 1961, Stegwee et al. 1963, de Kort 1969), as well as more recent studies (Lebenzon et al. 2020). After adult emergence flight muscles in nondiapausing beetles grow rapidly, reaching their maximum diameter in eight days, but flight muscles in diapausing beetles grow only slightly, then start to atrophy by day 4 and are represented thereafter by small myofibrils and few mitochondria. For beetles that are first reproductively active and then enter diapause, the transition from having fully developed flight muscles to having nonfunctional flight muscles is even more dramatic. Some diapausing adults such as the handsome fungus beetle *Stenotarsus rotundus* consistently have well-developed flight muscles prior to diapause, thus enabling the beetles to fly to their aggregation site, but the muscles undergo atrophy shortly upon arrival at the aggregation site (Wolda and Denlinger 1984). In both *L. decemlineata* and *S. rotundus* loss of muscle mass is a component of the diapause syndrome. Diapausing adults of *L. decemlineata* burrow underground for the winter, a site where flight muscles are clearly not needed, and adults of *S. rotundus* remain largely immobile within their aggregations. The muscle cells do not go through cell death but simply enter a phase of pronounced atrophy. This process in *L. decemlineata* is reminiscent of **sarcopenia**, the loss of muscle mass associated with human aging; in both cases the gene ***parkin*** appears to be a critical player (Lebenzon et al. 2020). The difference, of course, is that regeneration of aging human muscle mass can be reversed only modestly (with exercise), while insect flight muscles are fully restored at diapause termination.

Other species that remain active during adult diapause (*e.g.*, monarch butterflies and the mosquito *Culex pipiens*) retain fully developed flight muscles, but distinctions can sometimes still be noted in the fine details. For example, flight muscles in diapausing adults of the mosquito *Culex pipiens* express **beta-tubulin** at much lower levels and contain far fewer microtubules than their nondiapausing counterparts (Kim and Denlinger 2009). This disparity perhaps suggests that, although diapausing adults retain the ability to fly, their flight capacity is not on par with flight capacity of nondiapausing adults. During diapause in *C. pipiens*, two actin genes, ***actin 1*** and ***actin 2***, are highly upregulated, and there is a conspicuous redistribution of polymerized actin within the midgut. In nondiapausing females, polymerized actin remains clustered at intersections of muscle fibers that form the midgut musculature, while diapause and low temperature promote an even distribution of polymerized actin along the muscle fibers (Kim et al. 2006). The midgut is particularly vulnerable to low-temperature damage, and these alterations perhaps fortify the cytoskeleton for low temperatures of winter and contribute to the increased cold tolerance of diapausing mosquitoes (Rinehart et al. 2006). In the pharate first-instar diapause of the gypsy moth *Lymantria dispar*, abundance of actin is dramatically reduced in the brain at the onset of diapause, with levels rebounding quickly at diapause termination (Lee et al. 1998), suggesting that this cytoplasmic actin is essential during the post-diapause phase of development. Although its precise function related to diapause is unknown, the abundance profile of actin in *L. dispar* offers a reliable biomarker for diapause. Members of the actin gene family are downregulated during diapause in several additional species, including larvae of the leafcutter bee *Megachile rotundata* (Yocum et al. 2005) and adults of the carpenter bee *Ceratina calcarata* (Durant et al. 2016).

Large-scale –omics approaches commonly report differential expression of transcripts/proteins associated with the **cytoskeleton**. Perturbations in expression of both thick filaments (*myosin heavy chain, myosin light chain, myosin regulatory light chain, paramyosin*), thin filaments (*actin, troponin, tropomyosin*), members of the integrin signaling pathway, actin-binding proteins, tubulin, and other cytoskeletal-related genes and proteins are reported during diapause in a range of species including the twospotted spider mite *Tetranychus urticae* (Bryon et al. 2013, Zhao et al. 2017), an aphid *Acyrthosiphon pisum* (Le Trionnaire et al. 2012), parasitoids *Nasonia vitripennis* (Wolschin and Gadau 2009) and *Praon volucre* (Colinet et al. 2012), a solitary bee *Tetrapedia diversipes* (Santos et al. 2018), mosquitoes *Culex pipiens* (Robich et al. 2007, Zhang et al. 2019) and *Wyeomyia smithii* (Emerson et al. 2010), higher flies *Delia antiqua* (Hao et al. 2012), *Sarcophaga crassipalpis* (Rinehart et al. 2010), *Drosophila virilis* (Salminen et al. 2015), and *Bactrocera minax* (Wang et al. 2019a, 2020), moths and butterflies *Helicoverpa armigera* (Chen et al. 2010, Lu and

Xu 2010), *Danaus plexippus* (Iiams et al. 2019), *Bombyx mori* (Chen et al. 2017), and *Hyles euphorbiae* (Stuckas et al. 2014), and beetles *Dendroctonus ponderosae* (Bonnett et al. 2012)*, Colaphellus bowringi* (Tan et al. 2017b), and *Lissorhoptrus oryzophilus* (Xinxin et al. 2020). In some cases the transcripts or proteins are upregulated in association with diapause, while in others they are downregulated. For most of these species, we know relatively little about the downstream contributions of these genes to the diapause program, but from the abundance of reports showing differential expression of cytoskeletal genes during diapause it is apparent that major cytoskeletal realignments are a common feature of diapause.

Many studies cited above also report unique profiles of **cuticle proteins** or chitin-associated proteins associated with diapause. Distinct cuticle protein profiles and/or transcripts associated with chitin synthesis are noted during diapause in, among others, the aphid *Aphidius gifuensis* (Zhang et al. 2018a), beetle *Leptinotarsa decemlineata* (Yocum et al. 2009a,b), butterfly *Leptidea sinapis* (Leal et al. 2018), carpenter bee *Ceratina calcarata* (Durant et al. 2016), and the fly *Drosophila melanogaster* (Baker and Russell 2009, Kučerová et al. 2016). In the mosquito *Culex pipiens* a prominent cuticular protein is a reliable marker of early diapause (Li and Denlinger 2009). This protein contains an RR-2 consensus region associated with rigid cuticles, a feature that may be linked to the mosquito's enhanced stress resistance during diapause. Four hemolymph proteins annotated as cuticular proteins are elevated during overwintering diapause in queens of the bumble bee *Bombus terrestris*, one of which is elevated 478-fold (Colgan et al. 2019). In the leaf beetle *Gastrophysa atrocyanea* a chitinase (APAP I) is associated with feeding stage larvae and adults that have terminated diapause, and curiously suppression of this protein with RNAi prevents diapause termination by juvenile hormone (Fujita et al. 2006). Though we are not able to associate most cuticle proteins to specific functions at this time, it is likely that cuticular proteins with diapause-related profiles will ultimately prove to be contributors to desiccation resistance, other forms of protection, or special fortification of the cuticle during diapause.

Conspicuous structural features of **endocrine glands** sometimes distinguish diapausing from nondiapausing individuals. Size and shape of these glands frequently reflect activity levels. Cells of an inactive prothoracic gland from diapausing pupae of *Mamestra brassicae* are quite shriveled in comparison to cells of an active gland (Agui 1975). Although size of the corpora allata is not always a reliable indicator of activity, small corpora allata in diapausing adults commonly reflect an inactive gland, *e.g.*, adult diapause in *Culex pipiens* (Kang et al. 2014) and *Leptinotarsa decemlineata* (de Kort 1990). And, as one would expect with the cessation of reproduction that characterizes adult diapause, the ovaries and accessory reproductive glands are also typically diminished in size, as seen in Figure 7.1.

At the ultrastructural level, diapause distinctions are sometimes noted as well. For example, storage cells seldom seen in nondiapausing individuals of the horse chestnut leaf miner *Cameraria ohridella,* are abundant in diapausing pupae (Weyda et al. 2015). These circular cells contain abundant vacuoles packed with putative energy reserves needed uniquely for diapause.

7.8   Metabolic Restructuring

Not surprising, metabolic pathways deployed during diapause consistently differ from those seen in nondiapausing individuals, and fundamental shifts from early to late diapause are conspicuous as well. Based on Respiratory Quotients (RQ) and differing abundance of specific metabolites during diapause the uniqueness of diapause metabolism has been noted for many years, but the scale of change can now be appreciated more fully thanks to the growing number of transcriptomic, metabolomic, and proteomic studies comparing diapausing and nondiapausing individuals. Nearly 30% (41 out of 297) of protein differences between diapause and nondiapause in the cabbage beetle *Colaphellus bowringi* are involved in energy production and conversion, carbohydrate metabolism and transport, and lipid metabolism (Tan et al. 2017b). Metabolic functions frequently represent the largest functional category of differentially regulated genes in diapause, *e.g.*, twospotted spider mite *Tetranychus urticae* (Zhao et al. 2017), pea aphid *Acyrthosiphon pisum* (Le Trionnaire et al. 2012), migratory locust *Locusta migratoria* (Tu et al. 2015), pitcher plant mosquito *Wyeomyia smithii* (Emerson et al. 2010), onion fly *Delia antiqua* (Ren et al. 2018), cotton bollworm *Helicoverpa armigera* (Zhang et al. 2013), or at least this category is among the major functional categories, *e.g.*, the bumblebee *Bombus terrestris* (Amsalem et al. 2015).

Storage of energy reserves discussed in Section 6.2 indicates that many metabolic distinctions begin before the onset of diapause, but many interesting changes occur during diapause as well, especially shifts in utilization of specific reserves as diapause progresses. Direct measurements of reserves frequently portray major switches in the energy reserves used during diapause. Lipid levels drop in early pupal diapause of the flesh fly *Sarcophaga crassipalpis* but then remain stable from mid-diapause onward, whereas levels of carbohydrates and proteins, the non-lipid reserves, are stable initially but then decline during the last half of diapause, suggesting a shift from lipid utilization to utilization of non-lipid reserves at mid-diapause (Adedokun and Denlinger 1985). Similarly, lipid levels decline during the first two months of pupal diapause in the European cherry fruit fly *Rhagoletis cerasi* and then remain stable until diapause comes to an end (Papanastasiou et al. 2011). In the peach fruit moth *Carposina sasakii*, fats comprise approximately 30% of body content early in larval diapause (December and early January), but levels drop sharply to approximately 16% by the end of January and remain at that level for the remainder of diapause (Zhang et al. 2016a). In diapausing larvae of the blow fly *Calliphora vicina* fat content drops from 20% of dry weight to 8–10% by day 100 (Saunders 2000). Fat content in diapausing adults of the nymphalid *Euploea core corinna* drops from 24 to 18% between early June and mid-July in the Southern Hemisphere winter (Kitching and Zalucki 1981). The decline in lipid content during diapause in the Colorado potato beetle *Leptinotarsa decemlineata* is more modest, perhaps a reflection of the high degree of metabolic depression in this species or its greater reliance on non-lipid stores to fuel diapause (Lehmann et al. 2020).

7.8.1   Shifts in RQ Values

Major reliance on lipids or a combination of lipids and carbohydrates during diapause is suggested by RQ values, exemplified by an RQ of 0.78 in diapausing pupae of the silk moth *Hyalophora cecropia* (Schneiderman and Williams 1953), an RQ of 0.6 during the overwintering adult diapause of the linden bug *Pyrrhocoris apterus* (Koštál et al. 2008), and an RQ of 0.6–0.8 reported for diapausing adults of the solitary bee *Osmia lignaria* (Bosch et al. 2010). Though the RQ of *P. apterus* remains remarkably constant at 0.6 throughout 180 days of diapause (Koštál et al. 2008), RQ values frequently do not remain constant but instead change as the insect enters different phases of diapause, implying that the insect alters its reliance on specific energy reserves as diapause proceeds. In *O. lignaria* (Figure 7.7) RQ values are near 1.0 at the onset of diapause, drop to 0.6–0.8 for most of the winter and then rise to 1.0 in early spring (Sgolastra et al. 2010), suggesting reliance on carbohydrates early and late in diapause but utilization of proteins and lipids in mid-diapause. In early pupal diapause the RQ of the pierid butterfly *Pieris napi* is 0.75 and then after 10 days, drops to 0.55, indicating a shift from carbohydrates and protein in early diapause to lipids for the remaining months of diapause (Lehmann et al. 2016a). During larval diapause of the solitary bee *Megachile rotundata* RQ values are 0.6–0.8 during the winter months but increase to 1.0 in early spring (Yocum et al. 2005), a response suggesting movement toward strict carbohydrate reliance in late diapause, a trend similar to that seen in *O. lingaria* but opposite of that seen in *P. napi*. Although species differences can be expected, consistently there are distinctions between diapause and nondiapause as well as shifts during diapause.

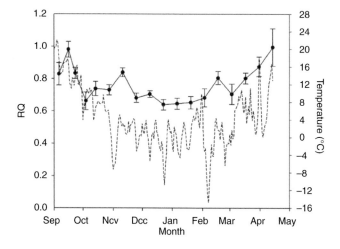

Figure 7.7 Change in RQ values during overwintering in diapausing females of the solitary bee *Osmia lignaria*. Dotted lines indicate mean daily temperature in Utah.
From Sgolastra et al. (2010), with permission from Elsevier.

7.8.2   Organ Distinctions

Fat body, the major center of insect metabolism, shows a more dramatic shutdown than other tissues such as the brain. In the brain of *Helicoverpa armigera* roughly the same number of genes are upregulated (27%) and downregulated (27%), while 47% are unchanged during pupal diapause, but in the fat body, the largest portion of genes are downregulated (60%), 7% are upregulated, and 33% show no change (Xu et al. 2012). This underscores the fact that a slowdown in metabolic activity in the fat body is especially noteworthy during diapause, a result consistent with the shutdown of the liver during mammalian hibernation (Williams et al. 2005).

7.8.3   Shift from Aerobic to Anaerobic Metabolism

Metabolomics studies reveal major shifts in metabolic pathways deployed during diapause, a shift favoring **glycolysis** and **gluconeogenesis** and reliance on anaerobic metabolism rather than the **tricarboxylic acid (TCA, citric acid) cycle.** Since diapausing insects almost exclusively depend on energy stores, the generation of new biosynthetic compounds relies heavily on **fatty acid β-oxidation**, the glyoxylate cycle, and gluconeogenesis. In the flesh fly *Sarcophaga crassipalpis* this shift is reflected in elevation of glucose, alanine, and pyruvate during pupal diapause. Concurrent depletion in pools of two aerobic metabolic intermediates, fumarate and citrate, indicates a reduction in TCA cycle activity (Michaud and Denlinger 2007). Glucose elevation is huge in this species, nearly 30-fold compared to nondiapausing pupae, thus greatly increasing the availability of substrate entering the glycolytic pathway, a feature essential for the generation of the fly's dominant cryoprotectants, glycerol, and sorbitol. Glucose, presumably derived from glycogen stores, is also elevated during adult diapause in the Colorado potato beetle, *Leptinotarsa decemlineata* (Lefevere et al. 1989), the tropical beetle *Stenotarsus rotundus* (Pullin and Wolda 1993) and *Drosophila melanogaster* (Kubrak et al. 2016). Pyruvate accumulation is noted in the rice stem borer *Chilo suppressalis* (Tsumuki and Kanehisa 1980), and elevated levels of alanine and other amino acids during diapause (Kukal et al. 1991, Rivers et al. 2000, Michaud and Denlinger 2007) are likely derived from pyruvate stores that accumulate during diapause. Pyruvate also offers a readily accessible energy source for the TCA cycle at the end of diapause, as proposed for embryonic diapause in the silk moth *Bombyx mori* (Kageyama and Ohnishi 1971). During larval diapause in the European corn borer *Ostrinia nubilalis* activity of lactate dehydrogenase, the enzyme that catalyzes conversion of pyruvate into lactate, is highly elevated (Uzelac et al. 2020), an indication that this insect also switches to anaerobic metabolism during diapause. By contrast, activity of citrate synthase, an enzyme indicative of an active TCA cycle, is low during diapause in *O. nubilalis* but high during nondiapausing stages, a result consistent with observations in several other species as well.

A shift toward anaerobic metabolism is also evident from transcriptomic datasets. Supporting the metabolomic results noted in *S. crassipalpis*, the transcriptome of this species displays major shifts toward anaerobic metabolism during diapause (Ragland et al. 2010). DAVID analysis comparing early diapause with nondiapause pupae shows diapause enrichment in carbohydrate and pyruvate metabolism, and GSA analysis reveals enrichment of glycerolipid metabolism, ether lipid metabolism, and pyruvate metabolism. Enriched gene lists include upregulation of key members of glycolysis and gluconeogenesis pathways, and although some members of the TCA cycle are upregulated, the initial steps in the TCA cycle involving conversion of pyruvate from glycolysis to acetyl-CoA are suppressed, a result concordant with pyruvate recycling into gluconeogenesis. At the end of diapause, transcript abundance for genes encoding key enzymes in aerobic metabolism and aerobic processing of anaerobic end products are elevated, accompanied by a simultaneous decline of expression for transcripts encoding enzymes involved in anaerobic metabolism.

The switch from **aerobic** to **anaerobic** metabolism during diapause and a return to aerobic metabolism at diapause termination is further supported by transcript profiling of embryonic diapause in the cricket *Allonemobius socius* (Reynolds and Hand 2009a), larval diapause in the fly parasitoid *Nasonia vitripennis* (Li et al. 2015) and alfalfa leafcutting bee *Megachile rotundata* (Yocum et al. 2015), pupal diapause in the apple maggot fly *Rhagoletis pomonella* (Ragland et al. 2011), aphid parasitoid *Praon volucre* (Colinet et al. 2012), butterfly *Pieris napi* (Lehmann et al. 2018), and cotton bollworm *Helicoverpa armigera* (Bao and Xu 2011, Xu et al. 2012, Wang et al. 2018), and adult diapause in the twospotted spider mite *Tetranychus urticae* (Zhao et al. 2017).

A metabolic shift is also evident from proteomics: Pupal diapause termination in the Chinese citrus fly *Bactrocera minax* is accompanied by a major shift in abundance of proteins involved in energy metabolism, underscoring a return to aerobic metabolism at the end of diapause (Dong et al. 2019). Similar protein shifts are noted in diapausing adults of the rice water weevil *Lissorhoptrus oryzophilus*; especially conspicuous is the lower abundance of proteins involved in the TCA cycle (Xinxin et al. 2020).

Elevation of glycerol in overwintering larvae of the mountain pine beetle *Dendroctonus ponderosae* also appears to be the consequence of glycogenolytic and gluconeogenic pathways (Fraser et al. 2017). Though alanine, glycerol, and trehalose are elevated during diapause in the sycamore lace bug *Corythucha ciliata* (Li et al. 2017a) and the parasitoid *Aphidius gifuensis* (Zhang et al. 2018a), transcripts and metabolites in glycolysis, gluconeogenesis, and the TCA cycle do not reflect the expected shift in metabolism, an observation that underscores the challenge of relying strictly on transcript levels and metabolites to designate pathways being utilized.

One recurring theme indicative of diapause in numerous species is the elevation of the transcript-encoding **phosphoenolpyruvate carboxykinase (*pepck*)**. Pepck is an anaerobic metabolic enzyme that catalyzes oxaloacetate into CO_2 and phosphoenol pyruvate, a metabolite that can be used in gluconeogenesis. *Pepck* is elevated ninefold during pupal diapause of *Sarcophaga crassipalpis* (Ragland et al. 2010). Elevation of the *pepck* transcript or the protein PEPCK is also reported for embryonic diapause in the Asian tiger mosquito *Aedes albopictus* (Poelchau et al. 2011), in ovaries of diapause egg producers of the silk moth *Bombyx mori* (Chen et al. 2017), larval diapause of the wheat blossom midge *Sitodiplosis mosellana* (Cheng et al. 2009), pitcher plant mosquito *Wyeomyia smithii* (Emerson et al. 2010), and mountain pine beetle *Dendroctonus ponderosae* (Fraser et al. 2017), pupal diapause of the apple maggot *Rhagoletis pomonella* (Ragland et al. 2011) and onion fly *Delia antiqua* (Hao et al. 2016), adult diapause of the bumblebee *Bombus terrestris* (Amsalem et al. 2015) and *Drosophila melanogaster* (Baker and Russell 2009, Kubrak et al. 2016), as well as the dauer state of the nematode *Caenorhabditis elegans* (Jeong et al. 2009) and during adult diapause in the copepod *Neocalanus flemingeri* (Roncalli et al. 2018). But, like most diapause responses, *pepck* upregulation is not universal. No transcript elevation is noted in larval diapause of the alfalfa leafcutting bee *Megachile rotundata* (Yocum et al. 2015) nor adult diapause of the twospotted spider mite *Tetranychus urticae* (Bryon et al. 2013), in spite of evidence that these species rely on glycolysis and gluconeogenesis. We cannot, of course, eliminate the possibility that PEPCK in these species is regulated at the protein level. In fact, there is good evidence that PEPCK is regulated posttranslationally by acetylation during diapause in the cotton bollworm, *Helicoverpa armigera* (Wang et al. 2018). Patterns of elevated *pepck* in numerous species and in diapauses of different stages suggest the dominance of anaerobic pathways during diapause. Although there are two isoforms of PEPCK, a cytosolic (PEPCK-C) and mitochondrial (PEPCK-M) form, most insect work does not distinguish between the variants. The two isoforms commonly respond differently to stresses such as cold shock and starvation, but both isoforms are elevated during pupal diapause in *S. crassipalpis* (Spacht et al. 2018).

The common metabolic shift noted during diapause, a decrease in aerobic metabolism and an accompanying increase in anaerobic metabolism, favors activity of glycolysis and gluconeogenesis, the pentose phosphate shunt, and the phosphoenolpyruvate carboxykinase (PEPCK)-succinate pathway to generate ATP and reduce equivalents (Hahn and Denlinger 2011). This shift is also commonly observed when animals are exposed to hypoxic or anoxic conditions, but during diapause this shift occurs not in response to hypoxia/anoxia but as a developmental response under normoxia.

Reliance on anaerobic metabolism is evident in diapauses of other invertebrates as well. One of the most profound examples is seen in the encysted embryos of the brine shrimp *Artemia franciscana*. This extreme anaerobic diapause, represented by a 97% drop in respiration rate (Clegg et al. 1996), is achieved by shifting energy metabolism from oxidative phosphorylation to glycolysis (Hand et al. 2011). Dauer larvae of the nematode *C. elegans* likewise have suppressed metabolic rates, accompanied by a 11-fold decline in activities of enzymes involved in the TCA cycle and elevated glycolytic and gluconeogenesis capacity, a shift that diverts carbon atoms from the TCA cycle to gluconeogenesis (O'Riordan and Burnell 1989, Wang and Kim 2003, Penkov et al. 2020). Interestingly, dauer larvae also utilize an alternative pathway of energy production involving fermentation, as indicated by the accumulation of ethanol and upregulation of alcohol fermentation genes (Glockner et al. 2008).

7.8.4   Lipid Management and Utilization

Another conspicuous metabolic switch observed in association with diapause involves management and utilization of lipid stores. Sequestering of fats and conserving their utilization is a vital component of successful overwintering (Sinclair and Marshall 2018). Pathways involved in **lipid storage** are likely to be more favored during diapause preparation and in early diapause than in nondiapausing individuals. By contrast, pathways involving **β-oxidation**, a pathway utilized to break down lipids for energy production, may be largely suppressed during certain phases of diapause. The changes noted above for shifts in RQ values underscore the dynamic nature of such changes during diapause as the insect shifts from reliance on one form of energy store to another. During the first half of pupal diapause the flesh fly *S. crassipalpis* relies on lipid reserves but then switches to utilization of protein and carbohydrate stores (Adedokun and Denlinger 1985). Just the opposite is noted in diapausing adults of the mosquito *Culex pipiens*: Females first deplete their glycogen stores and then switch to the use of lipids as shown by both radiotracer tracking (Zhou and Miesfeld 2009) and expression of β-oxidation genes (Sim and Denlinger 2009a). Among 31 genes related to fat metabolism in *C. pipiens*, 4 involved in fat accumulation are highly upregulated in early diapause while genes involved in fat utilization are downregulated. But, at the transition from early to late diapause genes involved in fat storage are downregulated while 19 transcripts involved in fatty acid utilization are upregulated as the diapausing mosquito begins to catabolize its lipid reserves as an energy source. Females subjected to knockdown of *fas-1* and other (*fas-3* and *fabp*) transcripts involved in fat storage fail to sequester the lipids needed for overwintering (Sim and Denlinger 2009a). Likewise, in the cabbage beetle *Colaphellus bowringi* knockdown of *fatty acid-binding protein*, a gene encoding a protein involved in fatty acid transport, or *fas-2*, a gene encoding an enzyme essential for lipid biosynthesis, results in failure of diapause-destined adult beetles to accumulate the lipid reserves required for successful overwintering (Tan et al. 2017a,b).

Transcript and metabolite profiles during embryonic diapause of the mosquito *Aedes albopictus* reflect distinctions in genes related to lipid sequestration and utilization (Reynolds et al. 2012, Batz and Armbruster 2018). Elevated expression of *lipid storage droplet protein 2* (*lsdp2*) likely contributes to the higher amounts of lipid found in diapausing eggs, and conservation of lipids during diapause is reflected in downregulation of numerous genes involved in lipid catabolism. Expression of *lsdp2* is also elevated during adult diapause in the Colorado potato beetle *Leptinotarsa decemlineata* (Güney et al. 2020), suggesting a possible common role for the protein it encodes in storage of lipids during diapause. Expression of genes related to lipid metabolism is overrepresented in diapausing adults of the small carpenter bee *Ceratina calcarata* (Durant et al. 2016). Changes in enzyme activity also provide insights on the dynamics of lipid utilization. Lipolysis enzymes increase in activity during overwintering in the gall moth *Epiblema scudderiana* (Rider et al. 2011), a response that is opposite that seen for the gall fly *Eurosta solidaginis* (Joanisse and

Storey 1996). Distinctions in lipid metabolism are also evident in proteomic signatures during diapause, as noted in the parasitoid *Nasonia vitripennis* (Wolschin and Gadau 2009).

7.8.5   A Role for Autophagy

In the absence of food, diapausing insects may also rely on autophagy to generate some of the energy supplies needed to survive winter. The fact that a protein associated with autophagy, gamma-aminobutyric acid receptor-associated protein, is highly elevated in the bean bug *Riptortus pedestris* when food is absent and also during late diapause suggests a possible reliance on autophagy in this bug, especially in the late phase of diapause when other reserves have been depleted (Tachibana et al. 2019). A proteomics profile of pupae of the fly *Bactrocera minax* also suggests a role for autophagy in diapause (Wang et al. 2020).

7.8.6   Changes Associated with Dietary Shifts

A particularly striking metabolic shift is noted in insects that switch food sources as a function of diapause. Nondiapausing adults of the mosquito *Culex pipiens* are ravenous blood-feeders, but those programmed for diapause feed exclusively on nectar. This is reflected in diapause downregulation of genes encoding enzymes needed to digest a blood meal (*trypsin* and *chymotrypsin-like protease*) and concurrent upregulation of *fatty acid synthase*, a gene associated with accumulation of lipid reserves (Robich and Denlinger 2005). As the end of diapause approaches, genes encoding the digestive enzymes begin to be expressed as the female prepares to take a blood meal upon the termination of diapause. And, as one might expect, diapausing adults such as the Colorado potato beetle cease feeding as they enter diapause, an event accompanied by downregulation of genes encoding digestive enzymes (Yocum et al. 2009a,b).

7.9   Protein Synthesis and Posttranslational Modifications

The overall rate of **protein synthesis** is greatly reduced during diapause. Whole body samples of diapausing pupae (Joplin and Denlinger 1989) as well as brains (Joplin et al. 1990) of *Sarcophaga crassipalpis* synthesize proteins at a lower rate than their nondiapausing counterparts, as indicated by pulse labeling. But, the rate is not uniform throughout diapause. Diapausing flesh flies have a cyclic pattern of O_2 consumption (Section 7.4) and incorporation of $[^{35}S]$ methionine is highest (5.4%) during O_2 consumption peaks and lowest (1.7%) during troughs of the cycle (Joplin and Denlinger 1989). Incorporation for the comparable nondiapause stage is eightfold higher than observed for the maximum diapause rate. Similarly, *in vitro* cultured brains that are pulse-labeled and analyzed using two-dimensional electrophoresis show fewer proteins being synthesized during diapause (Joplin et al. 1990). And, brains from older flies (60 days in diapause) synthesize slightly more proteins than brains from younger flies (14 days in diapause), again underscoring the dynamic nature of diapause. A total of 175 egg proteins differ in abundance between diapause and nondiapause in the migratory locust *Locusta migratoria*, and most (151) are less abundant in diapause samples (Cui et al. 2019). Although pulse labeling shows that rates of protein synthesis differ between diapause and nondiapause, the total number of proteins present in brain samples from the two categories is usually quite similar (Li et al. 2007, Pavlides et al. 2011), an observation also true for pupal brains of the cotton bollworm *Helicoverpa armigera* (Lu and Xu 2010).

Low presumptive rates of protein synthesis during diapause are also reflected in transcriptional datasets. Several transcriptomic studies show fewer genes being expressed during diapause, *e.g.*, among differentially expressed genes in brains of the small carpenter bee *Ceratina calcarata* 26% are upregulated while 76% are downregulated during adult diapause (Durant et al. 2016). A comparison of differentially expressed genes (DEGs) in diapause- and nondiapause-destined adults of the beetle *Colaphellus bowringi* shows 2,269 downregulated DEGs compared to 897 DEGs that are upregulated (Liu et al. 2017); in the tropical solitary bee *Tetrapedia diversipes*, 815 DEGs are downregulated and 352 are upregulated (Santos et al. 2018); in the sycamore bug *Corythucha ciliata*, 337 DEGs are downregulated and 149 are upregulated (Li et al. 2017); and in the fly *Chymomyza costata* 796 DEGs are downregulated and 517 are upregulated (Poupardin et al. 2015). Differences in down- and upregulation are rather small for pupae of the flesh fly, *Sarcophaga crassipalpis:* 368 DEGs are upregulated at least twofold and 336 are downregulated (Ragland et al. 2010). Diapause, however, does not consistently reflect more downregulated than upregulated transcripts, *e.g.*, among the DEGs in adults of the mosquito *Culex pipiens*, 241 are upregulated and 207 are downregulated in diapause (Kang et al. 2016). Diapause is, of course, a dynamic process and most comparisons of expression are based on a single time point. The progression of diapause in embryos of the mosquito *Aedes albopictus* shows fewer DEGs as the end of diapause approaches (Poelchau et al. 2013b). Different levels of up- or downregulation can also be expected for different organs, as mentioned above in Section 7.8.2.

In pupae of the cotton bollworm, *H. armigera*, gene expression patterns show upregulation of transcriptional repressors and concurrent downregulation of translation activators at diapause initiation (Bao and Xu 2011). Reduced expression of such genes as *serine protease* and *E3 ubiquitin* during adult diapause in the copepod *Neocalanus flemingeri* are also indicative of low protein turnover (Roncalli et al. 2018). The observed depression of RNA and protein synthesis during diapause suggests that turnover of macromolecules is similarly depressed (Hand et al. 2016), though few attempts have been made to measure turnover during diapause.

Posttranslational modification of proteins refers to the covalent and enzymatic modification of proteins that occur after the protein has been synthesized. Such modifications take numerous forms, including phosphorylation, glycosylation, ubiquitination, SUMOylation (Small Ubiquitin-related Modifier), and acetylation, among others. How such modifications are involved in diapause is still unclear, but the evidence available suggests that such modifications may emerge as important diapause regulators. Several diapause-specific phosphoproteins, including a few that show increased phosphorylation during diapause, are present in the brains of the flesh fly *Sarcophaga crassipalpis* (Pavlides et al. 2011). Although the overall level of protein phosphorylation in the brains of *Helicoverpa armigera* declines during pupal diapause, several specific proteins linked to phosphorylation and dephosphorylation show increased abundance (Lu and Xu 2010). **Protein phosphatase**, PP1c, is of special interest because it is quite abundant in the pupal brain of *H. armigera* during diapause and is known to regulate multicellular functions such as cell division, metabolism, protein synthesis, and cytoskeletal reorganization. In diapausing pupae of *H. armigera*, PP1c is thought to exert its effect in the brain through dephosphorylation. Another phosphatase, PP2A, is highly expressed during the photosensitive stage for diapause induction in *H. armigera* (Tian and Xu 2013), but PP2A shows reduced gene expression and enzyme activity during embryonic diapause in the silk moth *Bombyx mori* (Gu et al. 2017). Though the two phosphatases differ in response during diapause in these two species, the fact that both are altered during diapause underscores a potential role for phosphorylation in the diapause of these species. The transition from diapause to post-diapause development is accompanied by the elevation of several protein tyrosine phosphatases in *B. mori*, a result indicating that phosphorylation events are associated with the onset and progression of development once diapause has been completed (Gu et al. 2021).

Numerous enzymes are activated by phosphorylation. Utilization of glycogen relies on **glycogen phosphorylase**, an enzyme that exists in two interconvertible forms, an active phosphorylated *a* **form** and an inactive dephosphorylated *b* **form**. During pupal diapause in the flesh fly *Sarcophaga crassipalpis* the activated form of the enzyme is highest at the onset of diapause and then declines to fairly constant levels for the remainder of diapause (Chen and Denlinger 1990). Diapausing pupae held in the cold consistently maintain higher levels of the activated enzyme, a feature consistent with the generation of glycerol, the major cryoprotectant used by these flies during diapause. Activation of glycogen phosphorylase as well as phosphofructokinase, another enzyme in the pathway for glycerol formation, are not unique to diapause, but are responses that can be activated by low temperature in nondiapausing individuals as well (Storey and Storey 1991). Phosphorylation of **Glycogen synthase kinase-3** (GSK-3), a multifunctional protein kinase involved in regulating both glycogen synthesis and protein synthesis, declines abruptly at the onset of diapause in eggs of the silk moth *Bombyx mori*, a change consistent with a drop in glycogen levels and elevation of cryoprotectants (Lin et al. 2009). Transfer of eggs to low-temperature results in increased GSK-3 beta phosphorylation, thus again showing linkage to low-temperature responses, responses that are fundamental to winter survival but not necessarily restricted to the diapause program. Large-scale phosphorylation events noted during cold acclimation for *Drosophila melanogaster* are especially prevalent in proteins involved in cytoskeletal networks (Colinet et al. 2017).

A potentially important role for phosphorylation is also evident at diapause termination in embryonic diapauses of the silk moth *Bombyx mori* (Fujiwara et al. 2006a,b) and false melon beetle *Atrachya menetriesi* (Kidokoro et al. 2006b), as well as pupal diapause of the flesh fly *Sarcophaga crassipalpis* (Fujiwara and Denlinger 2007). The phosphorylated form of **Extracellular signal-regulated kinase** (ERK), a member of the MAPK family, is elevated at diapause termination, regardless of the mechanism (temperature manipulations, chemical shocks) used to break diapause. Phosphorylation of ERK may well emerge as a component of diapause termination common to numerous species (See also Chapter 9), although few species have been examined thus far. In *B. mori*, protein kinase C (PKC)-dependent phosphorylation of several unidentified proteins is considerably lower in diapausing embryos than in embryos that either do not enter diapause or have broken diapause (Gu et al. 2019, 2020), again suggesting an important, but still unclarified, role for phosphorylation as a potential diapause regulator.

Ubiquination, a modification that marks proteins for degradation, also may be involved in the diapause response. Ubiquitin-mediated proteolysis is blocked during diapause in the brine shrimp *Artemia franciscana* (Anchordoguy and Hand 1994) and the copepod *Neocalanus flemingeri* (Roncalli et al. 2018). In diapausing pupae of *Helicoverpa armigera* two ubiquitin-conjugate enzymes, E2M and effete, are less abundant at the initiation of diapause (Lu and Xu 2010).

The covalent linkage of SUMO to its substrates (**SUMOylation**) regulates numerous cellular processes including transcription, chromatin maintenance, cell cycle, and apoptosis. In diapause-destined pupae of *H. armigera*, the SUMO gene is upregulated, and Bao et al. (2011) provide evidence that SUMO binds to a Forkhead protein, causing downregulation in the expression of the gene encoding diapause hormone (DH), thus suggesting an important role for SUMOylation as a negative regulator of transcription, driving the DH titer to a low level in diapause-destined pupae and consequently promoting developmental arrest. In *H. armigera*, unlike *Bombyx mori*, a low DH titer is essential for diapause (See Section 9.1.2.3).

Acetylation and **deacetylation** events are also observed in association with diapause and are likely quite important as diapause regulators. Acetylation of histones and the enzymes central to acetylation are differentially regulated in diapausing pupae of the flesh fly *Sarcophaga bullata* and likely exert an important regulatory role (Reynolds et al. 2016), as discussed more fully in Chapter 9, Section 10.2. In pupae of the cotton bollworm *H. armigera*, as well as other species, pyruvate metabolism is central to regulation of diapause. Low pyruvate is essential for diapause induction, while high levels are associated with nondiapause development as well as resumption of development at diapause termination (Wang et al. 2018). Acetylation is a post-translational regulator of phosphoenolpyruvate kinase (PEPCK) and several other enzymes that are critical determinants of pyruvate levels.

Widespread roles for reversible protein phosphorylation and other forms of protein modification in regulatory schemes during development suggest that such modifications will emerge as key players in diapause regulation, as well as in other forms of arrested development across the animal kingdom (Storey and Storey 2004). Phosphorylation related to major signaling pathways such as FoxO and AMPK is discussed more fully in Chapter 9, as are other posttranslational modifications that are emerging as regulators of diapause.

7.10   Storage Proteins

As diapause approaches, many insects synthesize an abundance of storage proteins that accumulate in the hemolymph. These proteins, sometimes referred to as **diapause-associated proteins**, are typically much more abundant in hemolymph of diapausing insects than in their nondiapausing counterparts, as noted in Figure 7.8 for the boll weevil *Anthonomus grandis* (Lewis et al. 2002). Such proteins, first reported in diapausing larvae of the southwestern corn borer *Diatraea grandiosella* (Brown and Chippendale 1978), are now known from a number of diapausing larvae and adults, but they clearly are not a universal attribute of diapause. Many species do not produce such proteins. Storage proteins are synthesized primarily in the fat body and released into the hemolymph where they persist for the duration of diapause. These same proteins are frequently synthesized by nondiapausing individuals as well, but in that case they are quickly utilized as the insect proceeds with development. By contrast, storage proteins remain in the hemolymph of diapausing individuals until the end of diapause, at which time they quickly disappear. What is unusual is their abundance and persistence.

Several types of diapause-associated storage proteins have been reported. Most prominent are the **hexamerins**, proteins composed of six identical or nearly identical subunits with subunit weights of 70–80 kDa (Burmester 1999). Hexamerins can be further classified as either **arylphorins**, proteins containing a high proportion (18–25%) of aromatic amino acids, or **methionine-rich storage proteins**, a group that, in Lepidoptera, contain 4–11% methionine. Arylphorins are the dominant hexamerin in diapausing adults of the boll weevil *A. grandis* (Lewis et al. 2002), while a

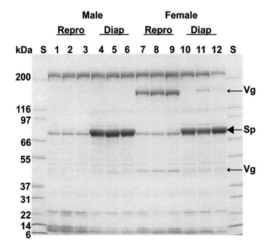

Figure 7.8 Coomassie blue-stained SDS-PAGE gel comparing hemolymph polypeptides from reproducing and diapausing adult males and females of the boll weevil *Anthonomus grandis.* Arrows indicate a storage protein (Sp) that is considerably more abundant in diapausing adults and vitellogenin (Vg), a protein present only in nondiapausing (reproducing) females.
From Lewis et al. (2002), with permission from Elsevier.

methionine-rich hexamerin dominates in diapausing larvae of the rice stem borer *Chilo suppressalis* (Sonoda et al. 2006). Diapause-associated proteins of relatively small size, *e.g.*, the 35 kDa diapause-associated protein in *D. grandiosella* (Dillwith et al. 1985), as well as proteins with much higher molecular weights are known from a number of species, *e.g.*, larvae of the tropical stem borer *Busseola fusca* (Osir et al. 1989) and pink bollworm *Pectinophora gossypiella* (Salama and Miller 1992).

Several forms of storage proteins may be present simultaneously. For example, 75 and 72 kDa hexamerins are present in larval hemolymph during diapause in the bamboo borer *Omphisa fuscidentalis* (Tungjitwitayakul et al. 2008); one is an aryl-phorin, the other a riboflavin-binding hexamerin. During larval diapause in the parasitoid *Nasonia vitripennis*, seven distinct storage proteins are present, six of which are altered in abundance during diapause: three hexamerins, two arylphorins, and one Glx storage protein (Wolschin and Gadau 2009). The gene encoding hexameric larval serum protein Lsp2 is prominently upregulated in preparation for adult diapause in *Drosophila montana* (Parker et al. 2016). In the migratory locust *Locusta migratoria* hexamerin-like protein 4 is more abundant during embryonic diapause, while hexamerin-like protein 1 is less abundant (Cui et al. 2019).

Storage proteins associated with diapause serve primarily as reservoirs of nitrogen and amino acids that will be used at diapause termination for generating new tissues during metamorphosis and/or for reproduction. Such a role is consistent with their rapid depletion at diapause termination. But, they may serve additional roles as well. Arylphorins bind and relocate lipids, a function especially important for lipid utiliza-tion during diapause, and hexamerins bind juvenile hormone (JH), suggesting a possible role for these proteins in regulatory schemes controlling diapause. The presence of several types of storage proteins in some species suggests distinct roles for specific proteins in the progression or maintenance of diapause.

Transcripts encoding these storage proteins reveal diverse temporal expression profiles. (1) Expression prior to diapause but not during diapause. In the spruce budworm *Choristoneura fumiferana* transcripts encoding two storage proteins are high during feeding phases of the first and second larval instars, but when the second instar enters diapause, synthesis ceases (Palli et al. 1998). Proteins transcribed from these transcripts, however, remain in the hemolymph throughout diapause and decline in abundance only when diapause is terminated. (2) Expression that persists during diapause. Though expression of RNAs encoding storage proteins usually begins prior to diapause, expression may also continue during diapause, as exemplified by larval diapause in *Galleria mellonella* (Godlewski et al. 2001) and adult diapauses of *Leptinotarsa decemlineata* (de Kort and Koopmanschap 1994, Koopmanschap et al. 1995) and *A. grandis* (Lewis et al. 2002). (3) Expression during diapause in response to cold. In the rice stem borer *C. suppressalis*, a hexamerin that is more abundant in nondiapausing than diapausing larvae at 20°C suddenly becomes more abundant in diapausing larvae when they are transferred to 5°C (Sonoda et al. 2006). When diapause is terminated naturally or by application of JH (adults) or ecdysteroids (larvae and pupae), one of the first obvious events is a decline in abundance of hemolymph titers of the diapause-associated proteins.

Abundance of storage proteins closely associated with diapause offers a valuable **biomarker** for determining diapause status in some species. Diapause in adults of the bollworm *A. grandis* is notoriously difficult to determine, especially in field-collected adults, but a diapause-associated protein present in both males and females offers an accessible diagnostic tool (Lewis et al. 2002). Similarly, hemolymph proteins have proven valuable markers for larval diapause in the codling moth *Cydia pomonella* (Brown 1980), the tropical stem borer *Busseola fusca* (Osir et al. 1989), and the pink bollworm *Pectinophora gossypiella* (Salama and Miller 1992).

7.11   Weight Loss during Diapause

Insects in diapause do not feed or feed very little, thus they are reliant on energy reserves sequestered prior to diapause entry. As one might expect, an insect not replenishing its energy reserves will undergo major weight loss. This is most commonly recorded as a drop in wet weight, but it presumably reflects a drop in dry weight as well. Between the end of December and early April, diapausing larvae of the stemborer *Busseola fusca* lose approximately 16% of their initial weight (Usua 1970). Diapausing larvae of the tropical moth *Omphisa fuscidentalis* lose nearly half their weight between September and May (Singtripop et al. 1999). Larvae of the blow fly *Calliphora vicina* lose 30% of their original weight after 100 days in diapause (Saunders 2000). From February to June, diapausing larvae of the chestnut weevil *Curculio elephas* lose 29% of their original weight (Soula and Menu 2005). Diapausing pupae of the swallowtail butterflies *Papilio glaucus* lose 10% of their weight between December and April, and loss for *P. canadensis* during this same period is 15% (Scriber et al. 2012). Adults of the Sunn pest *Eurygaster integriceps* lose 25% of their initial mass between the onset of diapause in July until the end of February (Hasanvand et al. 2020). Weight of adult nymphalid butterflies, *Inachis io* and *Aglais uritcae* drops 17–18% during 180 days in diapause in the laboratory at 4°C (Pullin 1987). The benefits of low winter temperatures in conserving energy reserves are particularly striking in diapausing adults of the solitary bee *Osmia lignaria* (Sgolastra et al. 2010). Overwintering bees lose 10–15% of their initial body weight when held at 0–7°C or if kept outdoors for up to 200 days, but when maintained at 22°C they lose 40% of their body weight within 100 days.

7.12   Water Balance

Water is at a premium during diapause, especially for embryos and pupae that lack access to drinking water. Drinking is an option for larvae and adults, but water resources may be frozen during winter and thus not accessible. Insects that diapause in hot, dry tropical environments or during summer, or those entering prolonged diapause confront a particularly daunting challenge. Because insects are small they have a large body surface area compared to body mass, a feature that further exacerbates the challenge of maintaining water balance. But, insects have some innate advantages as well. They are remarkably tolerant of water loss. While mammals encounter significant injury from loss of just a small amount of water, many insects can tolerate losses of 20–30% of their body water (Hadley 1994). Some, such as the Antarctic midge *Belgica antarctica*, readily tolerate losses as great as 70% (Hayward et al. 2007), while others, such as the African sleeping midge *Polypedilum vanderplanki*, survive in the nearly waterless state of **anhydrobiosis** (Cornette and Kikawada 2011). Suppressed metabolism during insect diapause also works to the insect's advantage. A low metabolic rate, as discussed above, reduces energy demands, slows the respiratory loss of water, and facilitates conservation of water resources stored in the body. Yet, the challenge remains. Although some leeway is possible, for an insect to maintain water balance, loss must be countered by water gain.

Insects in diapause sometimes contain less water than their nondiapausing counterparts (Danks 2000), a feature that may facilitate cold tolerance by lowering the risk of mechanical injury from ice formation and increasing concentrations of cryoprotectants (Section 7.13.6). Diapausing larvae of the parasitoid *Nasonia vitripennis* contain 10–25% less water than nondiapausing larvae (Yoder et al. 1994, Li et al. 2014). In the mosquito *Culex pipiens* water mass is the same in both diapausing and nondiapausing females, but because dry mass of diapausing females is greater, the percentage of water in diapausing females is 12% lower than in nondiapausing females (Benoit and Denlinger 2007). But, in some species no differences are apparent. Water pools do not differ between diapausing and nondiapausing pupae of the flesh fly *Sarcophaga crassipalpis* (Yoder and Denlinger 1991) and the corn earworm *Helicoverpa zea* (Benoit et al. 2015), nor for pupae of the Hessian fly *Mayetiola destructor* that are not in diapause, in summer diapause or in an overwintering diapause (Benoit et al. 2010b).

Diapausing insects are more **resistant to desiccation** than their nondiapausing counterparts, as exemplified for adults of the mosquito *Culex pipiens* (Rinehart et al. 2006). While no difference is detected in a saturated environment (100% relative humidity), at relative humidities of 85% and lower, diapausing adult females consistently survive longer. Interestingly, the limit of dehydration tolerance does not differ between the two types of mosquitoes: Neither can tolerate losing more than 40% of their body water (Benoit and Denlinger 2007). The difference is that diapausing mosquitos have in place enhanced mechanisms to prevent the loss of so much water. Water loss rates through **transpiration** are usually lower in diapausing insects, as shown in Figure 7.9 for pupae of the flesh fly *Sarcophaga crassipalpis* (Yoder and Denlinger 1991). Similar results are noted for diapausing pupae of the Hessian fly *Mayetiola destructor* (Benoit et al. 2010b) and corn earworm *Helicoverpa zea* (Benoit et al. 2015), diapausing larvae of the parasitoid *Nasonia vitripennis* (Yoder et al. 1994), and for diapausing adults of the mosquito *Culex pipiens* (Benoit and Denlinger 2007).

Water loss rates plotted as a function of temperature sometimes provide another indicator of unique water balance properties operating during diapause (Figure 7.10). For most insects, regardless of diapause status, two distinct rates of water loss are

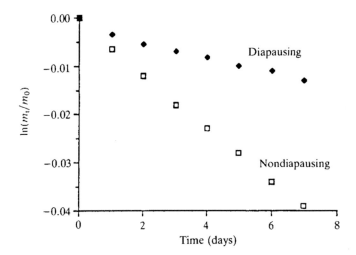

Figure 7.9 Net transpiration rates at 20°C in nondiapausing (open symbols) and diapausing (solid symbols) pupae of the flesh fly *Sarcophaga crassipalpis*. Semi-logarithmic plots of water mass ratio (m_t/m_0; mass at time t, m_t, divided by original mass at time 0, m_0 versus time describes transpiration rate as the negative slope of exponential decay.
From Yoder and Denlinger (1991).

detectable across a temperature range. The intercept of these two curves is referred to as the **Critical Transition Temperature** (CTT). As seen in Figure 7.10, the CTT for nondiapausing flesh fly pupae is 30.2°C, while for diapausing pupae the CTT is 39.1°C, a difference of 9°C (Yoder and Denlinger 1991). In the Hessian fly *Mayetiola destructor* the CTT is 26.3°C for nondiapausing flies and 31.7°C for diapausing pupae (Benoit et al. 2010b). The meaning of the CTT remains a bit controversial, but a conservative interpretation is that it reflects some sort of distinction in quality and/or quantity of cuticular hydrocarbons coating the integument. CTT differences are not always detectable between diapausing and nondiapausing individuals, *e.g.*, in *C. pipiens* CTTs for both types of adults are close to 41°C (Benoit and Denlinger 2007), yet diapausing adults of *C. pipiens* are clearly much more desiccation resistant. Though relatively few species have been examined, distinctions in water balance properties are likely a common theme of diapause and underscore the presence of special mechanisms employed during diapause for maintaining water balance.

The first line of defense against water loss is a **behavioral response**, selection of a protected diapause site, as discussed in Section 6.4. A site buffered from high winds reduces evaporative water loss, and a site with high humidity can ameliorate loss of water and may offer a source of drinking water. Diapausing adults of *C. pipiens* move within their underground sites to remain in a high humidity environment (Rinehart et al. 2006). Another behavioral trait used by some species to promote water conservation during diapause is to aggregate, as discussed in Section 6.5. A single adult of the tropical fungus beetle *Stenotarsus rotundus* loses water through transpiration

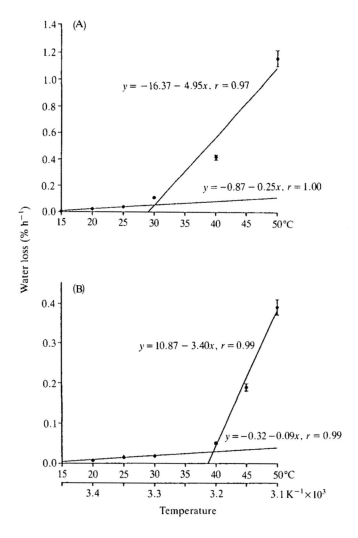

Figure 7.10 Arrhenius plot of water loss rates against temperature for (A) nondiapausing and (B) diapausing pupae of *Sarcophaga crassipalpis*. The intersection of the two lines is the critical transition temperature.
From Yoder and Denlinger (1991).

at a rate of 0.48%/hour, but the loss rate drops to less than half that amount (0.22%/hour) in aggregations of 250 diapausing beetles (Yoder et al. 1992b). An aggregation can be likened to a superorganism, reducing the surface area: volume ratio and hence reducing rates of water loss. The reduced metabolic rate noted in aggregating beetles also helps mitigate water loss. Clustering of diapausing eggs of the pierid butterfly *Neophasia terlooii* on living pine needles likely contributes to the eggs' high resistance to desiccation during winter (Halbritter 2020). More water, of course, is not always desirable in an overwintering site. Flesh fly larvae (genus *Sarcophaga*) seek dry sites for pupariation, thus presumably protecting vulnerable overwintering pupae from fungal and microbial attacks (Denlinger 1972b). But, selection of a dry environment calls for additional strategies to counter water loss.

On the plus side of the water balance ledger, many larvae and adult insects do indeed drink during diapause and thus can easily replace lost water when free water is available. Additionally, lipids can be utilized for the production of metabolic water. But, at least in diapausing pupae of the flesh fly *Sarcophaga crassipalpis* (Yoder and Denlinger 1991) and in diapausing adults of the mosquito *Culex pipiens* (Benoit and Denlinger 2007), metabolic water contributes little to this equation, at least in response to short bouts of dehydration. Although brief desiccation challenges may not generate metabolic water, long-term utilization of lipids during diapause can be expected to augment the water pool available during diapause.

Some diapausing individuals have an unusual ability to **absorb water vapor** from the atmosphere. At an R.H. of 85%, nondiapausing pupae of *S. crassipalpis* lose water (2% loss in 24 hours at 20°C), but diapausing pupae actually gain water under those same conditions (0.6%), suggesting a mechanism capable of absorbing water vapor from the air (Yoder and Denlinger 1991). Pupa must be alive for this to happen, suggesting an active process. Only when the entire puparium is coated with wax do diapausing pupae fail to absorb water from the atmosphere. Whether there is a specific site on the pupa (which resides within the puparium) responsible for water vapor uptake has not been determined. An even more striking case for water vapor uptake during diapause is noted during embryonic diapause in a tropical walking stick *Extatosoma tiaratum* (Yoder and Denlinger 1992). This species diapauses at two points during embryonic development: as an early embryo and then again as a pharate first-instar larva. Both stages are capable of absorbing water but the ability of young embryos to absorb water at an R.H. as low as 33% is particularly striking. A higher R.H. (60%) is required for water vapor absorption in diapausing pharate larvae. Uptake of water occurs over the entire surface of the chorion, as demonstrated by experimentally coating select portions of the chorion. As in *S. crassipalpis*, water vapor uptake does not occur in dead embryos, suggesting the possibility of an active process. The capacity for water vapor absorption is also evident in diapausing larvae of the parasitoid *Nasonia vitripennis* (Yoder et al. 1994), although no uptake is observed at an R.H. below 90%.

Water vapor absorption is certainly not universal for diapause. Diapausing adults of *C. pipiens* lose water at all humidities below saturation (Benoit and Denlinger 2007), as do diapausing pupae of *H. zea* (Benoit et al. 2015). Although failure to observe water uptake in dead individuals is often regarded as evidence for active uptake, such evidence is not conclusive. Passive sorption may very well be involved. A dead insect loses control of respiratory water loss, and in such cases gains from passive sorption are likely less than the loss of water through the respiratory system or integument. When considering competing fluxes of water gain and water loss, diapausing insects, compared to nondiapausing or dead individuals, possibly show water gain simply because they passively gain more than they lose. Relative contributions of active versus passive processes remain to be resolved, along with identification of sites of putative active uptake and the chemo-absorptive properties that possibly contribute to passive water gain.

Low metabolic demand during diapause translates into a **lower respiratory rate**, thus enabling the insect to capitalize on an obvious mechanism for enhancing water conservation during diapause. Some diapausing insects close their spiracles for long periods as discussed above (Section 7.5). The tracheal system is a major site of water loss, thus closing the spiracles offers a huge advantage for water retention.

The **integument** is the other major site of potential water loss. This route of loss is frequently diminished in diapause by enhanced waterproofing. Puparia of diapausing flesh flies (*S. crassipalpis*) are coated with nearly twice as much hydrocarbon as seen on puparia of nondiapausing flies (Yoder et al. 1992a). Interestingly, the additional hydrocarbons are packed on the inside of the puparium rather than on the exterior surface. Both types of puparia have an identical composition of saturated hydrocarbons with chain lengths of 25–33 carbons; the difference is simply that puparia of diapausing pupae contain nearly twice the amount of hydrocarbons seen in nondiapausing individuals (Yoder et al. 1995). Thus, the amount of hydrocarbons produced, rather than their composition, contributes to the lower transpiration rates in diapausing flesh fly pupae. The same is true for pupae of *Helicoverpa zea* and *Heliothis virescens*: Those in diapause have nearly double the amount of cuticular lipids seen in nondiapausing pupae, but the same components are present in both types and in roughly the same proportions (Buckner et al. 1996, Benoit et al. 2015).

Diapausing embryos of the Asian tiger mosquito *Aedes albopictus* are encased in a chorion coated with nearly 30% more hydrocarbons than found on the chorion of nondiapausing embryos, but composition does not differ (Urbanski et al. 2010b). Hydrocarbons extracted from both types of chorions contain the same 17 unsaturated hydrocarbons ranging in odd numbers from 19 to 51 carbons. The boost in hydrocarbon abundance nicely correlates with increased desiccation resistance in diapausing embryos and elevated expression of *fatty acyl coA elongase*, a gene encoding an enzyme involved in the synthesis of long-chain surface lipids known to mediate desiccation resistance in numerous insects. Compositions of cuticular lipids in diapausing and nondiapausing adults of the Colorado potato beetle *Leptinotarsa decemlineata* are also nearly identical (Yocum et al. 2011). Quantitative differences in abundance (twofold) are reported for diapausing, compared to nondiapausing, adults of *C. pipiens* (Benoit and Denlinger 2007), but lipid composition has not been investigated in this species.

Boosting abundance of **hydrocarbons** and other cuticular lipids, without changing composition, is a common response, but a few good examples show differences in composition as well. Hydrocarbon profiles from diapausing adults of the face fly *Musca autumnalis* differ considerably from hydrocarbon profiles of nondiapausing flies (Jurenka et al. 1998): Reproducing adults of both sexes have more alkenes and fewer methyl-branched alkanes than their diapausing counterparts. In *Drosophila montana*, diapausing adult cuticular hydrocarbons have chain lengths of C27 to C35, while chain lengths of C23 to C31 predominate in nondiapausing adults (Ala-Honkola et al. 2018). The long-chained alkanes and mono methyl-branched hydrocarbons on the cuticle of diapausing flies likely provide a more stable, protective barrier against desiccation. Concurrently, the shorter hydrocarbons coating nondiapausing adults likely function as volatile compounds used for recognition and elicitation of the mating response. Reproductively active males are attracted only to nondiapausing females producing short-chain hydrocarbons, further suggesting that the smaller hydrocarbons are involved as signaling compounds, while the long-chain hydrocarbons, found only on diapausing flies, function in desiccation resistance. Too few species have been examined to reliably predict a trend, but immature stages may rely more extensively on quantitative differences in cuticular lipids, while qualitative differences between diapausing and nondiapausing adults reflect distinctions in reproductive status and the exclusive reliance of nondiapausing flies on cuticular hydrocarbons as signaling compounds.

Cuticular proteins may also contribute to enhanced water retention. Some cuticular proteins are unique to diapause, as in *Culex pipiens* (Li and Denlinger 2009). The diapause-specific cuticular protein in *C. pipiens* contains an RR-2 motif frequently associated with rigid cuticles. Such proteins are candidates for desiccation resistance, but functional studies have not been completed. Transcripts encoding genes that produce constituents of the cuticle, as well as genes with chitin-binding activity, are overrepresented in numerous large diapause datasets, as discussed above under structural responses. A rich array of such transcripts and the proteins they encode suggest additional roles for the cuticle in desiccation tolerance. Increased melanization and thickening of the integument seen in some diapausing individuals (See Section 7.7) also likely improve water retention.

Other proteins, not necessarily associated with the cuticle, may contribute to desiccation resistance as well. Three diapause-associated transcripts are good biomarkers for diapause in the Colorado potato beetle *Leptinotarsa decemlineata* (Yocum 2003). One of the transcripts, DAT2, has high similarity to a desiccation-induced protein in the flour beetle *Tenebrio molitor*. Consistent with such a role is the fact that expression of DAT2 is elevated as body water content in diapausing beetles declines (Lehmann et al. 2014a). A diapause upregulated transcript in embryos of *Aedes albopictus*, *epithelial membrane protein* (EMP), also possibly contributes to desiccation resistance (Urbanski et al. 2010a). Though the correlation with desiccation stress is strong, how DAT2 and EMP function in response to diapause and desiccation is unknown. Two members of the *yellow* family are important contributors to egg desiccation resistance in *A. albopictus* (Noh et al. 2020), but whether expression of these genes differs between diapause and nondiapause is not known. **Heat shock proteins** (Hsps), discussed in Section 7.13.1, are best known for mitigating deleterious effects of heat and cold, but they too can be elicited by desiccation stress, as in *Sarcophaga bullata* (Tammariello et al. 1999). Hsps, especially Hsp70, are elevated in response to desiccation in several mosquito species (Benoit et al. 2010a). RNAi directed against Hsp70 does not alter the mosquito's water content or water loss rates, but dehydration tolerance is lowered, rendering the insect more vulnerable to water loss. Antifreeze proteins, best known for their role in cold tolerance, are also produced during the summer in some species and possibly contribute to dehydration resistance, thus they too may be added to the list of potential contributors to this important physiological response (Duman et al. 2010). **LEA proteins** (Late Embryogenesis

Abundant), critical for desiccation resistance in plants, are known from one insect, the chironomid *Polypodium vanderplanki*, as well as from an insect relative, a collembolan *Onychiurus arcticus* (Kikawada et al. 2006, Clark et al. 2007). Future work may reveal the presence of LEAs in additional insects.

High levels of the sugar **trehalose and sugar alcohols** such as glycerol, sorbitol, ribitol, or mannitol are routinely abundant in hemolymph of diapausing insects. These compounds, well known for their role as cryoprotectants (Lee 2010), are less appreciated for their potential role in helping to resist damage from desiccation (Holmstrup et al. 2010). As water is lost, sugars and polyols can function to slow the dehydration rate as well as enhance the stability of proteins and cellular membranes. Exposure to desiccating conditions (0% R.H.) for as little as 1 hour elicits a boost in hemolymph concentrations of glycerol in the flesh fly *Sarcophaga bullata*, and low levels of glycerol injected into flies can reduce water loss rates twofold (Yoder et al. 2006).

The insect's **renal system** is another potential site of water loss. The mechanics of excretion can be altered by diapause, as seen when a saline solution is injected into adults of the mosquito *Culex pipiens*. In response to a water overload, nondiapausing females excrete urine at a rate nearly triple that observed for diapausing females (Yang et al. 2017b), indicating a lower capacity for diuresis during diapause. Two aquaporins, Eglp1 and Aqp12L, are differentially expressed during mosquito diapause; the differences are not evident in the excretory system (Malpighian tubules and hindgut) but instead in midgut, ovaries, and abdominal wall, suggesting that aquaporins play an important, but unknown, non-excretory role unique to diapause. Expression of AQP1 is elevated during larval diapause in the stemborer *Chilo suppressalis* (Lu et al. 2013a), and expression of several aquaporins in larvae of the bark beetle *Dendroctonus armandi* peak in midwinter (Fu et al. 2019), but why this is so is not clear.

None of the –omics studies have thus far directly examined large-scale responses to dehydration stress during diapause, but we might expect to see some of the same major shifts that have been observed in nondiapausing insects subjected to desiccation. Comparative transcriptomics reveals a few common themes (Teets et al. 2012): Upregulation of genes encoding Hsps, especially Hsp40, ubiquitin-mediated proteasome, and GTPase involved in membrane trafficking, and downregulation of numerous metabolic genes. Yet, numerous genes differentially expressed in one species do not show the same distinctions in others, suggesting a diverse array of physiological responses that may have evolved to combat desiccation stress across taxa.

Other contributors to desiccation resistance, including those of maternal origin, are also likely. Diapausing embryos of the band-legged ground cricket *Dianemobius nigrofasciatus* are clearly more resistant to desiccation than their nondiapausing counterparts (Goto et al. 2008). The mother provisions the diapausing eggs with an unknown factor that increases both cold and desiccation tolerance, but the obvious candidates do not seem to be involved. Females of the mite *Phytoseiulus persimilis* respond to relative humidity by altering size and composition of their eggs (Le Hesran et al. 2019): Females reared at low humidity lay drought-resistant eggs that are larger and contain more amino acids, sugar alcohols, and saturated hydrocarbons than drought-sensitive eggs laid by females reared at high humidity. Though the experiments with *P. persimilis* do not pertain to diapause, the fact that a maternal effect is impacting drought resistance of the progeny suggests the interesting unexplored possibility of maternal contributions to desiccation resistance in association with diapause as well. How the mother senses relative humidity and responds by altering the properties of the eggs she deposits poses an intriguing, but unanswered, question.

In any one species, a **suite of responses**, rather than a single approach, combine to meet the water challenges of diapause. Multiple strategies are evoked simultaneously. For example, water retention during adult diapause in *Culex pipiens* is enhanced by flight to a high humidity environment, large size of diapausing females (reduction of the surface: volume ratio, thus lowering cuticular water loss), elevated levels of lipids, decrease in metabolic rate, decrease in respiration rate, addition of waterproofing hydrocarbons to the cuticle, decrease in the rate of renal excretion, possible contributions from a unique cuticular protein, and a boost in levels of Hsps (Benoit and Denlinger 2007, Li and Denlinger 2009, Sim and Denlinger 2009, Yang et al. 2017b). Similarly, in *H. zea*, pupation in a protected underground environment, large size of diapausing pupae, high fat content, reduced metabolic rate, and layering of additional lipids onto the surface of the integument combine with other likely unidentified features to offer high desiccation tolerance to diapausing pupae (Benoit et al. 2015). Many contributors to dehydration resistance during diapause (*e.g.*. sugars, polyols, Hsps) offer the added advantage of offering cross-protection against cold injury and other forms of environmental stress.

7.13   Enhanced Stress Responses

Common to most diapauses is a heightened responsiveness to environmental stresses such as temperature extremes or desiccation, as well as biotic stresses imposed by bacteria, fungi, or other microbes. Many of these defense responses are turned on as a component of the diapause program, rather than being activated directly by the stressor itself. This implies that the metabolically depressed diapausing insect is preemptively prepared to confront stresses it may encounter during diapause without *de novo* activation of these responses. Most evidence for upregulation of defense mechanisms during diapause is derived from transcriptomic and/or proteomic studies showing elevation of multiple stress responses during diapause. Somewhat fewer studies have actually tested the efficacy of these defense systems.

7.13.1   Heat Shock Proteins

One of the most conspicuous stress responses associated with diapause is the expression of transcripts encoding heat shock proteins (Hsps), also known as **stress proteins** (Denlinger et al. 2001, King and MacRae 2015). This highly conserved protein family consists of molecular chaperones, distinguished by their molecular masses. The most highly studied members include a cluster of small Hsps (smHsps) with molecular masses in the range of 20–30 kDa, Hsp60, Hsp70, and Hsp90. In addition to family members that are heat-inducible, some family members, the cognates, are expressed constitutively under nonstressed conditions, *e.g.*, heat shock cognate 70 (Hsc70). The smHsps are distinguished not only by their smaller size but by the fact that they are ATP-independent, while the larger Hsps are ATP-dependent proteins. Hsps were first identified as responders to heat stress, hence the name heat shock proteins, but we now know that expression can be elicited by a range of environmental and abiotic stresses (Feder and Hofmann 1999). These chaperone proteins are involved in protein folding and localization and can be expected to protect the integrity of critical enzymes and other proteins during diapause as well as to target select proteins for degradation. As King and MacRae (2015) suggest, during diapause, when ATP concentrations are low, they may alternatively function in sequestering rather than folding proteins.

Following discovery of the upregulation of an smHsp and Hsp70 during pupal diapause in the flesh fly *Sarcophaga crassipalpis* (Flannagan et al. 1998, Yocum et al. 1998, Rinehart et al. 2000), a more comprehensive survey of *S. crassipalpis* (Rinehart et al. 2007) revealed upregulation of at least 8 Hsp family members during diapause, downregulation of one Hsp (Hsp90) and no change in the cognate Hsc70 (Rinehart et al. 2007). Upregulation begins at the very onset of diapause, continues throughout diapause and post-diapause quiescence, but within hours after development is reinitiated, expression is turned off (Figure 7.11) (Hayward et al. 2005, Rinehart et al. 2007). Similar upregulation of Hsps is noted in a range of species including different diapause stages, *e.g.*, embryos of gypsy moth *Lymantria dispar* (Rinehart et al. 2007), silk moth *Bombyx mori* (Moribe et al. 2010), and katydid *Paratlanticus ussuriensis* (Shim et al. 2013), larvae of European corn borer *Ostrinia nubilalis* (Rinehart et al. 2007), corn stalk borer *Sesamia nonagrioides* (Gkouvitsas et al. 2008, 2009a), stem borer *Chilo suppressalis* (Lu et al. 2013a), wheat blossom midge *Sitodiplosis mosellana* (Cheng et al. 2016), solitary bee *Megachile rotundata* (Yocum et al. 2005), parasitoid *Trichogramma dendrolimi* (Zhang et al. 2020), pupae of onion maggot *Delia antiqua* (Chen et al. 2005b), walnut husk maggot *Rhagoletis suavis* (Rinehart et al. 2007), apple maggot *R. pomonella* (Rinehart et al. 2007), Chinese citrus fly *Bactrocerca minax* (Dong et al. 2019), tobacco hornworm *Manduca sexta* (Rinehart et al. 2007), cotton bollworm *Helicoverpa armigera* (Bao and Xu 2011), and adults of Colorado potato beetle *Leptinotarsa decemlineata* (Yocum 2001, Lehmann et al. 2014), carpenter bee *Ceratina calcarata* (Durant et al. 2016), *Drosophila montana* (Salminen et al. 2015), *D. suzukii* (Zhai et al. 2019) and in the CNS of adult females of the migratory locust *Locusta migratoria* programmed to lay diapausing eggs (Jarwar et al. 2019). Although expression of Hsps during diapause is quite widespread, it is not universal. There is no evidence for upregulation of Hsps during larval diapause in the blow fly *Lucilia sericata* (Tachibana et al. 2005) nor in the adult diapauses of *Drosophila triauraria* (Goto et al. 1998) or *Culex pipiens* (Rinehart et al. 2006). While most work has focused on transcripts, the Hsps encoded by these transcripts are indeed transcribed as indicated by several proteomics studies, *e.g.*, *S. crassipalpis* (Li et al. 2007). Transcriptional evidence for Hsp involvement in diapause goes beyond the identification of Hsp transcripts and proteins.

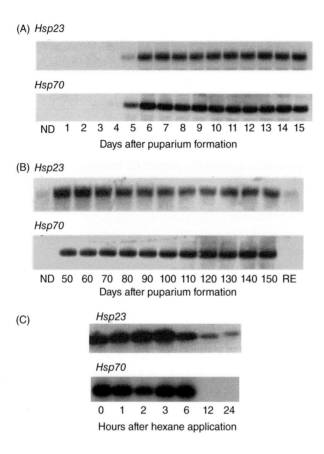

Figure 7.11 Expression patterns of heat shock proteins *Hsp23* and *Hsp70* in the flesh fly *Sarcophaga crassipalpis* show that Hsps are (A) turned on at the onset of pupal diapause, (B) continue to be expressed throughout diapause, and (C) are turned off within hours after diapause is terminated with an application of hexane.
From Rinehart et al. (2007).

In *S. crassipalpis*, transcripts encoding Hsf, DnaJ, and Hop, three proteins linked to Hsp70 function, are also elevated in diapausing pupae (Ragland et al. 2010).

The same Hsps are not consistently upregulated in all species. Reports of upregulation are most prevalent for Hsp70 and smHsps. By contrast Hsp90 is downregulated during diapause in several species, *e.g., Sarcophaga crassipalpis* (Rinehart and Denlinger 2000) and *Helicoverpa armigera* (Chen et al. 2010), but not in all, *e.g., Delia antiqua* (Chen et al. 2005b). Although Hsp90 is downregulated during diapause in *S. crassipalpis* it remains responsive to heat and cold shock: Both shocks evoke elevated expression (Rinehart and Denlinger 2000). Similarly, Hsp90 expression is boosted by environmental stress during diapause in *D. antiqua* (Chen et al. 2005b) and *Sesamia nonagrioides* (Gkouvitsas et al. 2009b). This is in contrast to the responses of Hsp70 and Hsp23 in *S. crassipalpis*: Both are maximally expressed during diapause, and further thermal shock fails to increase expression. Which Hsp is expressed may even differ within a single species, based on the season, as indicated in *Sitodiplosis mosellana* (Cheng et al. 2016). Hsp70 is highly expressed during summer diapause, while Hsp90 is more highly expressed during winter diapause. And, in this species, the transition from diapause to post-diapause quiescence is marked by elevation of Hsc70. Hsc70, however, shows no change in response to diapause or diapause termination in *S. crassipalpis* (Denlinger et al. 2001).

The above observations underscore several interesting features distinguishing the diapause Hsp response from the typical Hsp stress response. Upregulation of Hsps during diapause is programmed developmentally and occurs at temperatures that are quite moderate and insufficient to generate the typical stress-induced Hsp response. In the canonical heat shock response, Hsps are strongly expressed but synthesis of most other proteins ceases. By contrast, Hsps are synthesized concurrently with a whole suite of other proteins during diapause. This poses an intriguing question. Synthesis of Hsps is usually regarded as being incompatible with the synthesis of other proteins, but clearly the two occur simultaneously during diapause. In addition, stress such as heat shock typically prompts upregulation of all inducible Hsps, while the diapause Hsp profile shows selective upregulation of specific family members. For example, heat shock prompts simultaneous elevation of Hsp23, Hsp70, and Hsp90 in *S. crassipalpis*, but during diapause, only Hsp23 and Hsp70 are upregulated while Hsp90 is concurrently downregulated (Denlinger et al. 2001). Another oddity about the diapause Hsp response is that expression persists for such a long time, a period that may last 10 months or more, while the heat shock response is usually quite brief, quickly turned on when stress is presented and again quickly turned off when stress is removed.

Expression of Hsps during dormancy is not restricted to diapause in insects. Hsps are also strongly upregulated in forms of dormancies ranging from dauer formation in the nematode *Caenorhabditis elegans* (Cherkasova et al. 2000), encystment in the brine shrimp *Artemia francisana* (King and MacRae 2015, Rowarth and MacRae 2018a), quiescent stages of the freshwater sponge *Spongilla lacustris* (Schill et al. 2006), and mammalian hibernation (Eddy et al. 2005), as well as in seeds and other dormant plant tissues.

What functions do Hsps serve during diapause? With their multiple interactions with substrate proteins, they likely are vital components of the network maintaining cell homeostasis during diapause. When either Hsp23 or Hsp70 is suppressed by RNAi in diapausing pupae of *S. crassipalpis*, cold tolerance is dramatically lowered (Rinehart et al. 2007), suggesting Hsps are involved in maintaining the functional integrity of critical proteins needed at low temperature. Suppression of Hsps in flesh flies has no impact on either the decision to enter diapause or in determining diapause duration. Hsps may also provide a secondary mechanism guaranteeing that development will remain arrested. The presence of Hsps is usually incompatible with the progression of the cell cycle and consequent development, hence Hsps may also be helping to put the brakes on the cell cycle during diapause. Although Hsps do not appear to affect the diapause decision in *S. crassipalpis*, experiments with the brine shrimp *A. francisana* demonstrate an exciting role for Hsp40 in this crustacean's diapause (Rowarth and MacRae 2018b). When Hsp40 is knocked down, fewer than 50% of the nauplii enter diapause. This fascinating response provides the first example of a molecular chaperone that influences diapause entry.

In addition to Hsps, calcium-binding chaperones from the fat body may be implicated in diapause. Genes encoding two lectin-type chaperones with calcium affinity, calreticulin and calnexin, are differentially expressed in the fat body of the Colorado potato beetle *Leptinotarsa decemlineata* during adult diapause (Dogan et al. 2020). Calreticulin is downregulated during diapause in females, and both calreticulin and calnexin are elevated when diapause is broken. How they may contribute to the diapause response is unclear.

7.13.2   Innate Immunity

Immune responses continue to be fully operational during diapause. As demonstrated in the silk moth *Samia cynthia*, diapausing pupae readily respond to an immune system challenge (Nakamura et al. 2011). Injection of latex beads into diapausing pupae prompts **phagocytosis** and **encapsulation**, even at temperatures as low as 4°C. During the dormancy of *Drosophila melanogaster*, expression of immune genes *drosomycin* and *cecropin A1* are highly elevated (up to ninefold for *cecropin A1*) in response to injection of a cocktail of Gram-negative and Gram-positive bacteria (Kubrak et al. 2014). Diapausing larvae of the parasitoid *Nasonia vitripennis* appear to suppress overproduction of the symbiont *Arsenophonus nasoniae* by trapping the microbes in melanized nodules, a response that indicates an active immune response operating during diapause, but the presence of occasional necrotic larvae suggests that the immune response sometimes fails (Nadal-Jimenez et al. 2019). *Sarcotoxin II*, a member of the attacin family of antibacterial proteins, responds just as quickly (within 1h) to an immune challenge in diapausing pupae of the flesh fly *Sarcophga crassipalpis* as it does in nondiapausing pupae (Rinehart et al. 2003), but interestingly the response to physical injury of the body wall is a bit slower during diapause (6 vs. 3 hours). An enhanced immune response is sometimes noted not only during diapause but also in stages prior to and after completion of diapause: Larvae of the pierid butterfly *Pieris napi* destined for pupal diapause have higher phenoloxidase activity than their nondiapause-destined counterparts, and adults that have been through diapause show an enhanced encapsulation response (Prasai and Karlsson 2011).

Although a gene such as *sarcotoxin II* is not developmentally upregulated during diapause, a wide range of –omics studies show that many immune response genes are indeed diapause upregulated. Upon entry into diapause the **Toll pathway** is activated in adults of *D. melanogaster* (Kučerová et al. 2016), as evidenced by increased expression of the antifungal agent *drosomycin* and other *drosomysin-like* genes, canonical targets of Toll signaling. Additional targets of Toll signaling, *immune-induced molecule 1* and *immune-induced molecule 2*, are also strongly upregulated, as are several genes encoding peptidoglycan recognition proteins and Gram-negative bacteria-binding proteins. During larval diapause in the tropical bee *Tetrapedia diversipes* upregulated genes involved in the immune response include *toll, apidaecin,* and *peptidoglycan-recognition proteins SB1* (Santos et al. 2018). In diapausing pupae of the cotton bollworm *Helicoverpa armigera*, several proteins involved in the immune response are elevated: serine proteases, serpins, C-type lectins, and beta-1,3-glucan recognition protein (Zhang et al. 2013). C-lectins are also abundant in diapausing larvae of the aphid *Aphidius gifuensis*, as are the immune globulin hemocyanin and the mRNA encoding immune-related protein phenoloxidase (Zhang et al. 2018). Transcripts encoding Defensin 4 and a leucine repeat-rich protein are upregulated in diapausing adults of the sycamore lace bug, *Corythucha ciliata* (Li et al. 2017). In diapausing pupae of the flesh fly *Sarcophaga crassipalpis*, the immune signaling genes *cactus* and *dorsal* (an activator of the antimicrobial peptide diptericin), along with *defense*, an antimicrobial peptide, are elevated (Ragland et al. 2010). In queens of the bumble bee *Bombus terrestris* antimicrobial peptides become more abundant shortly after mating, and high levels persist throughout diapause (Colgan et al. 2019). Diapausing pupae of the onion maggot, *Delia antiqua*, express an abundance of *C-lectin, virus-induced protein dorsal,* and *dorsal-related immunity factor* (Hao et al. 2012). Similarly, transcripts involved in innate immune responses are among the top genes differentially expressed in diapausing pupae of the butterfly *Leptidea sinapis* (Leal et al. 2018).

Though numerous examples show elevated expression of genes encoding immune-related proteins or increased abundance of such proteins during diapause, the response is not universal. Several comprehensive –omics studies report no such boosts during diapause. Diapausing queens of the bumblebee *Bombus terrestris* are among those showing no elevation in immune responsive genes (Amsalem et al. 2015, but see Colgan et al. 2019). Two genes that might be expected to be elevated, *phenoloxidase subunit A3-like* (involved in wound healing) and *defensin* (antimicrobial peptide) are actually downregulated during diapause in *B. terrestris*. When the parasitic nematode *Sphaerularia bombi* infects a diapausing queen of *B. terrestris,* the nematode also enters a dormant state and relatively few changes of expression are noted within the queen, but within 6 hours after diapause is broken numerous immune genes, including members of the Toll signaling pathway, are upregulated (Colgan et al. 2020). In the Chinese citrus fly *Bactrocera minax* two genes linked to immunity, *Kunitz-type serine protease inhibitor* and a bacteriolytic lysozyme *LY21*, are downregulated, a response suggesting suppressed immunity during diapause in this species (Wang et al. 2019a). Rather than being diapause regulated, some immune-responsive genes are likely environmentally regulated in certain species, as suggested above for *sarcotoxin II* (Rinehart et al. 2003). And, within the same species, for example, *S. crassipalpis*, some immune-related genes may be elevated in response to the diapause program (*e.g., cactus, dorsal,* and *defense*), while others (*e.g., sarcotoxin II*) are elevated only in response to an immune challenge (Rinehart et al. 2003, Ragland et al. 2010). Levels of expression can also be responsive to environmental temperature (Ferguson et al. 2018a). In post-diapause quiescent larvae of the solitary bee *Megachile rotundata* immune-related transcripts are already high but are further elevated when bees are exposed to fluctuating temperatures (Torson et al. 2015). It thus appears that an elevated immune response is not consistently a component of the diapause response.

In some cases it is, but in other cases, the response is evoked only when the diapausing insect is under siege by bacterial or fungal pathogens or other environmental challenges.

Unique, but undefined, features in the milieu of diapausing stages sometimes exert dramatic effects on the microbes and parasitic nematodes they harbor. West Nile virus in the body of the female mosquito *Culex pipiens* (Nasci et al. 2001) and LaCrosse virus in embryos of *Ochlerotatus triseriatus* (Blitvich et al. 2001) cease to replicate during diapause, and eggs of the parasitic nematode *Heterotylenchus autumnalis* fail to hatch in diapausing adults of the face fly *Musca autumnalis* (Stoffolano 1967). Developmental regulation of this sort clearly offers a powerful way to synchronize the development of microbes and nematodes with the development of the host, but how this is achieved remains to be discovered.

7.13.3   Antioxidant Protection

Another stress response frequently elevated in diapausing insects is the antioxidant response, a response protecting against damage inflicted by **reactive oxygen species** (ROS), including superoxide anions, hydroxyl, alkoxyl, and peroxyl radicals, and hydrogen peroxide. Though low levels of ROS are important signaling molecules, in excess they can cause oxygen toxicity that potentially damages proteins, lipids, and nucleic acids (Dowling and Simmons 2009, Farooqui and Farooqui 2012). **Oxygen scavengers** (*e.g.*, glutathione and iron-binding proteins such as ferritin) and **antioxidant enzymes** (superoxide dismutase, glutathione peroxidase, glutathione reductase, and catalase) act together to protect against oxidative damage. Levels of **tocopherols** (Vitamin E), also recognized as antioxidants, are elevated in cell membranes during winter in diapausing adults of the linden bug *Pyrrhocoris apterus* (Koštál et al. 2013). Although less appreciated, carotenoids, hemocyanin, and other pigments also offer protection against ROS damage. Such biochemical defenses, combined with discontinuous gas exchange (Section 7.5), help mitigate injury against ROS during diapause.

There is still much we do not know about the aging process, but a recurring theory of aging purports that ROS damage is a major contributor to aging and ultimately to death. In this context, diapausing insects, by virtue of their longer lifespans, may be more vulnerable to accumulated oxidative damage than their short-lived nondiapausing counterparts. The protective mechanisms against ROS damage unique to many diapausing insects support the idea that an insect in diapause may help thwart the ravages of aging through such mechanisms.

Winter conditions of low temperature can also generate significant ROS damage, especially for insects that survive in the frozen state (Storey and Storey 2010). After a freeze/thaw event introduction of oxygen generates a burst of ROS generation that can trigger oxidative damage. Underground sites commonly used by diapausing insects are oxygen-limited, an effect exacerbated by periodic flooding that further reduces oxygen access (Woods and Lane 2016). **Anoxic bouts** are particularly stressful as insects shift between hypoxia to normoxia and are thus periodically exposed to a flood of oxygen and the accompanying boost in ROS production (López-Martínez and Hahn

2012, Ravn et al. 2019). The combination of a long dormancy, potential exposure to anoxia, and low-temperature damage can thus be expected to present a significant ROS challenge for many diapausing insects.

Manipulation of genes encoding the antioxidant enzymes **catalase** and **superoxidase dismutase** demonstrate the importance of these two protective enzymes in the overwintering adult diapause of the mosquito *Culex pipiens* (Sim and Denlinger 2011). Expression of both *catalase* and *superoxide dismutase-2* are considerably higher in young diapausing females than in their nondiapausing counterparts. Diapause specifically elevates *sod-2* but not the other *sod* transcripts present in *C. pipiens*. RNAi suppression of *catalase,* but not *sod-2*, results in increased damage to the ovaries, while suppression of either gene shortens the longevity of diapausing females. Both genes appear to be components of the pathway regulated by the forkhead transcription factor (*foxo*). In the migratory locust *Locusta migratoria* levels of both *catalase* and *sod* are lowered when one of the **peroxiredoxins**, *Prx6*, is knocked down, a result suggesting an upstream role for this peroxiredoxin in regulating the antioxidant response (Chen et al. 2020). Knockdown of *Prx6* also increases levels of FoxO phosphorylation and reduces the incidence of embryonic diapause in locusts, suggesting there may also be linkage with ROS and insulin signaling pathways. In diapausing pupae of the Tasar silk moth *Antheraea mylitta*, the fat body, a major site of metabolic activity, is especially vulnerable to oxidative damage as evidenced by accumulated tissue damage, as well as elevated levels of SOD, hydrogen peroxide, and glutathione (Sahoo et al. 2015), but no comparisons are available for fat body from nondiapausing stages.

Other evidence for elevated antioxidant responses during diapause is based on large transcriptomic or proteomic datasets. *Catalase* is elevated in diapausing adults of *Drosophila virilis* (Salminen et al. 2015), as is the protein catalase in diapausing pupae of the hawkmoth *Hyles euphorbia* (Stuckas et al. 2014). Both *catalase* and *Sod2* are elevated in dormant adults of *D. melanogaster* (Kučerová et al. 2016). Diapausing larvae of the mountain pine beetle *Dendroctonus ponderosae* have elevated levels of the antioxidant enzyme glutathione peroxidase as well as the oxygen scavenger **ferritin** (Bonnett et al. 2012). The protein ferritin or its transcript are elevated in diapausing copepods (Tarrant et al. 2008), diapausing larvae of the parasitoid *Nasonia vitripennis* (Wolschin and Gadau 2009), pupae of the flesh fly *Sarcophaga crassipalpis* (Rinehart et al. 2010), cotton bollworm *Helicoverpa armigera* (Bao and Xu 2011), onion maggot *Delia antiqua* (Ren et al. 2018), and adults of the rice water weevil *Lissorhoptrus oryzophilus* (Xinxin et al. 2020). In addition to ferritin, the antioxidant enzymes SOD and glutathione-s-transferase (GST) are more abundant during pupal diapause in *H. armigera* (Lu and Xu 2010). Expression of genes encoding peroxidase and GST are elevated during larval diapause in the solitary bee *Megachile rotundata* (Torson et al. 2015) and parasitoid *Trichogramma dendrolimi* (Zhang et al. 2020), as is the gene encoding GST in diapausing adults of *Drosophila suzukii* (Zhai et al. 2019) and the mosquito *Culex pipiens pallens* (Zhang et al. 2019). Seven antioxidant genes are upregulated during diapause preparation in embryonic diapause of the mosquito *Aedes albopictus* (Poelchau et al. 2013a), although none appear to be upregulated during diapause (Poelchau et al. 2013b). Hidden in supplementary tables is evidence for upregulation of several antioxidants during adult diapause of the beetle *Colaphellus bowringi* (Tan et al. 2017b) and larval diapause of the peach fruit moth *Carposina sasakii* (Cao et al. 2020).

Elevated antioxidant responses are not noted in all species. None of the antioxidant enzymes commonly associated with oxidative stress such as catalase, glutathione peroxidase, or superoxide dismutase are elevated during pupal diapause in *S. crassipalpis*, but levels of the oxygen scavenger ferritin and other metalloproteins are high, perhaps providing an alternative form of protection against oxidative stress (Ragland et al. 2010). No evidence of elevated ROS responses is seen during embryonic diapause in the silk moth *Bombyx mori* (Chen et al. 2017), diapause in larvae of the European corn borer *Ostrinia nubilalis* (Jovanović-Galović et al. 2007) or in adults of the Colorado potato beetle *Leptinotarsa decemlineata* (Krishnan et al. 2007), and in fact levels of some of the canonical antioxidant enzymes are even lower during diapause. Expression of genes encoding several key antioxidant enzymes (catalase, aldehyde dehydrogenase, GST) is upregulated during pupal overwintering in the spurge hawkmoth *Hyles euphorbiae*, but the response appears to be elicited by cold rather than the diapause program (Stuckas et al. 2014).

Among species that show elevated levels of antioxidants during diapause it is unclear whether the response is anticipatory of low oxygen or the consequence of an ROS signal generated by metabolic suppression. Both options would appear to be viable (Moreira et al. 2017), and as seen in so many diapause responses, there are quite likely species differences in this regard. In most cases, we cannot yet say with clarity which comes first, upregulation of the ROS response or metabolic depression acting as a trigger for the ROS response.

7.13.4   Defense against Toxins

Defenses against environmental and anthropogenic toxins are heightened in some diapausing insects but not all. The issue is complicated by the complex interactions between low winter temperatures, water availability, and bioavailability of pollutants as a function of the binding properties of toxins with environmental matrices (Holmstrup et al. 2010). Low metabolic activity and lack of feeding during diapause make a diapausing insect less vulnerable to toxins, but there are also examples of molecular mechanisms that presumably offer protection.

Most noticeable is elevation of a family of **mixed function oxidases** (cytochrome **P450s**) capable of oxidizing a wide range of endogenous substrates as well as xenobiotics. Transcripts encoding P450 are highly upregulated during embryonic diapause in the cricket *Allonemobius socius* (Reynolds and Hand 2009a), migratory locust *Locusta migratoria* (Tu et al. 2015), and silk moth *Antheraea yamamai* (Yang et al. 2008), in producers of diapause eggs in the silk moth *Bombyx mori* (Akitomo et al. 2017), in larval diapause of the parasitoid *Trichogramma dendrolimi* (Zhang et al. 2020), and during pupal diapause in the citrus fly *Bactrocerca minax* (Wang et al. 2019a). Among the 82 P450s annotated in the genome of the twospotted spider mite *Tetranychus urticae*, 28 are differentially expressed during adult diapause, the majority of which are diapause upregulated (Bryon et al. 2013). The transcript encoding aldehyde oxidase, an enzyme with high specificity for insecticide oxidation, is highly upregulated during adult diapause in the mosquito *Culex pipiens* (Robich et al. 2007), and the related enzyme aldehyde dehydrogenase, which is also involved in catalyzing oxidation of toxic compounds, is considerably more abundant in diapause of the aphid parasitoid *Praon volucre* (Colinet et al. 2012). Elevated selenoproteins and metalloproteins seen in several diapause examples, *e.g.*, pupae of *Sarcophaga crassipalpis* (Rinehart et al. 2010, Ragland et al. 2010) and adults of *C. pipiens* (Robich et al. 2007), can function not only in detoxification but also in the immune response. Though there are several good examples of elevated protection against toxins during diapause, this response appears to be not nearly as widespread as many of the other diapause-related stress responses.

In some cases, genes involved in cellular detoxification are actually downregulated. For example, *glutathione-S-transferase* and *UDP-glucuronosyl-transferase* are both downregulated during larval diapause in the tropical bee *Tetrapedia diversipes* (Santos et al. 2018) and the European corn borer *Ostrinia nubilalis* (Jovanović-Galović et al. 2007). This downregulation possibly reflects the fact that these diapausing larvae are no longer feeding and thus do not require protection from xenobiotics ingested with their food.

At the ecological level, exposure to insecticides may influence the propensity for diapause. Larvae of the leafroller *Choristoneura rosaceana* that diapause during summer under bark are presumably better protected and may escape exposure to insecticides (Carrière et al. 1995). This idea is supported by the observed selection for a higher summer diapause incidence in orchard populations exposed to insecticides. Similarly, pyrethroid applications to fields in Australia result in higher insecticide resistance in diapausing pupae of *Helicoverpa armigera*, a response that carries over to the following season (De Souza et al. 1995). A strain of *H. armigera* selected for resistance to *Bacillus thuringiensis* enters diapause more readily than an unselected, susceptible strain (Liang et al. 2007). Insecticide resistance can also influence diapause timing. Larvae of the codling moth *Cydia pomonella* homozygous for resistant alleles for P450 and glutathione-S-transferase show an altered critical photoperiod (Boivin et al. 2004). They respond to a longer critical daylength for diapause entry and hence enter diapause earlier than codling moths not homozygous for these traits. Thus, a number of interactions, some unexpected, link diapause and insecticide resistance.

But, the diapausing stage can also be highly vulnerable to insecticides, especially to applications received prior to diapause entry, as exemplified in overwintering bumble bee queens (Fauser et al. 2017). Subjecting the bumble bee *Bombus terrestris* to neonicotinoid insecticides prior to diapause not only increases overwintering mortality but also accelerates overwintering weight loss in queens.

7.13.5   Anoxia and Hypoxia

Winter habitats can be oxygen-limited, especially the underground sites favored by many species. Flooding after heavy rains and freezing of the substrate make the situation even worse. Likewise, aquatic habitats, high altitude sites, and habitats with decaying organic matter are confronted with significantly reduced oxygen availability (Hoback and Stanley 2001, Storey and Storey 2010, Cox and Gillis 2020). Days without an adequate oxygen supply present a physiological challenge. We know a fair amount about the effects of anoxia on metabolic pathways and the challenges posed for ATP, ion, pH, and water balance homeostasis, but most observations have focused on nondiapausing stages (Visser et al. 2018, Ravn et al. 2019). Though diapausing stages have received much less attention, we might anticipate higher resistance to anoxia damage in a diapausing insect since metabolism is already depressed and in many cases there is a switch from aerobic to anaerobic metabolic pathways. And, this expectation appears to be true. Diapausing larvae of the codling moth *Cydia pomonella* survive up to 42 days in 0.5% oxygen, but nondiapausing stages, eggs and adults, die after 1.25 hours in this low oxygen environment (Soderstrom et al. 1990). While nondiapausing pupae of the flesh fly *Sarcophaga crassipalpis* survive only one day of anoxia, diapausing pupae tolerate up to six days (Kukal et al. 1991).

Suppressed oxidative metabolism caused by anoxia induces production of the anaerobic metabolites glycerol and alanine in both diapausing and nondiapausing flesh fly pupae, but incorporation of labeled glucose is nearly twice as fast in nondiapausing pupae as in their diapausing counterparts (Kukal et al. 1991). Anoxia elicits synthesis of heat shock proteins (Hsps) in nondiapausing flesh flies (Michaud et al. 2011), and the fact that the same suite of Hsps synthesized in response to anoxia is normally synthesized in diapausing pupae (Section 7.13.1) suggests the possibility that Hsps present in diapausing flies offers preemptive protection against damage from anoxia. Anoxia elicits developmental stasis in nondiapausing flesh fly pupae, but this arrest is distinctly different from diapause: The arrest is permanent and cannot be broken with ecdysteroids, the hormonal trigger capable of initiating development in diapausing pupae. Interestingly, four days of anoxia exposure boosts the subsequent rate of oxygen consumption in diapausing flesh fly pupae and prompts diapause termination (Kukal et al. 1991). Anoxia also successfully terminates embryonic diapause in the false melon beetle *Atrachya menetriesi* if the eggs have already been chilled; unchilled eggs do not immediately terminate diapause, but diapause or post-diapause quiescence is shortened (Kidokoro and Ando 2006). Hypoxia treatment during post-diapause quiescence in larvae of the alfalfa leafcutting bee *Megachile rotundata* actually enhances survival to adulthood, but it exacts a toll on the adults, reducing feeding rates and shortening adult lifespan (Abdelrahman et al. 2014).

7.13.6   Relationship to Cold Hardiness

Low temperature is among the most significant environmental challenges confronting overwintering insects. As ectotherms insects are largely at the mercy of the environment, and temperatures that dip below the freezing point pose the potential for irreversible cell injury and death. As noted in a series of reviews (Denlinger and Lee 2010, Storey and Storey 2012, Teets and Denlinger 2013, Overgaard and MacMillan 2017, Toxopeus and Sinclair 2018), a suite of diverse mechanisms operates to mitigate cold injury (Figure 7.12). Many of the responses are noted during both the **seasonal cold hardening** that is associated with diapause as well as during **rapid cold hardening**, a cold hardening response that operates on a much shorter timescale (within minutes) and is seen in nondiapausing insects as well (Teets and Denlinger 2013). A few responses are restricted to one or the other form of cold hardening (*e.g.*, the heat shock response is well documented in seasonal cold hardening but not in rapid cold hardening). The pervasive impact of low temperature is highlighted by the fact that nearly one-third of the transcriptome and half the metabolome of *Drosophila melanogaster* are differentially regulated in response to cold (MacMillan et al. 2016).

A few insects tolerate freezing (**freeze-tolerant**), but most cannot (**freeze-intolerant or freeze-avoidant**). Diapausing insects are represented in both categories. For a freeze-intolerant species to survive a frozen environment, it must evoke mechanisms to avoid formation of extracellular ice. Commonly, this involves a boost in hemolymph osmolality, which concurrently lowers the body's **supercooling point** (SCP) to a temperature below the lowest temperature experienced in the insect's microhabitat. This is achieved by synthesizing polyhydric alcohols (glycerol, sorbitol, mannitol, ribitol), sugar alcohols (*myo*-inositol), sugars (trehalose, glucose), and/or amino acids (alanine, proline) that are pumped into the hemolymph and function colligatively to depress the SCP, in much the same way that antifreeze is added to an automobile's radiator to prevent freezing during winter. A select, but growing, group of insects also synthesize **antifreeze proteins** (AFPs) to enhance cold tolerance (Duman et al. 2010). AFPs, first reported in polar fish, function by binding in a non-colligative manner to ice crystals, retarding growth of the ice lattice and thus suppressing the SCP. Cryoprotectants and AFPs commonly lower SCPs to temperatures of -20 to $-30°C$ or lower (Turnock and Fields 2005). These additives enable many insects to tolerate winter temperatures down to the insect's SCP. SCPs frequently reflect the minimum temperature to which a population is exposed, *e.g.*, Figure 7.13 for the linden bug *Pyrrhocoris apterus* (Ditrich et al. 2018). But, the SCP is not consistently a reliable measure of an insect's lower lethal temperature. Many insects succumb at low temperatures well above the SCP (Lee 2010).

No single mechanisms or molecules are common to all freeze-tolerant species (Toxopeus and Sinclair 2018). Some mechanisms facilitate freezing through the use of **ice-nucleating proteins** or other **ice nucleators** (Ins) that promote ice formation at relatively high subzero temperatures. By organizing water into an ice-like structure, the embryo crystal, Ins promote freezing at high subzero temperatures, generating SCPs that are usually above $-10°C$. One possible advantage of this strategy is that ice formation reduces cuticular water loss in insects vulnerable to high rates of desiccation. A few freeze-tolerant species also produce proteins with ice-binding properties similar to AFPs, but how such proteins function in freeze-tolerant insects remains unknown (Duman et al. 2010). These proteins possibly serve an alternative function unrelated to cold tolerance. The fact that AFPs from the beetle *Dendroides canadensis* generate increased high-temperature survivorship in transgenic adults of *Drosophila melanogaster* suggests roles for AFPs that go beyond their well-documented low-temperature function (Vu et al. 2019).

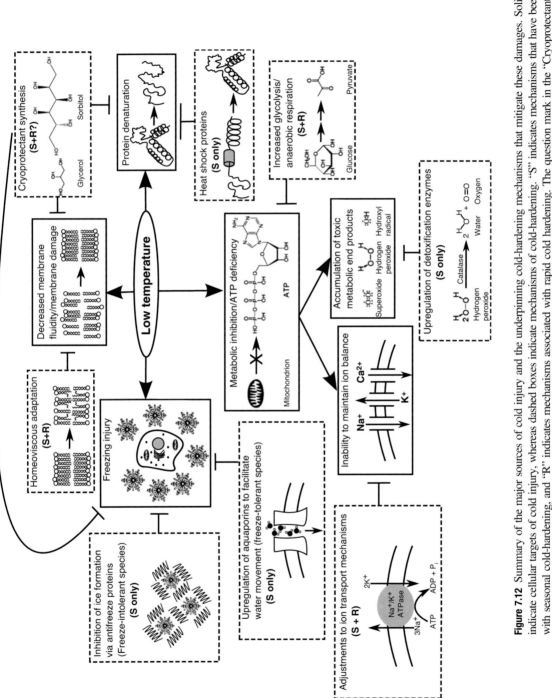

Figure 7.12 Summary of the major sources of cold injury and the underpinning cold-hardening mechanisms that mitigate these damages. Solid boxes indicate cellular targets of cold injury, whereas dashed boxes indicate mechanisms of cold-hardening. "S" indicates mechanisms that have been associated with seasonal cold-hardening, and "R" indicates mechanisms associated with rapid cold hardening. The question mark in the "Cryoprotectant Synthesis" box denotes that there is discrepancy over whether cryoprotectant synthesis is an essential component of rapid cold-hardening. From Teets and Denlinger (2013).

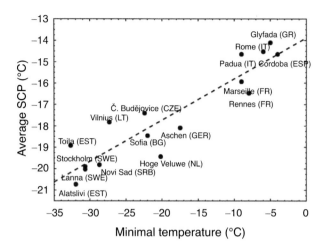

Figure 7.13 Supercooling points recorded for populations of the linden bug *Pyrrhocoris apterus* collected from various European localities with different climatic conditions (January–February temperatures).
From Ditrich et al. (2018), with permission from MDPI.

Distinctions between freeze-tolerant and freeze-intolerant can easily be made in dry laboratory conditions, but field conditions may blur the distinction. The linden bug *Pyrrhocoris apterus* is considered a classic freeze-intolerant species with a winter SCP in the range of −16 to −20°C, but in the humid leaf litter where it normally overwinters, it is highly susceptible to inoculative freezing by ice crystals at temperatures as high as −3°C (Rozsypal and Koštál 2018). Bugs inoculated through their body openings or across the cuticle survive freezing of up to 40% of their body water, thus showing that the simple dichotomy of freeze tolerance and intolerance may not adequately reflect the real world. The ability to survive some ice formation in the body is possibly widespread among species considered to be freeze-intolerant. While some freeze-intolerant species can block inoculative freezing (*e.g.*, the puparium of the flesh fly *Sarcophaga crassipalpis* effectively prevents transfer of external ice crystals to the pupa encased within [Kelty and Lee 2000]), many other species likely are unable to do so. Diapausing adults of the alder beetle *Agelastica alni* show remarkable flexibility and can be either freeze-tolerant or freeze-intolerant in northern Europe: Freeze-intolerant beetles have low SCPs (range of −12 to −15°C) and greater cold tolerance than freeze-tolerant beetles, which have SCPs in the −6 to −8°C range (Hiiesaar et al. 2018). Experimentally, freeze-intolerant larvae of *Drosophila melanogaster* can switch to being freeze-tolerant by feeding on a proline- or arginine-augmented diet (Koštál et al. 2012, 2016a), a demonstration that offers promise for exploiting this and similar techniques to generate freeze-tolerant insects for cryopreservation.

The rich literature on insect cold tolerance highlights not only cryoprotectants and proteins mentioned above, but also additional proteins including heat shock proteins (Section 7.13.1), aquaporins and dehydrins, as well as glycolipids and fatty acids. The cell membrane is another key site remodeled in response to cold (Koštál et al. 2003, Koštál 2010). Both phase and fluidity of the lipid bilayer are responsive to temperature, as manifested by a number of common alterations including increased desaturation, shortening of fatty acyl chain length, head group restructuring, and shifts in the glycerophospholipid/cholesterol ratio. Recent –omics approaches reveal numerous genes, proteins, and metabolites that are upregulated, phosphorylated, or differentially expressed in response to cold (Michaud and Denlinger 2010, Colinet et al. 2017), thus underscoring the breadth of impact elicited by low temperature.

But, the primary question of interest here is the relationship between diapause and cold hardiness. Many players and physiological processes are involved in generating both cold hardiness and diapause, but the literature is frequently vague about the relationship between the two. In many experiments, the studied insect is both in diapause and subjected to low temperature simultaneously, thus making it impossible to see which feature is responsible for the observed phenotype of enhanced cold tolerance. Are they distinct processes or is cold tolerance inherent to the diapause program? Examples of both possibilities exist in the literature: Cold tolerance can be either independent of diapause or a component of the diapause program. The relationship, as reviewed previously (Denlinger 1991), can be summarized briefly as follows. As shown in Figure 7.14, the two processes may be elicited independently or in association with each other, and when associated, the relationship may be either coincidental or linked.

Cold hardening without diapause is widespread. This is manifested in species that lack the capacity for diapause but are still capable of cold hardening. Cold hardening can also occur in nondiapausing stages of insects that do have the capacity for diapause, and the rapid cold hardening response that prevents cold shock injury is a

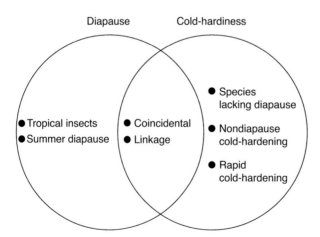

Figure 7.14 Relationship between diapause and cold hardiness. The two features may be expressed independently or in association with each other. When associated, the relationship may be coincidental or linked.
Adapted from Denlinger (1991).

ubiquitous response noted in all developmental stages and is not dependent on diapause. At the opposite end are tropical insects and those that diapause in summer: Diapause is evident but not associated with cold hardiness, although it is appropriate to note the caveat that cold hardiness has seldom been investigated in such species. At the center of the Venn diagram are species where the two are associated in time, but the association may be either coincidental or linked. I use the **linked** category to refer to cases where cold hardiness is a component of the diapause program. Inducing diapause results in elevated cold hardiness, even under warm conditions. This does not imply that diapausing cold-hardy insects cannot be made more cold-hardy by low temperatures but simply that cold hardening is already manifested by the induction of diapause. The **coincidental** category refers to cases where distinct environmental cues trigger the two events. For example, the insect may enter diapause in response to short daylength but not develop cold hardiness until exposed to a second trigger, low temperature.

Examples used previously to support these categories (Denlinger 1991, Hodková and Hodek 2004) are augmented here with several more recent examples distinguishing linked and coincidental responses. Diapause and cold tolerance appear to be linked in the butterfly *Pieris napi*, as evidenced by the enhanced cold hardiness and elevated sorbitol levels inherent to diapausing pupae (Lehmann et al. 2018). The programming of pupal diapause in the flesh fly *Sarcophaga crassipalpis* elicits production to two major cryoprotectants, glycerol and alanine, resulting in cold-hardy pupae that have not been exposed to low temperature (Michaud and Denlinger 2007). Likewise, in the Khapra beetle *Trogoderma granarium*, diapausing larvae are considerably more cold-tolerant than nondiapausing larvae exposed to the same low temperature (Wilches et al. 2017). Cold tolerance is a component of the nymphal diapause program in populations of the house spider *Achaearanea tepidariorum* from temperate latitudes but not in subtropical populations (Tanaka 1997). Diapausing larvae of the drosophilid *Chymomyza costata*, but not their nondiapausing counterparts, survive liquid nitrogen (Koštál et al. 2011), implying that the diapause program sets in place mechanisms for survival at this extremely low temperature. The critical cryoprotectant used by this species is proline, an amino acid that increases from 20 to 147 mM at diapause initiation. The sunn pest *Eurygaster maura* shows elevated expression of a gene encoding a putative AFP in late summer, during pre-diapause and early diapause, long before the adults migrate to their overwintering sites, suggesting that expression of this gene is a component of the diapause program (Guz et al. 2014). The boost in desaturated fatty acids, especially oleic acid, is a diapause-linked, cold-hardening response affecting cell membranes in pupae of the flesh fly *Sarcophaga crassipalpis* (Michaud and Denlinger 2006) and potato tuber moth *Phthorimaea operculella* (Hemmati et al. 2017). Linkage is also evident in expression of *Drosophila cold acclimation* and its paralog, *regucalcin*: The calcium-binding proteins encoded by these genes are associated with cold hardiness and are upregulated by the programming of adult diapause in *Drosophila montana* (Kankare et al. 2010, Vesala et al. 2012a) and other species in the *D. virilis* group (Salminen et al. 2015). The cryoprotectants glucose, trehalose, and proline are elevated in *D. montana* in the autumn, at the onset of diapause, but later, when winter conditions set in, there is a huge increase (400-fold) in *myo*-inositol (Vesala et al. 2012b), a result suggesting diapause-linked increases of certain cryoprotectants and then a further cold-hardening response (*myo*-inositol) triggered by the onset of low winter temperatures. Only diapausing pupae of the aphid parasitoid *Praon volucre* accumulate high levels of sorbitol, implicating the diapause program rather than low temperature as the trigger for synthesis of this cryoprotectant (Colinet et al. 2012). Fatty acid composition, an important feature for cold tolerance in diapausing larvae of the European corn borer *Ostrinia nubilalis*, is determined by the diapause program and is only modestly altered by cold acclimation (Vukašinović et al. 2017).

A coincidental relationship is suggested by numerous other examples. In the photoperiodically regulated adult diapause of the linden bug *P. apterus*, polyols ribitol and sorbitol begin to accumulate only in response to a low-temperature trigger, approximately 5°C (Koštál et al. 2004), thus implying that cold hardiness and diapause are coincidental in this species, but entering diapause is clearly an essential precondition for accumulation of the cryoprotectants. Likewise, in larvae of the rice stem borer *Chilo suppressalis*, the chief cryoprotectant, glycerol, is not accumulated in response to diapause itself but in response to the low temperatures that come with the onset of winter (Li et al. 2002b, Izumi et al. 2007, Lu et al. 2013a). The SCPs of the pentatomid bug *Halyomorpha halys* are high in early adult diapause but drop as diapause progresses, suggesting that cold tolerance is distinct from diapause but is acquired during the transition from autumn to winter (Cira et al. 2018). Cold hardening in diapausing larvae of the European corn borer *Ostrinia nubilalis* is distinct from entry into diapause and is attained only when diapausing larvae are subsequently exposed to low temperatures around 0°C, temperatures that activate the anaerobic enzymes needed to produce cryoprotectants (Uzelac et al. 2020).

It is not possible to distinguish between linked or coincidental responses in numerous other examples because the insects are both in diapause and simultaneously exposed to low temperatures, often under natural outdoor conditions, but such studies are helpful in revealing the types of cryoprotectants commonly used, results that underscore the diversity of cryoprotectants important for specific species. In the cotton bollworm *Helicoverpa armigera*, concentrations of trehalose, rather than polyols, are high in cold-acclimated diapausing pupae (Izumi et al. 2005). Diapausing embryos of the hemlock looper *Lambdina fiscellaria*, which have SCPs as low as −47°C in February, also appear to rely primarily on trehalose as their main cryoprotectant (Rochefort et al. 2011). Seasonal records for the crapemyrtle bark scale *Acanthococcus lagerstroemiae* show major winter elevation of the polyol mannitol (Wang et al. 2019b). Levels of glucose and myo-inositol are highly elevated in overwintering populations of the adult ladybird beetle *Hippodamia variegata* (Hamedi et al. 2013). In midwinter, glycerol levels are highly elevated in field populations of the bark beetle *Pityogenes chalcographus* (Koštál et al. 2014). In diapausing pupae of the fall webworm *Hyphantria cunea* the amino acid alanine peaks in midwinter (Li et al. 2001). Numerous −omics studies provide additional evidence for the presence of elevated levels of cryoprotectants during diapause.

For example, enzymes (or their transcripts) involved in glycerol synthesis are more abundant during larval diapause in the parasitoid *Nasonia vitripennis* (Wolschin and Gadau 2009) and mountain pine beetle *Dendroctonus ponderosae* (Bonnett et al. 2012, Fraser et al. 2017), and during adult diapause in the sycamore lace bug *Corythucha ciliata* (Li et al. 2017a). Genes encoding enzymes involved in trehalose synthesis are expressed more highly in diapausing larvae of the parasitoid *Aphidius gifuensis* (Zhang et al. 2018a), as are genes encoding enzymes involved in inositol synthesis in the twospotted spider mite *Tetranychus urticae* (Bryon et al. 2013). A cold tolerance candidate gene, *Frost,* shows increased expression in diapausing pupae of the flesh fly *Sarcophaga crassipalpis* (Ragland et al. 2010).

7.14   Does the Clock Stop?

A shutdown in clock gene expression is noted during hibernation in the European hamster (Revel et al. 2007), and in the liver (but not the hypothalamus) of the terai tree frog during hibernation (Borah et al. 2020). But, most research on clock cycling in relation to insect diapause has focused on the photosensitive stage prior to the onset of diapause, and little attention has been paid to the actual diapause stage. Yet, there is evidence that the clock shuts down during diapause in some species of invertebrates and expression patterns of clock genes may be altered in others. The clock shuts down during diapause in the copepod *Calanus finmarchicus*, a species that lives in cold water fjords of the North Atlantic. The canonical clock genes cycle during the active phase of development during the long photophases of summer but cease cycling in winter when fifth-stage juveniles migrate to deeper water and enter diapause (Hafker et al. 2018). Transcription of clock genes also appears to cease during embryonic diapause in the microcrustacean *Daphnia magna* (Schwarzenberger et al. 2020). In the linden bug *Pyrrhocoris apterus, cryptochrome* and *period* are abundant immediately after adult eclosion, but expression levels drop dramatically as the bugs enter diapause, remain low for the remaining 120 days of diapause, but rise again when the diapause is broken (Koštál et al. 2008). This same sort of shutdown in clock gene expression is thought to occur during pupal diapause in the flesh fly *Sarcophaga crassipalpis* (Short et al. 2016), but this could simply be a feature of the developmental stage of diapause because there is also no cycling observed in pupae that have not entered diapause.

One might assume that cycling would persist in species that retain some level of activity during diapause or in species that rely on a change in daylength to terminate diapause, and the scant evidence that exists on this topic suggests this is true. The mosquito *Culex pipiens* retains some activity during adult diapause and responds to long daylength for diapause termination, and in female brains the clock genes (*period, timeless, cycle,* cryptochrome2) continue to cycle throughout diapause (Meuti et al. 2015). Interestingly, the cycle of *timeless* expression is less robust during diapause in *C. pipiens*, and after three months, peak times of expression for both *period* and *timeless* are advanced considerably. The functional significance of such shifts remains unknown. Clock genes in diapausing eighth-instar larvae (55 days after egg hatch) of the moth *Sesamia nonagrioides*, a species that maintains some mobility during diapause, continue to cycle, but the oscillation peaks advance (Kontogiannatos et al. 2016, 2017), as noted in *C. pipiens*. Species that rely on photoperiodic cues for diapause termination, such as the linden bug *Pyrrhocoris apterus*, require a functional clock so they can respond to a switch from short days to diapause-terminating long days (Urbanová et al. 2016). The silk moth *Antheraea pernyi* is a species that breaks pupal diapause in response to long daylengths, and several clock genes, as well as the downstream gene encoding N-acetyltransferase, show cyclic patterns of expression at diapause termination (Mohamed et al. 2014), but earlier phases of diapause have not been rigorously examined. Is the clock functioning throughout diapause in such cases or is cycling simply initiated after some other diapause-terminating signal is received?

7.15   A Role for the Microbiome?

Only recently are we beginning to appreciate the enormous impact of microbiomes on physiological processes, behavior, and stress responses, and as an insect enters such a profoundly different state as diapause it seems likely that microbial composition could be impacted and possibly contribute to the diapause response. But, we are still in the beginning phase of recognizing a possible impact of the microbiome on diapause (Mushegian and Tougeron 2019). Does the microbial community survive unchanged during diapause, are components lost or gained, and do microbes remain active during diapause or do they also become dormant? Far too few cases have been examined to portray a complete picture, but several observations support the idea that the microbial community is reduced in diversity during diapause. This appears to be the case in the stink bug *Nezara viridula* (Medina et al. 2018) and the cockchafer *Melolontha hippocastani* (Arias-Cordero et al. 2012), as well as in the marine copepod *Calanus finmarchic* (Datta et al. 2018). Such differences could simply be the consequence of reduced feeding by the microbe's host during diapause.

Whether the microbial community continues to contribute to essential physiological functions during diapause has not been examined extensively, but several observations suggest that microbial activity is altered by diapause. The sycamore aphid *Drepanosiphum platanoidis* relies on the endosymbiont *Buchnera* to provide some of its essential amino acids, but during the aphid's reproductive diapause, such demands are reduced and symbiont activity is suppressed (Douglas 2000). Some microbes clearly remain critical for survival during diapause. During adult diapause the shield bug *Parastrachia japonensis* relies on *Erwinia*-like gut bacteria to produce uricase needed to convert uric acid into amino acids; without these bacteria the hosts do not survive (Kashima et al. 2006). In other cases, microbes may be providing antibiotic agents or other defensive agents that protect the insect during diapause. Such appears to be the case in the European beewolf *Philanthus triangulum*: Females culture *Streptomyces* bacteria within specialized antennal glands and then lace their brood cells with the bacteria prior to oviposition (Kaltenpoth et al. 2005). The bacteria are taken up by the larvae and incorporated into their overwintering cocoons to ward off fungal infections during diapause. In adults of *Drosophila melanogaster*, the presence of *Wolbachia* slightly decreases the likelihood of diapause (17% diapause in *Wolbachia*-infected females vs. 23% in uninfected females)(Erickson et al. 2020), but whether such a slight effect is biologically important is not clear.

Transovarial transmission, a common route for transferring microbes from mothers to offspring, can be impacted by diapause. *Regiella insecticola*, an intracellular symbiont of the pea aphid *Acyrthosiphon pisum*, is reliably transferred from partheno-genetic mothers to nondiapausing eggs, but this transfer does not always occur in sexually produced diapausing eggs (Moran and Dunbar 2006). And, in the parasitoid *Nasonia vitripennis* long cold storage of diapausing larvae can deplete the number of *Wolbachia* available for transovarial transmission (Perrot-Minnot et al. 1996).

A seasonal shift in the gut microbiome is noted in the field cricket *Gryllus veletis*, but in this case the *Wolbachia* titer actually increases during diapause, while abundance of *Pseudomonas* declines (Ferguson et al. 2018b). During diapause the cricket becomes more cold-tolerant, and although it retains the ability to clear bacteria from its hemolymph, the cricket becomes more vulnerable to fungal infections. Whether these features share a common basis with shifts in the microbiome remains a fascinating but unresolved question. Abundance of certain components of the gut microbiome in the adult cabbage beetle *Colaphellus bowringi* differ between diapause and non-diapause (Liu et al. 2016b): *Proteobacteria* and *Firmicutes* abundance is higher and *Bacteroidetes* is lower in diapause. Tantalizingly, these same positive and negative correlations are noted for human obesity, suggesting a possible contribution of the same microbes to the fat accumulation associated with diapause. The potential for involvement of the microbiome in executing a successful diapause and even for manipulating the diapause response is high and holds promise as an exciting focus for future research. Do the seasonal changes noted in the human gut microbiome (Smits 2017) foretell similar seasonal variation in insects?

7.16 Dynamics of Diapause

Though diapause has start and stop points, the events in between are far from static, and instead reflect a dynamic progression from establishing the developmental arrest (**diapause initiation**), to maintaining the arrested state (**diapause maintenance**), and eventually to the transition that ends diapause (**diapause termination**), as discussed in Chapter 1. In coining the term **diapause development**, Andrewartha (1952) conveyed the important concept that diapause represents a dynamic progression of subtle, often covert, developmental events that eventually culminate in bringing the arrest to an end. Equally important phases of post-diapause quiescence and resumption of development will be covered later in the discussion on diapause termination (Chapter 8). Thus, the emphasis here is strictly on the dynamics of the environmental and physiological responses that occur during diapause itself.

The dynamic nature of diapause is alluded to in many of the above descriptors of diapause that report changes in physiological and biochemical parameters during the course of diapause. At the onset of diapause, the diapause decision is rather labile and can be easily reversed, but during this initiation phase the diapause program becomes progressively more deeply entrenched and less vulnerable to termination by environmental or hormonal signals. Following diapause initiation, the insect enters the diapause maintenance phase, the long segment of diapause when the insect maintains the arrested state established during initiation. Diapause maintenance is a convenient catch-all term encompassing the interval between diapause initiation and termination, but it is the least well-defined phase and a bit artificial because it is not a static state but has phases that can clearly be distinguished, depending on the parameter observed. During this phase, diapause persists even if the insect is subjected to favorable environmental conditions. Finally, diapause termination refers to the phase when diapause comes to an end, allowing resumption of development. At this stage the insect is no longer refractory to reinitiating development and will resume development if presented the appropriate environmental conditions. As one might expect, there are species differences in how long various phases of diapause persist as well as differences in some of the characteristic changes that are noted with time.

Diapausing pupae of the flesh fly *Sarcophaga crassipalpis* illustrate the dynamic nature of diapause as witnessed by changes in their responses to exogenous hormones and stress. Four distinct stages of responsiveness are noted for the hormonal diapause terminator, 20-hydroxyecdysone (Denlinger et al. 1988b), as illustrated in Figure 7.15. In what is designated as Stage I, pupae are first highly sensitive to ecdysteroids but progressively become less sensitive by day 10. Stage II, extending from days 10–40, shows a reversal, a progressive increase in sensitivity. During Stage III, days 40–60, pupae become less sensitive to the ecdysteroid, as reflected in the increase in ED_{50}. In the final phase, Stage IV, pupae again become more sensitive to ecdysteroids as the end of diapause approaches. Responses to high-temperature stress (35°C) during diapause reflect similar phases (Denlinger et al. 1988b). During Stage I, the first 10 days of diapause, pupae become increasingly less susceptible to heat as a diapause terminator: On day 1 of diapause, a one-day exposure to 35°C is sufficient to avert diapause, but by day 10, 16 days are required to elicit the same response, suggesting that the diapause program is becoming progressively more entrenched during this time. The ET_{50} then progressively declines during Stage II, maintains a steady state during Stage III, and then declines further during Stage IV, at which point a 4 hour exposure to the high-temperature shock is sufficient to break diapause. Such variations in response are sometimes couched as variations in diapause intensity. One might anticipate a steady progression of declining intensity as diapause progresses (Hodek 2002), but this is not always the case, as exemplified in flesh flies.

Other diapause parameters also change throughout the diapause maintenance phase in the flesh fly. Lipids are used as the primary energy source during Stages I and II, but pupae switch to protein and carbohydrate usage during Stages III and IV (Adedokun and Denlinger 1985). Overall metabolic rate declines during Stages I and II, but gradually increases during Stages III and IV, a feature reflected in Figure 7.7, showing progressive elongation of O_2 consumption cycles toward mid-diapause and a shortening of the cycles toward the end of diapause (Denlinger et al. 1972). Cold tolerance increases during Stage I, remains stable during Stage II and progressively declines during Stages III and IV (Lee et al. 1987b). Concentrations of the cryoprotectant glycerol increase during Stages I and II, and then decrease during Stages III and IV (Lee et al. 1987b).

Changes in sensitivity to ecdysteroids for breaking diapause are also evident in pupae of the Bertha armyworm, *Mamestra configurata* (Bodnaryk 1977), but in this case, the ED_{50} drops during the first few days in diapause, increases over the next 30 days, remains constant for the next 40 days, and then declines toward the end of diapause. Experiments with diapausing pupae of the silk moth *Hyalophora cecropia* (Waldbauer et al. 1978) and the Australian bollworm *Heliothis punctiger* (Browning 1981) also show age-related changes in ecdysteroid sensitivity. Such changes in ecdysteroid sensitivity during pupal diapause are not universal: The ED_{50} remains remarkably constant throughout diapause in the tobacco hornworm *Manduca sexta* (Bradfield and Denlinger 1980).

During adult diapause in the linden bug *Pyrrhocoris apterus*, diapause intensity, measured as the time to initiate oviposition, increases at the beginning of diapause, reaching a peak at 31 days, gradually declines until 60–90 days, and then stabilizes at a steady state through day 180 (Koštál et al. 2008). Early in diapause the bugs are refractory to chilling as a stimulus for diapause termination, but not during later phases of diapause maintenance. Although the mechanism for this switch is not known, it has the obvious adaptive advantage of preventing premature egg maturation. Transcript levels of genes encoding aldose reductase and sorbitol dehydrogenase closely track changes in diapause intensity, thus indicating that certain transcript expression levels can be effective markers for stages of diapause maintenance. In *P. apterus* metabolic rate drops at the very beginning of diapause but then remains fairly constant for the remainder of diapause. The initiation phase of diapause in this species is a time of intense energy storage, increased polyol accumulation, and increased expression of clock genes *period* and *cryptochrome*, but also a time of diminished locomotion, feeding, and drinking, and a decrease in abundance of transcripts encoding rate-limiting enzymes involved in energy metabolism (Hodková et al. 2003, Koštál et al. 2004, 2008). In concordance with a number of other species, diapause initiation is a period of pronounced change as *P. apterus* switches from an active mode to the diapause program.

Dynamic changes in adult diapause of the solitary bee *Osmia lignaria*, reflected in changes of respiration rate, can be interpreted as changes in diapause intensity (Sgolastra et al. 2010). Diapause intensity increases (*i.e.*, respiration rate drops) during the first 30 days of diapause, is maintained at a plateau for the next 30 days, declines and then persists at another lower plateau for 40 days, before entering the period of post-diapause quiescence when exposure to high temperatures can reinitiate development. Changes in RQ values also systematically accompany changes in respiration rate.

Phases of diapause are sometimes reflected in overt developmental changes, as noted for males of the leaf beetle *Gastrophysa atrocyanea* (Ojima et al. 2015). In this

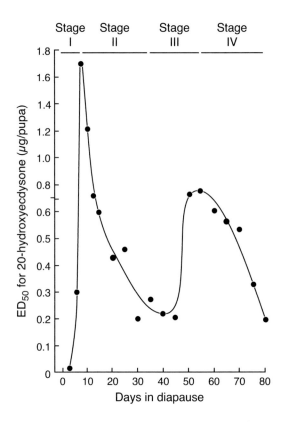

Figure 7.15 Developmental changes in sensitivity to 20-hydroxyecdysone for pupal diapause termination in the flesh fly *Sarcophaga crassipalpis*. Four distinct phases of sensitivity are noted.

From Denlinger et al. (1988b).

species the testes increase in volume during the first 60–70 days of diapause at 15 or 20°C, then decline until day 120 and remain at a constant size thereafter. Timing of peak size depends on temperature and is considerably later for beetles held at 25°C. As the testes shrink, spermatogenesis ceases, and sperm and their precursors are resorbed.

Transcript profiles are likely to change systematically throughout the course of diapause, but most diapause transcriptomic studies fail to reveal this because most are based on a single time point during diapause. Yet, several targeted gene studies have examined expression patterns of specific genes throughout diapause, and such studies underscore the dynamic nature of diapause (Denlinger 2002). In the example of the flesh fly *Sarcophaga crassipalpis*, illustrated in Figure 7.16 transcripts encoding certain genes are downregulated throughout diapause (PCNA, Hsp90), others are upregulated throughout (Hsp23, Hsp70, EcR), some are upregulated only in early diapause (pScD41), others are upregulated only in late diapause (USP), and still others (PO) are intermittently expressed in phase with oxygen consumption cycles in these flies.

In the mosquito *Culex pipiens*, targeted genes related to fat metabolism, examined at three time points, show high expression of genes related to fat synthesis (*e.g., fas*-1) in early diapause and concurrent suppression of genes related to fat metabolism (Sim and Denlinger 2009), while in late diapause, genes involved in fat synthesis are turned off and those involved in β-oxidation and other energy-generating steps are highly upregulated. In diapausing pharate larvae of the Asian tiger mosquito *Aedes albopictus*, RNA-seq, performed at 11, 21, and 40 days postoviposition (Poelchau et al. 2013b), shows the greatest difference between diapause and nondiapause at the earliest time point (day 11); thereafter patterns of gene expression in the two types of pharate larvae gradually converge as the end of diapause approaches. Functionally, early strong differences in genes related to metabolism gradually diminish over time, leaving only genes related to lipid metabolism as being distinct in late diapause.

The transition from diapause initiation to diapause maintenance has been tracked in a few species. In the Colorado potato beetle *Leptinotarsa decemlineata* this transition occurs approximately two weeks after adult eclosion, as indicated by a marked decline in metabolic rate, cessation of feeding and limited locomotion (Yocum et al. 2009b). The transition is accompanied by an abrupt shutdown in the expression of several transcripts, including genes with functions associated with protein synthesis, structural and metabolic functions, a retinol-binding protein, as well as genes encoding proteins of unknown function (Yocum et al. 2009a, 2009b). The fact that expression of these unidentified genes ceases within two weeks after adult emergence implies that their function is restricted to diapause initiation.

The phase transition between diapause maintenance and termination is of special interest because it is central to the timing of spring emergence in many species. PCA analysis of the metabolome of the butterfly *Pieris napi* reveals numerous differences distinguishing diapause maintenance from diapause termination (Lehmann et al. 2018). Pupal diapause in this species requires a cold signal for diapause termination, and several metabolites, including sorbitol, mannitol, and alanine, accumulate only when this diapause-terminating signal is present. In the solitary bee *Megachile rotundata* the phase transition from larval diapause maintenance to termination is marked by pronounced downregulation of the heat shock protein transcript *hsp70* (Yocum et al. 2005), however, in pupae of the flesh fly expression of genes encoding heat shock proteins persist beyond diapause termination, *i.e.*, expression continues through post-diapause quiescence and ceases only when development resumes (Hayward et al. 2005). The late diapause elevation in the expression of *Ultraspiracle*, seen in pupae of *Sarcophaga crassipalpis*, could play a critical role in the transition to diapause termination (Rinehart et al. 2001). Though the gene encoding Ecdysone Receptor (EcR), the dimerization partner of USP, is expressed throughout diapause, *Usp* is upregulated only near the end of diapause, a response that provides the diapausing pupa with a functional receptor for responding to ecdysteroids at the end of diapause. Expression patterns that change at diapause termination are good candidates for participating in signaling cascades essential for diapause termination.

A few custom array studies examine differences in all three major phases of diapause: initiation, maintenance, and termination. During diapause initiation in

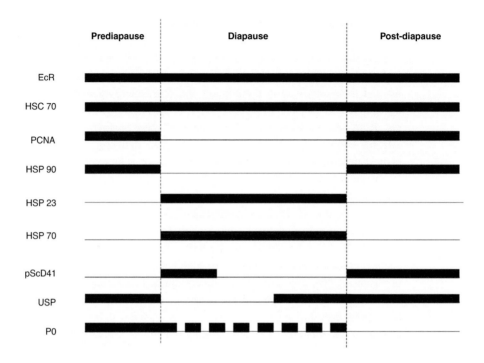

Figure 7.16 Expression patterns of select genes associated with different phases of pupal diapause in the flesh fly *Sarcophaga crassipalpis*.
From Denlinger (2002).

Drosophila montana, several genes, including those related to photoinduction, lipid metabolism, *couch potato*, and several cytoskeletal genes, are uniquely upregulated (Salminen et al. 2015). Genes uniquely expressed during the maintenance phase include those encoding ribosomal proteins, heat shock proteins, biogenic amines, and several clock proteins. During the diapause termination phase, the highest expression differences are the heat shock *Hsc70* gene and several related to phototransduction. Although many gene expression patterns are common to all three diapause phases, numerous distinctions in gene expression support the categorization of diapause into distinct phases. Another comprehensive study focuses on larval diapause in the drosophilid fly *Chymomyza costata* (Koštál et al. 2017). Armed with a custom microarray representing 1,042 genes, these authors document patterns of gene expression associated with distinct phases of diapause. PCA analyses clearly distinguish three phases of diapause. In these flies, the photosensitive period of diapause programming immediately precedes the initiation of diapause, and a gradual progression is noted between these two phases. Upon entering the maintenance phase, very few additional differences are noted early in the maintenance phase (only 0.7% differentially expressed sequences, DE), but the number of DE sequences increases markedly by the second month of diapause (6.3%), again reflecting the fact that the diapause maintenance phase is not static. Diapause in *C. costata* can be terminated by long

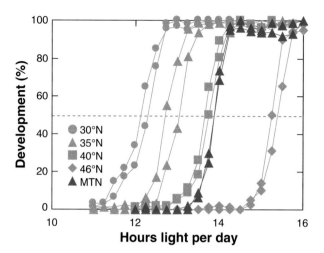

Figure 5.4 Geographic variation in photoperiodic response for larval diapause in populations of the pitcher plant mosquito *Wyeomyia smithii* from different localities in eastern North America. MTN (red) is a mountain population collected at 900–1,000 m, 35° N. Note the MTN shift in response from that noted in a low altitude population from the same latitude. Critical daylength is indicated by dashed line. (A black and white version of this figure will appear in some formats.)

From Bradshaw and Hozapfel (2010a), with permission from Annual Reviews.

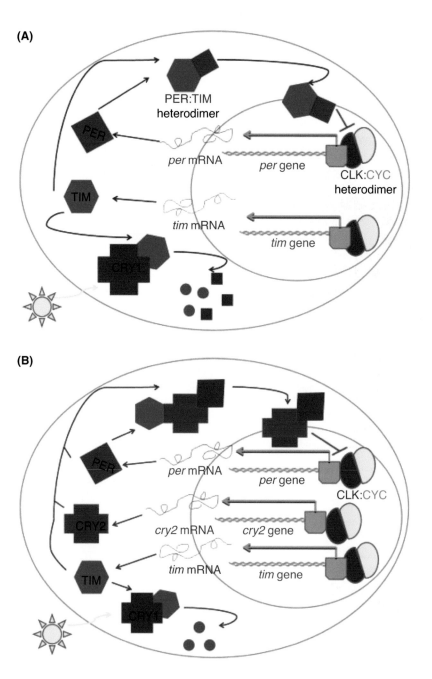

Figure 5.13 The circadian clock model for (A) *Drosophila melanogaster* and (B) most other insects. In the nucleus of *D. melanogaster* circadian clock cells, CLOCK (CLK) and CYCLE (CYC) proteins form a heterodimer that acts as a transcription activator by binding to the E-box promoter region of the *period* (*per*) and timeless (*tim*) genes. *per* and *tim* mRNA are translated in the cytoplasm of the cell. PER and TIM proteins then form a heterodimer and translocate back into the nucleus where PER inhibits the action of CLK:CYC, thereby suppressing the transcription of *per* and *tim*. CRYPTOCHROME1 (CRY1) protein degrades TIM and itself in the presence of light. This results in increasing levels of *per* and *tim* mRNA throughout the day when CLK:CYC activity is uninhibited, and decreasing levels of *per* and *tim* mRNA during the night. (B) Circadian clocks in most other insects differs in that they possess a light-insensitive CRY2 protein that acts as the major negative transcriptional regulator of the core circadian clock. In this case, PER appears to assist CRY2 in nuclear translocation, whereas TIM helps to stabilize PER and CRY2. (A black and white version of this figure will appear in some formats.)
From Meuti and Denlinger (2013).

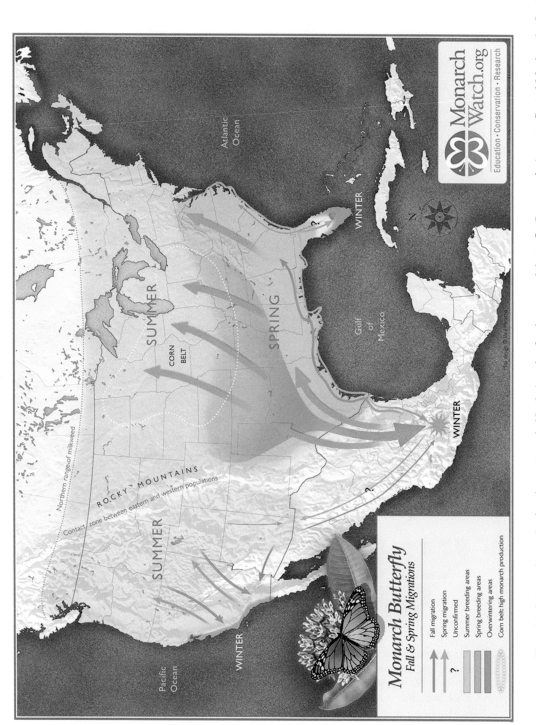

Figure 6.3 Autumn (orange) migration patterns in North America for the monarch butterfly *Danaus plexippus* to Central Mexico, the Southern California coast, and southern Florida, and the spring (green) northward return. (A black and white version of this figure will appear in some formats.)

Courtesy of Monarch Watch (monarchwatch.org).

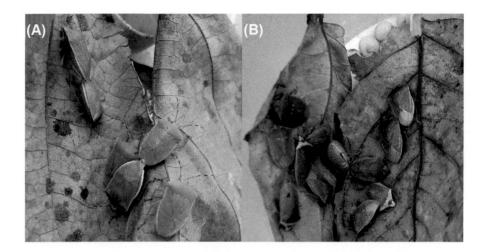

Figure 6.5 Diapause is associated with a color change in the stink bug *Nezara viridula*. (A) Green adults are not in diapause and (B) brown adults are in diapause. (A black and white version of this figure will appear in some formats.)
Photographs courtesy of Dmitry Musolin (St. Petersburg State University).

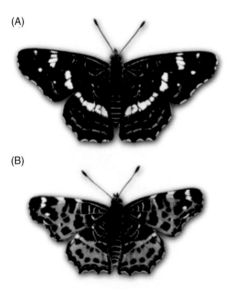

Figure 6.6 Seasonal polyphenism in the butterfly *Araschnia levana*. (A) Adults from a summer generation that have not experienced pupal diapause emerge with black and white coloration. (B) Those emerging in spring after overwintering in pupal diapause are predominantly orange. (A black and white version of this figure will appear in some formats.)
From Morehouse et al. (2013), with permission from John Wiley & Sons.

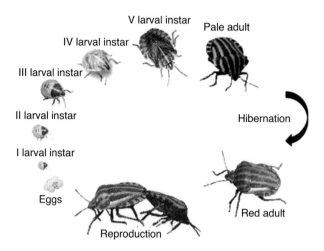

Figure 6.7 Life cycle of the striated stink bug *Graphosoma lineatum* showing the transition from a pale and black striated adult that enters diapause in the autumn to a red and black adult that breaks out of diapause in the spring. (A black and white version of this figure will appear in some formats.)

From Johansen et al. (2010), with permission from John Wiley & Sons.

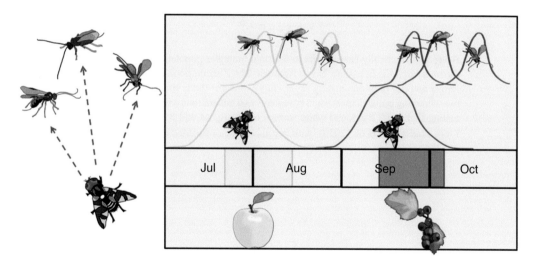

Figure 8.1 Allochronic speciation in *Rhagoletis* fruit flies driven by differences in seasonal phenology of the ancestral hawthorn host plant (red) and introduced domestic apples (yellow). Radiation of the univoltine flies onto apples also created a new temporal niche allowing sequential speciation of a community of univoltine parasitoid wasps that attack the flies. Both the flies and the parasitoids have shifted timing of their diapause to synchronize hosts and consumers across trophic levels, driving genetic divergence by allochronic isolation. (A black and white version of this figure will appear in some formats.)

From Denlinger et al. (2017).

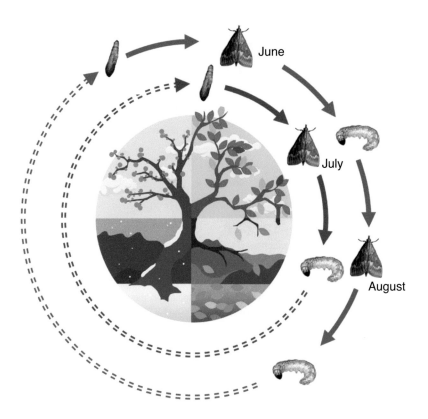

Figure 8.2 Genetically based shifts in seasonal timing in populations of the European corn borer *Ostrinia nubilalis* in eastern North America. The Z strain population (red) remains in diapause longer and has only one mating flight (July), whereas the E strain (blue) has a shorter overwintering pupal diapause and completes two generations with early (June) and late (August) mating flights. (A black and white version of this figure will appear in some formats.) From Wadsworth et al. (2020), with permission from BioMed Central.

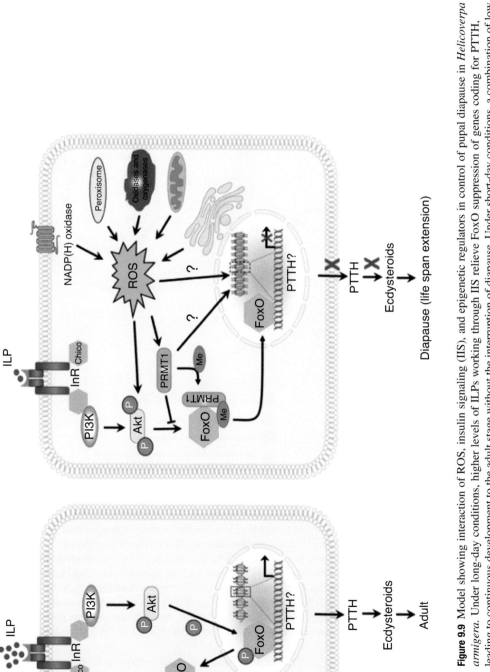

Figure 9.9 Model showing interaction of ROS, insulin signaling (IIS), and epigenetic regulators in control of pupal diapause in *Helicoverpa armigera*. Under long-day conditions, higher levels of ILPs working through IIS relieve FoxO suppression of genes coding for PTTH, leading to continuous development to the adult stage without the interruption of diapause. Under short-day conditions, a combination of low ILP levels, higher levels of ROS and PRMT1 block FoxO phosphorylation, promoting FoxO nuclear localization and suppression of genes coding for PTTH, resulting in diapause. (A black and white version of this figure will appear in some formats.) From Palli (2017) based on Zhang et al. (2017b), with permission from National Academy of Sciences, USA.

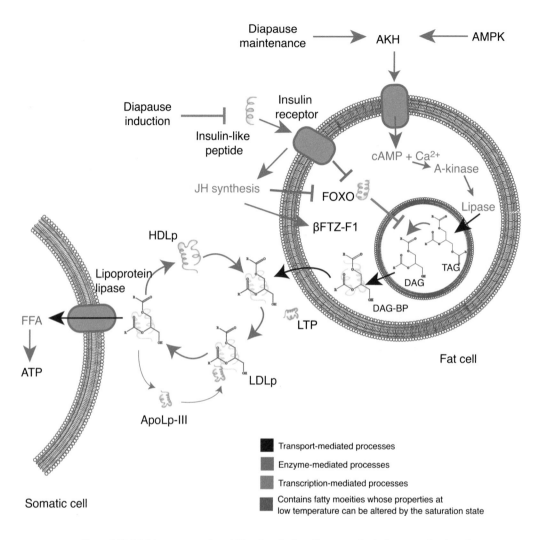

Figure 9.11 Lipid storage and mobilization during diapause, depicting contributions by adipokinetic hormone (AKH) and insulin-like peptides (ILPs). Diapause induction involves inhibition of ILPs, which removes the inhibiting effect of the insulin receptor on FoxO, allowing accumulation of lipids. During diapause maintenance, AKH is produced in response to accumulation of AMP-activated protein kinase (AMPK), resulting in stimulated production of diacylglycerol (DAG) from triacylglycerol (TAG) via a cAMP and Ca^{2+} signaling cascade. DAG is then exported from the lipid droplet into the hemolymph by binding to high-density lipophorin (HDLp) with the help of a lipid transport particle (LTP), forming low-density lipophorin (LDLp). ApopLp-III, apolipophorin 3; DAG-BP, diacylglycerol binding protein; FFA, free fatty acid; JH, juvenile hormone. (A black and white version of this figure will appear in some formats.)

From Sinclair and Marshall (2018), with permission from The Company of Biologists.

daylengths, and the photoperiodic response is accompanied by another boost in the number of DE sequences (8.2%). Such differences again reinforce the idea that specific patterns of gene expression are associated with distinct phases of diapause.

The functional basis for many of the changing responses during diapause is not always clear, but they underscore the dynamic nature of diapause and convey the concept that an insect in diapause does not simply stop developing and then start again at some later point. Many processes are underway during diapause, ultimately leading to resumption of a developmental trajectory similar to that seen in their nondiapausing counterparts. Deciphering and integrating the developmental processes and signaling pathways that foster progression through diapause remains an exciting challenge in diapause research.

8 Ending Diapause and Reinitiating Development

Ending diapause, remaining in the holding pattern of post-diapause quiescence, and resuming development are three distinct developmental landmarks. The **end of diapause** is marked by the insect's capacity to resume development if conditions are favorable. Prior to that time development is arrested and cannot yet be reinitiated even under favorable conditions. This represents a period of fixed latency, a refractory period that, for winter diapause, prevents the overwintering insect from developing in autumn when temperatures would be permissive.

Once diapause has ended, most insects enter **post-diapause quiescence**, a phase in which the insect is capable of responding to favorable environmental conditions by resuming development but does not do so due to the prevailing environmental conditions. The transition from diapause to post-diapause quiescence is a covert transition, marked only by the newly acquired capacity to proceed with development if the right conditions are present. Insects in diapause and post-diapause quiescence are morphologically identical, metabolic depression persists (Denlinger et al. 1972), and gene expression patterns remain largely unchanged (Hayward et al. 2005). Once favorable conditions return, the insect in post-diapause quiescence is primed to respond by initiating development.

From an ecological perspective what is probably most important is the timing of events that culminate in resumption of the next active phase of the lifecycle. To reach that point, diapause must be broken, and upon the return of favorable conditions, the insect must reinitiate development and proceed until it reaches the next active phase of its life cycle. For an adult that has been in diapause or a larva that diapauses in an early instar the transition can be rapid, but for a diapausing pupa to attain the active life of an adult it must first complete pupal and pharate adult development. A larva that diapauses in its final larval instar requires even more time to reach adulthood, and a significant interval may be required for a diapausing embryo to complete embryogenesis and emerge as a feeding larva. Under field conditions, completion of development after diapause has been terminated may take months rather than days.

As discussed below (Section 8.6), a common pattern for insects in temperate latitudes is for the diapause period to be initiated in the autumn and persist until around the time of the winter solstice, at which point diapause is completed and the phase of post-diapause quiescence begins. During post-diapause quiescence the insect is fully competent to develop but fails to do so because temperatures are too low to allow development. Only when temperatures rise later in the spring will development be initiated and proceed until completion.

8.1   Diapause Duration

How long an insect remains in diapause under fixed environmental conditions is referred to as **diapause duration.** Terms **diapause intensity** or **depth of diapause** are sometimes used interchangeably with diapause duration, but I would argue that these are distinctly different diapause parameters. Duration is an unambiguous term that simply defines how long the insect is in diapause. In contrast, the terms diapause intensity or depth of diapause imply how easily the diapause can be broken. A diapause could be of long duration but not necessarily of great intensity. Or, a diapause could have a short duration but still be quite intense. Measures of intensity can be derived from levels of metabolic rate or responsiveness to various diapause-terminating stimuli, as discussed in the section on diapause dynamics (Section 7.16). Although responsiveness to external stimuli may vary at different times during diapause, (*e.g.*, Figure 7.15), the general trend is for the intensity of diapause to decline toward the end of diapause, making it easier to terminate a diapause that has already been underway for some time. Unlike diapause induction, an all-or-none event, diapause duration is highly variable within a species, frequently depending on environmental conditions that prevail either before or during diapause. As nicely summarized by Masaki (2002), this variation in diapause duration is seen both within and across species.

Just as photoperiodic response curves show major clinal variation, latitudinal gradients are evident for diapause duration as well. **Latitudinal clines** of diapause duration either decrease or increase from south to north and are noted for both winter and summer diapauses. Differences can be quite pronounced (Masaki 2002), *e.g.*, summer pupal diapause duration for the cabbage moth *Mamestra brassicae* on the Japanese Islands varies from 130 days at 28° N to 30 days at 43° N, and duration of overwintering embryonic diapause in the Emma field cricket *Teleogrylllus emma* is 150 days and highly variable in a southern population (33° N) but lasts approximately 90 days and is less variable in a northern population (43° N). Reverse tendencies are noted in the pitcher plant mosquito *Wyeomyia smithii*, a species in which diapause duration is fivefold longer at 60° than at 40° N (Bradshaw and Lounibos 1977), and in the blow fly *Calliphora vicina*, larval diapause is completed in 60 days for a population in Italy (44° N), but some larvae remain in diapause more than 100 days in Finland (65° N) (Saunders 2000). Altitude has a similar impact: Although a highland (1,050 m) population of the cherry fruit fly *Rhagoletis cerasi* is less than 1° latitude from a coastal (20 m) population in Greece, the coastal population has a much shorter diapause (Papanastasiou et al. 2011).

Diapause duration is subject to selection, as demonstrated by success in altering the length of diapause by artificial selection, *e.g.*, embryonic diapause of the ground cricket *Dianemobius fascipes* (Masaki 2002). Interestingly, the incidence of diapause in *D. fascipes* is also altered (decreased) by selection for increased duration, a feature not universally observed. In the tobacco hornworm *Manduca sexta*, the opposite relationship is noted: A low incidence of pupal diapause is associated with diapause of long duration, a response allowing the few pupae that enter diapause early in the season to bridge the long interval before the onset of winter (Denlinger and Bradfield 1981).

In its extreme, high variability for the trait of diapause duration is sometimes reflected in **bimodal patterns** of adult emergence, as noted for diapausing pupae of the silk moth *Hyalophora cecropia* in central Illinois: An early emerging cohort of adults appears in late May, followed by a second cohort emerging in late June (Waldbauer and Sternburg 1986). Termination of pupal diapause and adult emergence is unimodal for Illinois populations of the apple maggot *Rhagoletis pomonella*, but in Texas the pattern is bimodal, with a second peak occurring approximately 55 days after the first (Lyons-Sobaski and Berlocher 2009). After completion of pupal diapause in the turnip root fly *Delia floralis*, adult emergence is bimodal, with roughly half the flies emerging 10–20 days after diapause termination and the other half emerging 40–60 days after diapause has been terminated (Biron et al. 2003). A bimodal emergence pattern is also noted for nondiapausing flies of this species, thus suggesting that the pattern is not based on diapause duration but rather on different rates of pharate adult development. Similar bimodal patterns of emergence are noted in the closely related cabbage maggot *D. radicum* and onion maggot *D. antiqua* (Lepage et al. 2014). A polymorphism is associated with vernal development in larvae of the midge *Chaoborus americanus* (Bradshaw 1973). Both morphs are responsive to long daylengths and food for breaking diapause, but the morphs differ in speed of their response and hence in emergence time. The fast-responding morph predominates in a spring that is continuously warm, but the slow-responding morph is dominant in a spring with repeated thaws and refreezing events. Such polymorphisms likely represent bet-hedging tactics enabling the species to maintain different adaptive phenotypes that may be more or less successful in specific years.

The precise timing of adult emergence following pupal diapause depends not only on the duration of pupal diapause but also on the duration of post-diapause quiescence, as well as the time required to complete pharate adult development. Theoretically, diapause duration may be identical in two populations but thresholds needed to prompt adult development or to complete adult differentiation may differ. For example, a higher temperature threshold for development in one subset of a population would translate into delayed adult eclosion. Variation in attributes of post-diapause quiescence and the triggers required to initiate adult development, as well as the time required to complete pharate adult development could be just as effective in altering the timing of spring emergence as diapause duration. Relatively few experiments distinguish between these options. As discussed below, variability in these responses provides the phenotypic plasticity for species to switch to new host plants that display a different seasonal phenology.

8.2   Skipping Post-diapause Quiescence

A period of post-diapause quiescence, between completion of diapause and resumption of development, appears to be the most common progression, but there are species in which this phase appears to be skipped, and the insect goes directly from completion of diapause to resumption of development. A prime example is the apple maggot *Rhagoletis pomonella* (Dowle et al. 2020). The host race that feeds on apple emerges in August, while the race that feeds on hawthorn emerges in September. Variation in timing of adult emergence (98% of the variation) is accounted for by variation in the time diapause is terminated (Powell et al. 2020). The time required for the completion of pharate adult development accounts for only 2% of the variation in emergence time, thus differences in emergence times for the two races are based almost solely on when diapause is completed. Temperatures are obviously high enough to sustain development in spring and early summer, but the flies remain locked in pupal diapause. This is in stark contrast to the more common pattern that uses the cold of winter to block the onset of development and then uses the seasonal rise in temperature to synchronize emergence with the early flush of spring resources. Interestingly, not all species of *Rhagoletis* lack post-diapause quiescence. The European cherry maggot *R. cerasi* exhibits a classic post-diapause quiescence to synchronize its emergence in Greece with early fruiting cherries in May (Papanastasiou et al. 2011, Moraiti et al. 2020). Thus, it appears that post-diapause quiescence is skipped only in those species that emerge late in the summer. How common it may be to skip post-diapause quiescence is unknown, but it is likely prevalent among species that synchronize emergence with late season hosts.

An alternative is that an environmental factor other than low temperature could keep the insect in post-diapause quiescence, just as some tropical species complete diapause and then remain in post-diapause quiescence until the rains return, as discussed in Section 8.8. Another alternative is to skip post-diapause quiescence and progress with development but do so very slowly. This scenario seems to be operating in the green oak tortrix *Tortrix viridana* (Du Merle 1999). Embryonic diapause in this species is terminated in late autumn/early winter, and the embryos then develop slowly but continuously until egg hatch in the spring.

8.3   Why Is Timing So Important?

Though the onset of overwintering diapause in a population frequently occurs over a period of months in late summer and early autumn, spring emergence after completion of diapause is remarkably well synchronized, regardless of the date of diapause entry. This fact underscores the utility of diapause in generating a large pool of potential mates present at the same time in the spring and allows overwintering, plant-feeding insects to finely coordinate spring emergence with host phenology.

Flowering and bud burst are both notoriously brief events of early spring, and if an insect is to successfully exploit these resources it must make its appearance at the right time! Ideally, leaf-feeding insects will be present just after buds open so they can take advantage of the nutritious young growth. Development outside this optimal period has severe fitness consequences, and matching of larval hatching with bud burst can be expected to be under strong selective pressure. Similarly, fruiting and seed production occur during narrow seasonal windows, requiring insects that exploit such resources to coordinate activity with the brief interval that the resource is available. For such phenological specialists, diapause offers both a mechanism for surviving when food is not present as well as a mechanism to precisely time its activity to optimally exploit a seasonal food source (Tauber and Tauber 1981, van Asch and Visser 2007).

An insightful comparison of two species of pierid butterflies nicely demonstrates how specialization can influence seasonal timing (Posledovich et al. 2015). *Pieris napi* is a multivoltine, phenological generalist that consumes green leaves of various species of Brassicaceae throughout the warm season in Sweden, while *Anthocharis cardamines* is a univoltine phenological specialist that occupies the same geographic area but feeds on flowers and siliquae of Brassicaceae, resources that are available for approximately one month (May–June). Both species overwinter in pupal diapause and eclose as adults in the spring. Diapause duration is shorter in *A. cardamines* and remains unaltered by longer cold periods, while the longer diapause of *P. napi* is shortened by low temperatures and shows a more variable response. The more stringent pattern observed for *A. cardamines* is likely critical for synchronizing emergence within the narrow temporal window of its food source, while such precision is less important for the generalist species *P. napi*. The rapid thermal response involved in reinitiating development in *A. cardamines* enables this species to track its food source, even in a warming environment.

8.4   Seasonal Timing as a Driver of Speciation

The timing of diapause termination and spring emergence can be remarkably precise, but there is population variation, and the extremes provide the grist allowing sympatric populations to diverge based on the timing of emergence. This, in turn, can allow exploitation of new host plants, the development of host races, and eventually the reproductive isolation required for the evolution of new species. Speciation, such as this, that occurs in response to a phenological shift, rather than geographic isolation, is referred to as **allochronic speciation** (speciation based on temporal isolation).

The most extensively studied taxon in this regard is the apple maggot *Rhagoletis pomonella* (Feder et al. 2005). As proposed in pioneering work by the prominent North American entomologist Benjamin D. Walsh (1867), introduction of apples to North America set the stage for a host plant shift by the fly from native hawthorn to apple. Making this shift requires an earlier end to the fly's obligate pupal diapause (Figure 8.1), shifting adult emergence from late September (hawthorn fruiting) to late

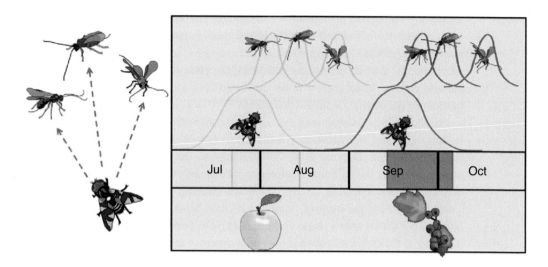

Figure 8.1 Allochronic speciation in *Rhagoletis* fruit flies driven by differences in seasonal phenology of the ancestral hawthorn host plant (red) and introduced domestic apples (yellow). Radiation of the univoltine flies onto apples also created a new temporal niche allowing sequential speciation of a community of univoltine parasitoid wasps that attack the flies. Both the flies and the parasitoids have shifted timing of their diapause to synchronize hosts and consumers across trophic levels, driving genetic divergence by allochronic isolation. (A black and white version of this figure will appear in some formats. For the color version, please refer to the plate section.)
From Denlinger et al. (2017).

July–early August (apple fruiting). Colonizing apples presumably was possible as a result of natural variance in diapause duration and selection that favored flies with a shortened pupal diapause. This shift in timing has repercussions across trophic levels. The apple maggot *R. pomonella* hosts several parasitoids which also have an univoltine life cycle, and interestingly, these parasitoids have likewise now made the shift to a shorter diapause (Figure 8.1) in coordination with the apple maggot (Hood et al. 2015). This segregation into distinct host races, an event that has taken place in less than 150 years, appears to be an early step leading to speciation (Egan et al. 2015, Doellman et al. 2018). Allozyme alleles *Dia-2, Me,* and *Had 100* are among the alleles associated with developmental time and energy reserves, factors that may be components of the regulatory scheme separating the two host races of *R. pomonella* (Feder et al. 2010). Diverse approaches including genetic association, quantitative trait loci, and common-garden studies indicate that timing is highly heritable in *R. pomonella* (Egan et al. 2015). A detailed evaluation of transcriptional changes during diapause suggests that multiple genes, each with small effects, contribute to the rate of diapause development, especially during the onset and final phases of diapause (Dowle et al. 2020). The shift to a seasonally earlier fruit (apples) appears to be driven by more rapid diapause development in the early fly population. A related species, the blueberry maggot *R. mendax,* also appears to be in an early phase of isolation into two temporally segregated populations (Teixeira and Polavarapu 2005). A late blueberry cultivar recently introduced to New Jersey hosts a late-emerging population of *R. mendax*. Rather than breaking diapause and emerging in late June/early July to match the traditional fruiting period, the late-emerging cohort emerges in September when the late cultivar bears fruit, a delay attributed to a longer period of post-diapause quiescence (or possibly a longer diapause).

Duration of larval diapause in the European corn borer *Ostrinia nubialis* is a central feature separating two genetically distinct populations (E- and Z-pheromone strains) in North America (Dopman et al. 2005, 2010). As the strain names imply, the two populations also use different pheromone blends, but the E-strain terminates larval diapause earlier, allowing adults to emerge in early spring, and completes two generations during the summer, while the Z-strain breaks diapause nearly one month later and completes only a single generation in New York (Figure 8.2). This difference in timing at the end of diapause is the major driver of reproductive isolation for these

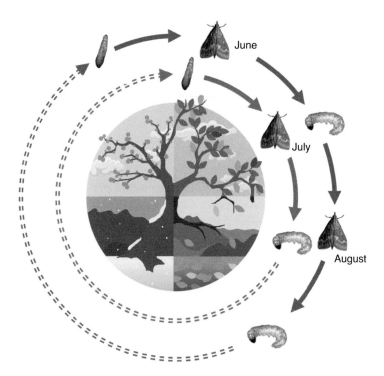

Figure 8.2 Genetically based shifts in seasonal timing in populations of the European corn borer *Ostrinia nubilalis* in eastern North America. The Z strain population (red) remains in diapause longer and has only one mating flight (July), whereas the E strain (blue) has a shorter overwintering pupal diapause and completes two generations with early (June) and late (August) mating flights. (A black and white version of this figure will appear in some formats. For the color version, please refer to the plate section.)
From Wadsworth et al. (2020), with permission from BioMed Central.

two populations (Wadsworth et al. 2013). Because it emerges earlier and pupates later in the season, the E-strain is also subjected to lower temperatures than the Z-strain during its active phase, a feature reflected in higher cold tolerance attributes of the E-strain (Wadsworth et al. 2020). Whole genome approaches suggest distinctions in two circadian clock genes, *period* (*per*) and *pigment-dispersing factor receptor* (*Pdfr*), may underlie this polymorphism (Kozak et al. 2019). *Per* and *Pdfr* are within interacting quantitative trait loci and differ in allele frequency among early- and late-breaking diapause individuals, suggesting yet another link between the circadian clock and the seasonal photoperiodic clock (Gotthard and Wheat 2019).

Other similar examples of potential incipient speciation are emerging in the literature. The generalist moth *Operophtera brumata* thrives on several tree species, including *Prunus padus* and *Quercus robur*, species that represent extremes in timing of spring bud burst (Tikkanen and Lyytikäinen-Saarenmaa 2002). Spring egg hatch in populations of *O. brumata* closely matches the timing of bud burst in trees from which they are collected, suggesting a highly heritable trait. Transfer of eggs from one tree to the other results in a mismatch between egg hatch and bud burst, as well as decreased growth rates and increased dispersal rates for both groups of larvae. Populations of the green oak tortrix *Tortrix viridana* feed on a variety of oak species with different times of bud burst, but there is remarkable synchrony with the time of egg hatch on each host (Du Merle 1999), suggesting phenological polymorphism leading to hatch dates precisely tailored to each oak species and hence to potential speciation. The stem borer *Chilo suppressalis* is another species with two host-plant-associated populations: one associated with rice and the other with water-oat (Zhou et al. 2018). Reproductive isolation of these two populations is at least partially due to differences in their diapause responses. Critical daylength, larval development time, duration of the photosensitive period, and diapause duration differ between the two populations, as well as gene expression profiles, suggesting that differences in diapause induction and termination are promoting divergence in the two populations of *C. suppressalis*. The browntail moth *Euproctis chrysorrhoea* is a highly polyphagous species capable of feeding on a range of deciduous and evergreen trees in Spain (Frago et al. 2010). Though larvae reared on both tree types enter diapause at the same time (October), those reared on the evergreen strawberry tree *Arbutus unedo* exploit the availability of evergreen leaves and start feeding after two months in diapause, grow considerably during this winter feeding period, and consequently emerge as adults much earlier than cohorts overwintering on deciduous hosts. Although this system has not been examined extensively, these observations suggest the potential for allochronic speciation based on the availability of winter feeding.

8.5   How Is Synchronous Spring Emergence Achieved?

Seasonal dynamics of development in the flesh fly *Sarcophga bullata* illustrate the scatter in timing of autumn entry into pupal diapause and the contrasting precise timing of adult eclosion in the spring (Denlinger 1972a). In central Illinois, a few

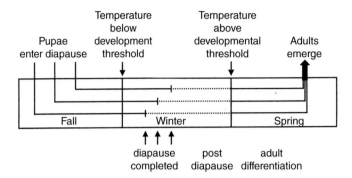

Figure 8.3 The role of low temperature in synchronizing post-diapause development in the spring, based on the flesh fly *Sarcophaga bullata* (Denlinger 1972a,b). Pupae enter diapause over a period of several months in the autumn. The latency inherent to diapause prevents further development during the remainder of the autumn. Pupae will complete diapause at different times during the winter, but they remain locked into post-diapause quiescence by the low temperatures that prevail during winter and then simultaneously initiate adult differentiation when higher temperatures return in the spring. Thus, pupae that enter diapause at quite different times in the autumn can complete development and emerge as adults at the same time in the spring.
From Denlinger (2001).

pupae enter diapause in July (2%), more do so in August (20%), and by September entry into diapause is 99%, thus diapause entry is spread over at least three months, and probably longer because adult flies can be collected as late as mid-October. By contrast, the timing of spring adult eclosion is nearly identical for pupae that enter diapause at different times. Flies pupariating in both August and September have identical spring median emergence dates of May 11 (range May 8–15 for August flies and May 9–17 for September flies). This sort of precision is achieved by a remarkable but simple mechanism. As illustrated in Figure 8.3, pupae enter diapause at different times in the summer and autumn. The fixed latency period of diapause is the same for all pupae, regardless of pupariation time. This means that the earliest pupae entering diapause complete diapause earlier than late pupariating flies, but because the cold of winter prevails, none of the pupae are able to immediately reinitiate development toward adulthood. For the early pupariating flies, diapause may be completed by early December whereas diapause may not be completed in later pupariating flies until mid-January or later. The cold serves as a mechanism for stockpiling all of these pupae in post-diapause quiescence, essentially stockpiling pupae that are now all fully competent to start developing synchronously when temperature rises above a certain threshold in the spring. For *S. bullata* in central Illinois, the first signs of adult differentiation (migration of antennal discs) are noted in the field in early April, but development progresses slowly due to the prevailing low temperatures, culminating in synchronous adult eclosion in mid-May.

For many insects diapause itself is completed rather early. In the example given above for *S. bullata*, most pupae have completed the diapause phase of overwintering by early January, and they then simply wait in post-diapause quiescence until higher temperatures in the spring permit resumption of development (Denlinger 1972b). In the Colorado potato beetle *Leptinotarsa decemlineata* diapause in the field is completed after three months, but adult emergence from the soil in the Netherlands does not occur until late in spring when soil temperature exceeds 11°C (Lefevere and de Kort 1989). In central Europe adult diapause in the linden bug *Pyrrhocoris apterus* is completed in the field by late December, but the bugs remain in post-diapause quiescence until the rise of temperatures in April prompts reproductive development (Hodek 1971b). In the Netherlands, adults of the Asian ladybird beetle *Harmonia axyridis* enter diapause in late October and shift to post-diapause quiescence in mid-December (Raak-van den Berg et al. 2013). Larvae of a Greek strain of the leafroller *Adoxophyes orana* terminate diapause in late January but fail to develop further until temperatures rise in the spring (Milonas and Savopoulou-Soultani 2004). Diapausing larvae of the wheat blossom midge *Sitodiplosis mosellana* in northern China complete diapause in early December and then transition into post-diapause quiescence, with adults emerging in late April (Cheng et al. 2017). The transition from diapause to post-diapause quiescence occurs in mid-late January for diapausing larvae of the peach fruit moth *Carposina sasakii* in northern China (Zhang et al. 2016a). Embryonic diapause in the European red mite *Panonychus ulmi* is completed in northern Greece by early February, and egg hatching occurs in mid-April, upon accumulation of 155-degree days above a 7.4°C threshold (Broufas and Koveos 2000).

Counter to the above examples that report diverse times of diapause entry but synchronous resumption of development, there are also a few examples in which spring emergence is not synchronous but instead correlates with the date of diapause onset. For example, diapausing eggs of the parasitoid *Diachasma alloeum* that are laid early also emerge early, rather than in synchrony with eggs that are deposited later (Forbes et al. 2010). This "first in-first out" pattern is unusual and the adaptive basis for this pattern is not immediately obvious.

8.6   Diapause Development

The fixed latency of diapause, the fact that a certain interval must elapse before diapause can be terminated, implies that the diapausing insect cannot normally break out of developmental arrest immediately after diapause has been initiated. As discussed above, this is, of course, a safeguard preventing the untimely initiation of development when conditions are or soon will be prohibitive. The fact that diapause can eventually be terminated implies that a progression of molecular and physiological events must be completed to bring diapause to an end. Although the term is perhaps an oxymoron, **diapause development** (Andrewartha 1952) captures the idea that events transpiring during diapause lead to its eventual termination. Other terms commonly used for these same processes include **diapause-ending processes, reactivating processes,** or **conditioning** (Waldbauer 1978). Although each of these terms has drawbacks, the underlying connotation is that diapause is a dynamic process, and certain physiological changes must be completed before the insect can break diapause. Prevailing environmental conditions frequently dictate how rapidly these processes are completed.

8.6.1   Environmental Influences

A range of environmental parameters influences how long diapause will last, or stated differently, how rapidly diapause development proceeds. Many of the influences are species-specific. For some species, chilling is required, but not for others. Daylength, the dominant environmental signal for diapause induction, is less important in determining diapause duration, but for some species it is the major factor.

Chilling is essential for some species. Classic experiments with *Hyalophora cecropia* demonstrate that termination of pupal diapause in this giant silk moth requires a 10-week period of chilling at 5°C before diapause can be broken, and it's not the whole body that needs to be chilled, just the brain (Williams 1946). Like *H. cecropia*, termination of embryonic diapause in the silk moth *Bombyx mori* requires a two-month exposure to 5°C (Yaginuma and Yamashita 1999). Adults of the bluegrass weevil *Listronaotus maculicollis* collected from the field in September fail to break diapause if they are not chilled but do so readily if allowed to remain in the field until early January (Wu et al. 2017). Adults of the solitary bee *Osmia lignaria* also require a period of chilling, but this period can be quite brief, as short as one week at 4°C (Sgolastra et al. 2010). Optimal survival and highly synchronous spring emergence of this important pollinator are achieved with a chilling period of 210 days at 4°C (Bosch and Kemp 2003). The chilling temperature required to break embryonic diapause in the hemlock looper *Lambdina fiscellaria* is rather high, 15°C (Delisle et al. 2009), thus a chill does not necessarily require temperatures near the freezing point. Larval diapause in the langriid beetle *Lagria hirta* cannot be broken at constant temperatures ranging from 5 to 30°C; only if larvae are chilled for three months at 5°C and then transferred to 15–25°C can the diapause be ended (Zhou and Topp 2000). Larval diapause of the wheat blossom midge *Sitodiplosis mosellana* also requires a period of low temperature to break diapause; optimum chilling conditions are 60–90 days at 4–8°C (Cheng et al. 2017). Without chilling, the green oak tortrix *Tortrix viridana* remains locked in an embryonic diapause that may last a year or more at 20°C. The embryos eventually die without developing unless they are chilled (Du Merle 1999). Populations of the orange tip butterfly *Anthocharis cardamines* from northern Europe require cold exposure to break pupal diapause: 30 days at 2.4°C is inadequate but a 60-day chill is sufficient to elicit synchronous spring emergence (Stålhandske et al. 2015). Embryos of the spider mite *Eotetranychus smithi* are capacitated to break diapause by a 30–90 day chilling period at 5°C (Takano et al. 2017).

But, as argued (Hodek and Hodková 1988, Hodek 2002), the need for chilling has been overemphasized in the diapause literature, possibly due to the early popularity of the papers by Williams on pupal diapause in *H. cecropia*. In numerous cases diapause duration is shorter at higher temperatures and no chilling is needed, but chilling often reduces diapause duration, even if it is not a requirement for breaking diapause.

Chilling is not needed to end adult diapause in the Colorado potato beetle *Leptinotarsa decemlineata*; diapause development can proceed at 4, 17, or 25°C, and it proceeds more quickly at higher temperatures (Lefevere and de Kort 1989). Diapausing pupae of the flesh fly (several *Sarcophaga* species) likewise do not require chilling. Diapause at 17°C lasts nearly twice as long as it does at 25°C, but transferring pupae from 17 to 25°C further hastens the process of diapause development (Denlinger 1972b). Chilling accelerates the breaking of summer diapause, as well as winter diapause, in the onion maggot *Delia antiqua*; brief periods (one to five days) at temperatures ranging from 2 to 21°C break summer diapause in pupae of this species (Ishikawa et al. 2000). Pupal diapause in the western cherry fly *Rhagoletis indifferens* eventually ends at constant temperatures of 23–26°C, but termination is considerably more rapid and synchronous after a 20-week period of chilling at 3°C (Neven and Yee 2017). Pupal diapause in the fall webworm *Hyphantria cunea* does not require chilling to break diapause, but chilling at 5°C for 90–120 days significantly accelerates diapause termination (Chen et al. 2014a). Likewise, termination of larval diapause in the wheat blossom midge *Sitodiplosis mosellana* is hastened by chilling (Cheng et al. 2017). Temperatures both above and below 25°C hasten termination of the pseudopupal diapause of the blister beetle *Epicauta gorhami*, but the optimum is in the 10–15°C range (Terao et al. 2012). Though chilling is not essential for breaking diapause in the Chinese rice grasshopper *Oxya chinensis*, it does hasten completion of diapause: Optimal chilling condition is 60 days at 8°C (Zhu et al. 2009). The impact of cold exposure is, of course, a function of both temperature and time, as well as when the chilling period is executed during diapause.

The Mormon cricket *Anabrus simplex*, a species with an embryonic diapause, offers yet another scenario demonstrating a manner in which temperature can influence diapause duration (Srygley and Senior 2018). Culturing embryos in a daily cycling temperature regime that is constant in both amplitude and thermoperiod yields a diapause duration of 128 days, while culturing in a cycling temperature patterned after natural fluctuations in temperature generates a much shorter 42-day diapause, a result that underscores the diversity of seasonal features that may contribute to diapause regulation.

In the cabbage butterfly, *Pieris melete*, temperatures experienced prior to the onset of pupal diapause determine the duration of summer diapause (rearing at 22°C produces a shorter diapause than rearing at 20 or 18°C), but interestingly, such an effect is not observed for winter diapause in this species (Xiao et al. 2008b). In a closely related species, *P. napi*, a chilling temperature of 2–4°C is optimal for breaking pupal diapause: Lower temperatures result in higher mortality and higher temperatures retard the rate of diapause development (Lehmann et al. 2017b). In the predatory mite *Euseius finlandicus* a one to four week period at 5°C fails to accelerate diapause termination compared to mites held continuously at 20°C (Broufas et al. 2006). Numerous additional examples compiled by Hodek and Hodková (1988) and Hodek (1971b, 2002) provide ample proof that many species do not require low temperatures to break diapause, but chilling can influence the rate of diapause development.

As pointed out in Section 7.16, diapause is a dynamic process showing systematic developmental changes to various stimuli throughout diapause. Thus, we can assume that responses to temperature and other influential environmental factors are also not constant throughout diapause. This feature is evident in the temperature response during pupal diapause in the onion maggot *Delia antiqua*, a species showing two distinct phases of sensitivity to low temperature (Nomura and Ishikawa 2000). During the first phase, lasting approximately 60 days, low temperature (5.6°C) has no effect on diapause development, but after 60 days, low temperatures retard the rate of diapause development. A similar pattern is noted for embryonic diapause in the gypsy moth *Lymantria dispar* (Gray et al. 2001). Subjecting chilled (5°C) diapausing gypsy moth eggs to brief pulses (1–75 minutes) of different temperatures (−5 to 25°C) shows that the response is not constant throughout diapause, but is better characterized as a two-phase response. Early in diapause chilling has little impact on diapause duration, but late in diapause the effect is pronounced, resulting in extremes of 30 and 126 days in diapause.

Several examples show that high temperature can increase diapause duration, a response especially common among insects that have a summer diapause (Masaki 1980). More recent examples also exist. Rearing eggs and larvae of the grape berry moth *Lobesia botrana* under a range of temperatures (12–30°C) at LD 12:12 generates pupal diapauses of different durations, with the highest duration observed at high temperatures (Roditakis and Karandinos 2001). A similar response is noted for the linden bug *Pyrrhocoris apterus*. Higher adult diapause duration at short daylengths is seen when nymphs are reared at 25 rather than 20°C, and an even higher duration is evoked by coupling photoperiod with a 25/15°C thermoperiod (Kalushkov et al. 2001). High temperature (26 vs. 20°C) increases diapause duration in larvae of the mango blossom gall midge *Procontarinia mangiferae*, and switching larvae to the lower temperature brings diapause to a halt (Amouroux et al. 2014).

In some species, photoperiod experienced during diapause influences the rate of diapause development. A period of low temperature is not required for completion of larval diapause in the fruit moth *Adoxophyes orana*, but photoperiod is influential: Short days maintain the diapause state and long days lead to diapause termination (Milonas and Savopoulou-Soultani 2004). In the linden bug *P. apterus* photoperiodic cues are used to terminate adult diapause, but the bug is responsive to these cues only at temperatures above 15°C, and morphogenesis cannot proceed unless food is available (Hodková and Hodek 1987). Daylength, along with temperature, impacts diapause duration in *Pieris melete*, and summer and winter diapause responses differ: Summer diapause is longer at long daylengths, while winter diapauses are longer at short daylengths (Xiao et al. 2006). Long daylengths, rather than low temperatures, accelerate adult diapause termination in the green stink bug *Nezara viridula*: At 25°C, diapause lasts 83 days at LD 13:11, 126 days at LD 12:12, and 154 days at LD 10:14 (Musolin et al. 2007). Short daylengths in late autumn and early winter help to maintain the second-instar larval diapause of the peach twig borer *Anarsia lineatella* in northern Greece, but the photoperiod effect is diminished by midwinter and a period of at least 30 days of chilling at 4°C is needed to synchronously terminate diapause

(Damos and Savopoulou-Soultani 2010). In the cherry fruit fly *Rhagoletis cerasi* long days during the chilling period result in a longer pupal diapause, even though pupae would normally be underground at this time (Moraiti et al. 2020). Considerable work has focused on photoperiodic mechanisms regulating diapause induction, but the role of photoperiod in diapause termination has been examined less extensively. The evidence that is available suggests that the same timekeeping mechanisms operate to control both diapause induction and termination. Experiments with the zygaenid moth *Pseudopidorus fasciata*, employing classic approaches including night interruptions, resonance experiments, and the Bünslow protocol suggest that a long-night measuring hourglass is involved in both diapause induction and termination in this species (Li et al. 2003).

Photoperiod and temperature experienced during the programming phase of diapause influence diapause duration in some species. Figure 5.8, shown in Chapter 5, demonstrates such a relationship for pupal diapause in the tobacco hornworm *Manduca sexta*: Exposure to a few short days during larval life produces a low incidence of diapause but a diapause of long duration, while exposure to many short days yields a high diapause incidence but a diapause of short duration. This enables those entering diapause early in the summer (exposed to only a few short days) to bridge the longer time interval before the onset of winter (Denlinger and Bradfield 1981). Change to a longer or shorter photophase during larval development or larval diapause in the tiger moth *Cymbalophora pudica* causes a significant extension (up to 138 days) or shortening (down to 10 days) of summer diapause, respectively, a response not seen if photophase remains constant (Koštál and Hodek 1997). The shortened diapause caused by a switch to short photophase allows the moth to complete an additional generation in late summer. Duration of larval diapause in the pine caterpillar *Dendrolimus puncatatus* also depends on the rearing photoperiod: Photophases near the critical daylength yield a diapause of short duration, while rearing at a shorter daylength results in a longer diapause and one that cannot be terminated unless larvae are transferred to long daylengths (Zeng et al. 2013). Summer diapause in prepupae (just prior to pupation) of the zygaenid moth *Pseudopidorus fasciata* is also influenced by photoperiods received during larval rearing: Duration is nearly twice as long at LD 15:9 than at LD 11:13 or 13:11, and transferring pupae from long to short days hastens termination (Wu et al. 2006). In the longicorn beetle *Psacothea hilaris* duration of larval diapause is influenced by both photoperiod and temperature experienced prior to diapause, generating 40–140 day extremes of diapause duration, with the longest diapause induced by long days, 20°C, and the shortest diapause by short days at 20°C (Asano et al. 2004). Both photoperiod and temperature also impact pupal diapause duration in the swallowtail *Sericinus montelus*: At 25°C, LD 16:8 diapause lasts 83 days but the duration increases to 130 days at LD 12:12; dropping the temperature to 15°C, LD 16:8, increases diapause duration to 173 days (Wang et al. 2009). Transferring larvae of the Mediterranean corn borer *Sesamia nonagrioides* from short days to longer days prompts either immediate diapause termination or a reduction in diapause duration, while a transfer from long days to short days results in a slight increase in diapause duration (Fantinou et al. 2003);

chilling does not appear to hasten diapause termination in this species (Eizaguirre et al. 2008). Duration of the obligate pupal diapause in the Chinese citrus fruit fly *Bactrocera minax* is influenced both by larval rearing temperature and pupal weight (Zhou et al. 2020): Larvae reared at 8°C remain in pupal diapause 193 days at 15°C, while those reared as larvae at 20°C remain in diapause approximately 230 days, and diapause is slightly longer in heavier pupae. In the fall webworm *Hyphantria cunea* short daylengths of 8, 10, or 12 hours during larval development evoke a longer pupal diapause than daylengths of 13 or 14 hours (Chen et al. 2014a). Adult diapause duration in the bean bug *Riptortus clavatus* is influenced by photoperiod acting during pre-diapause development as well as after adult eclosion (Nakamura and Numata 2000). In this case, the response is quantitative, with daylengths near the critical daylength generating a diapause of shorter duration than observed at shorter daylengths.

Environmental information determining how long diapause will last is sometimes channeled through the mother, as discussed for maternal effects operating on the programming of diapause (Section 5.13). Cases in point include the walking stick *Ramulus irregulariterdentatus* which enters a longer embryonic diapause if the mother is exposed to long days (Yamaguchi and Nakamura 2015) and the mite *Halotydeus destructor*, a species that enters a longer embryonic diapause when parents are reared in a hot, dry environment (Cheng et al. 2019).

Feeding opportunities can also influence length of the overwintering period by altering either duration of diapause or post-diapause quiescence. As indicated above (Section 8.4), larvae of the browntail moth *Euproctis chrysorrhoea*, when overwintering on an evergreen tree break diapause after two months, feed on the available leaves during winter, grow and reach maturity several months earlier than larvae that overwinter on deciduous trees (Frago et al. 2010). A likely scenario is that diapause in larvae reared on either type of tree is terminated after two months, and those overwintering on deciduous trees remain in post-diapause quiescence while those on evergreen exploit the availability of leaves by resuming development. In the Mississippi Delta adults of the tarnished plant bug *Lygus lineolaris* overwintering on common henbit, a plant that blossoms in early December, terminate diapause in December, but those overwintering on plant species that blossom later terminate diapause nearly a month later (Snodgrass et al. 2012). The presence of food augments the effect of long daylengths in promoting diapause termination in larvae of the midge *Chaoborus americanus* (Bradshaw 1973). Food is required for resumption of development in diapausing adult females of the rice bug *Leptocorisa chinensis*, but not for males (Tachibana and Watanabe 2008). Rainfall, long daylength, and high temperature all promote diapause termination in adults of the tropical Australian nymphalid butterflies *Euploea core* and *E. sylvester*, but only when these signals are complemented with a food source (nectar) is there 100% diapause termination (Canzano et al. 2003). This fail-safe combination of cues may be especially important to the nymphalids for preventing untimely diapause termination. The cabbage bug *Eurydema rugosum* has two types of diapause: One induced by short days with either rapeseed or leaves as the food source and the other induced by long days with rapeseeds as the food (Ikeda-Kikue and Numata 1994). The short-day-induced diapause can be broken after a period of chilling, regardless of the food source, but long-day bugs reared on seeds require food to terminate diapause. Regardless of photoperiod, diapausing adults of the seed-sucking bug *Dybowskyia reticulata* refrain from reproduction until they are offered seeds from their carrot host, a plant that does not produce seeds until early summer in Osaka, Japan (Numata 2004). Starvation thus delays the onset of reproduction and synchronizes onset of oviposition in the population.

Depletion of energy reserves may also prompt diapause termination. If sufficient reserves are not available, breaking diapause early may be the only viable option (Hahn and Denlinger 2007). This may be occurring in the brown marmorated stink bug *Halyomoropha halys* (Skillman et al. 2018). Adults that depart their overwintering shelters early are consistently lighter in weight and have fewer remaining glycogen and sugar reserves than adults that remain sheltered in diapause.

Males and females may interpret the environmental signals involved in resumption of development differently. There are examples of both males and females emerging earlier, but when there is a difference, males are more frequently the first to emerge. In the rice bug *Leptocorisa chinensis* males terminate diapause earlier than females, even though they enter diapause at the same time (Tachibana and Watanabe 2007). Males resume post-diapause development without food being available, but females do so only in the presence of food. Males of the univoltine parasitoid wasp *Diachasma alloeum* complete embryonic diapause and emerge five days earlier than females (Forbes et al. 2010). Adults males of the tarnished plant bug *Lygus lineolaris* also terminate diapause earlier than females (Snodgrass et al. 2012). Overwintering males of the weevil *Sternechus subsignatus* emerge a few weeks earlier than females (Socías et al. 2016). In the ladybird beetle *Cheilomenes sexmaculata* females tend to emerge slightly earlier than males following both winter or summer diapause (Kawakami et al. 2017), a response possibly driven by the need of the female to feed before becoming sexually mature, but differences are slight, seldom more than a day or so.

8.6.2   What Physiological Processes Dictate Completion of Diapause?

As the above discussion indicates, we know quite a bit about environmental conditions that influence the rate of diapause development, but we still have a poor understanding of underlying molecular and physiological mechanisms governing this diapause-terminating process. The capacity to end diapause is a function of the rate of diapause development. In the apple maggot *Rhagoletis pomonella* this rate is the consequence of numerous genes, each with small effects, that collectively dictate how rapidly the process advances (Dowle et al. 2020). Some sort of threshold event is likely to be ultimately involved in bringing diapause to an end. We can imagine either the gradual erosion of a factor that maintains diapause and/or the gradual accumulation of a factor that reaches a threshold, triggering completion of diapause. It seems likely that any such factors would be integrated with a nutrient-sensing mechanism capable of monitoring nutrient reserves and thus be capable of providing feedback about the quantity of reserves available. This feature is essential if the insect is to retain sufficient reserves for completion of development once diapause has been terminated.

A few examples suggest how such a mechanism could work. Embryonic diapause in the silk moth *Bombyx mori* is induced and maintained by the presence of sorbitol in the yolk (Horie et al. 2000), and a period of chilling is needed to terminate diapause. Once sorbitol surrounding the embryo is removed, development can proceed, thus a process, perhaps directed by a low-temperature-activated enzyme that removes sorbitol, could indeed lead to diapause completion. As diapause progresses, sorbitol is gradually converted to glycogen, thus removing the developmental block on the diapausing embryo. Several players may be involved in the pathway capacitating *B. mori* to end its diapause. A novel cold-inducible gene, *Samui*, activated in response to 5°C, is elevated shortly after oviposition and possibly plays a role in receiving the chill signal (Moribe et al. 2001). Elevation of *Samui* correlates with a gradual increase in abundance of heat shock protein Hsp70a (Moribe et al. 2010), suggesting that these two gene products may work in concert as components of the signaling mechanism culminating in diapause termination. Extracellular signal-regulated kinase/mitogen-activated protein kinase (**ERK/MAPK**) is another strong contender for regulating developmental competence in *B. mori* (Iwata et al. 2005a, Fujiwara et al. 2006a,b). ERK/MAPK is activated by low temperature, and after 45 days at 5°C, a point coinciding with the chilling period needed to break diapause, the activated form of ERK/MAPK is abundant in yolk cells surrounding the embryo. The progression of development is inhibited by a MAPK-ERK kinase inhibitor, further suggesting a critical role for ERK/MAPK in attaining competence to end diapause and begin development, quite likely through activating transcription of enzyme genes involved in ecdysteroid and sorbitol metabolism. Chilling at 5°C is more effective than chilling at higher or lower temperatures, a feature consistent with temperatures that optimally activate ERK (Fujiwara and Shiomi 2006). Another possible contributor may involve activation of a metallo-glycoprotein ATPase, with high homology to members of the superoxide dismutase family (Isobe et al. 2006). This enzyme reliably marks the completion of diapause development in *B. mori* (Kai et al. 1987, 1995, Tani et al. 2001, Ti et al. 2004). Described as a **Time-Interval-Measuring Enzyme** (TIME), it is activated in response to a well-defined period of chilling at 5°C, a process that can occur either *in vivo* or *in vitro*. But, whether TIME actually contributes to diapause development or simply marks its completion is not yet clear (Hirai et al. 2008), and a link between sorbitol removal and TIME activation is not established. Thus, several factors appear to respond to the mandatory chilling required for breaking diapause in *B. mori*, but exactly how these agents work together to achieve competence still remains unknown.

A role for ERK/MAPK signaling in prompting termination of diapause (*i.e.,* developmental competence) is proposed not only for embryonic diapause of *B. mori* but also for embryonic diapause in the false melon beetle *Atrachya menetriesi* (Kidokoro et al. 2006b) and pupal diapause in the flesh fly *Sarcophaga crassipalpis* (Fujiwara and Denlinger 2007), as discussed in Section 9.1. Although association with resumption of development is clear in each of these cases, results with *A. menetriesi* suggest a further link to regulation of diapause duration. Activation (phosphorylation) of ERK appears to influence how long the beetle embryos remain in diapause. As in many other species, diapause in this beetle is terminated by cold. **Phospo-ERK** (p-ERK), the activated form of ERK, is high during pre-diapause but declines upon entry into diapause (Kidokoro et al. 2006b). Exposure of diapausing embryos to 7.5°C, an optimal temperature for terminating diapause, boosts levels of p-ERK to a consistently high level. By contrast, environmental conditions that result in intensification of diapause (0°C or early transfer to a high temperature) correlate with a decline in levels of p-ERK. These results imply that chilling conditions promoting diapause termination are consistently linked to elevated levels of p-ERK, while conditions that delay diapause termination are associated with low p-ERK. Determining how cold influences levels of p-ERK and what downstream responses are elicited by the activation of ERK offer promise for deciphering a vital link in the diapause regulatory pathway. Genes associated with calcium signaling and cold hardening are proposed as determinants of diapause duration in the monarch *Danaus plexippus* (Green and Kronforst 2019), but this conjecture still lacks functional evidence.

Although most work on diapause termination has focused on the role of temperature, a possible molecular mechanism for terminating diapause in response to daylength is discussed in Section 5.7 in the context of a counter mechanism proposed for the silk moth *Antheraea pernyi*, a mechanism that relies on the dynamics of **N-acetyltransferase** to convey a signal to the brain for releasing PTTH (Mohamed et al. 2014, Wang et al. 2015a,b). Another potentially interesting contributor to daylength-mediated diapause termination is *diapause response gene 1*, a transcript that is strongly upregulated in the mosquito *Wyeomyia smithii* when diapausing larvae are transferred to long daylengths that terminate diapause (Emerson et al. 2010).

Embryonic diapause of the gypsy moth *Lymantria dispar* offers another scenario explaining how erosion of a diapause-promoting factor could allow diapause development to be completed. The developmental brake in this case is applied by the presence of maternally provisioned ecdysteroids in the yolk; removal of yolk-bound ecdysteroids ends diapause (Lee and Denlinger 1997, Lee et al. 1997). Degradation of the ecdysteroids during a period of chilling could permit completion of diapause development, although at this point we lack evidence if and how chilling exert such an effect. A contrasting possibility involving ecdysteroids is suggested by the gradual accumulation of ecdysteroids in diapausing pupae of the cabbage army moth *Mamestra brassicae* in response to chilling (Yamada and Mizoguchi 2017). Unchilled pupae do not release ecdysone, but upon transfer to 4°C pupae begin secreting ecdysone within one week and continue to do so throughout the chilling period, but development is not initiated by high temperature until the pupae have been chilled for two months. The low level of ecdysteroid release during the chilling period likely capacitates the pupa for completion of diapause, possibly through some covert action of the hormone.

Another way to determine the diapause end point would be through an **energy-sensing mechanism** capable of monitoring energy reserves and then using this information to end diapause when a critical amount of reserves have been depleted

(Hahn and Denlinger 2007, 2011). Such a system, although not well defined, is absolutely critical for determining when reserves approach a level that cannot sustain completion of the post-diapause development needed to reach the next feeding stage. In many cases the insect must retain sufficient reserves to fuel the costs of post-diapause metamorphosis and perhaps to migrate to a distant destination. If an energy-sensing mechanism is involved one would predict that diapause duration could be altered by manipulating rates of metabolism during diapause. Higher metabolic rates should translate into a shorter diapause. This is precisely what is seen for pupal diapause in the flesh fly *Sarcophaga crassipalpis* and related species (Denlinger et al. 1972, 1984, 1988b). Metabolic rates are elevated at high temperatures, and application of juvenile hormone also boosts the metabolic rate during diapause. Cumulative metabolism (measured as total oxygen consumed) expended by the end of diapause is rather constant for low temperature, high temperature, and in response to JH application, even though the duration of diapause in each of these cases varies considerably. This suggests that a fixed amount of metabolic expenditure signals the time to complete diapause, an observation that implies feedback between an energy-sensing mechanism, located perhaps in the fat body, and the coordinating centers in the brain. Small larvae of the blow fly *Calliphora vicina* either avert larval diapause or terminate diapause earlier than well-fed larvae (Saunders 1997), again suggesting some sort of energy sensor. Likewise, smaller larvae of the rice stem borer *Chilo suppressalis* break diapause earlier than larger larvae (Xu et al. 2011), as do smaller pupae of the Chinese citrus fruit fly *Bactrocera minax* (Zhou et al. 2020).

Where would such a sensor be located? The fat body is the logical site, especially since recent mammalian and insect findings point to adipocytes as active participants in regulating metabolism through both localized subcellular mechanisms as well as endocrine interactions between the gut and brain (Baker and Thummel 2007, Arrese and Soulages 2010). Experiments with *Drosophila melanogaster* identify two sets of neurons that, when silenced, produce very fat flies and when overactivated produce very lean flies (Al-Anzi et al. 2009), a result indicating that the insect brain has distinct groups of neurons that regulate feeding and nutrient homeostasis similar to the vertebrate brain-hypothalamic axis. Although the basis for such an energy-sensing mechanism is strong, few experiments thus far have examined this possibility in relation to insect diapause.

One set of experiments suggesting such a feedback mechanism focuses on **cross talk** between the **fat body** and **brain** regulating termination of pupal diapause in the boll-worm *Helicoverpa armigera* (Xu et al. 2012). Gene expression is strongly suppressed in the fat body during diapause, resulting in low levels of tricarboxylic acid (TCA) intermediates circulating within the blood. At diapause termination, the fat body is activated, releasing an abundance of TCA intermediates that act on the brain to prompt development by stimulating release of prothoracicotropic hormone and subsequently ecdysteroid production. This conclusion is supported by the fact that diapause can also be broken by injection of TCA intermediates and upstream metabolites. Although the evidence is somewhat limited, the results support the general idea that a flux of metabolites could provide information used to fine-tune diapause progression.

As discussed in Chapter 9, a number of signaling pathways involved in insect diapause, including insulin signaling, Target of Rapamycin (TOR) signaling, adipokinetic hormones, AMP-activated protein kinase (AMPK), invertebrate neuropeptide F (NPF), among others (Hahn and Denlinger 2011) could provide links between energy sensing and diapause regulation.

8.7   Molecular Signatures Marking Transitions to Post-diapause Quiescence and Resumption of Development

Because diapausing and post-diapause quiescent insects must both remain in a developmental hiatus, maintain suppressed metabolism and combat the same environmental challenges, molecular differences distinguishing these two states may be subtle or not apparent at all. For example, conspicuous signatures of diapause in pupae of the flesh fly *Sarcophaga crassipalpis* include upregulation of genes encoding two heat shock proteins (*hsp23 and hsp70*)and concurrent downregulation of the cell cycle regulator *pcna* (Hayward et al. 2005). No changes are noted in expression patterns of these genes during the transition from diapause to post-diapause quiescence, but this should be expected because the cold protection offered by heat shock proteins and suppression of the cell cycle are needs that persist into post-diapause quiescence.

Several species, however, do reveal signs of molecular changes associated with the transition from diapause to post-diapause quiescence. During diapause in the solitary bee *Megachile rotundata*, respiratory quotients remain around 0.7 until May but then increase to 0.8 in June, a time point thought to mark the transition to post-diapause quiescence (Yocum et al. 2006). This change is accompanied by decreased expression of *hsp70*, but not of the cognate *hsc70*. A similar shift is noted for larvae of the wheat blossom midge *Sitodilposis mosellana*, but in this case it is the cognate form, *hsc70*, that changes: Expression is low during diapause but rises approximately fourfold at the beginning of post-diapause quiescence (Cheng et al. 2016). Though the basis for the change in *hsc70* expression remains unclear, it offers a useful transition marker in this species. Activation of an ATPase, dubbed TIME (Time-Interval-Measuring Enzyme, see Section 8.6.2), nicely coincides with the completion of diapause development in embryos of the silk moth *Bombyx mori* (Kai et al. 1995, Ti et al. 2004). Quite likely additional biomarkers of this type will be found, but few researchers have sought such genes.

Many floodwater mosquitoes, *e.g.*, *Aedes albopictus*, that overwinter in embryonic diapause complete diapause and then enter a period of post-diapause quiescence. What is somewhat unusual is that a quiescent stage identical to post-diapause quiescence is also entered into by nondiapausing individuals. Both use quiescence to await a flooding event that will trigger hatching (Denlinger and Armbruster 2014). Pharate larvae are unable to respond to the flooding stimulus during diapause and become responsive only after they have entered post-diapause quiescence. By contrast, those not entering diapause develop quickly to the pharate larval stage and, until flooding occurs, enter a quiescent phase identical to that seen in post-diapausing pharate larvae.

Transcript differences between diapause and nondiapause gradually decline throughout diapause, and by the beginning of post-diapause quiescence few distinctions remain (Poelchau et al. 2013b). Although changes progress through diapause, few diapause-specific distinctions remain once the quiescent stage is reached. It hasn't been explicitly tested, but the presumption is that both types of quiescent individuals will show the same major shifts in transcription once development is triggered, as seen below for numerous other species.

Changes in gene expression at the resumption of development are huge and occur rapidly. Expression patterns for the above-mentioned signature genes that remain unchanged during the transition from diapause to post-diapause quiescence in *S. crassipalpis* are quickly altered at resumption of development. The heat shock protein gene *hsp70* ceases to be expressed within 12 hours after development is stimulated (Rinehart et al. 2000), and a dramatic boost in expression of the cell cycle regulator *pcna* is evident within this same timeframe (Tammariello and Denlinger 1998), as shown in Figure 7.14. A transcriptomic analysis of *S. crassipalpis* shows dramatic changes in patterns of gene expression across a broad spectrum of functional categories within 24 hours after diapause is broken, a change that closely aligns the new profile with patterns observed in nondiapausing pupae (Ragland et al. 2010).

Within 16 hours after the transfer of diapausing pupae of the sweet potato hornworm *Agrius convolvuli* to a diapause-terminating temperature, cytochrome c oxidase activity and the mRNA encoding COX1 are highly elevated in the brain (Uno et al. 2004), a response consistent with the anticipated rise in metabolic demand as development resumes.

8.8   Triggering Resumption of Development

Following post-diapause quiescence, development resumes in response to ecologically appropriate environmental conditions that foretell the end of the unfavorable season. For many overwintering insects, the **rise in spring temperature** prompts the beginning of development, although quite a few days may elapse between resumption of development and its completion. For example, the transition from post-diapause pupal quiescence to the onset of pharate adult development for the flesh fly *Sarcophaga bullata* in central Illinois occurs in early April, but it takes until mid-May for the flies to complete pharate adult development and emerge as adults (Denlinger 1972a). In this case, as in many others, it is the rise in temperature that permits resumption of development. In diapausing adults of the bluegrass weevil *Listronotus maculicollis* 62 degree days above a threshold temperature of 7.8°C must be reached to initiate post-diapause development (Wu et al. 2017). In the deer ked *Lipoptena cervi*, chilling is not essential for breaking pupal diapause, but low temperatures in the northern boreal zone suppress development until higher temperatures permit adult emergence in mid-August (Härkönen and Kaitala 2013). Soil temperature must exceed 11°C for post-diapausing adults of the Colorado potato beetle

Leptinotarsa decemlineata to emerge from their underground overwintering site (Lefevere and de Kort 1989).

Food availability and **moisture levels** also feature prominently as cues for resumption of development in some species. Though some species or individuals carry over sufficient reserves to begin reproducing without further feeding, many rely on a spring food source to initiate reproduction, *e.g.*, Colorado potato beetle *L. decemlineata* (Lefevere and de Kort 1989) and the linden bug *Pyrrhocoris apterus* (Hodek and Hodková 1986). As mentioned earlier, when the tarnished plant bug *Lygus lineolaris* is denied food or offered a less-favored host plant diapausing adults of a Mississippi population terminate diapause in January, but when a highly favored food plant, henbit, is present they end diapause and reinitiate development in December (Snodgrass et al. 2012), suggesting that not only is food presence a factor but also food quality. As diapause progresses, intensity declines in the tropical bruchid *Bruchidius atrolineatus*: Flowers or pods of the host plant *Vigna unguiculata* elicit no effect on diapause termination in early diapause, but in late diapause their presence promotes resumption of reproductive development (Glitho et al. 1996).

The importance of water and other forms of moisture in relation to diapause is nicely summarized in a review by Hodek (2003). Water appears not to be particularly important for diapause development but is sometimes a cue for resumption of development. The importance of water uptake for resumption of development was recognized in early experiments with diapausing embryos of the grasshopper *Melanoplus differentialis*: Removal of the egg's waxy coating with xylol or other solvents permits absorption of water, triggering resumption of development (Slifer 1946).

Numerous tropical species rely on the **onset of the rainy season** for resumption of development (Denlinger 1986). Just as insects from temperate latitudes await the rise in spring temperatures to resume development, many tropical species complete diapause long before they reinitiate development and remain in post-diapause quiescence until the return of the rains, a cue that reliably signals the advent of new plant growth, thereby facilitating insect-host plant synchrony. For *Stenotarsus rotundus*, an endomychid beetle from Central America, adult diapause is initiated in July/August, and in response to the slight increase in daylength after the winter solstice, adults gradually attain reproductive competence and redevelop their flight muscles over the next few months but, although primed to reproduce, they remain in post-diapause quiescence until the onset of the rains in late April (Wolda and Denlinger 1984, Tanaka et al. 1987a). A similar scenario seems to operate for diapausing adults of the red locust *Nomadacris septemfasciata*; adults are in a stage of post-diapause quiescence prior to the onset of the rainy season in Madagascar and respond quickly by becoming reproductively active when the rains arrive (Franc and Luong-Skovmand 2009). The sorghum stem borer *Busseola fusca* bridges the dry season in Kenya in larval diapause; early in diapause larvae are unresponsive to rainfall but in late diapause, after they have experienced a dry season, they respond quickly to rainfall by terminating diapause and pupating (Okuda 1991). Likewise, the galling insect *Eurytoma* sp. diapauses within dead leaves that have dropped to the forest floor during the dry season, but simulated rainfall prompts adult eclosion (Barbosa and Fernandes 2019).

Adults of the weevil *Sternechus subsignatus* spend the dry Argentina winter inside an underground hibernaculum, but respond to spring rains by emerging from the soil, an event that occurs with a delay of approximately 15 days after a critical rainfall threshold of 80 mm (Socías et al. 2016). In cases such as this, moisture likely not only signals new plant growth but also softens the hard upper crust of the soil, making it easier for adults to reach the surface. Resumption of development in diapausing embryos of the balsam twig aphid *Mindarus abletinus* is prompted by both elevated temperature and water absorption (Doherty et al. 2018). Unless eggs are soaked in water few successfully hatch; a mere increase in relative humidity does not suffice. Embryonic development in the tropical grasshopper *Oedaleus senegalensis* resumes after rain elevates moisture levels in the soil at the beginning of the rainy season or in response to elevated relative humidity in the laboratory (Gehrken and Doumbia 1996). Post-diapause development in the wheat blossom midge *Sitodiplosis mosellana* is triggered by both optimal temperature (22–26°C) and soil moisture (20–60% dry weight); levels of moisture below or above these optima dramatically reduce successful adult emergence (Cheng et al. 2017).

A rise in **humidity** alone can also be used as a trigger or accelerant of development. Direct contact with rainfall may not be needed. High humidity (70–80% RH) augments effects of short daylength and the presence of a host plant for termination of adult diapause in the tropical bruchid *Bruchidius atrolineatus* (Glitho et al. 1996). The tropical braconid parasitoid *Microplitis demolitor* also responds to high humidity for diapause termination and/or resumption of development after post-diapause quiescence (Seymour and Jones 2000). Diapausing larvae maintained at 48% RH require 92 days for adult emergence, those transferred to 81% RH emerge in 38 days, while those transferred to 5% RH emerge in 166 days. Removal of the larval cocoon reduces diapause duration from 62 to 23 days, presumably because moisture can more easily reach the diapausing larva. In pupae of the cherry fruit fly *Rhagoletis cerasi*, humidity has no effect on diapause duration, but high RH accelerates the rate of post-diapause development and can thus influence the timing of spring emergence (Moraiti et al. 2020).

A fascinating final step triggering egg hatch in the migratory locust *Locusta migratoria* is **vibration** (Tanaka et al. 2018). Eggs from the same pod that are separated hatch sporadically, but two eggs held in contact with each other hatch synchronously, suggesting there is some type of communication between eggs that prompts synchronous hatching. Eggs held separately but connected by a wire hatch synchronously, as do eggs offered vibrational stimulation by music! All evidence supports the hypothesis that locust embryos use vibrational stimuli from neighboring eggs to synchronize hatching. Although this would presumably not be a response used exclusively by diapausing eggs, it underscores a two-step process of being primed to respond but waiting to do so until a final signal, in this case vibration, is provided. Rainfall, food availability, or more commonly a rise in temperature frequently provide the final permissive step breaking post-diapause quiescence and allowing progression of development.

Another interesting twist on emergence time is seen in the solitary bee *Megachile rotundata*. Eggs are deposited in brood cells constructed linearly, one on top of the other, thus posing an interesting challenge for those deposited first. The solution is to regulate emergence so that the last ones deposited emerge first. Several processes contribute to achieving this end. One of the contributors, noted for males, is that the position of the brood cell impacts post-diapause development rate, with those near the entrance developing more quickly than those at the bottom (Yocum et al. 2014).

Numerous factors thus dictate when diapause ends and development resumes. Together with the timing of diapause induction, the timing of these events at the end of diapause is absolutely critical for the precise coordination of insect seasonal cycles with favorable seasons. Selection for strict adherence to seasonal constraints offers a well-honed mechanism for surviving until the next favorable season returns and for assuring synchrony of development within populations at the end of diapause. Yet, there is also sufficient geographic and within-population variation in diapause duration, as well as attributes of post-diapause quiescence and rates of post-diapause development to facilitate shifts to new host plants and for adjusting to a changing climate and to accommodate migration into new geographic areas.

9   Molecular Signaling Pathways that Regulate Diapause

The endocrine system finely orchestrates the progression of development from embryonic stages, through larval and pupal development, to adulthood and finally to reproductive maturation. The halt in development characteristic of diapause implies a block in the endocrine signaling pathway that, in time, is countered to again reinitiate development. Development is marked by progressive rises and falls in hormone titers, especially titers of the ecdysteroids and juvenile hormones, as well as upstream neuropeptides that preside over the synthesis and release of these growth hormones. Both the rise and decline of hormone titers are critical for developmental progression, and development can be halted by either failing to elevate a specific hormone or by maintaining a hormone titer at a high level, that is, preventing its decline. Both mechanisms are observed in certain diapauses.

Insect endocrinology has deep roots in the diapause literature. Pioneering work by S. Fukuda and K. Hasegawa on embryonic diapause of the silk moth *Bombyx mori* culminated in the discovery of the brain's role in regulating embryonic diapause through the production of the neuropeptide diapause hormone (Fukuda 1951, Hasegawa 1951). Experiments by Carroll Williams using the giant silk moth *Hyalophora cecropia* revealed the critical role of the brain-prothoracic gland axis governing the onset and termination of pupal diapause (Williams 1946, 1952). And, experiments from Jan de Wilde's laboratory on the Colorado potato beetle *Leptinotarsa decemlineata* defined the role of the corpora allata and its product, juvenile hormone, in regulating adult diapause (de Wilde and de Boer 1961). Diapause offered a powerful tool allowing these early workers to identify endocrine glands and hormones that were blocked during developmental arrest. Their insightful conclusions proved useful not only for understanding diapause but for identifying the hormones used to coordinate the grand schemes of insect development.

This chapter explores the roles of the major insect hormones in diapause, as well as upstream connections to the diapause programming centers in the brain and downstream gene networks used to evoke diverse features of the diapause program. Previous reviews provide a historic context for understanding hormonal control of diapause (Denlinger 1985, 2002, Schiesari et al. 2011, Denlinger et al. 2012) and set the stage for emerging developments that define the complexity of gene networks presiding over the diapause response.

9.1 Hormonal Control Systems

This section examines the classic insect hormones and how they contribute to diapause regulation. Though species differences are apparent for each of the stages, regulation of larval and pupal diapauses are fairly similar to each other but distinct from regulatory schemes operating in embryos and adults, thus the three major categories examined are embryonic, larval and pupal diapause, and adult diapause.

9.1.1 Embryonic Diapause

The small size of insect embryos makes it considerably more difficult to perform the sorts of glandular ablations, transplants, and hormone applications that have proven so useful for probing diapauses occurring at later stages of development. Progress is further hampered by the paucity of information concerning the role of the endocrine system in directing events of embryogenesis. Numerous checkpoints exist during embryonic development, a feature underscored by the diverse stages of embryogenesis in which diapause can be manifested (Section 1.2.1). What is especially striking about embryonic diapause is the rich and diverse control mechanisms that prevail, a feature that is all the more intriguing because so few species have actually been examined.

9.1.1.1 Diapause Hormone

Diapause in *Bombyx mori* is dictated by long daylength and high temperature received by the mother when she herself is an embryo and young larva. If she receives those cues early in life, as an adult she will synthesize and release the neuropeptide diapause hormone (DH) that acts on her ovarioles to determine the diapause fate of her own embryos (Yamashita 1996). In the presence of DH, the developing oocytes incorporate sorbitol and the embryos enter diapause. It was originally assumed that the role of sorbitol was simply to enhance cold hardiness for overwintering survival, and that likely is an important contribution, but sorbitol appears to also bring about the developmental arrest (Horie et al. 2000, Iwata et al. 2005b). Diapause-programmed embryos cultured in the absence of sorbitol avert diapause and nondiapause-programmed embryos cultured in a medium supplemented with sorbitol halt development.

DH is a 24-amino acid member of the **FXPRL-amide peptide family** (Figure 9.1). It shares the same C-terminus five amino acid sequence with several other neuropeptides including pheromone biosynthesis activating neuropeptide (PBAN) and pyrokinen (Imai et al. 1991, Sato et al. 1993). DH, PBAN, and three FXPRL neuropeptides of unknown function are encoded in a single reading frame (Figure 9.1), and the resulting prohormone is posttranslationally processed to excise the active peptides, a feature stressing the importance of posttranslational control of hormone release. A single copy of DH-PBAN is present in the *Bombyx* genome, and linkage analysis suggests the locus is on chromosome 11 (Pinyarat et al. 1995). Knockout of DH and the DH receptor genes by TALEN-based mutagenesis results in females that produce

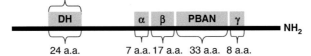

TDMKDESDRGAHSERGALWFGPRI-NH₂

Figure 9.1 Amino acid sequence of diapause hormone (DH) from the silk moth *Bombyx mori* and organization of the precursor polyprotein including locations of DH, PBAN, and three additional subesophageal ganglion neuropeptides.
Adapted from Sato et al. (1993), with permission from National Academy of Sciences, USA.

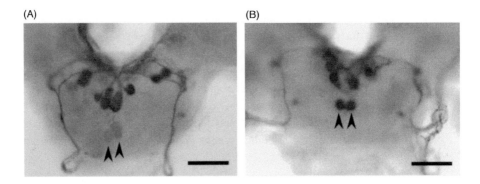

(A)   (B)

Figure 9.2 Localization of neurosecretory cells containing diapause hormone (DH) in the subesophageal ganglion of adult silk moths (*Bombyx mori*) that will produce either (A) diapause or (B) nondiapause eggs. Release of DH leads to a decrease in staining in the diapause-type females. Arrows indicate somata containing DH.
From Sato et al. (1998), with permission from Elsevier.

only nondiapausing eggs (Shiomi et al. 2015), further underscoring the critical role for DH in eliciting the diapause response.

The environmental programming of diapause in *B. mori*, completed during embryonic and early larval development, is encoded within the brain, which in turn directs, via circumesophageal connectives, synthesis and storage of DH within the subesophageal ganglion. A set of seven neurosecretory cells within the subesophageal ganglion (Figure 9.2) is the exclusive site of DH synthesis and storage (Sato et al. 1998, Morita et al. 2003), and axons projecting from those cells pass through the brain into the corpus cardiacum (Ichikawa et al. 1995), the site of DH release. Levels of DH-PBAN mRNA in the brain (Figure 9.3) show elevated expression in diapausing larvae beginning in the fourth larval instar and continuing through to adult eclosion (Xu et al. 1995).

The neurotransmitter **dopamine** plays a role in regulating DH transcription (Noguchi and Hayakawa 2001): Dopamine levels are higher in diapause egg producers during the time DH is being synthesized, DOPA injections boost DH mRNA levels, and feeding DOPA to nondiapause-type larvae or injection of DOPA into

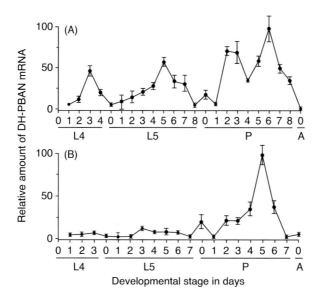

Figure 9.3 Levels of DH-PBAN mRNA in (A) diapause and (B) nondiapause-type females of *Bombyx mori* during the course of development. L4, L5, fourth- and fifth-instar larvae; P, pupa; A, adult.

Adapted from Xu et al. (1995), with permission from Elsevier.

pupae results in the production of diapause eggs by nondiapause-type females. Transcription factors including a member of the pituitary homeobox gene family (Shiomi et al. 2007), as well as a least two members of the POU transcription factor family (Zhang et al. 2004a), also participate in regulating DH transcription. Release of DH from the subesophageal ganglion is controlled by GABAergic neurons (Shimizu et al. 1997). Culturing isolated brain-subesophageal ganglion complexes with the inhibitory neurotransmitter **GABA** reduces release of DH, while incubation with picrotoxin, a GABA inhibitor, promotes release.

The Kosetsu strain of *B. mori* is sensitive only to temperature for the programming of diapause and thus offers a way to uniquely probe temperature's role in diapause initiation (Tsuchiya et al. 2021). Temperatures above 20°C yield diapause in the female's progeny, regardless of photoperiod, while lower temperatures result in nondiapausing progeny. The results identify an important interaction between GABA and **corazonin**, an undecapeptide neurohormone. The temperature effect is thought to be modulated by temperature-dependent expression of a plasma membrane GABA transporter, that, when present, prompts elevation of GABA, thereby inhibiting release of corazonin, an action that prevents corazonin from prompting release of DH. Conversely, in the absence of the GABA transporter, low levels of GABA prevail, allowing release of corazonin and consequently release of the DH needed to evoke the diapause response. The regulatory scheme (Figure 9.4) leading to the release of DH thus involves not only GABA but also an upstream GABA transporter and the downstream action of corazonin.

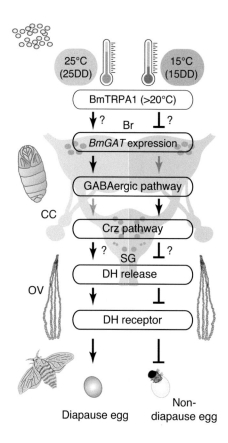

Figure 9.4 Schematic representation of the pathway from maternal reception of temperature cues by embryos of the silk moth *Bombyx mori* through GABAergic and corazonin (Crz) pathways to the release of diapause hormone (DH) and deposition of diapausing eggs by adult females. Br, brain; CC, corpus cardiacum; SG, subesophageal ganglion; TRPA1, thermosensitive channel; GAT, plasma membrane GABA transporter; OV, ovaries.
Adapted from Tsuchiya et al. (2021), courtesy of K. Shiomi (Shinshu University), with permission from National Academy of Sciences, USA.

The ovaries appear to be the sole target of DH in females destined to produce diapausing eggs, but late in the fifth larval instar expression of the DH receptor is also noted in the prothoracic gland, suggesting DH may play an additional role promoting ecdysteroidogenesis late in larval life (Watanabe et al. 2007). The DH receptor is a 436 amino-acid, G-protein-coupled receptor that shows affinity for several FXPRL-amides but exhibits highest affinity for DH, and expression of the gene encoding the receptor nicely coincides with the DH-sensitive stage (Homma et al. 2006). Several key components of the signaling network have been identified. DH prompts upregulation of the enzyme **trehalase** (Su et al. 1994), a protein encoded by both *treh-1* and *treh-2* (Kamei et al. 2011). The rise in trehalase activity correlates with a 1,000-fold elevation in *treh-2*, suggesting that transcription of *treh-2* is largely responsible for the DH-induced increase in trehalase activity. Increase in trehalase activity is accompanied by an influx of Ca^{2+} (Ikeda et al. 1993), resulting in conversion of glycogen to sorbitol, which in turn is incorporated into the egg, where it acts as the developmental arrestant (Horie et al. 2000). DH also reduces the concentration of cyclic GMP in diapause-destined eggs through regulation of guanylate cyclase activity (Chen and Yamashita 1989).

At diapause termination, an **ERK/MAPK**-mediated pathway signals reconversion of sorbitol to glycogen as well as release of ecdysteroids that prompt initiation of development (Fujiwara et al. 2006a,b). Ecdysteroids of maternal origin appear to remain conjugated to yolk proteins throughout diapause, but at diapause termination the conjugates are hydrolyzed, freeing ecdysteroids to initiate post-diapause development (Sonobe and Yamada 2004). **Protein kinase C (PKC)-dependent phosphorylation** is considerably higher in nondiapausing- than diapausing-embryos of *B. mori* (Gu et al. 2020), suggesting this signaling pathway may also be important for diapause and the subsequent processes of diapause termination and resumption of development. Activation of several **protein tyrosine phosphatases** (PTPs) is also linked to the onset of embryonic development once diapause has been broken (Gu et al. 2021), a response suggesting a critical role for PTP signaling in coordinating the progression of development upon termination of diapause.

A huge number (38) of insulin-like peptides, the **bombyxins**, are known from *B. mori*, and at least one, bombyxin-Z1, has an expression profile unique to diapause (Gu et al. 2019). Bombyxin-Z1 appears to be linked to PI3K/Akt signaling, and its expression profile suggests a connection to embryonic diapause, possibly through phosphorylation of glycogen synthesis kinase. Suppression of the **insulin signaling pathway** and elevation of **reactive oxygen species** (ROS) are prominent responses noted during diapause entry in *B. mori* (Gong et al. 2020), suggesting possible commonality with insulin and ROS signaling in regulation of pupal diapause in *Helicoverpa armigera* (Section 9.1.2.4). Transcriptomes of ovaries from diapause- and nondiapause-producing females identify 183 genes with higher expression and 106 genes with lower expression in diapause producers (Chen et al. 2017), indicating that numerous genes are differentially expressed in response to DH or other components of a diapause-inducing environment.

The hormonal scheme regulating embryonic diapause in *B. mori*, fascinating as it is, likely is not common to many other species, but a similar response may operate in some closely related Lepidoptera. The white-spotted tussock moth, *Orgyia thyellina* (family Lymantriidae), also appears to use DH for inducing embryonic diapause, and interestingly, in this species all four FXPRLa neuropeptides encoded in the same reading frame (Figure 9.1) are capable of eliciting the diapause response (Uehara et al. 2011). DH may also contribute to embryonic diapause induction beyond the Lepidoptera, as noted in a species of Orthoptera, *Locusta migratoria* (Hao et al. 2019). Several lines of evidence support this contention: DH-like neuropeptides present in *L. migratoria* boost the incidence of embryonic diapause when injected into fifth-instar nymphs reared under short days, mRNAs encoding DH-like neuropeptides are expressed at higher levels in short-day females, and RNAi directed against DH-like genes greatly reduces diapause incidence in the female's progeny.

DH, documented in numerous Lepidoptera species, has also been implicated in controlling pupal diapause but as a diapause terminator rather than an inducer of diapause (See Section 9.1.2.3). Interestingly, DH may also be involved in the formation of diapausing cysts in the brine shrimp *Artemia*; a DH receptor-like gene is present in *Artemia*, it is expressed during the formation of diapausing cysts, and RNAi knockdown of the receptor gene prevents diapause (Ye et al. 2017). No other embryonic diapauses have been examined nearly as extensively as *B. mori*, but it is evident that very different control mechanisms are at play in some other species.

9.1.1.2   Ecdysteroids

The gypsy moth *Lymantria dispar* diapauses as a pharate first-instar larva, a larva that is fully developed but remains encased within the chorion of the egg. At diapause termination, the pharate larva consumes the remaining yolk, breaks through the chorion, and begins larval life. Several lines of evidence indicate that this diapause is induced and maintained by an elevated titer of ecdysteroids. KK-42, an imidazole derivative thought to block ecdysteroid synthesis, averts diapause in *L. dispar* (Suzuki et al. 1993, Lee and Denlinger 1996), but diapause can be reinstated in such embryos with ecdysteroids (Lee and Denlinger 1997). Similarly, after a period of chilling, diapause can readily be terminated by transferring eggs to high temperature, but this diapause-terminating response can also be blocked with ecdysteroids. Experiments with a nondiapausing mutant of the gypsy moth also support a role for ecdysteroids: The diapause response is restored in such mutants by artificially boosting ecdysteroid levels (Lee et al. 1997). Synthesis of the immune protein **hemolin** (55 kDa), a reliable gut biomarker for diapause in the gypsy moth (Lee et al. 2002), is blocked when a ligature is placed behind the prothoracic gland, and co-culturing the gut with an active prothoracic gland promotes hemolin synthesis, as does an ecdysteroid injection (Lee and Denlinger 1996, 1997). Collectively, this evidence suggests that diapause in *L. dispar* is induced and maintained by a high ecdysteroid titer.

Ecdysteroids of maternal origin are well known to be incorporated into eggs in other species (Lagueux et al. 1981), but at the late stage of embryonic diapause in *L. dispar* (pharate first-instar larva), the ecdysteroids present are likely derived from the pharate larva's prothoracic gland. The prothoracic gland possibly synthesizes ecdysteroids throughout diapause and then shuts down production as a first step in diapause termination. A diapause-promoting effect of ecdysteroids may appear paradoxical, especially considering that lack of ecdysteroids is a dominant feature of most larval and pupal diapauses (Section 9.1.2.1). But, it is well known that a drop in the ecdysteroid titer is an essential component of the signal permitting progression of development (Sláma 1980, Ždárek and Denlinger 1987), and the gypsy moth appears to have exploited this feature to bring about its developmental arrest. Ecdysteroids also possibly direct entry into embryonic diapause in the ground cricket *Allonemobius socius*, as suggested by prominent mRNA signatures of ecdysteroidogenesis noted during pre-diapause (Reynolds and Hand 2009b).

But, interestingly, a low ecdysteroid titer appears to elicit embryonic diapause in some other species. In both the Australian plague locust *Chortoicetes terminifera*

(Gregg et al. 1987) and migratory locust *Locusta migratoria* (Tawfik et al. 2002b), nondiapausing eggs contain approximately three times more ecdysteroids than diapausing eggs. In these species, diapause occurs early in embryonic development, and the ecdysteroids present are likely of maternal origin. The ecdysteroid titer in nondiapausing eggs increases after the stage at which diapause occurs, and the titer also rises when diapause is averted by high temperature (Gregg et al. 1987). A rise in ecdysteroids is noted at diapause termination in *L. migratoria* (Tawfik et al. 2002b), and exogenous ecdysteroids can be used to terminate the diapause (Kidokoro et al. 2006a), thus the locust response consistently points to diapause being caused by the absence of ecdysteroids, similar to the regulatory scheme directing larval and pupal diapause in many other species. Like diapause in the silk moth, termination of embryonic diapause in the migratory locust is linked to activation of ERK (Kidokoro et al. 2006b). The diapause fate of embryos laid by females of *L. migratoria* also appears to be influenced by ROS and insulin signaling pathways linked downstream to phosphorylation of FoxO (Chen et al. 2020).

9.1.1.3   Diapause-Promoting Factor

Diapause in pharate first-instar larvae of the silk moth *Antheraea yamamai* presents yet another intriguing regulatory scenario (Suzuki et al. 1990, 1993). In this species, **KK-42**, the imidazole that blocks ecdysteroid synthesis, is highly effective in terminating diapause in intact pharate first-instar larvae as well as in headless pharate larvae, but diapause persists in an isolated head-thorax preparation, suggesting the presence of a repressive factor in the thorax. Isolated abdomens break diapause whether or not KK-42 is added. These results are interpreted by Suzuki et al. (1990) as evidence for a diapause-promoting factor produced in the mesothorax and a development-promoting factor produced in the midgut region of the second to fifth abdominal segments, a region that is a particularly rich source of peptide hormones (An et al. 1998). During diapause the diapause-promoting factor prevails, halting development. A palmitoyl conjugate of the pentapeptide **Yamamarin** (Figure 9.5), isolated from diapausing pharate first instars of *A. yamamai*, is the potential diapause inducer (Yang et al. 2007b, Sato et al. 2010). It is capable of eliciting a reversible cell cycle arrest and suppressing metabolic rate, thus eliciting a diapause-like state. Intriguingly, Yamamarin is capable of not only arresting proliferation of insect cells but is a powerful arrester of mammalian cell proliferation as well.

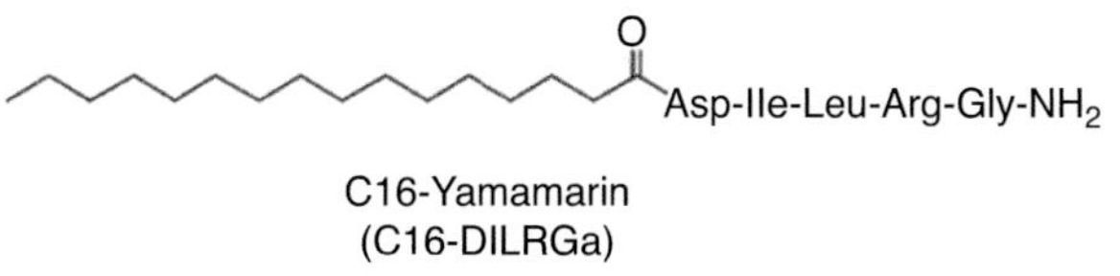

C16-Yamamarin
(C16-DILRGa)

Figure 9.5 Structure of yamamarin, a diapause-inducing compound isolated from diapausing pharate first-instar larvae of *Antheraea yamamai*.
From Sato et al. (2010), with permission from Elsevier.

9.1.1.4 Juvenile Hormones

The pharate first-instar larval diapause of the mosquito *Aedes albopictus* appears to be induced by the absence of juvenile hormone (Batz et al. 2019). While titers of ecdysteroids present in eggs of diapausing and nondiapausing eggs do not differ, the abundance of JH III is considerably lower at the onset of diapause. The argument for JH being lower in diapausing eggs is strongly supported by transcript data showing downregulation of genes in the JH biosynthesis pathway as well as genes that are JH-inducible. A gene encoding JH esterase (involved in JH degradation) is elevated in early diapause (Batz et al. 2019) and remains elevated throughout diapause (Poelchau et al. 2013b), suggesting a mechanism for keeping JH III at a low level throughout the course of diapause in this species. The JH analog pyriproxyfen can terminate diapause in *A. albopictus* (Suman et al. 2015), further supporting the idea that this diapause is maintained by low JH.

Five species examined most extensively, *B. mori, L. dispar, L. migratoria, A. yamamai*, and *A. albopictus* represent divergent hormonal control mechanisms. Diapause of *B. mori* intercedes early in embryogenesis and is dictated by secretion of the neuropeptide diapause hormone, diapause in *L. dispar* occurs late in pharate larval development and is maintained by high ecdysteroid levels, the early embryonic diapause of *L. migratoria* appears to result from an ecdysteroid deficiency, the pharate first-instar larval diapause of *A. yamamai* is regulated, at least in part, by a unique peptide called Yamamarin, and the pharate larval diapause in *A. albopictus* seems to be the consequence of a JH deficiency. The fact that so much diversity is evident in these few species that have been examined and the fact that countless other embryonic diapauses occur in diverse stages of embryogenesis suggest a rich variety of control mechanisms remain to be discovered for embryonic diapause.

9.1.2 Larval and Pupal Diapause

Locking a larva or pupa into developmental arrest implies the hormonal signal for advancing to the next stage has not been received. This suggests failure of the prothoracic gland (PG) to produce ecdysteroids needed to initiate the next molt, and indeed a shutdown in the brain-prothoracic gland axis is usually a conspicuous feature of larval and pupal diapause. Juvenile hormones can also play an important role, especially for larvae that undergo stationary molts.

9.1.2.1 Role of the Brain-Prothoracic Gland Axis

The older literature is replete with experiments showing that brain activation is essential for terminating diapause. Classic experiments by Carroll Williams (1946, 1952) demonstrated that the brain of the giant silk moth *Hyalophora cecropica*, when activated by chilling, is capable of terminating diapause when implanted into a brainless pupa. We now know it is not the whole brain that is required but just a pair of neurosecretory cells in the pars intercerebralis of each brain hemisphere that produce **prothoracicotropic hormone** (PTTH), the neuropeptide that prompts the prothoracic gland (PG) to produce ecdysteroids (Agui et al. 1979, Sauman and Reppert 1996, Wei et al. 2005).

Intracellular electrical recordings from cells that produce PTTH in the tobacco hornworm *Manduca sexta* show that these cells are silent during pupal diapause (Tomioka et al. 1995). Differences between these cells in diapausing and nondiapausing pupae are evident within one day after pupation, and only after a period of chilling do the cells gradually display more spontaneous action potentials and eventually more excitatory postsynaptic potentials characteristic of active cells.

The classic method for monitoring PTTH activity is to culture PG *in vitro* and monitor ecdysone production when brains or brain extracts are added to the medium. With this protocol, low PTTH activity during diapause is commonly noted, for example, larvae of the southwestern corn borer *Diatraea grandiosella* (Yin et al. 1985) and European corn borer *Ostrinia nubilalis* (Gelman et al. 1992) and pupae of the cabbage armyworm *Mamestra brassicae* (Endo et al. 1997). But, this is not always the case. PTTH activity in the brains of diapausing larvae of the wax moth *Galleria mellonella* is as high as in nondiapausing larvae (Muszyńska-Pytel et al. 1993), as also noted for pupae of *M. sexta* (Bowen et al. 1984) and the flesh fly *Sarcophaga argyrostoma* (Richard and Saunders 1987). Larvae of *S. peregrina* destined for pupal diapause actually contain higher PTTH levels than those not destined for diapause (Moribayashi et al. 1992). These observations suggest that PTTH is present during diapause but simply not released. In *G. mellonella* severance of the nerve tract leading from the brain to the PG prompts release of PTTH, thus suggesting the PG fails to synthesize ecdysteroids during diapause due to nervous inhibition from the brain (Muszyńska-Pytel et al. 1993). With recent molecular advances, it is possible to examine the role of PTTH with more precision, as elegantly demonstrated for pupal diapause in *M. brassicae* (Mizoguchi et al. 2013). PTTH is abundantly released into the hemolymph of nondiapausing individuals, followed by a major surge in ecdysteroids (Figure 9.6), whereas very little PTTH or ecdysteroid is detected in hemolymph of individuals destined for pupal diapause. Injection of PTTH into young diapausing pupae averts diapause in *M. brassicae*, consistent with expectations from earlier models for pupal diapause. In this species, PTTH content in the brain of diapausing pupae is higher than in nondiapausing pupae, indicating that the diapausing pupae have sufficient PTTH but that it is not released, a view consistent with other reports cited above but inconsistent with earlier experiments with this species that were based on less reliable *in vitro* experiments (*e.g.*, Endo et al. 1997, cited above).

In *Heliothis virescens* (Xu and Denlinger 2003), *Helicoverpa armigera* (Wei et al. 2005), and *H. zea* (Zhang and Denlinger 2012), the PTTH transcript is expressed at relatively high levels during larval development, but in larvae destined for pupal diapause expression drops at the end of the feeding phase (onset of wandering behavior) and remains low throughout pupal diapause, while expression remains high in larvae and pupae destined for nondiapause. These observations suggest that, in these species, PTTH is regulated at the level of transcription. Thus, there are likely species differences in mechanisms of PTTH regulation. A halt in either PTTH

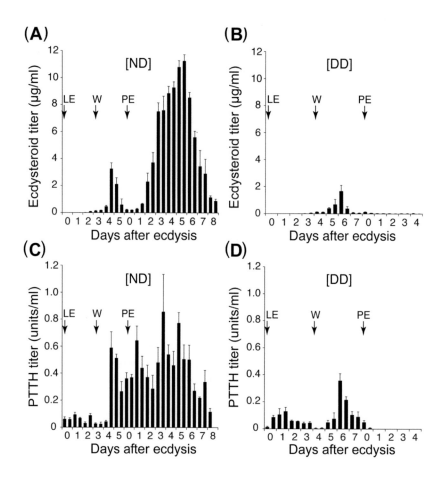

Figure 9.6 Developmental changes in (A and B) ecdysteroid and (C and D) PTTH titers in the hemolymph (A and C) nondiapausing and (B and D) diapausing pupae of the cabbage armyworm *Mamestra brassicae*. Arrows indicate times of final larval ecdysis (LE), onset of wandering (W) and pupal ecdysis (PE).
From Mizoguchi et al. (2013), with permission from PLOS.

synthesis or a failure to release PTTH elicits the same developmental hiatus, and there is no reason to assume that control mechanisms are identical in all species.

The location of PTTH cells within a cluster of **lateral neurosecretory cells** in the brain places them in close proximity to circadian clock cells that are critical for the photoperiodic response in both *Antheraea peryni* (Sauman and Reppert 1996) and *Manduca sexta* (Shiga et al. 2003b), as discussed in Section 5.4. This proximity provides ample opportunity for interactions between these two types of cells, interactions that likely dictate the timing of PTTH synthesis and release. PTTH has another interesting link to light perception: Through its receptor, **Torso**, PTTH acts on two light sensors, **Bolwig's organ** and two light-sensitive neurons in the central brain, to direct larvae to avoid light and seek a dark site for pupariation (Yamanaka et al. 2013b).

A mechanism of this sort is likely especially important for diapausing larvae and pupae that must find an overwintering site that is well concealed and protected.

Aminergic neurotransmitters, including **dopamine** and **serotonin**, have been implicated numerous times in pupal diapause. High dopamine levels in the brain are critical for diapause induction, possibly by preventing release of PTTH, as suggested for diapause-destined larvae of the fly *Chymomyza costata* (Koštál et al. 1998), pupae of the moths *Mamestra brassicae* (Noguchi and Hayakawa 1997) and *Antheraea pernyi* (Matsumoto and Takeda 2002) and butterfly *Pieris brassicae* (Isabel et al. 2001), and adults of the butterfly *Leptidea sinapis* (Leal et al. 2017). A drop in brain dopamine levels directly precedes diapause termination in the silk moth *A. pernyi* (Matsumoto and Takeda 2002). An agonist of the D2-like dopamine receptor delays timing of adult emergence under long daylengths in *A. pernyi*, while an antagonist terminates pupal diapause (Wang et al. 2015b), results consistent with a role for dopamine in locking pupae into diapause, an action possibly linked to preventing adenylate cyclase from generating cyclic AMP. Some of the most compelling evidence suggesting a role for dopamine in diapause induction is the observation that feeding L-DOPA to last-instar larvae of *M. brassicae* reared under long daylengths (not diapause-inducing) elicits a diapause-like state in the resulting pupae (Noguchi and Hayakawa 1997). Among the genes upregulated by both short daylengths and the feeding of L-DOPA is a gene with high identity to a receptor for activated protein kinase C (RACK) (Uryu et al. 2003). As a member of the **protein kinase C** (PKC) signaling cascade, RACK is a potential contributor to the signaling pathway leading to diapause, possibly by maintaining high levels of dopamine, which in turn prevents release of PTTH (Noguchi and Hayakawa 1997). One of the serotonin receptors (5HTR$_\beta$) present on PTTH cells in *A. pernyi* also has the potential for blocking PTTH synthesis or release (Wang et al. 2013). When this receptor is knocked down using RNAi pupae terminate diapause early, and PTTH accumulates in the cell, suggesting that this serotonin receptor plays a role in regulating PTTH synthesis or release, thus serving a role in maintaining diapause.

Not only is PTTH not synthesized and/or released during diapause but its target, the PG, is also refractory during this time, thus providing double assurance that the neuroendocrine system will not generate the ecdysteroid signal needed to promote development. Assays for PTTH activity indicate that PG dissected from diapausing individuals are refractory to stimulation, for example, in larval diapause of the blow fly *Calliphora vicina* (Richard and Saunders 1987) and European corn borer *Ostrinia nubilalis* (Gelman et al. 1992), and pupal diapause of the tobacco hornworm *Manduca sexta* (Bowen et al. 1984) and flesh fly *Sarcophaga argyrostoma* (Richard and Saunders 1987). Loss of competency can be rapid. In *S. argyrostoma* the PG loses competency to produce ecdysone within one to two days after the onset of pupal diapause. The loss is more gradual, occurring over a six-day period, at the onset of larval diapause in *C. vicina*, and even during diapause the gland retains some ability to synthesize ecdysone. When diapausing larvae of *C. vicina* are transferred from 11 to 25°C, the PG regains its competency to produce ecdysone within 24 hours (Richard and Saunders 1987). The neuropeptide **myosuppressin** appears to play an important role in inactivating the PG at the onset of diapause, at least in *Mamestra brassicae*

(Yamada et al. 2017). The PG becomes highly refractory to stimulation by PTTH within a day after pupation, a time point coinciding with a rise in the hemolymph titer of myosuppressin, and PTTH-stimulated activation of the PG *in vitro* can be effectively countered by myosuppressin. Thus, not only is PTTH not produced, but the PG is also actively suppressed during diapause.

How reactivation of the brain-PG axis is achieved remains unknown, but in diapausing pupae of *M. brassicae* ecdysteroid synthesis and expression of the gene encoding PTTH are noted three to five weeks after the onset of chilling, although this initial signal is insufficient to elicit development (Yamada and Mizoguchi 2017). By contrast pupae that remain unchilled fail to show these signs of brain-PG activation. A low-level release of ecdysone prompted by chilling may ultimately capacitate the pupa to terminate diapause. Could such a model offer a timing mechanism for diapause termination in this species? Possibly so, but evidence from other species does not suggest that such a model is widespread as a determinant of diapause duration.

PTTH stimulation of the PG in nondiapausing insects is mediated by **cyclic AMP** as a second messenger, but diapausing glands of *Manduca sexta* (Smith et al. 1986) and *Sarcophaga argyrostoma* (Richard and Saunders 1987) fail to respond to cyclic AMP or agents known to elevate cyclic AMP, implying that the block in gland function occurs beyond the stage of cyclic nucleotide action, possibly the consequence of active suppression of the PG by myosuppression and/or additional regulatory molecules.

Cells of the PG are considerably smaller during diapause in many Lepidoptera species; both cytoplasm and nuclei are reduced in size, and mitochondria take on a swollen configuration (Denlinger 1985). In *Sarcophaga crassipalpis*, there is a slight reduction in cell size within the PG component of the ring gland, but this difference is not as pronounced as in Lepidoptera (Joplin et al. 1993). Curiously, an ultrastructural examination of the ring gland in *S. crassipalpis* reveals extensive arrays of rough endoplasmic reticulum, suggesting that the gland remains active in protein synthesis during pupal diapause. The function of this activity remains unknown but appears to be unrelated to ecdysone synthesis since all evidence indicates that this function is shut down during diapause.

Whether synthesis or release of PTTH is regulated appears to vary with species, but the net effect, a shutdown in the synthesis of ecdysteroids, is consistent. Numerous reports document low or undetectable ecdysteroid titers in larvae or pupae during diapause and an elevation at diapause termination. Examples include larval diapause of the European corn borer *Ostrinia nubilalis* (Gelman and Woods 1983, Peypelut et al. 1990), bamboo borer *Omphisa fuscidentalis* (Singtripop et al. 2002), seabuckthorn carpenterworm *Holcocerus hippophaecolus* (Zhou et al. 2016), and yellow-spotted longicorn beetle *Psacothea hilaris* (Munyiri and Ishikawa 2004), and pupal diapause of the Bertha armyworm *Mamestra configurata* (Bodnaryk 1985), cabbage moth *M. brassicae* (Mizoguchi et al. 2013, Figure 9.6), tobacco hornworm *Manduca sexta* (Friedlander and Reynolds 1992), and flesh flies *Sarcophaga argyrostoma* (Richard and Saunders 1987) and *S. peregrina* (Moribayashi et al. 1988), among others.

Consistent with the model showing absence of ecdysteroids to be central to maintaining diapause is the observation that exogenous ecdysteroids or their analogs are highly effective in terminating larval and pupal diapause. The literature reports numerous examples of this response, including termination of larval diapause in the European corn borer *Ostrinia nubilalis* (Gadenne et al. 1990) and pine sawfly *Diprion pini* (Hamel et al. 1998), and pupal diapauses in several Diptera, including *Sarcophaga crassipalpis* (Ždárek and Denlinger 1975), and Lepidoptera, including *Antheraea mylitta* (Mishra et al. 2008), *Mamestra configurata* (Bodnaryk 1985), *Pieris brassicae* (Pullin and Bale 1989) and *Manduca sexta* (Friedlander and Reynolds 1992, Champlin and Truman 1998). Initiating pharate adult development requires sustained elevation of 20-hydroxyecdysone persisting over several days, thus the diapause-terminating effect of a single ecdysteroid injection likely exerts its effect by stimulating the PG to produce its own ecdysone, perhaps through a feedback loop acting directly on the PG or PTTH cells in the brain (Ždárek and Denlinger 1975). This view is supported by the observation that a small elevation of ecdysone in the hemolymph elicits PTTH release in *Bombyx mori* (Mizoguchi et al. 2015).

Cytochrome P450 enzymes encoded by the **Halloween genes** catalyze a series of hydroxylation steps resulting in the production of 20-hydroxyecdysone. These genes are characteristically expressed at low levels during diapause but show a boost in expression at the break of diapause, as seen in larval diapause of the seabuckthorn carpenterworm, *Holcocerus hippophaecolus* (Zhou et al. 2016). Involvement of the ecdysone signaling pathway is also indicated by experiments with the imidazole derivative KK-42, a synthetic growth regulator thought to block ecdysteroid biosynthesis. KK-42 dramatically delays pupal diapause termination in *Antheraea pernyi* and *Helicoverpa zea* and boosts pupal diapause incidence in *Sarcophaga crassipalpis* (Liu et al. 2015), results consistent with low ecdysteroids being the cause of pupal diapause.

Shutdown of the brain-prothoracic gland axis is an attractive model for larval and pupal diapause regulation. In a simplified version of this model, the PG fails to produce ecdysone because it has not been stimulated by PTTH from the brain. Diapause in *Hyalophora cecropia* nicely reflects this scenario. Chilling of the brain is essential for the release of PTTH, and this in turn stimulates the PG to synthesize the ecdysone needed to break diapause and initiate development. Without a brain, the pupa is locked into a permanent diapause. But not all species conform to this rather simple model (Denlinger 1985). In *Manduca sexta, Pieris rapae*, and *Antheraea polyphemus*, removal of the brain during the first month or so of pupal diapause results in a permanent diapause, but at later stages removal of the brain has no impact (Figure 9.7). The extreme case is noted in *Helicoverpa zea*: Brain removal within the first 4 hours after pupation results in permanent diapause, but within 24h the diapausing pupa is already independent of the brain (Meola and Adkisson 1977). Stationary molts of diapausing larvae of the moth *Sesamia nonagriodes*, as well as pupal and adult molts can also proceed without a brain (Pérez-Hedo et al. 2010, 2011). These results suggest the possibility that PTTH exerts its effect on the PG very early, and after receiving the PTTH signal, the PG assumes the role of the regulatory organ that responds to high temperature by initiating development.

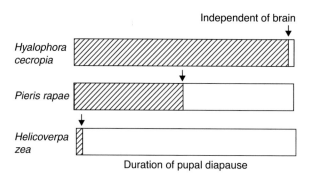

Figure 9.7 Time during pupal diapause at which several species are no longer dependent on the brain for initiation of adult development.
Adapted from Denlinger (1985).

An alternative interpretation, and perhaps the more likely scenario, is that our traditional view of PTTH playing the sole role in promoting ecdysteroidogenesis is not correct. Experiments with *Drosophila melanogaster* show that ablation of PTTH-producing cells has no effect on molting or metamorphosis (McBrayer et al. 2007), and further *Drosophila* experiments indicate that multiple factors, in addition to PTTH, act on the PG to regulate ecdysteroidogenesis (Marchal et al. 2010, Yamanaka et al. 2013a). Although PTTH plays a central role, it is not the only developmental signal triggering the onset of ecdysone biosynthesis. Additional trophic signals, especially nutritional signals, reflect the state of general metabolism and act directly on the PG, a result implying that multiple static regulators act to suppress ecdysteroidogenesis. One particularly intriguing observation is the presence of a receptor for **pigment dispersing factor** (PDF) in the PG of *Bombyx mori* and the responsiveness of the PG to PDF stimulation (Iga et al. 2014). This feature provides a direct conduit between circadian clock genes and ecdysteroidogenesis. The PG emerges as the prime decision-maker, responding to both stimulation and suppression. Although these conclusions are largely derived from studies of *Drosophila* development, complexity of the PG response is likely equally valid for diapause control and may help explain some of the anomalies in PG regulation reported in the diapause literature. In spite of the fact that diapausing pupae of *Helicoverpa zea* and some related species can terminate diapause without a brain, intact pupae of *H. zea* show a conspicuous rise in PTTH mRNA when diapause is terminated (Zhang and Denlinger 2012). PTTH retains its role as a major player in the regulation of larval and pupal diapause through its action on the PG, but there are supplementary pathways that also contribute to PG regulation.

Other intriguing results suggest the PG may be activated differently in nondiapausing individuals than at termination of diapause. In *Sarcophaga crassipalpis*, cyclic AMP levels in the brain and ring gland are high in nondiapausing pupae and low in diapausing pupae at the time of diapause onset, and diapause can easily be averted in diapause-programmed pupae if cyclic AMP levels are artificially boosted at the time of diapause entry (Denlinger 1985), an observation compatible with other evidence suggesting that cyclic AMP is used as a second messenger for PTTH. One might thus assume that cyclic AMP would be effective in breaking diapause. Surprisingly, this is not the case. Cyclic AMP injected along with ecdysteroids actually retards the onset of development. By contrast, **cyclic GMP**, which is ineffective in averting diapause, can readily break diapause by itself and will enhance the diapause-breaking effect of ecdysteroids (Denlinger and Wingard 1978). It is not clear why cyclic GMP works to break diapause while only cyclic AMP can avert diapause. A minimal interpretation is that alternative, perhaps unique, prothoracicotropic factors contribute to activation of the PG at the termination of diapause, a conclusion compatible with multiple stimulatory and suppressive input signals.

One component of the signaling pathway leading to ecdysone synthesis is extracellular signal-regulated kinase (**ERK**), a member of the mitogen-activated protein kinase (**MAPK**) family. When pupal diapause is terminated in the flesh fly *Sarcophaga crassipalpis* by either high temperature or hexane one of the first responses noted is activation (phosphorylation) of ERK (Fujiwara and Denlinger 2007). Injection of ecdysteroids does not activate ERK, nor is ERK activated within the ring gland, suggesting that ERK elicits its effect through stimulation of the brain, quite possibly as a component of the cascade transmitting environmental information to PTTH cells within the brain. **Torso**, a tyrosine kinase receptor, functions as the PTTH receptor within the PG, and its activation by PTTH is thought to stimulate ERK phosphorylation, leading eventually to ecdysone synthesis (Rewitz et al. 2009). Thus, one might anticipate seeing ERK activation in the ring gland of *S. crassipalpis* at diapause termination, but instead ERK activation is not seen in the ring gland, only in the brain, epidermis, and fat body (Fujiwara and Denlinger 2007). What is clear is that PTTH release and ERK activation are central to resumption of development but their precise relationship awaits clarification.

The timing of diapause termination could also be influenced by the presence or absence of the receptor for the active form of ecdysone, 20-hydroxyecdysone. Two proteins, **Ecdysone receptor** (EcR) and **Ultraspiracle** (USP), dimerize to form the functional receptor. 20-Hydroxyecdysone binds to this complex, which in turn binds directly to ecdysone response elements to elicit specific gene action that prompts the onset of development. In the absence of a functional receptor the tissues cannot respond, thus having a receptor in place is essential for initiating development. During pupal diapause in *Sarcophaga crassipalpis* expression of the mRNA encoding EcR persists throughout diapause, but expression of USP gradually declines at the onset of diapause, remains undetectable throughout mid-diapause, and again appear in late diapause, at a time coinciding with increased sensitivity of pupae to injected ecdysteroids (Rinehart et al. 2001). This suggests that reappearance of USP is a preparatory step, perhaps a critical one, leading to the eventual termination of diapause. In the second-instar larval diapause of the spruce budworm *Choristoneura fumiferana*, transcripts encoding both isoforms of EcR and USP are present throughout diapause, but several ecdysone-inducible transcription factors only reappear at diapause termination (Palli et al. 2001). In *Manduca sexta*, the transcript encoding

EcR disappears at the onset of diapause but reappears when diapause is terminated (Fujiwara et al. 1995), a pattern also observed for larval diapause in the midge *Chironomus tetans* (Imhof et al. 1993), the bamboo borer *Omphisa fuscidentalis* (Tatun et al. 2008) and the seabuckthorn carpenterworm *Holcocerus hippophaecolus* (Zhou et al. 2016). Patterns of expression of EcR and USP during diapause do not appear to be consistent among the few species that have been examined, but the presence or absence of these receptor proteins is likely an important component of the regulatory scheme for larval and pupal diapause.

9.1.2.2   Role of the Corpora Allata

As summarized (Denlinger et al. 2012), early views of larval and pupal diapause focused strictly on the brain-prothoracic gland axis, leaving little reason to suspect the involvement of juvenile hormones (JH), but this view was challenged by a set of experiments on larval diapause of the southwestern corn borer *Diatraea grandiosella* (Yin and Chippendale 1973, Chippendale 1977). What launched their work was the surprising discovery that injection of ecdysteroids into diapausing larvae of *D. grandiosella* prompts a stationary larval molt rather than pupation. A **stationary molt** suggests the presence of JH because the larva would have been expected to pupate in the absence of JH. This suspicion of JH involvement was followed by verification that the hemolymph titer of JH remains high and the corpora allata (CA) are active during diapause. At the end of diapause, the JH titer drops, an important signal permitting release of the ecdysone that triggers pupation, as illustrated in Figure 9.8 (Yin and Chippendale 1973). A similar regulatory scheme is evident in *Chilo* stem-borers, another taxon that undergoes stationary molts during diapause. High JH titers reported for larval diapause of the Mediterranean corn borer *Sesamia nonagrioides* (Eizaguirre et al. 2005) and the yellow-spotted longicorn beetle *Psacothea hilaris* (Munyiri and Ishikawa 2004) suggest that they too have a diapause maintained by JH. Knocking down the transcription factor *FoxO* in diapausing nymphs of the brown planthopper *Laodelphax striatellus* results in a reduced JH titer and a concomitant reduction in duration of nymphal diapause (Yin et al. 2018), indicating that JH may also maintain this diapause. Though high JH esterase (JHE) activity involved in JH degradation distinguishes nondiapausing larvae of *P. hilaris* from their diapausing counterparts (Munyiri and Ishikawa 2004), no such differences are noted in *S. nonagrioides* (Schafellner et al. 2008), suggesting that, as known from many studies, the JH titer is regulated by several synthetic and degradation mechanisms that may vary in importance among species. In the cricket *Modicogryllus siamensis* short daylengths that elicit extra nymphal molts characteristic of diapause appear to be driven by photoperiodic upregulation of *juvenile hormone O-methyl transferase*, a gene encoding the enzyme that converts an inactive JH precursor to active JH (Miki et al. 2020).

Species that do not undergo stationary molts presumably do not need to maintain a high JH titer during larval diapause, a conclusion supported by evidence from several species of Lepidoptera, for example, European corn borer *Ostrinia nubilalis* and codling moth *Cydia pomonella* (Denlinger 1985), and the beet webworm *Loxostege sticticalis* (Jiang et al. 2011). In these species, JH is actually high at the onset of

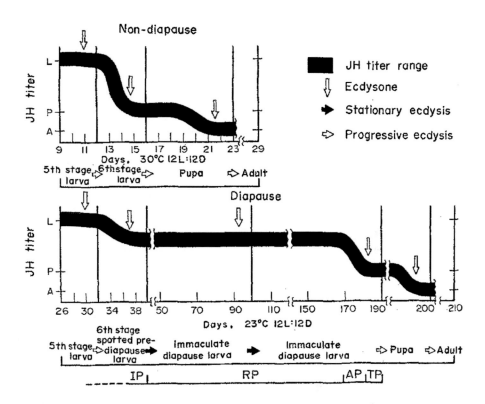

Figure 9.8 Scheme illustrating the titers and timing of juvenile hormone (JH) and ecdysone secretion in nondiapausing and diapausing larvae of *Diatraea grandiosella*. L, P, and A represent the JH thresholds for larval, pupal, and adult development. IP, RP, AP, and TP indicate the inductive, refractory, activated, and termination phases of diapause. From Yin and Chippendale (1973), with permission from Elsevier.

diapause, but the titer drops to low levels shortly after diapause is initiated, and an edysteroid injection elicits pupation, not a stationary molt. Why JH is high early in diapause in these species is not clear, but it does not appear to be involved in the induction process, as evidenced by the fact that application of JH to nondiapause-destined larvae does not result in diapause.

Application of JH can terminate larval diapause in the bamboo borer *Omphisa fuscidentalis* (Singtripop et al. 2000, Tatun et al. 2008), a response reported as well for a number of other larval and pupal diapauses (Denlinger 1985). JH presumably exerts this effect by stimulating the PG to produce ecdysone (Singtripop et al. 2008). Expression of the transcript encoding JH binding protein is elevated by either a JH analog or 20-hydroxyecdysone, and simultaneous application of the two hormones results in an additive response (Ritdachyeng et al. 2013), suggesting that both hormones may be involved in activating the PG at diapause termination. JH analogs can also stimulate the PG in debrained larvae of the moth *Sesamia* nonagrioides (Pérez-Hedo et al. 2011), again reinforcing the point that numerous agents appear capable of activating the PG. Whether JH plays a role in naturally breaking diapause in these two moth larvae remains unclear. The possibility of a combined role for JH and ecdysteroids is, however, likely in some species. In flesh flies, JH by itself has no effect in breaking pupal diapause, but dramatic synergism is noted when JH is applied in concert with a small dose of ecdysteroid (Denlinger 1979). When diapause is terminated naturally in flesh flies, a small peak of JH activity precedes the sustained elevation in the ecdysteroid titer, thus the combined application of JH and ecdysteroid likely mimics the natural reactivation process.

The contribution of JH to larval diapause goes beyond simply locking the larva into a stationary molt. In *Diatraea grandiosella*, application of a JH analog to a nondiapausing larva induces a diapause-like state and prompts accumulation of hexamerin within the fat body, a response normally only seen in diapausing larvae (Brown and Chippendale 1978). Conversely, allatectomy of diapausing larvae suppresses accumulation of the protein. In pre-diapausing larvae this storage protein reaches a maximum titer at the onset of diapause, is gradually utilized during diapause, with only trace amounts remaining by the end of diapause. JH also contributes to the regulation of certain metabolic events occurring during pupal diapause. In flesh flies (*Sarcophaga* sp.), JH is involved in the regulation of metabolic cycles that persist throughout diapause (Denlinger et al. 1984). In these flies, the metabolic rate is not constant but is periodically elevated, displaying cycles with a periodicity of four to six days (Section 7.4). These cycles are driven by JH. The hemolymph titer progressively increases during the trough of the cycle, reaches a critical point, triggers a bout of high oxygen consumption, and then drops precipitously. The rapid decline in JH titer is aided by a rise in JH esterase (Denlinger and Tanaka 1989). Removal of the CA component of the ring gland causes diapausing pupae to become acyclic (Denlinger et al. 1988b). Application of JH to pupae just before they enter diapause does not alter the diapause fate of the pupae, but such pupae fail to exhibit oxygen consumption cycles and instead consume oxygen at a high rate throughout diapause. Pupae treated this way have a much shorter diapause, presumably because their high metabolic rate prematurely exhausts their energy reserves. This conclusion assumes an operative mechanism used by pupae to monitor and respond to depleted energy reserves (Hahn and Denlinger 2011). Few other diapausing pupae exhibit such metabolic cycles, thus this particular role for JH during diapause may be restricted to flesh flies and a few others.

9.1.2.3   A Role for Diapause Hormone

Though diapause hormone (DH) is well known as an inducer of embryonic diapause in the silk moth *Bombyx mori* (Section 9.1.1.1), attempts to induce diapause with DH in most other species have failed. Yet, a gene encoding a DH-like neuropeptide is widely distributed within the Lepidoptera, including species that lack embryonic diapause (Denlinger 1985). This suggests that DH may have some other primary function and was simply captured by *B. mori* as a diapause regulator.

Somewhat surprising, DH also plays a role in pupal diapause, but the effect is just the opposite of what is seen in *B. mori*. Rather than promoting diapause induction, DH breaks pupal diapause in species within the *Heliothis/Helicoverpa* complex, including

H. virescens (Xu and Denlinger 2003), *H. armigera* (Zhang et al. 2004b,c), and *H. zea* (Zhang et al. 2008), as summarized (Zhang et al. 2015a). The DH titer is high in larvae and pupae not destined for diapause, but in diapause-destined individuals the titer drops at the onset of diapause and remains low until diapause ends, further supporting the idea that low DH is essential for the manifestation of pupal diapause. When diapausing pupae of *H. zea* are transferred to high temperatures to terminate diapause, the mRNA encoding DH-PBAN is quickly elevated (Zhang and Denlinger 2012). In *H. virescens*, DH-PBAN mRNA is most highly expressed in the subesophageal ganglion complex (Xu and Denlinger 2003). The hemolymph titer of DH is much higher in nondiapausing pupae than in diapausing pupae. Injection of DH into diapausing pupae prompts a major increase in hemolymph ecdysteroids four to eight days after DH injection (Zhang et al. 2004b). DH stimulates ecdysteroidogenesis in the PG both *in vivo* and *in vitro*, and labeled DH binds to the PG membrane, all of which suggest that DH breaks diapause by activating the PG to synthesize ecdysone (Zhang et al. 2004c). DH appears to exert its stimulatory effect on the PG through the diazepam binding inhibitor/acyl-CoA-binding protein signaling pathway (Liu et al. 2005). The transcript encoding the DH receptor in *H. zea*, expressed at low levels throughout diapause, increases only when adult development is initiated (Zhang et al. 2014), consistent with a role for DH in diapause termination. DH-PBAN gene expression is regulated upstream by a **sumoylation** event: High expression of the SUMO gene in *H. armigera* reduces the activity of the transcriptional factor forkhead (FoxA), which in turn downregulates DH-PBAN gene expression, thus contributing to the low DH titer associated with diapause (Bao et al. 2011). All evidence from these noctuid moths consistently indicates that a drop in DH is essential for pupal diapause induction and that DH elevation acts as a trigger for the ecdysteroid surge needed for diapause termination.

That both DH and ecdysteroids can break diapause in heliothine pupae begs the question of how and why two distinctly different hormones elicit the same response. The effect of these two hormones is, however, a bit different and may offer an important clue about their relationship (Zhang and Denlinger 2012). Though ecdysteroids can break diapause at temperature of either 18 or 21°C, DH can do so only at the higher temperature. This suggests a possible scenario linking DH to the temperature-sensing mechanism used to terminate diapause: When the proper temperature message is received DH is released and, perhaps in concert with PTTH, triggers synthesis of ecdysone by the PG. As we know from *Drosophila* (Yamanaka et al. 2013a), stimulation of the PG is not the exclusive domain of PTTH. Involvement of both DH and PTTH in activating the PG is suggested by the simultaneous elevation of PTTH and DH mRNAs in response to diapause-terminating stimuli (Liu et al. 2005, Zhang and Denlinger 2012).

The sequencing of DH in *Heliothis/Helicoverpa* made it possible to develop a number of highly potent agonists and at least one antagonist of DH (Zhang et al. 2009, 2011a). The 24 amino acid DH sequence contains an active core of 7 amino acids located at the C terminus of the native DH, and using this information it was possible to develop rationally based **agonists** and an **antagonist**. One of the DH agonists is at least 50× more potent in breaking diapause than native DH, some agonists also can be injected into larvae to prevent diapause, and one of the agents actually blocks termination of diapause (Zhang et al. 2011a). Such agents offer potential tools for disrupting diapause, but delivery across the insect integument or into the food source remains a formidable challenge (Zhang et al. 2015b). DH can also be delivered to larvae by having them feed on transgenic tobacco plants that have been modified to express the peptide, but attempts thus far to disrupt the diapause response with this method have not succeeded (Zhou et al. 2016).

The ubiquity of the DH-PBAN gene within Lepidoptera, including many species with a larval or pupal diapause, raises the specter that DH could be influencing diapause in species beyond the heliothine moths. In diapausing pupae of the tobacco hornworm *Manduca sexta*, the expression pattern of the transcript encoding DH-PBAN within the subesophageal ganglion is distinctly different from the pattern noted in nondiapausing pupae (Xu and Denlinger 2004). A conspicuous drop in expression is seen in diapausing pupae nine days after pupation, and at earlier stages (wandering larvae and day 3 pupae) expression is high in diapause-destined individuals but low in those not destined for diapause. The DH-PBAN transcript is also expressed at low levels in the subesophageal ganglion during the early phase of larval diapause (October to December) in the bamboo borer *Omphisa fuscidentalis* but increases in January, reaching a peak in March, just before diapause is terminated (Suang et al. 2015). Such differences suggest possible, but untested, contributions of DH to diapause regulation in additional species of Lepidoptera.

9.1.2.4   Insulin Signaling

The efficacy of bovine insulin in breaking pupal diapause in the butterfly *Pieris brassicae* was the first hint suggesting a role for insulin signaling in diapause (Arpagaus 1987), a discovery that preceded documentation that **insulin-like peptides** (ILPs) do indeed exist in insects. Though mammals have relatively few ILPs, many ILPs are now known from insects (Wu and Brown 2006). *Bombyx mori*, for example, has more than 30 ILPs. Yet, in spite of this abundance of ILPs, all or most ILPs appear to exert their effect through a single receptor. Elegant work with the nematode *Caenorhabditis elegans* identified many key genes linked to dauer formation (the nematode equivalent of diapause) as orthologs of genes in the insulin signaling pathway (Fielenbach and Antebi 2008). A role for insulin signaling is emerging as a common theme among diverse forms of animal dormancy (Hand et al. 2016). A link between ILPs and insect diapause has been examined most extensively for adult diapause, but some information is available for pupal diapause as well. The most extensively studied system in this regard is pupal diapause in the cotton bollworm *Helicoverpa armigera* (Li et al. 2017, Zhang et al. 2017b), but several transcriptomic studies also suggest the involvement of insulin signaling in larval and pupal diapause.

Higher ILP levels are present in the hemolymph of nondiapausing than in diapausing pupae of *H. armigera* (Li et al. 2017, Zhang et al. 2017b). Low ILP levels would traditionally be thought to downregulate Akt, leading to activation of the transcription

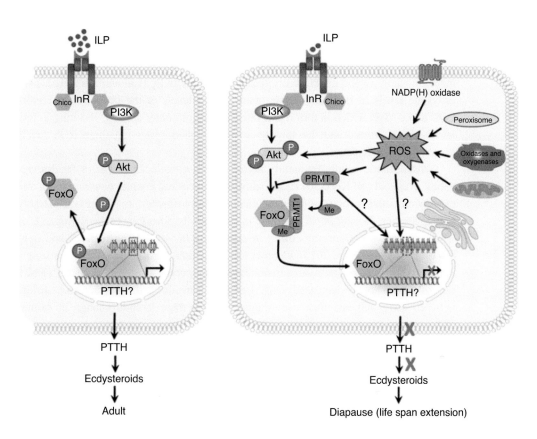

Figure 9.9 Model showing interaction of ROS, insulin signaling (IIS), and epigenetic regulators in control of pupal diapause in *Helicoverpa armigera.* Under long-day conditions, higher levels of ILPs working through IIS relieve FoxO suppression of genes coding for PTTH, leading to continuous development to the adult stage without the interruption of diapause. Under short-day conditions, a combination of low ILP levels, higher levels of ROS and PRMT1 block FoxO phosphorylation, promoting FoxO nuclear localization and suppression of genes coding for PTTH, resulting in diapause. (A black and white version of this figure will appear in some formats. For the color version, please refer to the plate section.)

From Palli (2017) based on Zhang et al. (2017b), with permission from National Academy of Sciences, USA.

factor FoxO, which in turn promotes features of the diapause phenotype, but that is not the path of action here. Akt, p-Akt, and FoxO are all higher in brains of *H. armigera* during diapause, a result not expected to be consistent with the observed low ILP levels and negative regulation of FoxO by p-Akt. Surprisingly, in this case reactive oxygen species (ROS), not ILPs, elicit high expression of Akt and abundant p-Akt in the brain and elevate arginine methyltransferase 1 (PRMT1), thus blocking phosphorylation of FoxO and promoting its accumulation in the brain, as schematically presented in Figure 9.9. The results suggest a mechanism by which the brain regulates diapause through a distinct **ROS-mediated insulin signaling pathway** and highlights an unexpected role for physiological levels of ROS as a regulatory signal (Palli 2017, Zhang et al. 2017b). Activity of hexokinase, the first rate-limiting step in glycolysis, is also a central player in this response; downregulating hexokinase activity reduces metabolic activity and concurrently increases ROS activity (Lin and Xu 2016). Low ecdysone levels in short-day pupae result in low levels of the transcription factors POU and c-Myc, which in turn repress hexokinase promoter activity, resulting in low levels of hexokinase and the subsequent low metabolic rate and high ROS levels associated with diapause.

Transcriptomic results frequently reveal diapause-distinct patterns of insulin signaling (Ragland and Keep 2017), but some differences run counter to expectation, and expression patterns for specific components of the pathway may differ considerably among species. During pupal diapause in the flesh fly *Sarcophaga crassipalpis*, 9 of 12 insulin pathway transcripts differ in expression between early diapause and nondiapause (Ragland et al. 2010). Counter to expectation, eight of those nine transcripts are relatively increased compared to nondiapausing pupae, whereas the other, PTEN, a negative regulator, is decreased. Only 1 of the 12 insulin pathway transcripts differs between early and late diapause, suggesting that there is little change in insulin pathway transcripts throughout diapause. Upon diapause termination, expression patterns of all nine transcripts flip, matching the nondiapause profile. Transcript abundance alone does not indicate whether activity of the insulin pathway is increased or suppressed, and possibly the unexpected high expression of insulin transcripts reflects a readiness to respond, perhaps in preparation for triggering the boost in ecdysteroids needed to terminate diapause.

During larval diapause in the fly *Chymomyza costata*, relatively few changes in insulin signaling transcripts are noted throughout the maintenance phase of diapause, but shifts in both positive and negative regulators are noted at diapause termination (Koštál et al. 2017). Several members of the insulin growth factor pathway are altered in expression as a result of diapause in the European corn borer *Ostrinia nubilalis* (Wadsworth and Dopman 2015). Differences in the insulin signaling signature, noted for the insulin receptor *InR* and *Tsc1* (mediates cross talk between the insulin and TOR pathways), are evident between apple and hawthorn races of the fly *Rhagoletis pomonella* (Meyers et al. 2016), suggesting that the insulin signaling pathway may contribute to differences in pupal diapause duration between these two races. Though insulin signaling has been examined less thoroughly in larval and pupal diapause, it is likely to be just as important for these stages as it is for adult diapause.

9.1.2.5   Other Players?

Quite likely additional hormones contributing to pupal diapause await discovery. Among such candidates is **neuropeptide-like precursor 4** (Nplp4), a peptide precursor with no known function but with a strong association with diapause in the flesh fly *Sarcophaga crassipalpis* (Li et al. 2009a). The open reading frame of this cDNA encodes a 61-amino acid residue precursor protein containing a predicted 18 residue signal peptide, one 22-amino acid and one 2-amino acid propeptide, and a 19-amino acid neuropeptide. Nplp4 mRNA levels are quite low in nondiapausing pupae but are highly upregulated throughout diapause and then drop two days after diapause is terminated. This close association with diapause suggests a potential role for Nplp4 in initiating and maintaining diapause, but its precise function remains unknown. **Inos,** *myo*-inositol-1-phosphate synthase, is another compound of potential interest for diapause in *S. crassipalpis* (Li et al. 2009b). Abundance of the protein Inos is low in the brain during diapause but the protein becomes quite abundant in the brain within 24 hours after diapause termination. The fact that Inos plays a role in a number of signal transduction pathways in mammals suggests it may contribute to early events associated with diapause termination.

In addition to DH, do other members of the FXPRL family of peptides also influence the diapause response? As shown in Figure 9.1, a single mRNA encodes not only DH but also pheromone biosynthesis activating neuropeptide (PBAN) as well as three additional neuropeptides (*a*-SGN, β-SGN, γ-SGN) of unknown function. Though DH has the highest diapause terminating activity in pupae of *Helicoverpa armigera*, the other four neuropeptides are cross-reactive for diapause termination (Zhang et al. 2004b,c). Whether they actually play a role in regulating diapause is not known.

Numerous transcripts and proteins of unknown function emerge from -omics studies, suggesting a rich array of additional factors associated with diapause, many of which may prove to contribute to its regulation.

9.1.3   Adult Diapause

Adult diapause implies a shutdown in reproduction. The most conspicuous feature in females is the arrest of oocyte development, and in males the most obvious attribute is the failure to mate with a receptive female. The status of the testes is not consistently a reliable indicator for diapause in males (Pener 1992): In some species testes are underdeveloped, while in others the testes are replete with mature sperm cysts. Accessory glands in both females and males are usually reduced in size. Flight muscles, especially in some of the Coleoptera, may degenerate during diapause and then regenerate when diapause comes to an end. Much of the early work on adult diapause focused on the role of juvenile hormone (JH), and that still remains the central focus for adult diapause, but more recent roles have been identified for ecdysteroids, insulin-like peptides (ILPs), as well as other neuropeptides.

9.1.3.1   Role of the Brain

As with larval and pupal diapause, the brain is the central control center, dictating activation of the major endocrine organs, and in the case of adult diapause it is control of the corpora allata (CA) that has received the most attention. The JH signaling pathway (Jindra et al. 2013) is central to regulation of all adult diapauses. The brain directs its control over the CA through both neural connections as well as neuropeptides that either promote (**allatotropins**) or suppress (**allatostatins**) activity.

But, the CA apparently can be regulated through additional channels as well. Surprisingly, removal of the pars intercerebralis (PI) or pars lateralis (PL) from females of the black blow fly *Protophormia terraenovae* does not alter JH production, implying that neurons from these regions of the brain are not absolutely essential for mediating changes in JH synthesis (Shiga et al. 2003a). These regions of the brain are, however, important for regulation of diapause: Removal of the PI blocks ovarian development, as in diapause, and PL removal prevents entry into diapause (Shiga and Numata 2000), suggesting that the PL normally provides some inhibition to ovarian development during diapause. In adults of the bug *Plautia stali* the mid-region of the brain produces an agent that inhibits JH biosynthesis in cultures of a corpus cardiacum-corpus allatum complex, an inhibition that is suppressed under long-day conditions (Matsumoto et al. 2013).

As noted for larval and pupal diapause, **aminergic signaling** within the brain exerts a key role in determining how and when environmental signals are relayed to the relevant endocrine cells, as shown by genetically manipulating aminergic neurotransmitter signaling in *Drosophila melanogaster* (Andreatta et al. 2018). High levels of serotonin or dopamine promote diapause, while downregulation of these signaling pathways leads to reduced diapause. Serotonin appears to exert its effect on diapause by inhibition of Insulin-like peptide (ILP) production and release. When Dopamine Receptor (DopR1), present on the CA and fat body, is silenced, diapause incidence drops, suggesting that signaling through DopR1 contributes to the regulation of adult diapause. Exactly how aminergic signaling relates to other known regulators of the CA and insulin cells remains unknown, but the results highlight another important link within the brain's control system. Although there may be variation in how the CA is regulated, consistent evidence points to a shutdown of the CA as the critical endocrine response resulting in diapause.

Among approximately 150 pacemaker neurons in the brain of *D. melanogaster* a subset of 6 pairs of neurons produce allatostatin C (Zhang et al. 2021). Interestingly, these neurons make inhibitory connections to insulin-producing cells, and those cells do indeed have allatostatin receptors, thus providing a direct connection from clock-containing neurons to the cells that produce insulin. Though the functional experiments in this case focus on the circadian pattern of oogenesis, rather than diapause, it seems likely that such a network is also engaged in diapause regulation.

9.1.3.2   Role of the Corpora Allata

Among the best-studied systems probing hormonal control of adult diapause are the Colorado potato beetle *Leptinotarsa decemlineata*, the northern house mosquito *Culex pipiens*, and the linden bug *Pyrrhocoris apterus*. Pioneering work in the 1950s by Jan de Wilde and his colleagues in the Netherlands revealed the central role of JH in regulating diapause in *L. decemlineata* (de Kort 1990). They documented a low JH titer in diapausing, short-day reared adult beetles and its subsequent rise at diapause termination and reported that diapause could be terminated with exogenous JH. Histological evidence shows the CA to be small and inactive during diapause, and surgical removal of the CA from long-day (nondiapause) beetles induces a diapause-like state. Chemical destruction of the CA with precocene II elicits the same diapause-like state in long-day beetles (Bowers 1976). The JH hemolymph titer is a function of

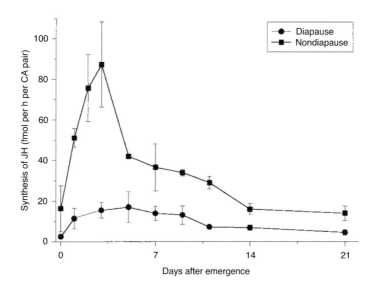

Figure 9.10 *In vitro* activity of the corpora allata (CA) from diapausing and nondiapausing females of the mosquito *Culex pipiens* on different days following adult emergence. Adapted from Readio et al. (1999), with permission from Elsevier.

not only JH synthesis but also its degradation. **JH esterases** (JHE) that degrade JH play a central role in the diapause of *L. decemlineata*. High JHE activity at the onset of diapause contributes to the low JH titers that prevail at this time (Vermunt et al. 1999). A substantial body of information on adult diapause in *L. decemlineata* and other species with adult diapause conforms to this model depicting adult diapause as a JH-deficiency syndrome (Denlinger 1985, Denlinger et al. 2012).

Early results arguing a role for JH in regulating adult diapause in the mosquito *Culex pipiens* are based on the ability of a JH application to terminate diapause in females, but a series of more recent experiments provide ample additional evidence for JH involvement. Cultured CA dissected from diapausing females secrete little JH (Figure 9.10), whereas those dissected from nondiapausing females are quite active, especially during the first week after adult eclosion (Readio et al. 1999). Females that are allatectomized and then transferred to diapause-terminating conditions remain in a diapause-like state and refuse to take a blood meal. Size of the CA is not consistently a sign of low activity, but in *C. pipiens* the small size of the CA from diapausing females is consistent with low synthetic activity (Kang et al. 2014). The argument for JH involvement is further strengthened by gene silencing experiments (Sim and Denlinger 2008). Knockdown of two ribosomal genes, S2 and S3a, causes a diapause-like state which can be reversed with application of JH III (Kim and Denlinger 2010, Kim et al. 2010). RNAi-induced suppression of the gene encoding allatotropin, the neuropeptide that stimulates the CA, results in a diapause-like state in females not programmed for diapause, a response that can be reversed with application of JH III (Kang et al. 2014). The mRNA encoding allatotropin is also expressed at low levels during diapause, further supporting a role for allatotropin in regulating the

CA. By contrast, no difference in expression of allatostatin, a neuropeptide that inhibits JH biosynthesis, is seen between the two types of females. This suggests that, for *C. pipiens,* it is suppression of allatotropin that is the critical link in regulating a shutdown of the CA during diapause.

Research on control of diapause in the linden bug *Pyrrhocoris apterus* from the laboratory of Magdalena Hodková and her colleagues at the Czech Academy provides a rich database for regulation of adult diapause, especially details outlining the brain's role in regulating the CA, as well as regulation of a parallel, nondiapause-starvation-induced reproductive arrest. Figure 7.1 (Chapter 7) shows the conspicuous differences in reproductive systems dissected from diapausing and nondiapausing males and females of *P. apterus.* This species conforms to the expected suspension of JH synthesis during diapause and resumption of development upon application of JH analogs (Hodková 1976, Smykal et al. 2014). The natural JH in this species is JH III skipped bisepoxide. Information from the brain reaches the CA in this species by nervous connections (Hodková and Socha 2006). The CA are inhibited from secreting JH under both diapause-inducing short days and in response to starvation, but the diapause-associated inhibition is much stronger (Hodková 1992, Hodková and Okuda 2019). The short-term inhibition by starvation is readily reversed by disconnection of the CA from the brain, but even if nervous connections to the brain are severed in diapausing females, the CA continue to be inhibited when cultured *in vitro.* JH exerts its effect during metamorphosis by binding to the **JH receptor**, Methoprene-tolerant (Met), and its binding partner **Taiman** (Tai), which in turn upregulates a key JH-inducible protein, **Krüppel-homolog 1** (Kr-h1). To test whether this same JH signaling pathway is linked to breaking diapause in *P. apterus*, these key genes were functionally evaluated by knockdown experiments using RNAi (Smykal et al. 2014). Loss of Met and Tai matched the effects of CA ablation, indicating that the JH receptor (Met-Tai complex) is required to stimulate vitellogenesis at the break of diapause. RNAi directed against Kr-h1, however, had no effect, suggesting that downstream paths for JH may differ for metamorphosis and the onset of vitellogenesis. Among the most conspicuous diapause traits in males of *P.apterus* are reduced synthesis of accessory gland proteins and enhanced synthesis and release of hexamerin into the hemolymph. Knockdown of Met and Tai in nondiapausing males reduces the amount of accessory gland protein and elevates hexamerin levels, consistent with the expectation of JH deficiency characteristic of diapause (Hejnikova et al. 2016). Interestingly, this same treatment does not seem to alter male locomotory activity or mating (Urbanová et al. 2016). A naturally occurring diapause (a natural absence of JH) is clearly more effective in shutting down reproduction than either ablation of the CA or knockdown of genes encoding the JH receptor, suggesting that diapause goes beyond a simple shutdown of these contributing factors.

A sampling of additional research indicates a role for JH in regulation of adult diapause, for example, the blow fly *Protophormia terraenovae* (Shiga et al. 2003a), black rice bug *Scotinophara lurida* (Cho et al. 2007), stink bug *Plautia stali* (Kotaki et al. 2011, Matsumoto et al. 2013), bean bug *Riptortus pedestris* (Dong et al. 2021), plum curculio *Conotrachelus nenuphar* (Hoffmann et al. 2007), leaf beetles

Gastrophysa atrocyanea (Ojima et al. 2015) and *Galeruca daurica* (Ma et al. 2021), cabbage beetle *Colaphellus bowringi* (Liu et al. 2016a, 2017, Guo et al. 2021), moth *Caloptilia fraxinella* (Evenden et al. 2007), and solitary bee *Osmia rufa* (Wassielewski et al. 2011), among others. Experiments with the bean bug *R. pedestris* also confirm that the photoperiodic response regulated by JH operates downstream of JH through action of the transcription factor Krüppel-homolog 1 (Dong et al. 2021), as discussed above for *P. apterus*.

Although experiments consistently point to a role for JH, differences are apparent in how the CA is regulated as well as how the JH titer is maintained. Nervous control of CA activity by the brain is evident in diapauses of *Leptinotarsa decemlineata* (Khan 1988), *Protophormia terraenovae* (Matsuo et al. 1997), *Pyrrhocoris apterus* (Hodková et al. 2001), *Locusta migratoria* (Okuda and Tanaka 1997), among others. Hormonal control of the CA is also evident in numerous systems, including stimulation by allatotropins and inhibition by allatostatins. The contribution of JHE in reducing the JH titer may also differ among species. While JHE is important in suppressing the hemolymph JH titer in *L. decemlineata*, there are no differences in JHE titers between diapausing and nondiapausing adults of the locusts *Nomadacris japonica* and *L. migratoria* (Okuda et al. 1996). Genes encoding three JH degradation enzymes, JH epoxide hydrolase 1 (JHEH1), JHEH2, and JH diol kinase in the cabbage beetle *Colaphellus bowringi* are more highly expressed in diapause-destined beetles, and their knockdown results in a higher JH titer and suppression of some key diapause traits such as fat storage and stress tolerance but does not affect ovarian development (Guo et al. 2019). This observation suggests the JH degradation pathway contributes to the diapause phenotype but does not dictate the ovarian shutdown characteristic of diapause.

The role of the CA may differ between males and females. In some species, for example, the grasshopper *Anacridium aegyptium*, an inactive CA adequately explains the cause of male diapause as manifested by lack of mating behavior (Greenfield and Pener 1992). In others, allatectomy has little effect on male behavior, but females are profoundly affected by CA removal, for example, *Pyrrhocoris apterus* (Hodková 1994, Urbanová et al. 2016) and *Protophormia terraenovae* (Tanigawa et al. 1999). In both of these species, males respond to photoperiod and display mating behavior in the absence of a CA, yet accessory gland development and hexamerin synthesis appear to be robust components under the control of the CA. This disparity suggests that, at least in males, components of the diapause response may be controlled by multiple regulatory pathways. This point is further amplified in diapause of the leaf beetle *Gastrophysa atrocyanea* (Ojima et al. 2015): JH stimulates egg maturation in females but has a limited effect on males. JH accelerates testes development up to the point of spermiogenesis but does not lead to the formation of cysts with mature sperm. An unknown factor is required for maturation of sperm cysts. As this distinction between males and females suggests, absence of JH is likely not the exclusive cause of diapause induction and maintenance, even for females. JH application, at best, can stimulate only a short burst of oviposition in diapausing females of *Leptinotarsa decemlineata* and other species, suggesting that the actual termination of diapause requires a sustained stimulation of the CA from the brain (Denlinger 1985). A minimal interpretation is that, at the end of diapause, the brain must become reactivated to promote JH synthesis by the CA. The brain may independently direct specific events associated with diapause, and some of these events may occur prior to CA activation, as suggested by the observation that adults of *L. decemlineata* dig their way out of the soil prior to activation of the CA (Lefevere 1989). This raises the interesting possibility that activation of the CA may be a consequence of diapause termination, rather than the cause.

JH also is involved in the autumn switch by the pea aphid *Acyrthosiphon pisum* to its sexual mode of reproduction, essential for the production of overwintering diapause eggs (Section 6.9). JH application prevents aphids from making the switch to the sexual mode (Hardie 2010), and several transcriptional signatures of JH synthesis and signaling are conspicuously downregulated in autumn aphids that have already made the switch (Le Trionnaire et al. 2012).

Adult diapause in social insects is also linked to JH. As discussed further in Section 11.3, accumulation of a storage protein, hexamerin 1, influences the JH titer, thereby determining the diapause fate of emerging females (Hunt et al. 2007). If further verified, this scenario is indeed the reverse of the usual assumption that it is JH and ecdysteroids that drive production and accumulation of hexamerins, not that hexamerins drive synthesis of JH (Denlinger et al. 2005).

9.1.3.3   Role for Ecdysteroids

Ecdysteroids, once thought to be restricted to pre-metamorphic life stages, are also present in adults. Rather than being produced by the prothoracic gland, a structure that has degenerated by the time the insect reaches adulthood, ecdysteroids in adult females are derived primarily from the ovaries. Although ecdysteroids have received much less attention than JH in regulation of adult diapause, several studies indicate that they too may contribute. Among species using ecdysteroids to regulate vitellogenesis, a role for ecdysteroids in diapause is perhaps a foredrawn conclusion.

Some of the most compelling evidence for ecdysteroid involvement comes from the *Drosophila* literature. Developmental arrest in this species occurs at the pre-vitellogenic stage of ovarian follicular development (Saunders et al. 1989). **Yolk proteins** (YPs) are present in the female hemolymph, but oocytes fail to incorporate the YPs (Saunders et al. 1990). Ecdysteroids are proposed to activate endocytosis at the termination of dormancy, allowing initiation of vitellogenesis (Richard et al. 1998). During dormancy the ovaries produce low levels of ecdysteroids. The rate of synthesis increases within 4 hours after transfer of flies to dormancy-terminating conditions of 25°C and reaches maximum levels by 12 hours (Richard et al. 1998). Injection of 20-hydroxyecdysone terminates the dormancy as reflected by ovarian maturation and increased abundance of nurse cell proteins involved in endocytosis (clathrin, adaptin, YP receptor)(Richard et al. 2001a,b). Interactions between ecdysone and insulin signaling pathways are also quite possible, especially in light of the observation that insulin receptor mutants in *D. melanogaster* show reduced synthesis of ecdysone (Tu et al. 2002).

How can these results with ecdysteroids be reconciled with the huge body of data suggesting that vitellogenesis in higher Diptera is mediated by JH? JH can indeed break the dormancy of *D. melanogaster* (Saunders et al. 1990), but this response is thought to be an indirect effect of JH triggering the ecdysteroid-mediated initiation of vitellogenesis (Richard et al. 1998, 2001b). When flies are transferred to reproduction-permissive conditions, it is the ecdysteroid titer that goes up quickly (within hours), while the JH titer remains low.

Ecydsteroids have been implicated in adult diapauses of several other species as well. In another fly, *Protophormia terraenovae*, levels of ecdysteroids are considerably lower in hemolymph and ovaries of diapausing females (Tanaka et al. 2013). The source of ecdysteroids in nondiapausing females remains a bit of a puzzle since high ecdysteroid titers persist even after extirpation of the ovaries, one of the likely sources of adult ecdysteroids. Removal of the pars intercerebralis, a potential source of neuropeptides controlling ecdysteroid synthesis, also does not influence the ecdysteroid titer, but surgical removal of the pars intercerebralis does suppress expression of yolk proteins. Ecdysteroid titers are also lower in diapausing than in nondiapausing females of *Locusta migratoria* (Tawfik et al. 2002a). JH application induces ovarian maturation, a response that correlates with elevated levels of ovarian and hemolymph ecdysteroids, similar to levels seen in reproductively active females. A role for ecdysteroids is also proposed for the linden bug *Pyrrhocoris apterus* (Sauman and Sehnal 1997). Under long daylengths maturation of male accessory glands and testes occur during the fourth nymphal instar, but these organs remain immature under short daylengths. An ecdysteroid injection into short-day males induces sexual development, suggesting that low pre-diapause ecdysteroid levels are a prerequisite for diapause in males. That suppression of ecdysone signaling is essential for the reduced JH signaling at the onset of diapause is also elegantly demonstrated in a series of experiments on the cabbage beetle *Colaphellus bowringi* (Guo et al. 2021).

But, not all species show low levels of ecdysteroids in adult diapause. In fact, just the opposite is noted in the Colorado potato beetle *Leptinotarsa decemlineata* (Briers and de Loof 1981). In these beetles, ecdysteroid levels are high in diapause-destined beetles shortly after adult eclosion, but the titer drops within the first week, whereas the ecdysteroid titer is initially low in nondiapausing adults and then increases in females at the onset of reproduction. During diapause the ecdysteroid titer gradually rises (Briers et al. 1982), a feature that may be essential for diapause termination. A combination of JH and ecdysteroids is more effective in breaking diapause than either hormone by itself (Lefevere 1989). Thus, both ecdysteroids and JH likely contribute to the diapause response in this well-studied beetle.

One would assume that adults, lacking a prothoracic gland, would have no role for prothoracicotropic hormone (PTTH), the neuropeptide that plays such a central role in regulating molting and earlier stages of diapause. But, curiously, mRNA encoding PTTH in diapausing adults of the mosquito *Culex pipiens* is expressed at levels nearly double that in nondiapausing mosquitoes during the first month after adult eclosion (Zhang and Denlinger 2011a). Whether the observed mRNA patterns reflect similar differences in synthesis and release of the neuropeptide remains unknown, and what function PTTH might have at the onset of diapause is not at all clear. Does it participate in stimulating ecdysone synthesis or does it take on a completely different role in the adult?

Adults of the nymphalid *Polygonia c-aureum* fail to break diapause in response to an ecdysteroid injection, and no differences are noted in ecdysteroid titers of diapausing and nondiapausing adults (Hiroyoshi et al. 2017). This diapause is strictly associated with a JH deficiency, with no evidence of ecdysteroid involvement.

9.1.3.4   Insulin Signaling

A solid body of information is available supporting a role for insulin signaling in adult diapause, especially for the dormancy of *Drosophila melanogaster* and, to a somewhat lesser extent, diapause in the mosquito *Culex pipiens* and a few other species (Sim and Denlinger 2013a). Although it remains debatable whether reproductive dormancy of *D. melanogaster* is quiescence or diapause (see Section 1.1.3), the power of this model system makes it possible to dissect the role of insulin signaling in ways not available in other systems, and the results likely serve as a valuable template for diapause in other species. Eight **insulin-like peptides** (ILPs), but only one receptor, have been identified in *D. melanogaster*. The different ILPs are independently regulated transcriptionally and appear to have distinct functions, although there is also some redundancy (Nässel et al. 2013).

The first indication suggesting a role for insulin signaling in the overwintering adult dormancy of *D. melanogaster* is the link proposed between overwintering and the insulin-regulated phosphatidylinositol 3-kinase (**PI3-kinase**) gene, *Dp110* (Williams et al. 2006). Using transgenic flies, these authors elegantly demonstrated that deletions of *Dp110* increase the proportion of flies in dormancy, while augmenting expression in the nervous system decreases dormancy incidence. Their data is consistent with a model in which reduced signaling via the insulin/PI3-kinase pathway increases diapause.

Genetic dissection of the insulin signaling pathway provides substantial support for involvement of ILPs as regulators of dormancy in *D. melanogaster* (Kubrak et al. 2014, Schiesari et al. 2016). Ablation or reducing excitability of the **insulin-producing cells** (IPCs) in the brain that produce ILPs-2,3,5, employing a *chico* mutation to impair insulin signaling, or inhibiting ILP2 and 5 in the hemolymph all dramatically boost the incidence of dormancy. By contrast, overexpression of *ilps-2/5* or enhanced excitability of the IPCs results in no dormancy, results consistent with dormancy being invoked by reduced insulin signaling. A transcriptional reporter assay reveals that IPC activity is reduced when female flies are subjected to dormancy-inducing conditions (Ojima et al. 2018). The photoperiodic response providing input to the IPCs appears to be mediated by *pigment dispersing factor* expressing neurons, with output from the IPCs to the corpora allata (CA). Upstream control of the IPCs is under serotonergic control that exerts an inhibition on production and release of the ILPs (Andreatta et al. 2018), as discussed in Section 9.1.2.1. A link between the CA and insulin signaling is also evident from other lines of experimentation with *D. melanogaster* (Mirth et al. 2014). The effect of JH, at least for larval growth, is dependent on

forkhead box O transcription factor (FoxO), which is negatively regulated by the insulin signaling pathway.

ILP-1 is a rather poorly known ILP, but it has an interesting expression pattern in association with dormancy in *D. melanogaster*. The mRNA encoding ILP-1 is expressed in brain IPCs from early in the pupal stage through to the first few days of adult life (Liu et al. 2016c). In females programmed for dormancy, expression persists for at least nine weeks, with levels influenced by feedback regulation from ILP2,3, and 5. Female *ilp1* mutants fail to deposit eggs, similar to the diapause syndrome. Expression levels are influenced by short neuropeptide F and JH, suggesting a potentially important role for this particular ILP in the dormancy of *D. melanogaster*. ILP-1 is also associated with starvation, possibly linking diapause, a nonfeeding stage, to other perturbations of feeding known to be regulated by insulin signaling (Erion and Sehgal 2013). Genome-wide transcriptome changes show upregulation of *ilp2 3,5* and *6*, but downregulation of *ilp8*, a known coordinator of growth and development, and no change in the insulin receptor during dormancy (Kučerová et al. 2016). In spite of upregulation of several *ilps*, a feature also noted during pupal diapause in *Sarcophaga crassipalpis* (Ragland et al. 2010), expression of downstream genes is consistent with attenuation of insulin signaling during diapause. Reduced insulin signaling impacts processes that would need to be adjusted in a fly entering diapause (Allen 2007, Sim and Denlinger 2013b, Sharma et al. 2019), thus making this signaling pathway a particularly attractive regulatory target for diapause.

Males of *D. melanogaster* have received much less attention, but short days and low temperatures also bring about a developmental halt in males: Development of the testes and accessory glands is arrested, as is spermatogenesis (Kubrak et al. 2016). As in females, stores of lipids increase at the onset of dormancy and then decline as they are utilized, expression of genes associated with stress responses and immunity are elevated, and males do not attempt to mate. Like females, dormant males of *D. melanogaster* show evidence of suppressed insulin signaling.

Several lines of evidence also support a role for insulin signaling, as well as JH signaling, in regulating adult diapause in the mosquito *Culex pipiens* (Sim and Denlinger 2008). Knockdown of the insulin receptor in long-day females halts ovarian development in a stage comparable to diapause, and this arrest can be reversed by application of JH, thus suggesting that insulin signaling may be involved in the pathway leading to JH synthesis. A role for insulin signaling is further suggested by experiments focusing on knockdown of the transcription factor **FoxO**. When the insulin signaling pathway is activated, FoxO is normally suppressed, thus the likely scenario is that FoxO is suppressed in long-day females that develop without diapause. This scenario is consistent with the effect observed when diapause-programmed mosquitoes are injected with dsRNA directed against FoxO. Such females fail to accumulate the huge lipid stores characteristic of diapause, a result consistent with a model that short days lead to a shutdown in the insulin signaling pathway and thereby release the repression of FoxO, leading to the diapause phenotype. Transcripts encoding ILP-1 and ILP-5 are expressed at lower levels in diapausing females, but knockdown of only ILP-1 results in a cessation of ovarian development akin to diapause (Sim and Denlinger 2009b). The RNAi-induced ovarian arrest can be reversed with application of JH, showing a link between insulin signaling and JH production. Levels of FoxO are much higher in the fat body of diapausing females than in their nondiapausing counterparts. Application of JH to diapause-destined females suppresses FoxO and the accumulation of lipid is reduced (Sim and Denlinger 2013b), again supporting the presence of a link between JH synthesis and FoxO that is critical for expression of the diapause phenotype.

Other experiments examining insulin signaling in relation to diapause are rather limited but frequently point to a role for this signaling pathway. Rearing the pea aphid *Acyrthosiphon pisum* under short-day conditions of autumn results in the female laying diapausing eggs (see Section 6.9). Such females express genes encoding ILP-1, ILP-4, and insulin receptor at lower levels than their long-day counterparts (Barberà et al. 2019). Scattered transcriptome analyses also reveal transcript expression patterns suggestive of perturbations in insulin signaling, and likely more such evidence will emerge when this pathway is explicitly examined in additional species.

9.1.3.5   Other Players?

Adult diapause is likely not based strictly on a JH (or ecdysteroid) deficiency. Most examples showing a role for ecdysteroids also imply a role for JH, suggesting roles for both hormones in orchestrating regulation of adult diapause. But, several additional pieces may be missing. The absence of JH is clearly one of the big factors regulating adult diapause, but roles are also evident for insulin and ecdysone signaling in some cases. A suggested role for **ecdysis-triggering hormone** (ETH) in diapause (Guo et al. 2021) is just one such example. The full diapause response can seldom be explained on the basis of what we currently know. Termination of diapause and resumption of reproduction cannot usually be fully initiated with the simple addition of JH. A brief burst of egg-laying, for example, may be prompted, but that differs from the sustained reproductive activity seen in a female that has truly broken diapause. This suggests that additional regulatory paths contribute to bring into play the full response for both entering and exiting diapause. For example, JH may cause a short burst in egg-laying but the brain likely must also be fully engaged to orchestrate completion of diapause and resumption of development. Examples below of cross talk between diverse signaling systems provide hints of the likely complexity integral to the natural response.

9.2   Cross Talk among Signaling Pathways

Though we have a good body of information explaining how the major insect hormones are involved in the diapause response, many underlying gene networks are still poorly understood. Diverse responses are sometimes elicited by the same hormone and in some cases different hormones elicit the same response, features that are suggestive of coordination or cross talk among signaling systems. Maintaining a condition of homeostasis demands cross talk between systems to allow the feedback and fine-tuning needed to maintain stability while continuously adjusting to both external and dynamic internal changes. The complexity of the diapause response, as evident from Chapter 7, also imposes a need for coordination among multiple systems. How can a simple environmental cue be translated into hormonal action that results in the manifestation of so many diverse traits? All of this requires sophisticated communication and interacting gene networks that communicate and amplify a simple response to elicit multiple effects. The discussion below highlights several signaling systems implicated in nutrient regulation, homeostasis, and other physiological and developmental decisions relevant to diapause. Unlike stage-specific hormonal regulation schemes discussed above, many of the signaling pathways described below transcend stage specificity, impacting multiple stages of diapause. The fact that most of these signaling systems are interconnected implies they do not operate independently but act in concert to facilitate a coordinated response.

9.2.1   Insulin Signaling

As suggested from some of the above examples, involvement of insulin signaling appears to be a widespread, if not universal, component of diapause regulation (Sim and Denlinger 2013a). This signaling pathway, well known for its role in regulating carbohydrate and fat metabolism in mammals, is potentially involved in many aspects of the diapause phenotype. It is a promising candidate for growth regulation and for regulating reserves during diapause. Regulatory roles for insulin signaling during diapause are documented for embryonic (Section 9.1.1), larval (Section 9.1.2), pupal (Section 9.1.2) and adult (Section 9.1.3) diapause, and thus will not be discussed extensively here. Many fascinating questions about the insulin signaling system in insects remain unanswered. Especially bewildering is the huge number of insulin-like peptides (ILPs), as many as 38 in the silk moth *Bombyx mori* (Wu and Brown 2006). What does it mean to have so many different ILPs? Although it was originally thought that insects had only a single insulin receptor, two to three receptors are now known for some insect species (Smykal et al. 2020), but we are still at an early stage in understanding the functional significance of these diverse components of the signaling system. Insects are not alone in exploiting insulin signaling for diapause regulation. Insulin signaling features prominently in the dauer state of the nematode *Caenorhabditis elegans* (Fielenbach and Antebi 2008), diapause in the copepod *Calanus finmarchicus* (Christie et al. 2016), and embryonic diapause of the annual killifish *Austrofundulus limnaeus*, as well (Woll and Podrabsky 2017).

The insulin signaling pathway can prompt the PG to initiate ecdysone synthesis (Smith et al. 2014) and the CA to produce JH (Jindra et al. 2013), and downstream, the pathway connects to forkhead transcription factor (FoxO), a transcription factor with numerous targets relevant to diapause. The effects of JH on growth rates are dependent on FoxO, which is negatively regulated by insulin and ecdysone signaling pathways (Mirth et al. 2014), thus multiple interconnections are evident.

The insulin signaling pathway, along with a sister pathway, Target of Rapamycin (TOR), also participates in mediating the utilization of amino acids in protein synthesis and growth (Nijhout and Callier 2013) and interacts with members of the Ras/Raf-mediated MAPK cascades involved in regulating cellular growth. Developmental events of these types are central to the diapause response, as evidenced by experimental results discussed above for diverse stages of diapause.

9.2.2   TOR Signaling

Target of rapamycin (TOR) signaling, like insulin signaling, participates in regulating body size and nutritional status (Grewal 2009), and because the two pathways share many key elements the two are sometimes collectively referred to as the insulin/TOR pathway. In larvae of *Drosophila melanogaster* uptake of amino acids during feeding boosts TOR activity and stimulates larval growth. It is unknown how the fat body interprets and transduces amino acid availability into enhanced TOR signaling, but several fat body amino acid transporters have been implicated in activating TOR signaling, suggesting they are components of a nutrient-sensing mechanism (Colombani et al. 2003). The role played by TOR in nutrient sensing suggests it could be a key participant in diapause decisions such as accumulating additional reserves prior to diapause and utilizing nutrients during and after diapause. Cross talk between insulin and TOR pathways is facilitated by *Tsc1*, hence the difference noted in expression levels of *Tsc1* among populations of *Rhagoletis pomonella* may be linked to this cross talk and possibly contribute to the different durations of diapause in host races of this species (Meyers et al. 2016).

The TOR pathway has not been examined extensively in relation to diapause, but one elegant example examines the extended nymphal growth period associated with overwintering in the cricket *Modicogryllus siamensis* (Miki et al. 2020). While nymphs held at 25°C under long-day conditions emerge as adults in 60 days, those reared under short days undergo additional molts and require up to 140 days to reach adulthood. At a higher temperature (30°C), development proceeds more quickly, but short- and long-day differences still persist. Thus, the slow growth of overwintering nymphs is regulated by both temperature and photoperiod. Based on transcript abundance and RNAi experiments directed against key members of the signaling pathways, it appears that low temperature downregulates expression of ILP and TOR genes to slow the growth rate without affecting the number of molts. The number of molts is regulated by photoperiod, independent of temperature, through the JH signaling pathway. Thus, the JH signaling pathway and the insulin/TOR pathway cooperate to regulate slow development of overwintering nymphs.

Though few experiments have specifically probed the role of TOR signaling in relation to diapause, several transcriptome results suggest a role for TOR signaling in diapause, for example, perturbations in the insulin/TOR signaling pathway are consistently noted in a comparative dataset based on 11 species of Diptera, Lepidoptera, and Hymenoptera (Ragland and Keep 2017).

9.3   The Multiplying Effect of FoxO

Numerous downstream effects of insulin signaling are mediated by the forkhead transcription factor (FoxO), including diapause-relevant processes such as regulation of the cell cycle, development, lifespan extension, suppressed metabolism, fat hypertrophy, enhanced stress responses, among others (Sim and Denlinger 2013b). FoxO is suppressed in the presence of insulin, and when insulin is absent FoxO is activated by phosphorylation and translocated from the cytoplasm into the nucleus where it prompts a number of responses. Knockdown of FoxO in diapause-destined females of the mosquito *Culex pipiens* results in failure to accumulate the huge fat reserves typical of diapause (Sim and Denlinger 2008). ChiP sequencing, used to identify direct targets of FoxO, reveals numerous targets in *C. pipiens* with direct relevance to diapause, including stress tolerance, metabolic pathways, lifespan extension, cell cycle and growth regulation, and circadian rhythms (Sim et al. 2015). Knockdown of three known targets of FoxO (*glycogen synthase, atp-binding cassette transporter, low-density lipoprotein receptor chaperone*) suppresses glycogen and lipid biosynthesis, resulting in reduced accumulation of these vital energy stores needed for diapause, a result that further underscores FoxO's critical role in regulating features of the diapause phenotype (Olademehin et al. 2020). FoxO thus acts as a **master switch** in generating the diapause phenotype. Rather than a multitude of genes acting independently to produce diapause, many diapause characteristics may be prompted by one or more transcription factors such as FoxO that activate diverse networks acting collectively to generate the diapause phenotype. Though FoxO appears to play a central role in diapause, not all features of diapause are elicited through this pathway. FoxO is involved in both fat accumulation and suppression of ovarian development in *C. pipiens*, but these two features are not always linked: The circadian transcription factor *Vrille* selectively influences egg maturation while *Pdp1* influences the pathway affecting fat accumulation (Chang and Meuti 2020), an observation suggesting both complex interconnections as well as independence of various transcription factors in generating components of the diapause program.

Multiple diapause-related features also appear to be under control of FoxO during nymphal diapause in the planthopper *Laodelphax striatellus* (Yin et al. 2018). RNAi directed against FoxO results in decreased JH titers, lower levels of cold tolerance, as well as perturbations of carbohydrate, amino acid, and fatty acid metabolism. In brains of the cotton bollworm *Helicoverpa armigera* elevation of protein arginine methyltransferase 1 (**PRMT1**) blocks **FoxO phosphorylation**, resulting in accumulation of FoxO in brains of diapause-destined pupae, leading to symptoms of diapause (Zhang et al. 2017b). FoxO also appears to regulate the storage of energy reserves and stress tolerance responses during embryonic diapause in the locust *Locusta migratoria* (Chen et al. 2020). DAF-16, a forkhead transcription factor in the nematode *C. elegans* exerts a similar broad impact on dauer formation, as indicated by numerous targets identified by ChiP sequencing, including extended life span, fat accumulation, and other dauer-related functions (Oh et al. 2006).

Several transcriptome comparisons also reveal diapause-related differential expression for components of the FoxO signaling pathway, for example, ovaries of *Bombyx mori* programmed to produce diapausing eggs (Chen et al. 2017), as well as embryos of *B. mori* at the onset of diapause (Gong et al. 2020) and adult dormancy in *D. melanogaster* (Kučerová et al. 2016), but FoxO is activated by phosphorylation, a posttranslational event, thus roles for FoxO can better be probed by examining protein or activity levels rather than transcript abundance.

9.4   Adipokinetic Hormones

Adipokinetic hormones (AKHs), a family of small neuropeptides, eight or sometimes nine amino acids in length, are synthesized in the corpora cardiaca and function to mobilize lipids, glycogen, and occasionally amino acids (Lorenz and Gäde 2009, Gäde et al. 2020). These neuropeptides were originally linked to mobilization of flight fuels, but AKHs are also important for **nutrient homeostasis**, including accumulation and mobilization of lipid and glycogen reserves (Hahn and Denlinger 2011). Functionally, AKHs are similar to the vertebrate hormone glucagon. Functions ascribed to AKH suggest a possible role in diapause.

As summarized by Sinclair and Marshall (2018) and depicted in Figure 9.11, AKH binds to receptors on the surface of fat body cells, boosting intracellular cAMP and Ca^{2+} cascades that activate triacylglycerol lipases, which cleave a fatty moiety from a triacylglcerol, leading eventually to the production of the free fatty acids (FAA) used to generate ATP. As depicted in Figure 9.11, signaling pathways involving JH, insulin, FOXO, among others (Toprak 2020) are involved in the network of signals directing the events of lipid storage and mobilization.

Evidence for AKH involvement in diapause is currently limited, but a connection has been proposed for the linden bug *Pyrrhocoris apterus* (Socha and Kodrik 1999). Diapausing adult females of *P. apterus* mobilize approximately twice as much lipid as nondiapausing females when injected with AKH, a result suggesting that diapausing females have higher sensitivity to AKH for lipid mobilization into the hemolymph. In *D. melanogaster*, a transcript encoding AKH is elevated more than fourfold during the first three weeks of dormancy and remains elevated approximately twofold for the remainder of the dormancy period (Kubrak et al. 2014), a response indicative of a shift in regulation of metabolic homeostasis during diapause.

The large AKH superfamily also includes **corazonin**, a neuropeptide that may also have diapause functions. Corazonin elicits several diverse effects in different species, including promoting pigmentation, a feature that is important in producing dark coloration in locusts (Tanaka 2001) and perhaps in other diapausing insects as well (Section 6.7). Corazonin co-localizes with the clock gene PER in lateral neurosecretory cells within the brain of the tobacco hornworm *Manduca sexta* (Shiga et al. 2003b), suggesting a possible role in timing processes. Ablation of these cells suppresses the diapause response. And, as discussed above for *Bombyx mori*

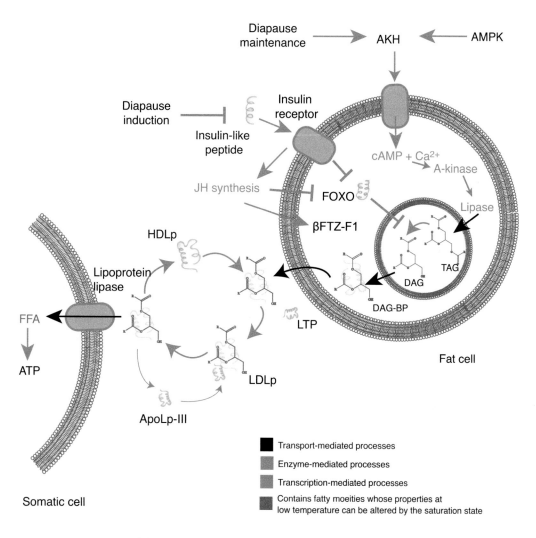

Figure 9.11 Lipid storage and mobilization during diapause, depicting contributions by adipokinetic hormone (AKH) and insulin-like peptides (ILPs). Diapause induction involves inhibition of ILPs, which removes the inhibiting effect of the insulin receptor on FoxO, allowing accumulation of lipids. During diapause maintenance, AKH is produced in response to accumulation of AMP-activated protein kinase (AMPK), resulting in stimulated production of diacylglycerol (DAG) from triacylglycerol (TAG) via a cAMP and Ca^{2+} signaling cascade. DAG is then exported from the lipid droplet into the hemolymph by binding to high-density lipophorin (HDLp) with the help of a lipid transport particle (LTP), forming low-density lipophorin (LDLp). ApoLp-III, apolipophorin 3; DAG-BP, diacylglycerol binding protein; FFA, free fatty acid; JH, juvenile hormone. (A black and white version of this figure will appear in some formats. For the color version, please refer to the plate section.)
From Sinclair and Marshall (2018), with permission from The Company of Biologists.

(Section 9.1.1.1), corazonin participates in the signaling system prompting release of DH (Tsuchiya et al. 2021).

9.5   Wnt Signaling

Among genes commonly downregulated during diapause are members of Wnt signaling pathways (Ragland and Keep 2017), highly conserved pathways that regulate gene transcription and cell proliferation (the canonical Wnt β-catenin-dependent pathway), and noncanonical pathways involved in cell polarity and intracellular calcium regulation. Downregulation of most members of this pathway during diapause is consistent with a halt in development that characterizes diapause. Wnt pathways are interconnected with ecdysone, JH, insulin, and TOR signaling pathways as well, suggesting numerous possible links to diapause regulation.

Among eleven species examined, most transcripts in this signaling pathway are downregulated during diapause (Ragland and Keep 2017), but there is one anomaly: *Spen* is almost consistently upregulated. This gene acts pleiotropically to repress **Notch signaling**, a pathway that participates in regulation of tissue proliferation. Interestingly, high expression of *spen* is also associated with longer diapause duration, suggesting that this pathway may contribute to the differences in seasonality noted among populations of *Rhagoletis pomonella* (Meyers et al. 2016).

As might be anticipated for a signaling pathway involved in growth and cell proliferation, shifts in expression of genes associated with Wnt signaling are recorded at diapause termination or shortly before. Somewhat surprisingly, upregulation of both positive regulators of Wnt signaling (*basket* and *c-jun*) and a negative regulator (*shaggy*) are seen in diapausing larvae of the fly *Chymomyza costata* at this time (Koštál et al. 2017). In diapausing pupae of *R. pomonella*, several members of the calcium-dependent Wnt pathway are upregulated late in diapause, consistent with a function in reinitiating development (Ragland et al. 2011).

Experiments with the cotton bollworm *Helicoverpa armigera* functionally evaluate the role of Wnt signaling in diapause. Ecdysone upregulates the transcription factor c-Myc through the Wnt β-catenin pathway (Chen and Xu 2014). During diapause, when ecdysone is absent, the Wnt pathway is repressed in the pupal brain. Blockage of Wnt β-catenin signaling with an inhibitor simulates the diapause response. Though few other experiments have examined the role of Wnt signaling in diapause, the combined transcriptomic results and the work with *H. armigera* consistently suggest that a shutdown of this pathway is integral to diapause. And, like the other signaling pathways discussed in this section, the role of Wnt transcends the developmental stage of diapause as well as taxonomic classification: Perturbations in Wnt signaling are noted in larval, pupal, and adult diapauses, as well as in species from different orders, including Diptera, Lepidoptera, and Hymenoptera. And, in the mosquito *Culex pipiens*, differences in Wnt signaling are evident prior to the onset of diapause as well (Hickner et al. 2015).

9.6   TGF-β and BMP Signaling

The transforming growth factor-β (TGF-β) and bone morphogenic protein (BMP), polypeptide members of the cytokine superfamily, contribute to embryogenesis, cell growth, cell differentiation, apoptosis, neuronal wiring, synapse remodeling, and cellular homeostasis through signaling cascades that interact with Wnt signaling to phosphorylate **SMADs**, a family of proteins which, in turn, regulate transcription (Massaqué 2012). Thus, in common with Wnt signaling, one might anticipate roles for TGF-β and BMP signaling pathways in the developmental shutdown of diapause, as well as in processes leading to resumption of development at the end of diapause.

Among its diverse roles, the TGF-β signaling pathway contributes to JH signaling and thus has considerable potential as a contributor to diapause regulation. The final reaction in production of JH is conversion of an inactive JH precursor to active JH. This reaction is controlled by JH acid O-methyltransferase (Jhamt), a rate-limiting enzyme encoded by the gene *jhamt*. A member of the TGF-β family, *myoglianin*, downregulates expression of *jhmat*, consequently reducing the JH titer (Ishimaru et al. 2016). This action has a direct connection to nymphal diapause in the cricket *Modicocgryllus siamensis* (Miki et al. 2020), as mentioned in Section 9.2.2, thus underscoring interconnections of multiple signaling systems, in this case JH signaling, insulin/Tor signaling, and the TGF-β signaling pathway.

An interaction between TGF-β and insulin signaling is also evident for pupal diapause regulation in the cotton bollworm *Helicoverpa armigera* (Li et al. 2017b). A high level of phosphorylated Akt (p-Akt) is associated with low insulin levels and the manifestation of diapause. **Protein phosphatase 2A** (PP2A), an upstream regulator of the TGF-β pathway, binds to and dephosphorylates p-Akt, thus the TGF-β pathway negatively regulates the insulin signaling pathway to promote the diapause response. The BMP signaling pathway, a sister branch of the TGF-β superfamily, also participates in regulation of diapause in *H. armigera* (Li et al. 2018a). Both TGF-β and BMP signals are weaker in brains of diapause-destined pupae and downstream targets including p-Smad, POU, and TFAM are less abundant as a consequence. In *H. armigera* TGF-β and BMP appear to activate a common target, Smad1. Smad1 binds to the POU promoter, regulating its expression by boosting expression of TFAM, which in turn regulates metabolic activity. When an inhibitor is used to suppress TGF-β and BMP signals, downstream targets are less abundant and development is delayed. The results are consistent with downregulation of TGF-β and BMP signaling as a component of the diapause program, and activation, which leads to biosynthesis of ecdysone (Gibbens et al. 2011), is thus expected to be critical for post-diapause development.

Transcriptomic datasets also support a role for TGF-β and BMP signaling. Among the most dramatic differences between pupae of *Culex pipiens* destined for adult diapause or nondiapause are transcripts in the TGF-β and BMP pathways, many of which are upregulated in diapause-programmed pupae (Hickner et al. 2015). This species diapauses as an adult, thus the differences noted here occur before the onset of diapause, when diapause is being programmed by environmental cues. Though we might expect these transcripts to be downregulated during diapause, apparently during the programming phase of diapause transcripts in these two pathways are more abundant. One caveat is that both temperature and photoperiod were altered in these experiments, thus some of the differences could be due to the different temperature regimes used. But, temperature and photoperiod are normally both involved in the diapause response thus altering both variables is ecologically relevant.

Involvement of the TGF-β pathway in dormancy goes beyond the insect literature. SMADs are among the TGF-β targets associated with dauer formation in *Caenorhabditis elegans* (Inoue and Thomas 2000), and certain daf genes, for example, *daf-7*, have TGF-β ligands.

9.7   Couch Potato

The gene *couch potato* (*cpo*) has a potentially interesting but ambiguous role in diapause. *cpo* encodes an RNA binding protein and is expressed in diverse tissues including the brain and larval ring gland of *Drosophila melanogaster*. The fact that it has upstream ecdysone response elements and linkage to insulin signaling suggests it may contribute to the gene network regulating diapause, and the hypoactivity associated with the mutated *cpo* gene is reminiscent of diapause. An intriguing set of experiments with *D. melanogaster* first suggested a role for *cpo* in regulation of adult dormancy (Schmidt et al. 2008). Populations from North America with a high dormancy incidence express *cpo* at lower levels than populations with a low dormancy incidence, and allele frequency drops in summer when the dormancy incidence is low (Cogni et al. 2013). Variation in this response is suggested to be the consequence of an amino acid substitution (Lys/Ile), and manipulation of the *cpo* locus by complementation analysis predictably alters the dormancy response. But, curiously, a recent genome-wide association study does not show an effect of *cpo* on diapause in North American populations of *D. melanogaster* (Erickson et al. 2020). A relationship between *cpo* and diapause clines of *D. melanogaster* does not appear to hold for either Australia (Lee et al. 2011) or Europe (Zonato et al. 2016), and new evidence reporting 12 isoforms of *cpo*, some of which have opposing regulatory functions in the ovary (Zhao et al. 2016), suggests the story is more complicated than initially thought.

The work on *cpo* in *D. melanogaster* prompted exploration of this gene in other species. In *D. montana*, another drosophilid with an adult diapause, *cpo* is upregulated throughout diapause (Kankare et al. 2010, Salminen et al. 2015). Deletion and reduced expression of *cpo* does not alter the incidence of adult diapause, but it does shorten the developmental time from egg hatch to adult eclosion (Kankare et al. 2012). In *Sarcophaga crassipalpis*, a fly with a pupal diapause, a *cpo* homolog is downregulated during diapause (Ragland et al. 2010). In adult diapause of the mosquito *Culex pipiens* transcript abundance of *cpo* is nearly the same for both diapausing and nondiapausing females at adult eclosion, but within a week after adult eclosion, *cpo* is much more highly expressed in diapausing females (Zhang and Denlinger 2011b). When diapause is terminated naturally or by an application of JH, expression of *cpo* declines. Comparing *cpo* expression in different populations of *D. melanogaster* is fundamentally different than comparisons between diapausing and nondiapausing individuals within a population, as reported for *D. montana*, *S. crassipalpis*, and *C. pipiens*, but in all of these cases there are perturbations of *cpo*, even if the direction of change is not consistent. Exactly what the role of *cpo* may be in diapause remains unanswered, but it remains a gene with connections to signaling systems of interest for diapause regulation.

9.8 Other Signaling Pathways

In addition to signal transduction pathways mentioned above, pathways including **Notch, Jak-Stat, Hippo, Hedgehog,** among others, are also likely involved in evoking the diapause response. Members of some of these pathways appear in transcriptomic studies but have thus far not been subjected to experimental scrutiny. Several other signaling pathways of demonstrated relevance to diapause have already been discussed in the context of larval/pupal diapause (see Section 9.1.2). These include **Protein kinase C** (PKC) and extracellular signal-regulated kinase (**ERK**), a member of the mitogen-activated proteins kinase (**MAPK**) family. A role for ERK signaling is evident for diapause termination in both embryonic and pupal diapause, suggesting that this pathway is not restricted to taxa or developmental stage of diapause.

The fact that cold adaptations are frequently a component of the diapause program or may be evoked independently in response to cold suggests further linkage with neuropeptide and biogenic amine signaling systems that contribute to mitigation of cold stress (Lubawy et al. 2020). Maintaining homeostasis in a cold environment requires numerous specific adjustments in metabolism, ion regulation, muscle function, and other physiological processes, many of which are vital for winter survival during diapause.

Another intriguing hormone signaling system that may coordinate the integration of energy homeostasis and the timing of diapause events is corazonin. Though a role for corazonin in diapause was discussed above, this neuropeptide plays an additional interesting role in coordinating seasonality in the marine bristleworm *Platynereis dumerilii* (Andreatta et al. 2020). Sexual maturation in this annelid is tightly regulated by both metabolic state and the lunar cycle, and this interaction is coordinated through corazonin signaling. Whether a similar role for corazonin in coordinating energy resources and developmental timing operates during insect diapause has not been explored.

9.9 Cross Talk between Organs

It is imperative that organs communicate to generate a coordinated response. One conspicuous mechanism requiring cross talk is **energy sensing**, the monitoring of energy reserves (Hahn and Denlinger 2011). It is critically important to assess levels of reserves in the body and dictate appropriate action if these reserves become dangerously low. Prior to diapause entry, many insects "decide" to accumulate extra reserves, and during diapause they may switch from one type of reserve to another, and finally, diapausing insects have the capacity to monitor their reserves and end diapause if reserves dip to a certain critically low point. Depletion of reserves without allowing sufficient resources for diapause termination and completion of post-diapause development would, of course, be lethal. This implies a mechanism for monitoring reserves and signaling the brain to take action. Most prominent is a need for communication between the site of energy storage (fat body) and the central coordinating center (brain), but cross talk between other organs may be equally important.

The **fat body** is much more than a deposit of accumulated reserves. Adipocytes that make up the fat body actively participate in regulating metabolism both through localized subcellular mechanisms as well as through endocrine interactions with the gut and brain. The brain remains central, but it must receive input from other organs to inform its response. **Insulin signaling** and its sister pathway, **target of rapamycin** (TOR), are among the most prominent pathways likely involved in energy sensing and monitoring of nutritional status (Hahn and Denlinger 2011). Though not experimentally linked to diapause, release of insulin-like peptides in the brain of *Drosophila* is dependent on activation of the brain's *Methuselah* receptor, and this receptor is triggered by two peptides encoded by the *stunted* gene in the fat body and released into the hemolymph where they act on insulin-producing cells in the brain (Delanoue et al. 2016). This form of organ cross talk modulates insulin levels in response to nutrients and may provide circuitry critical for energy-sensing during diapause as well.

In the context of diapause, cross talk between the fat body and brain is best documented for regulation of pupal diapause in the cotton bollworm *Helicoverpa armigera* (Xu et al. 2012). Gene expression in the fat body is strongly suppressed during diapause, resulting in low levels of TCA intermediates circulating in the hemolymph, but at diapause termination the fat body is activated, releasing an abundance of TCA intermediates into the hemolymph that, in turn, act on the brain to prompt release of PTTH and the subsequent production of ecdysteroids by the PG. This model is supported by the observation that injecting a mixture of TCA intermediates and upstream metabolites (glucose and pyruvate) results in elevation of PTTH within the brain and the subsequent breaking of diapause. Ecdysteroid injection results in elevation of COX activity and ATP content in the fat body (*i.e.,* activation of the fat body) and causes a switch in hemolymph metabolite levels from the diapause to nondiapause profile. Elevation of the **glucose transporter** (Glut) in the brain of diapausing pupae of *H. armigera* (Zhang et al. 2017b) appears to facilitate sensing and absorption of low glucose levels present in the hemolymph during diapause. Together, these results are consistent with cross talk between biochemical processes in the fat body and the brain, components that engage insulin, ROS, PTTH, and ecdysone signaling pathways.

Shifts in metabolic pools at diapause termination in other species, for example, pupal diapause in the tephritid *Bactrocera minax* (Dong et al. 2019, Wang et al. 2020), are consistent with the above results for *H. armigera,* but functional experiments have not tested whether the *H. armigera* observations are broadly applicable.

Cross talk among other organs has received less attention in the context of diapause, but the gut, often underappreciated as an endocrine center, is a major hub of endocrine activity and could very well contribute to the unique feeding and nutrient features of diapause. Unique microbiota associated with the diapausing gut, for example, cabbage beetle *Colaphellus boweringi* (Liu et al. 2016b), could contribute to features of the diapause phenotype such as fat accumulation, perhaps communicating their effects through the gut's endocrine cells, but this connection remains unexplored.

9.10 Epigenetic Mechanisms Implicated in Diapause

Epigenetic processes add a whole new level of complexity to regulation of insect diapause. The term implies features that are "on top of" the traditional basis for inheritance, mechanisms that modify gene expression rather than altering the DNA sequence (Deichmann 2016). The major epigenetic processes, DNA methylation, histone modifications, and small RNA interference, have all been implicated in diapause processes (Reynolds 2017), as might be expected for mechanisms capable of switching gene activity on or off. There are, of course, many protein modifications associated with diapause, as discussed in Section 7.9, but not all represent epigenetic processes. For example, some histone modifications are indeed linked to epigenetic processes, but most probably are not because they lack long-term "memory," a response perpetuated in the absence of the original signal (Ptashne 2013). Since we are still in an early phase of this work with insect diapause adequate experimentation frequently does not allow us to make this distinction. Much of the current evidence for epigenetic involvement in insect diapause is based on transcriptomic findings, thus the results can best be considered putative evidence for epigenetic involvement. C.H. Waddington's use of the term epigenetics is considerably broader and refers to gene-environment interactions that influence the organism's developmental trajectory (Waddington 1952), a definition compatible with many of the processes discussed below.

9.10.1 DNA Methylation

DNA methylation, the addition of methyl groups to cytosine nucleotides, is a widely studied mechanism of chromatin remodeling, but it has received somewhat less attention in the diapause literature, perhaps because Diptera models for diapause lack robust systems of DNA methylation. DNA methylation possibly participates in regulation of embryonic diapause in *Bombyx mori*: A methyltransferase gene is upregulated in diapausing embryos of *B. mori* (Sasibhushan et al. 2013), but a DNA methylation inhibitor fails to alter the diapause incidence (Egi et al. 2016), thus functional evidence is lacking. Evidence is more robust for larval diapause of the jewel wasp *Nasonia vitripennis* (Pegoraro et al. 2016). Rearing *N. vitripennis* under diapause- and nondiapause-inducing photoperiods reveals distinct methylation profiles, and RNAi knockdown of *DNA methylase* or chemical inhibition of the enzyme causes females to produce diapausing eggs regardless of photoperiod, thus providing strong evidence implicating DNA methylation in the induction of diapause in this species. The fact that DNA methylation also is an important component of photoperiod-driven seasonal cycles of reproduction in Siberian hamsters (Stevenson and Prendergast 2013) and in diapause of the annual killifish (Toni and Padilla 2016) suggests a potentially wide-ranging, but still underappreciated, role for DNA methylation in regulating seasonality.

9.10.2   Histone Modification

Histones are highly basic proteins that **remodel chromatin** by acting as spools to order DNA into structural units called nucleosomes. Some histones (H3 and H4) have tails that can be covalently modified to increase or decrease interactions between the DNA strand and the nucleosome, thereby altering transcription at a specific DNA site. Increased interaction generally leads to a compact chromatin structure that is transcriptionally silent, while decreased interaction between histones and DNA yields relaxed chromatin that is transcriptionally active. Histone modification refers to the posttranslational modification of histone proteins by methylation, phosphorylation, acetylation, ubiquitylation, or sumoylation, all of which modify gene expression by altering chromatin structure. Histone methylation and acetylation have thus far been linked to diapause responses.

Transcriptome profiles comparing diapausing and nondiapausing larvae of the fly *Chymomyza costata* reveal a dramatic and rapid downregulation of *dpy-30* associated with diapause (Poupardin et al. 2015). This gene encodes a subunit of an enzymatic complex important for H3 (H3K4) histone methylation, a methylation event usually associated with transcriptional activation of genes involved in cellular growth and proliferation, cell cycle, DNA replication, and other processes that can be expected to be integral to diapause. Genes encoding two histone demethylases, *LSD1* and *Su(var) 3-9*, are upregulated in photosensitive larvae prior to the onset of pupal diapause in the flesh fly *Sarcophaga bullata* (Reynolds 2017), suggesting they may have roles in the programming phase of diapause.

Histone modification by **methylation** may be a feature shared among dormant states. As in *S. bullata,* both *LSD1* and *Su(var)3-9* are involved in regulating the dauer stage of the nematode *Caenorhabditis elegans* (Katz et al. 2009, Meister et al. 2011), and a histone methyltransferease SET-4 represses an insulin-like gene (*ins-9*) that contributes to detection of environmental cues that trigger dauer entry (Delany et al. 2017).

Acetylation is the other common form of histone modification, and transcript expression profiles from several species suggest involvement of acetylation in the diapause response. Prior to diapause entry, *reptin,* a gene encoding a portion of the Tip60 histone acetylation complex, is twofold upregulated in embryos of the cricket *Allonemobius socius* but is then downregulated once diapause has been entered (Reynolds and Hand 2009a,b). The Tip60 histone acetylation complex can silence chromatin through acetylation of histone H4, suggesting it may be playing a role in the initiation phase of diapause. In adult diapause of the mosquito *Culex pipiens,* genes encoding NuRD and Sin3 are upregulated in association with diapause (Hickner et al. 2015): NuRD promotes chromatin condensation and silences gene transcription, and the Sin3 complex includes histone deacetylase 1 (HDAC 1) and HDAC 2, enzymes that repress transcription. Pax-interacting protein (Pax) is also upregulated during diapause in *C. pipiens* (Sim and Denlinger 2008). This protein interacts with Sirtuin 1 (Sir2), an NAD-dependent HDAC, as well as other silencing proteins.

Histone **deacetylation** is commonly associated with gene silencing, and a nearly 75% reduction in the total amount of acetylated histone H3, noted during pupal diapause in *S. bullata* (Reynolds et al. 2016), is suggestive of large-scale down-regulation of gene transcription. The decrease in total histone H3 acetylation is accompanied, somewhat surprisingly, by a decrease in enzyme activity of HDACs and decreased expression of several genes encoding HDACs and histone acetyltrans-ferases (HATs). *hdac3, hdac6, sirt1,* and *sirt2,* along with *gnc5,* a gene that encodes a HAT, are underexpressed during pupal diapause. These observations suggest that histone deacetylation likely does not occur during diapause but during the earlier photosensitive larval stage, as reflected in the upregulation of many relevant genes during that time.

9.10.3   Polycomb Group Proteins and Histone Methylation

The Polycomb group (PcG) proteins work cooperatively to silence gene expression by remodeling chromatin structure. Two PcG complexes prominent in insects include Polycomb Repressive Complex 1 (**PRC1**), which ubiquitinates histone 2A, and **PRC2**, which trimethylates histone H3 at lysine 27 (mark H3K27me3). Many genes silenced by PRC2 through methylation are involved in developmental processes, a function anticipated to be prominent in diapause.

A role for PRC2 is noted in regulating pupal diapause in the cotton bollworm *Helicoverpa armigera* (Lu et al. 2013b). Expression of **Extra sex combs** (Esc), a component of PRC2, is elevated under long (diapause averting) daylengths, resulting in increased levels of H3K27me3. Somewhat surprisingly, the increase in H3K27me3 activates *ptth* expression rather than suppressing it, resulting in diapause termination. RNAi directed against *esc* or chemical inhibition of PRC2 reduces both H3K27me3 methylation and PTTH production, resulting in a developmental arrest in pupae programmed for continuous development. Thus, in *H. armigera*, H3K27me3 appears to stimulate rather than represses PTTH production and channels pupae toward continuous development rather than diapause. A role for PRC2 in insect diapause is also suggested by transcript profiles of the alfalfa leafcutter bee *Megachile rotundata* (Yocum et al. 2015). Five genes encoding members of the PRC2 complex are upregulated at larval diapause termination, a response suggesting a role for the PRC2 complex in repressing gene activity that prevails during diapause.

Interestingly, the most highly upregulated genes in diapause of the African killifish *Nothobranchius furzeri* include PcG members (Hu et al. 2020). Presence of the chromatin mark regulated by Polycomb, H3K27me3, is central to maintenance of diapause in this vertebrate species. Cold-induced vernalization in *Arabadopsis* is also achieved epigenetically by methylation of histones in the PcG complex (Yang et al. 2017a). The fact that a specific chromatin regulator is involved in diapause of both an insect and a fish, as well as the seasonal vernalization response of plants, suggests a prominent role for the PRC2 complex in various forms of seasonal response.

9.10.4   Heterochromatin Protein 1

The heterochromatin protein 1 (HP1) family is a highly conserved group of epigenetic factors that contribute to formation of heterochromatin domains (*e.g.*, telomeres and centromeres) through interaction with trimethylated lysine 9 of histone H3 (H3K9me3), and through interactions with euchromatin HP1 in activating or repressing transcription. A link with dormancy was first reported in dauer larvae of the nematode *Caenorhabditis elegans* (Meister et al. 2011). A homolog of HP1, HPL-2, interacts with TGF-β and insulin signaling pathways to regulate the dauer state. Loss-of-function mutants lacking HHPL-2 are more likely to enter the dauer stage than wild-type worms. HPL-2 mutants remain in the dauer stage longer than nonmutant counterparts, and the addition of HPL-2 facilitates exit from the dauer stage.

This protein family has not been examined extensively in insects, but *Hp1a* is downregulated approximately 50% in diapausing pupae of *S. bullata* (Reynolds and Denlinger, unpublished data), but whether this downregulation contributes to the diapause phenotype remains unclear.

9.11   Small Noncoding RNAs

Like epigenetic processes, small noncoding RNAs (sncRNAs) are emerging as major posttranscriptional gene regulators, involved in a wide range of critical biological processes including cell cycle regulation, cell proliferation, developmental timing, metabolism, stress resistance, and immune responses. Whether sncRNAs should be included as epigenetic processes is debated (Deans and Maggert 2015), but it is clear that they can influence the phenotype without altering the DNA sequence, thus they function in much the same way as the epigenetic processes discussed above. Many processes controlled by sncRNAs are highly relevant to diapause, thus they are prime candidates as regulators of insect diapause (Reynolds 2017) as well as other forms of animal dormancy (Reynolds 2019). Although sncRNAs are thought to be "untranslated" some do indeed encode functional peptides (Wei and Guo 2020).

The scnRNAs are divided into three classes: **microRNAs** (miRNAs), **small-interfering RNAs** (siRNAs), and **piwi-associated RNAs** (piRNAs). These classes are categorized according to size, origin, structure, localization, and mode of action. The miRNAs and siRNAs are 20–25 nucleotides in length and originate from double-stranded RNA precursors processed by Dicer1 and Dicer2, respectively. These short fragments are loaded into RNA-induced silencing complexes (RISCs) and aid in degradation or translational repression of their targets. miRNAs typically associate with RISCs that contain Argonaute-1 (Ago-1) at their catalytic center, and siRNAs bind to RISCs containing Ago-2. piRNAs, the largest class of scnRNAs, are 23–30 nucleotides long, they originate from single-stranded RNAs and negatively regulate expression and movement of transposable elements (TEs). An RNA helicase, Spindle E, is involved in piRNA biogenesis, and mature piRNAs are loaded into a RISC-containing Piwi, Aubergine (Aub), or Ago-3.

Transcripts involved in synthesis or processing of all three classes of small RNAs are differentially expressed in association with diapause in several insect species, but the miRNAs have received the most attention. Photoperiodically sensitive larvae of the flesh fly *Sarcophaga bullata* upregulate components of the siRNA pathway, including *ago-2* and *R2D2* (Reynolds et al. 2013). The piRNA pathway is also implicated: Genes encoding Piwi and Spindle-E are both upregulated more than twofold in photosensitive larvae. Within the nucleus, Piwi-piRNA complexes can regulate gene expression through interaction with HP1, which is encoded by a gene that is also upregulated during the photosensitive stage. Although this has not been tested, upregulation of *piwi* and *spin-E* could be expected to boost the proportion of heterochromatin in diapause-destined larvae of *S. bullata* and consequently initiate preparation for entry into pupal diapause. *Piwi* and *spin-E* transcripts are also more abundant during embryonic diapause in the band-legged ground cricket *Dianemobius nigrofasciatus* (Shimizu et al. 2018b). Differential expression of *ago-2* in relation to larval diapause in the leafcutter bee *Megachile rotundata* suggests involvement of miRNAs in this diapause as well (Yocum et al. 2015).

Distinct differences of expression in transcripts involved in miRNA synthesis (Figure 9.12A) are noted between diapausing and nondiapausing pupae of *S. bullata* (Figure 9.12B), as well as in response to the transition between diapause and post-diapause (Figure 9.12C) (Reynolds et al. 2013). Generally, high abundance of an miRNA implies that downstream targets are repressed, while low abundance is associated with target activation. *Ago-1*, associated with miRNA-mediated silencing, is upregulated more than twofold during diapause in pupae of *S. bullata* (Reynolds 2013) and is accompanied by changes in abundance of 10 additional miRNAs (Reynolds et al. 2017). Putative targets of these miRNAs in *S. bullata* include numerous diapause-relevant genes and pathways, such as *phosphoenolpyruvate* (*pepck*), a gene encoding a key metabolic enzyme that is upregulated during diapause in numerous species (see Section 7.8.3). The upregulation of *pepck* is consistent with the noted downregulation of miR-9c. Downregulation of several miRNAs at diapause termination, including targets in the Wnt and TOR pathways (see Sections 9.5 and 9.6), is consistent with activation of signaling pathways involved in diapause termination. An miRNA overexpressed in diapausing pupae of *S. bullata*, miR-289-5p, is also highly expressed in diapause-destined larvae of the butterfly *Araschnia levana*, but no difference in miR-289-5p expression is noted during the butterfly's actual diapause stage, the pupa (Mukherjee et al. 2020). Overall, miRNA profiles from *A. levana* demonstrate major differences during diapause, as well as during the pre-diapause larval stage. During summer diapause in pupae of the chrysomelid beetle

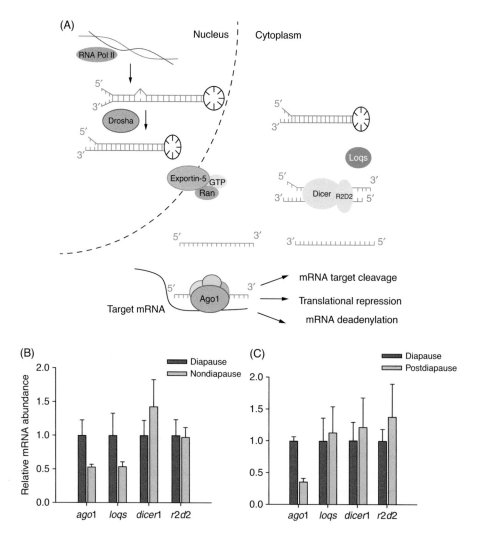

Figure 9.12 MicroRNA involvement in pupal diapause of the flesh fly *Sarcophaga bullata*. (A) Pathway for miRNA biosynthesis. miRNA expression comparisons (B) between diapausing and nondiapausing pupae and (C) between diapause and post-diapause stages. Adapted from Reynolds et al. (2013).

Galeruca daurica one miRNA, mir-1, is elevated during adult diapause (Duan et al. 2021), as it is during pupal diapause in the corn earworm *Helicoverpa zea* (Reynolds et al. 2019), and as in other studies, more miRNAs are downregulated (20) during diapause in *G. daurica* than are upregulated (5). Two downregulated miRNAs common to both diapausing pupae of *S. bullata* and diapausing adults of *G. daurica* include *let-7*, an miRNA thought to be involved in developmental timing, and *miR-100,* an miRNA involved in cell cycle progression.

In *Culex pipiens*, a mosquito with an adult diapause, miRNA profiles also change from pre-diapause to diapause and show some parallels to pupal diapause in *S. bullata*: In both species, mir-13b-3p, miR-275-3p, and miR-305-5p are under-expressed during diapause (Meuti et al. 2018). MicroRNA expression profiles for the mosquito *Aedes albopictus* indicate that the mother does not provide her eggs with a unique set of miRNAs, but distinctions are evident later, during the pharate larva's period of diapause maintenance (Batz et al. 2017). Several differentially regulated miRNAs in *A. albopictus* have targets relevant to diapause traits. In response to diapause termination by HCl in embryos of *Bombyx mori*, 23 miRNAs are upregulated and 38 are downregulated (Fan et al. 2017). Particularly striking in the *B. mori* results is the fact that miR-2761 is fivefold more abundant in diapausing embryos and drops sharply at diapause termination. miR-2761 negatively regulates expression of *sorbitol dehydrogenase* (*sdh*), the gene encoding a key enzyme that converts sorbitol to glycogen, an enzymatic reaction previously associated with diapause termination in silk moths. miR-2761 inhibition of *sdh* thus keeps sorbitol levels high, ensuring that embryos remain in diapause (see Section 9.1.1). Pupal diapause in *H. zea* can be broken with ecdysone, diapause hormone, or a diapause hormone analog, and injection of any of these diapause terminators results in downregulation of one miRNA, miR-277, an miRNA linked to the insulin/FOXO signaling pathway and regulation of metabolism, suggesting that miR-277 may play an important role in restarting metabolism at diapause termination (Reynolds et al. 2019). All three hormones also promote upregulation of *spook* and *iswi*, genes involved in ecdysone signaling. The responses of several other miRNAs vary depending on the agent used to terminate diapause, thus suggesting that different pathways can lead to the same final developmental response.

Results on hand suggest there may be some commonality to miRNA expression profiles related to diapause in different species and stages, but it remains too early to know for certain. The fact that some miRNAs are upregulated in association with diapause while others are downregulated underscores the idea that some processes are turned on specifically during diapause while others are turned off, and the fact that miRNA expression patterns can be different prior to diapause, during diapause, and at diapause termination underscores the dynamic nature of diapause and the contributions of miRNAs.

A **maternal effect** in *Sarcophaga bullata* prevents pupal diapause in progeny of females that have themselves been in diapause (see Section 5.13.2), and this effect may also be regulated by miRNAs (Reynolds et al. 2017), a possibility suggested by the fact that three miRNAs (let-7-5p, miR-125-5p, miR-190-5p) are more abundant in flies that have experienced diapause than in flies that have not been through diapause. Genes targeted by these miRNAs are possibly important for establishing the diapause block in the next generation. Six miRNAs are differentially regulated in developing pupae that are unable to enter diapause because of the maternal effect, and some (*let-7, miR-100, miR-125*) are tied to PTTH and ecdysone production, likely targets for promoting continuous development. Epigenetic processes appear to be involved in parental effects operating not only in invertebrates but also in fish, reptiles, birds, and mammals (Ho and Burggren 2010).

Involvement of miRNAs in **circadian clock regulation** in *Drosophila melanogaster* suggests parallel involvement in the photoperiodic clock mechanism. The miRNA **let-7** is implicated as a regulator of circadian rhythms: Overexpression of let-7 in clock neurons lengthens circadian period, and its deletion attenuates the morning activity peak (Chen et al. 2014c). At least six miRNAs show cyclic expression within the pigment dispersing factor (PDF) neurons, and one of them, mir-92a, peaks during the night and is responsive to light pulses and phase shifts (Chen and Rosbash 2016). Mir-92a modulates PDF neuronal excitability by rhythmically suppressing SIRT2 levels. Knocking down the miRNA biogenesis pathway severely affects behavioral rhythms in *D. melanogaster* and implicates three clock genes (*clock, vrille, clockworkorange*) as being under miRNA-mediated control (Kadener et al. 2009). Though none of these experiments relate directly to diapause, involvement of circadian clocks in diapause (see Section 5.6) provides a strong rationale for proposing that miRNAs contribute in a similar fashion to the clock mechanisms regulating diapause.

Other components of the diapause program, such as **cold tolerance**, also rely on miRNA regulation. The goldenrod gall fly *Eurosta solidaginis* is a freeze-tolerant species that alters its miRNA profile in response to freezing (Courteau et al. 2012, Lyons et al. 2016). Expression profiles of 24 miRNAs are altered in *E. solidaginis* by freezing, including several that are significantly reduced by freezing and others that are elevated. Target predictions for several of these miRNAs suggest regulation of translation and the TCA cycle. A freeze-avoiding species, the wood-boring beetle *Monochamus aeternatus*, upregulates miR-31-5p as larvae prepare for winter diapause and become cold-acclimated (Zhang et al. 2020a). Mechanistically, miR-31-5p appears to negatively target the rate-limiting enzyme acyl-CoA oxidase by promoting production of an **ascaroside** (asc-C9) thought to serve as a signal for metabolic depression and the accumulation of cryoprotectants. These results suggest that ascarosides, small signaling molecules identified in nematodes, may also be involved in regulating metabolic functions in insects. Metabolic depression, starvation resistance, and responses to anoxia and other stresses also commonly evoke miRNA responses in mammals and invertebrates alike (Zhang et al. 2011b, Biggar and Storey 2018), thus suggesting miRNA involvement in numerous physiological and biochemical adaptations incorporated into the diapause syndrome.

The **siRNA pathway** is possibly involved in the decision by the fly *Chymomyza costata* to enter larval diapause: *Ago-2* is downregulated in larvae during the photosensitive stage (Poupardin et al. 2015). Although the functional role for *ago-2* has not been evaluated, a contribution to diapause is likely. Interestingly, *ago-2* is downregulated in *C. costata* prior to larval diapause but is upregulated prior to pupal diapause in *Sarcophaga bullata*. A link between the siRNA pathway and diapause is also suggested for the leafcutter bee, *Megachile rotundata*: Alterations in expression of *ago-2, drosha,* and *pasha* are noted during the transition from diapause to post-diapause quiescence (Yocum et al. 2015). We remain at an early stage in evaluating functional roles for these small RNA pathways in association with diapause, but the transcriptional results suggest a rich potential for sncRNAs as influential players in diapause regulation.

A role for sncRNAs in control of dormancies goes beyond insect diapause. A role is well documented in dauer formation in the nematode *Caenorhabditis elegans*. A piwi mutant develops into an adult with the reproductive shutdown characteristic of adult diapause, and like a conventionally induced reproductive diapause in *C. elegans*, the diapause can be terminated when a food source is provided (Heestand et al. 2018). *C. elegans* also has a well-studied larval diapause, and this too appears to utilize scnRNAs, through interactions of the let-7 miRNA family with the DAF-12 receptor and transcription factor Hunchback-like-1 (*hbl-1*) (Hammell et al. 2009). In brief, DAF-12 receptor and its steroid ligand stimulate the production of let-7 family miRNAs, which repress *hbl-1*, promoting progression of larval development. In the absence of the DAF-12 ligand, production of let-7 is halted, allowing expression of *hbl-1* and consequently dauer formation. Together, DAF-12, let-7, and *hbl-1* form a hormone-coupled switch regulating dauer formation. Might a similar switch be operating in insect diapause?

Both histone modification and miRNAs are linked to embryonic diapause in brine shrimp in the genus *Artemia* (Dai et al. 2017). Several histone marks are more abundant during diapause than in stages prior to or after diapause and are thought to contribute to cell cycle arrest. Over 100 miRNAs, including some contributors to cell cycling, are differentially expressed during diapause in *A. parthenogenetica* (Zhao et al. 2015), suggesting wide-ranging involvement of miRNAs in the dormancy of these crustaceans. Several sncRNAs are also uniquely expressed in diapausing embryos of the annual killifish *Austrofundulus limnaeus* (Romney and Podrabsky 2017), and a growing list of mammals appear to rely on sncRNAs for regulating hibernation (Arfat et al. 2018). Parallels extend to plant seasonal cycles as well: Vernalization pathways in the perennials *Arabis alpina* (Bergonzi et al. 2013) and *Cardamine flexuosa* (Zhou et al. 2013) are coordinated through pathways regulated by miR156 and miR172.

Though connections between insect diapause and sncRNAs remain largely correlative at this point, it seems likely that sncRNAs will continue to emerge as important links in regulatory schemes for diapause. But, the field is fraught with technical challenges. Each miRNA can target numerous mRNAs, sometimes more than 100, making it especially difficult to link miRNAs with target mRNAs of interest for diapause, and the current inability to easily manipulate sncRNA abundance with specific inhibitors, sponges, or related tools makes it challenging to certify function. But, recent progress in the development of small RNA-Seq libraries and other advances will hopefully enable the community to make strides toward understanding the roles played by these potentially important gene regulators.

Recent decades have revealed layers of complexity to the simple regulatory schemes we once envisioned for diapause, yet the early models based on contributions from brain neuropeptides, the prothoracic gland, and corpora allata prevail and remain central to our understanding of the mechanisms of diapause regulation. What have been added are intricacies contributing upstream and downstream components to diapause pathways evoked largely by diapause hormone, ecdysteroid, and juvenile hormones. Some important observations remain isolated and are yet to be fully integrated into a grand regulatory scheme. The challenge now is to further assemble the many pieces of the puzzle to make a more complete pathway leading from reception of environmental cues to full execution of the diapause program. As work progresses, it will be exciting to witness the continued emergence of regulatory elements common to the diapauses of diverse species. The final product, however, will not be a single pathway but multiple alternative routes approaching the same developmental endpoint.

10 Genetic Control of Diapause

A genetic basis for diapause is suggested by the huge variability in the diapause response within a population (Chapter 3) and the well-documented geographic variation in responses to environmental cues (Section 5.1). Variation is also reflected in the decision to enter diapause, what cues to rely upon to interpret seasonal changes, and how long to remain in diapause. The sexes do not necessarily respond the same way (Section 5.1.2), further underscoring variation within a species. All these features are under strong genetic control, subject to selection. Genetic crosses between diverse strains of the commercial silk moth *Bombyx mori*, summarized by Kogure (1933) and Tazima (1964), provided the earliest experimental evidence demonstrating heritability of diapause traits. Strains of *B. mori* range from being univoltine (obligate diapause), to bivoltine (facultative diapause) to multivoltine (no diapause), and genetic crosses between these different strains revealed a rather complicated pattern of inheritance for diapause (or voltinism) involving three pairs of autosomal alleles and one sex-linked gene with three alleles. Other early experiments from the Russian scientist A.S. Danilevskii (1965) and his colleagues probed the genetic basis for diapause by crossing geographic populations showing different diapause traits, such as critical photoperiod. Excellent summaries of previous genetic experiments are provided by Tauber et al. (1986) and Danks (1987).

My goal here is to briefly summarize the early findings, supplemented with additional information on more recent experiments and new genetic approaches probing heritability of the diapause response. It is clear that there are no consistent patterns of inheritance. Some reflect classic Mendelian (monogenetic) patterns, most seem to be polygenic, both dominance and recessive genes may be involved, and some patterns are sex-linked.

Within a single genus or between different populations of the same species, both monogenic and polygenic patterns may occur, as noted for different species of spider mites (genus *Tetranychus*) as well as different populations of *T. urticae* (Bryon et al. 2017b). Likewise, QTL-mapping in the mosquito *Wyeomyia smithii* shows that different modes of inheritance can exist in different populations (Bradshaw et al. 2012). There is no reason to assume that all nondiapause mutants within the same species result from a mutation at the same locus. Diapause is the consequence of a multistep process involving photoreception, transduction of the photoperiodic signal to the brain, storage of this message within the brain, and then ultimately engagement of the hormonal signaling system that generates the diapause syndrome. This complicated process resulting in diapause can be derailed at any step along the way, thus suggesting a wealth of potential targets that could produce mutants affecting the diapause response.

In some cases, diverse traits associated with diapause are inseparable, but in other cases, different attributes of diapause appear to be **modular**, that is, inherited independently. For example, in sympatric hawthorn- and apple-infesting host races of the fly *Rhagoletis pomonella,* diapause intensity and the timing of diapause termination appear to be separate genetic modules (Ragland et al. 2017). The same appears to be true for diapause determination and duration in species as diverse as the blow fly *Calliphora vicina* (McWatters and Saunders 1996, 1997) and the cotton bollworm *Helicoverpa armigera* (Chen et al. 2012). The sorts of differences noted between and within species underscore the rich repertoire of existing inheritance patterns and imply an equally rich diversity in the evolution of genetic structures supporting the diapause response.

10.1   Selection Experiments

Diapause is more easily lost than gained, although success has been recorded in both directions. Numerous laboratory colonies founded by insects with a diapause lose diapause after being in the laboratory for some time. Optimal laboratory rearing conditions, frequently without cycling temperature and light cues, may simply remove the selective forces maintaining diapause. The rapid loss of diapause traits in the laboratory suggests that ecologically relevant experiments on diapause should be carried out soon after bringing wild populations into culture. Inbreeding, a common feature in laboratory cultures that originate from a small starter population, can also reduce the full scope of natural variance found in wild populations, as demonstrated by reduced variance in diapause duration of inbred lines of the Kanzawa spider mite *Tetranychus kanzawai* (Ito et al. 2012).

The rapidity of change is illustrated with experiments on various flesh fly species. In our laboratory's experiments with the flesh fly *Sarcophaga crassipalpis,* diapause under short-day conditions was high (85%) at 25°C for the first few years after introduction to the laboratory, but within 10 years, the flies had to be reared at temperatures of 20°C or lower to obtain the same high diapause incidence. By selecting for nondiapause in *S. bullata,* pupal diapause incidence dropped from 93% to nearly 0% in five generations (Henrich and Denlinger 1983). In a tropical flesh fly, *Poecilometopa spilogaster,* diapause incidence fell from 80% to less than 20% in five generations of selection (Denlinger 1979), and selection over five generations was sufficient to generate a nondiapausing variant of *S. similis* (Goto 2009). Without any attempts at selection, a long-standing colony of *S. bullata* (in the laboratory >20 years) lost its diapause response, even at low temperatures and short daylength, although the wild-type population from which it was derived had a robust pupal diapause (Goto et al. 2006). Since the many traits of diapause are related, it is often possible to select on the basis of one diapause-related trait and influence other diapause traits. For example, selection for late pupariation in *S. bullata*, a trait associated with entry into pupal diapause, resulted in both a higher incidence of pupal diapause as well as diapause of greater duration (Henrich and Denlinger 1982b). Attempts to boost the diapause incidence in sarcophagids by selecting for diapause have been modestly successful, but it appears to be not nearly as easy as selecting for the nondiapause phenotype.

A striking success in **selecting for nondiapause** is seen for the gypsy moth *Lymantria dispar*. This is a species with an obligate pharate larval diapause, but a few individuals escape diapause and develop without interruption. Hoy (1977) was able to exploit this feature by selecting an exclusively nondiapausing strain of the gypsy moth, an attractive experimental resource. For larval diapause in the fly *Chymomyza costata* selection resulted in a decline of the diapause response from nearly 100% to nearly 0% within five generations (Riihimaa and Kimura 1988). Selection for nondiapause in adults of the cabbage beetle *Colaphellus bowringi* yielded a steady decline in diapause incidence with successive generations, ending with nearly no diapause by the tenth generation (Xue et al. 2002). Starting with conditions that yielded an adult diapause incidence of approximately 50% in *Drosophila littoralis*, selection almost completely eliminated diapause in six to seven generations (Oikarinen and Lumme 1979).

Selection in the opposite direction, **selection for diapause**, is reported for a number of species. Diapause incidence in larvae of the blow fly *Calliphora vicina* reared under long days increased from under 10% to nearly 100% within five to six generations (Saunders and Cymborowski 2003). In the Indianmeal moth *Plodia interpunctella*, selection for larval diapause is remarkably rapid: Within two generations diapause incidence increased from 26 to 91% (Wijayaratne and Fields 2012), and selection for larval diapause in the southwestern corn borer *Diatraea grandiosella* raised the diapause incidence from 56 to 100% in three generations (Takeda and Chippendale 1982).

When subject to bi-directional selection, adults of the chrysomelid beetle *Ophraella communa* shifted from a 30–60% diapause incidence to 100% diapause in six to seven generations and to 0% diapause in five to eight generations (Tanaka and Murata 2017). Adults of the rice leaf bug *Trigonotylus caelestialium* can switch from laying diapausing (short days) or nondiapausing (long days) eggs to the opposite type of egg in response to photoperiod, and within 5–10 generations selection generated lines with longer or shorter response times required for this switch (Shintani 2009b).

Just as selection for a diapause-associated trait (long wandering period) in *S. bullata* (discussed above) increased the incidence of pupal diapause, selection for nondiapause under short days in *Drosophila montana* altered several diapause-associated traits, including critical daylength and cold tolerance (Kauranen et al. 2019b), underscoring the regulatory linkage sometimes evident among diverse diapause traits.

Selection can also alter the stage of diapause. Larval diapause in the burnet moth *Zygaena trifolii* occurs naturally in the third or any subsequent larval instar, but a third-instar diapause is most prevalent (87%). Selection for a fourth-instar diapause over six generations increased the incidence of fourth-instar diapause from 13 to 86% (Wipking and Kurtz 2000).

But, selection for nondiapause raises concerns about loss of genetic diversity, as demonstrated in a colony of the western corn rootworm *Diabrotica virgifera virgifera* selected for nondiapause (Kim et al. 2007). Though this colony offers considerable research utility, the colony, after approximately 190 generations in the laboratory, lost considerable (15–39%) genetic diversity compared to wild populations. Likewise, the nondiapause-selected line of the gypsy moth *Lymantria dispar* (Hoy 1977) appears to be less fit than the wild population and has been difficult to maintain long term in the laboratory.

10.2   Inheritance Patterns

The rich variety of selection experiments that have been performed, as well as natural variation in diapause responses found in geographic populations, offer possibilities for constructing genetic crosses that can reveal inheritance patterns of diapause traits. Clearly, there is no single pattern of inheritance that operates in all species, and it is also apparent that different features of the diapause response may be inherited in different manners.

10.2.1   Mendelian Inheritance

Though Mendelian-type inheritance patterns, also known as **monogenetic** inheritance patterns, appear to be in the minority, several good examples in the diapause literature conform fairly closely to a Mendelian pattern. Crosses between sibling species of *Chrysoperla carnea* and *C. downesi* suggest a single pair of alleles at two unlinked autosomal loci regulate adult diapause in these lacewings (Tauber et al. 1977). Crosses between populations of the mosquito *Ochlerotatus triseriatus* with different diapause capacities point to a single major recessive factor that regulates this embryonic diapause (Shroyer and Craig 1983). A simple Mendelian inheritance pattern appears to operate in the spider mite *Tetranychus pueraricola* (Suwa and Gotoh 2006), the linden bug *Pyrrhocoris apterus* (Doležel et al. 2005), and in the flesh fly *Sarcophaga bullata* (Han and Denlinger 2009b). In each of these examples, the segregation rate is close to 3:1, implying a major role for one gene or a small gene cluster in diapause induction. A comparison of pupal diapause incidences in *S. bullata* with theoretically predicted values from a Mendelian inheritance model (Figure 10.1) shows close alignment of experimental observations with predicted outcomes for crosses and back-crosses. Two parallel sets of crosses between a nondiapausing line and lines with either a low or high diapause incidence yield similar results: (1) diapause incidences are nearly the same in both directions, suggesting no role for sex linkage, (2) the nondiapausing phenotype is almost completely dominant, (3) the diapause incidence in the F_2 generation is nearly a quarter of that in the diapausing strains, and (4) backcrosses show no diapause in the BcN backcrosses and reduced diapause in BcD backcrosses.

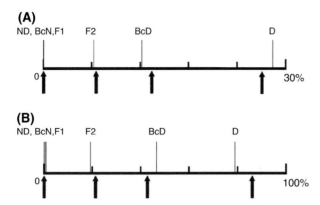

Figure 10.1 Comparisons of pupal diapause incidences in the flesh fly *Sarcophaga bullata* with theoretically predicted values in a Mendelian inheritance model for crosses between (A) a nondiapausing line (0% diapause) and a line with a low diapause incidence (28%) and (B) the nondiapausing line and a line with a high diapause incidence (78%). The four arrows (from left to right) indicate theoretically predicted diapause incidences of the nondiapause strain (N), F_2, Backcross (Bc), diapause strain (D). BcN and F_1 share the same predicted diapause incidence as N. Vertical lines represent observed diapause incidences.
Adapted from Han and Denlinger (2009b).

These results are consistent with a Mendelian inheritance model in which the non-diapause phenotype is a single autosomal dominant trait. Earlier experiments that crossed two different selected lines of *S. bullata* (Henrich and Denlinger 1983) also point to a relatively small number of gene loci controlling diapause.

More sophisticated analysis involving **Quantitative Trait Loci** (QTL) may very well reveal that some cases thought to follow a monogenetic inheritance pattern involve a more complex pattern of inheritance. Early experiments suggested that natural variation of larval diapause duration in the European corn borer *Ostrinia nubilalis* could be attributed to a Mendelian pattern of inheritance, but more recent QTL estimations indicate involvement of at least two genes within two adjacent, interacting QTLs (Kozak et al. 2019). A small number of loci may be involved in many inheritance patterns for diapause that were originally thought to be Mendelian.

10.2.2   Polygenic Inheritance

Polygenic inheritance patterns have been identified more frequently than simple Mendelian (or monogenetic) inheritance, suggesting this is the most prevalent form of diapause inheritance. Having numerous genes involved facilitates the types of quantitative changes common to geographic adaptation, as pointed out by Beck (1980) and others. Strong evidence for polygenic inheritance is evident from crossing experiments between northern and southern European populations of the Colorado potato beetle *Leptinotarsa decemlineata*: In this case photoperiodic induction of diapause is strongly heritable, with an additive polygenic autosomal background (Lehmann et al. 2016b). In the cotton bollworm, *Helicoverpa armigera*, crosses between northern and southern populations show that both the photoperiodic induction of diapause and its duration are under polygenic control (Chen et al. 2012), as also demonstrated for crosses between geographic populations of the cabbage beetle *Colaphellus bowringi* (Lai et al. 2008), ladybird beetle *Harmonia axyridis* (Reznik et al. 2017), and chrysomelid beetle *Ophraella communa* (Tanaka and Murata 2017). Inheritance patterns in hybrids of two closely related swallowtail butterflies, *Papilio glaucus* and *P. canadensis* indicate likely involvement of multiple genes, rather than just a single Mendelian locus (Ryan et al. 2018). Crossing experiments between a diapausing population of the mosquito *Culex pipiens* and a closely related nondiapausing strain of *C. quinquefasciatus* reveals four QTLs, located on three chromosomes, that affect diapause (Mori et al. 2007). A QTL analysis of the critical photoperiod in another mosquito, *Wyeomyia smithii*, compares populations from Florida and Alberta, Canada, and reports nine QTLs that explain 62% of the variation (Mathias et al. 2007). One QTL in *W. smithii* also affects the stage of diapause (third vs. fourth-instar larva). Selection for nondiapause in adults of the northern fly *Drosophila montana* reveals divergence from the control line in 16 gene clusters, most of which are not randomly distributed on all chromosomes but are enriched on chromosomes 3 and 4, a result similar to differences noted in latitudinally divergent populations (Kauranen et al. 2019b). Variation in the diapausing stage of *Zygaena trifolii*, noted above, is heritable and also reflects polygenic inheritance (Wipking and Kurtz 2000). A polygenic pattern of inheritance is also apparent as a determinant of pupal diapause duration in the apple maggot *Rhagoletis pomonella*; numerous genes, each exerting a small effect appear to dictate the rate of diapause development (Dowle et al. 2020). The same appears to be true for *Drosophila melanogaster*, where hundreds of alleles, each with a small effect, contribute to the diapause response (Erickson et al. 2020).

When multiple genes are involved in the diapause response, we can expect both additive effects, where two or more genes contribute equally to the response, or epistasis, where the effect of one gene dominates the other.

10.2.3   Dominant or Recessive Trait?

Traits attributed to a dominant allele will always be expressed even if a recessive allele is present, while traits associated with a recessive allele will be expressed only when both alleles present are recessive. While clear patterns of dominance may appear under a Mendelian pattern of inheritance, many diapauses are under polygenic control, involving numerous genes. Thus, simple declarations of dominance can be expected to be rare. Incomplete dominance is displayed when the presence of two different alleles of a gene generates an intermediate response with neither allele's effect dominating. Across a broad range of insect species, examples of each of these patterns are noted by Danks (1987), as well as in the more recent literature cited below.

Crosses between populations of the dung fly *Scathophaga stercoraria* from Switzerland and Iceland reveal that the incidence of pupal diapause at 12°C is **dominant** (Demont and Blanckenhorn 2008), as is adult dormancy in *Drosophila melanogaster* (Schmidt et al. 2005). Diapause in adults of the Asian ladybird beetle *Harmonia axyridis* also appears to be a dominant trait, as revealed by crossing experiments (Reznik et al. 2017).

Other crossing experiments reveal **incomplete dominance**, as shown by the F_1 displaying a diapause incidence that is significantly lower than in the high diapause strain but significantly higher than in the low diapause strain. This is seen in the cabbage beetle *Colaphellus bowringi* (Kuang et al. 2011), the flesh fly *Sarcophaga similis* (Goto 2009), and the cotton bollworm *Helicoverpa armigera* (Chen et al. 2012). Similarly, crosses between two populations of the Asian cornborer *Ostrinia furnacalis* that have different critical daylengths produce progeny with an intermediate critical daylength for larval diapause (Huang et al. 2013).

Inheritance of diapause traits using **recessive alleles** is more rare, but a few examples are known. Summer diapause in pupae of the pierid butterfly *Pieris brassicae* is one such example (Held and Spieth 1999). Crosses between a population from southern Spain, which has a summer diapause, with a population from Denmark that lacks summer diapause indicates an inheritance pattern with summer diapause as a recessive trait. F_1 hybrids completely lack the summer diapause-associated features of the Spanish population, and their photoresponse curve is identical to that of the population from Denmark. But, interestingly, crosses between northern populations of *P. brassicae* indicate that winter diapause in this species is inherited as a dominant trait. Results such as this point to the rich variety of inheritance patterns that can exist, even within different populations of the same species.

10.2.4 Sex Linkage or Autosomal?

Sex linkage is evident in some species but not in others. And, as cited above for *Bombyx mori*, inheritance patterns are not necessarily exclusively one type or the other. Loci on both autosomal and sex chromosomes may be involved within the same species. There is an advantage in having a beneficial mutation located on a sex chromosome. Selection can act more swiftly on a sex chromosome because the heterogametic sex (sex which has sex chromosomes that differ) exposes a mutation to selection more readily than when the mutation is hidden as a recessive locus on an autosome (Charlesworth et al. 1987, Pruisscher et al. 2017). This attribute allows an advantageous sex-linked genetic variant to spread through a population more rapidly than a mutation on an autosome (Meisel and Connallon 2013). This may be especially important to quickly evolve new diapause traits in a changing climate or for invasion of new geographic regions. Indeed, sex chromosomes appear to host genes implicated in diverse population differences within the Lepidoptera (Söderlind and Nylin 2011), including features of the diapause response.

Sex linkage is evident in crosses between a Spanish population of the green-veined white butterfly *Pieris napi*, which has a facultative pupal diapause with a precise critical photoperiod, and a Swedish population, which enters diapause regardless of photoperiod (Pruisscher et al. 2017). The inheritance pattern for diapause is strongly sex-linked in this species. Although the major component determining diapause in *P. napi* resides on the Z-chromosome (males are the homogametic sex with two Z-chromosomes, whereas females are heterogametic with either ZW or ZO chromosomes), there are minor effects from autosomes as well. Similarly, inheritance of critical photoperiod for pupal diapause in the speckled wood butterfly *Parage aegeria* has a strongly sex-linked component (*period*) but also relies on an autosomal gene (likely *timeless*) as well as other minor loci that exert small effects (Pruisscher et al. 2018). Whether pupal diapause is obligate or facultative in the tiger swallowtail *Papilio glauscus* is dependent on the subspecies. In the northern subspecies, *P.g. canadensis*, diapause is obligate, and the southern subspecies, *P.g. glaucus*, has a facultative diapause induced by short daylengths. Crosses between these two subspecies show that hybrids inherit the diapause response of their fathers (Rockey et al. 1987). Inheritance of the photoperiodic response to daylength in larvae of the Asian corn borer *Ostrinia furnacalis* is also influenced more strongly by the male parent (Huang et al. 2013). In adults of the cabbage beetle *Colaphellus bowringi* it is the female, not the male, that has the major influence in determining the critical photoperiod, although both sexes contribute (Lai et al. 2008). Post-diapause development time in the European corn borer *O. nubilalis*, a feature critical for distinguishing univoltine from bivoltine populations, is inherited as a sex-linked trait associated with the molecular marker *triose-phosphate isomerase* (Dopman et al. 2005).

Autosomal inheritance of the critical photoperiod is evident for adult diapause in *Drosophila littoralis* (Lumme and Oikarinen 1977). Populations of *D. littoralis*, present across a geographic range from 42 to 69° N, display critical daylengths ranging from 13.0 to 17.6 hours, and crosses between populations show inheritance of critical daylength to be due to an incompletely dominant allele on an autosomal chromosome. Likewise, inheritance of pupal diapause between two selected lines of the flesh fly *Sarcophaga bullata* appears to be strictly autosomal (Henrich and Denlinger 1983). Adult diapause in the Colorado potato beetle *Leptinotarsa decemlineata* also appears to be autosomal, based on crosses between northern and southern populations (Lehmann et al. 2016b). Genetic determination of critical daylength in crosses between Spanish and Swedish populations of the comma butterfly *Polygonia c-album* is mainly autosomal, with contributions from some minor sex-linked effects (Söderlind and Nylin 2011).

Some features of the diapause syndrome within a species may be sex-linked, while others are not. For example, inheritance of photoperiodic induction of larval diapause in the blow fly *Calliphora vicina* relies almost exclusively on the mother, whereas both parents contribute to diapause duration (McWatters and Saunders 1997). In the cotton bollworm *Helicoverpa armigera*, pupal diapause determination and duration also follow different inheritance patterns: The male parent plays a more important role in the decision to enter diapause, but both sexes influence the duration of diapause (Chen et al. 2012).

10.2.5   Linkage and Pleiotropy

Linkage and pleiotropy reflect distinct mechanisms, but the two are sometimes challenging to distinguish. Genes located in close proximity within a chromosome may be linked, that is, consistently inherited together as a single unit. Such inheritance patterns appear to be monogenic (Mendelian-like) in crossing experiments. Pleiotropy, in contrast, refers to a single gene influencing multiple phenotypic effects. Both linkage and pleiotropy can be expected to be common in relation to diapause, a complex trait that is expressed as a syndrome encompassing multiple phenotypic traits. Diapause not only is reflected in a developmental arrest at a precise stage, but also by many additional trait distinctions, as discussed in Chapters 6 and 7, including pre-diapause rates of development, altered behavior, accumulation of energy stores, and enhancement of various forms of stress tolerance. Are these diverse traits genetically linked and/or do they reflect some form of pleiotropy? It is not always easy to make the distinction.

An early report of **linkage** is based on regulation of critical daylength in adult diapause of the northern fly *Drosophila littoralis*: This response appears to be controlled by a cluster of closely linked alleles inherited as a single autosomal segregating unit (Lumme and Oikarinen 1977). The fact that some features of the diapause response are consistently inherited as a unit suggests some form of linkage. In the flesh fly *Sarcophaga bullata*, selection for late pupariation, as mentioned above, results in both a higher incidence of pupal diapause and a diapause of longer duration (Henrich and Denlinger 1982b), a result that may be the consequence of pleiotropic effects of genes associated with late pupariation or from one or more genes associated with late pupariation being closely linked to genes that affect diapause. QTL analysis for diapause and three other life-history traits (wing length, wing venation, developmental time) in two sibling species within the *Culex pipiens* complex show that individual QTLs across the traits commonly map to similar genome locations, suggesting these diverse traits are controlled by genes with pleiotropic effects or multiple tightly linked genes (Mori et al. 2007). Overlapping QTL peaks for several diapause-related life-history traits in *D. montana* also likely reflect either pleiotropy or gene linkage (Tyukmaeva et al. 2015). Selection for insecticide resistance in the leafroller *Choristoneura rosaceana* results in a higher proportion of larvae entering diapause, a correlation thought to be due to pleiotropic effects of the resistance allele(s) (Carrière et al. 1995).

One of the notable debates on the role of **pleiotropy** centers on the role of clock genes (Emerson et al. 2009b). As discussed in Section 5.6.2, clock genes are a critical component of the diapause timekeeping mechanism, but they also serve non-clock functions, making it challenging to distinguish a perturbation of diapause caused by an altered clock from some other function, such as gut activity (Bajgar et al. 2013b) or cold tolerance (Poikela et al. 2021) that also engage clock genes but in a non-timekeeping role. For example, reproduction could cease because a malfunctioning clock gene halts reproductive processes, thus mimicking the diapause syndrome, or the halt could be the consequence of a malfunctioning clock function that disables the timekeeping mechanism governing diapause. The distinction can be challenging to resolve.

Elegant experiments with the bean bug *Riptortus pedestris* evaluate the role of clock genes on both the diapause decision as well as the circadian pattern of cuticle deposition (Ikeno et al. 2010, 2011c). Disruption of clock genes results in disruption of both the diapause response as well as the circadian rhythm, suggesting it is the functioning of the clock (the timekeeping function) that is disrupted, not some pleiotropic effect of the clock genes. In the European corn borer *Ostrinia nubilalis* two strains (E and Z) differ in the time of day that they mate, as well as the seasonal timing in which they break diapause and initiate mating (Levy et al. 2018). Different circadian clock genes are associated with these two timing events, suggesting the two traits are not maintained by pleiotropy.

10.3   Power and Pitfalls of QTL Analysis

Mapping of quantitative trait loci (QTL) to identify regions of the genome that contribute to diapause offers a powerful tool, and as discussed in several of the above sections, this approach has proven useful for estimating sites and numbers of regions contributing to the diapause response. But, as pointed out by Bradshaw et al. (2012), several problems do limit its usefulness. QTLs represent regions of the genome, not individual genes, thus a single QTL can be expected to include several genes. Unless a large number of markers are used and the number of individuals genotyped and phenotyped is large, a QTL analysis overestimates the effect size of the QTL and underestimates the number of sites contributing to genetic divergence of the trait. And, if divergence between populations involves the cumulative effect of many genes of small effect, this approach is likely to miss some regions of the genome that are potentially important. But, in spite of these challenges, our understanding of diapause through QTL analysis has benefitted significantly from the recent growth in the number of species subject to this approach.

QTL mapping has been used to analyze diapause in several mosquito species. Exploiting the wide variation of photoperiodic responses in geographic populations of the pitcher plant mosquito *Wyeomyia smithii* Mathias et al. (2007) reported a range of interesting QTL associated with the larva's critical daylength: one that is unique, one that is sex-linked, one that overlaps with QTL for the stage of diapause, and one that interacts epistatically with the circadian rhythm gene *timeless*. Further work with *W. smithii* used this approach to develop QTL maps for recently diverged populations and more ancient populations (Bradshaw et al. 2012), results that highlight the complexity of diapause and pinpoint numerous sites on all three chromosomes with **single nucleotide polymorphisms** (**SNPs**) that contribute to the diapause response. QTL analysis was also used to identify loci responsible for adult diapause in *Culex pipiens*, as discussed above (Mori et al. 2007). A sibling species, *C. quinquefasciatus*, lacks diapause, and a comparison of these two sibling species has identified QTL associated with the diapause phenotype as well as other life-history traits that distinguish these two members of the *C. pipiens* complex. The majority of QTL in this study show intermediate effect magnitudes, suggesting that the traits examined reflect complex interactions among several genes rather than, as often noted in QTL studies, a single major effect QTL accompanied by several minor effect QTL. High-density linkage mapping combined with a bulk-segregant analysis (a QTL mapping technique for identifying genetic markers associated with a mutant phenotype) of hybrid lines created by crossing a tropical (nondiapause) population and a temperate (diapause) population of the Asian tiger mosquito *Aedes albopictus* identified 268 diapause-associated SNPs (Boyle et al. 2021). These SNPs are found at numerous locations throughout the genome but are largely clustered toward the beginning of chromosome 1, the middle of chromosome 2, and the end of chromosome 3. The large number of diapause-associated SNPs and broad chromosomal distribution support the hypothesis of a polygenic basis for diapause in this species. Advances in genomics and fine-scale genetic mapping can be expected to further enhance prospects for identifying genes within QTL that regulate the diapause phenotype.

QTL mapping of adult diapause in *Drosophila montana* indicates a major QTL for diapause induction on the X chromosome, as well as several QTL on the fourth chromosome (Tyukmaeva et al. 2015). Related traits, including cold tolerance, development time, and larval body weight show QTL on the second, third, and fifth chromosome, some with overlapping peaks, indicating that these traits share some QTL, likely reflecting either pleiotropic effects or multiple tightly linked genes. In another fly, *Rhagoletis pomonella*, loci associated with timing of diapause termination are mainly observed in SNPs mapping to chromosomes 1–3, while SNPs involved in the initial intensity of diapause are present on all five chromosomes, indicating that separate genetic modules likely control these two distinct aspects of diapause (Ragland et al. 2017). Although some gene regions on chromosome 2 affect both traits, most regions involved in the two traits are different.

QTL studies have also proven useful in identifying key genes likely implicated in diapause. QTL mapping of early- and late-emerging populations of the European corn borer *Ostrinia nubilalis* reveal that two clock-related genes *period* and *pigment dispersing factor receptor* are located within interactive QTL on the Z chromosome, and differences in allele frequency of the two genes undergird the tracking of winter length, accounting for a three-week difference in emergence time and ultimately to the temporal segregation of these two populations (Kozak et al. 2019). QTL analysis of two geographically distinct populations (France and Finland) of the parasitoid *Nasonia vitripennis* point to genomic regions on chromosomes 1 and 5 associated with diapause, and interestingly, these chromosome regions contain the clock genes *period* and *cycle* (chromosome 1) and *cryptochrome* (chromosome 5) (Paolucci et al. 2016). QTL mapping also proved useful in identifying *couch potato* (see Section 9.7), a gene encoding an RNA binding protein, as a putative regulator of clinal variation of adult dormancy in *Drosophila melanogaster* (Schmidt et al. 2008).

What all of these QTL studies elegantly demonstrate is both complexity of the diapause response and involvement of numerous chromosomal sites that have both major and minor impacts on diapause. Diapause traits that appear to be Mendelian may very well prove to involve multiple minor effect loci when examined under the scrutiny of QTL analysis. This approach is proving to be a powerful method for evaluating modularity of diverse diapause traits, as well as locations of specific loci and interactions between loci. The utility of QTL analysis for diapause has also proven useful beyond insects, as demonstrated for the crustacean *Daphnia magna* (Czypionka et al. 2018).

Recent technological breakthroughs foretell possibilities for rapid advancement in our understanding of the genetic basis for diapause. For species amenable to the technology, CRISPR/Cas9 editing makes it possible to directly test the roles of specific genes. High-throughput, next-generation sequencing technology makes it feasible to now score thousands of genetic polymorphisms using methods such as restriction site-associated DNA (RAD tag) markers. That, along with increasing availability of whole-genome sequences, present new opportunities for dissecting the genetics of diapause in non-model species. Studies with *Rhagoletis pomonella* (Ragland et al. 2017) nicely demonstrate the power of genotyping-by-sequencing methodologies to survey large numbers of SNPs in populations with distinct diapause features, and experiments on clinal variation in the speckled wood butterfly *Parage aegeria* (Pruisscher et al. 2018) reveal the combined power of whole-genome sequencing, genome-wide association, and analysis of allele-frequency clines to identify genomic regions with large and small effects on diapause induction. A novel hybrid swarm-based genome-wide association study deployed in *Drosophila melanogaster* successfully analyzed 2,800 flies, revealing hundreds of small-effect SNPs associated with dormancy in this model species (Erickson et al. 2020). As argued by Ragland et al. (2019) opportunities to integrate forward and reverse genetics with physiological and evolutionary studies offer a rich potential for enhancing our understanding of how various features of the complex diapause response are regulated and evolve.

11 Evolution of Diapause

Several themes relevant to diapause evolution have been discussed previously. Chapter 3 discussed variation, a key component providing the grist for evolutionary change, Section 8.4 discussed the impact diapause timing can have on speciation, and Chapter 10 discussed genetic control of diapause, highlighting selection experiments that can alter the diapause response. This chapter addresses three additional evolutionary questions. What is the origin of diapause? How does evolution shape the diapause response in space and time, that is, across geographic ranges and in response to climate change? Of these two questions, the second is more easily addressed, but the first question, although perhaps unanswerable, offers intriguing possibilities for consideration. The third question probes how the evolution of diapause may have had a downstream impact on the evolution of insect sociality.

11.1 Origin of Diapause

As pointed out by Tauber et al. (1986), the earliest thoughts on diapause evolution converged on the idea that diapause had its origin in **quiescence** (*e.g.*, Mansingh 1971, Müller 1976). The fact that the transcriptional profile for *Drosophila melanogaster* in quiescence (Baker and Russell 2009, Kubrak et al. 2014, 2016, Lirakis et al. 2018) shares many of the signatures of diapause adds credence to this idea. Food restriction also generates a transcriptome profile similar to diapause (Kubrak et al. 2014), suggesting diapause shares a molecular signature similar to several general **stress responses**. But, one of the key elements of diapause is its anticipatory nature, a feature not present in quiescence or food deprivation (Tauber et al. 1986). Timekeeping mechanisms are ancient and well represented in single-cell organisms, fungi, nematodes, crustaceans, and other organisms more ancient than insects. Presumably the first insects to emerge during the Devonian already had an intact clock mechanism capable of at least monitoring and regulating a daily, and perhaps a seasonal cycle of physiology and behavior. Regrettably, the fossil record offers few insights, although it's not inconceivable that diapause-specific morphological features may sometime be detected in the fossil record.

11.1.1   Insect Development Is Ideal for Stops and Starts

The progression of insect development is a sequence of internally programmed stops and starts, making it ideal for imbedding a period of arrested development at numerous points along the way. These stops and starts are especially pronounced throughout embryonic development, at molts from one larval instar to the next, at the onset of metamorphosis, and after adult eclosion but before reproduction begins. This progression reflects pulses of hormone release, and development can be halted either by blocking this release or by preventing the decline of the hormone at the end of a pulse. For example, as discussed in Section 9.1, insects will not progress to the next larval instar or proceed with metamorphosis unless the steroid hormone ecdysone is released, thus one conspicuous checkpoint is a halt in release of this molt-inducing hormone. And, in adults, juvenile hormone (JH) is commonly needed to prompt reproduction. Paradoxically, development can also be halted if a hormone continues to be present at the "wrong" time. For example, the JH titer must drop at the end of larval life to trigger release of ecdysone. When this doesn't happen, the larva remains a larva, thus setting the stage for a persistent larval diapause or a molt that is stationary (larva-larva) rather than progressive (larva-pupa). Similarly, persistent elevation of the ecdysone titer in some embryos prevents hatching, locking the insect into a holding pattern as a pharate first-instar larva. Thus, progression of development without diapause requires both rising and falling phases of hormone titers, and this uninterrupted progression can be hijacked at numerous points to place development on hold. This is presumably why diapauses invariably occur at the beginning or end of an instar or commonly prior to the onset of reproduction, rather than midway through an instar or in the middle of metamorphosis. Embryonic diapauses defy this generalization and seem to occur at nearly any time during embryogenesis, a response that possibly reflects the fact that embryonic development contains many stop and start points that can be exploited for insertion of diapause. The net result is that pauses are a natural component of insect development, and evolutionarily it may not be all that difficult to extend the duration of these pauses into periods of developmental arrest.

The evolutionary foundation for diapause that is inherent in the stops and starts characteristic of insect development is augmented by natural variation in **developmental timing**. By this I mean variation in the precise timing of hormonal release that dictates when the insect progresses to the next stage. There is considerable natural phenotypic variation within a population for the duration of each developmental stage (*i.e.*, developmental timing), and selection can be expected to exploit this genetic variation to generate phenotypes that develop at different rates. Variation in developmental timing is evident across taxonomic groups, as discussed in Chapter 3. An especially rich variation is evident among crickets and grasshoppers (*e.g.*, Masaki 1978, Miki et al. 2020), taxa in which both the number of nymphal molts as well as the total time required to go from egg hatch to adulthood varies considerably. Such variation is prevalent among other taxonomic groups as well, for example, neotropical flesh flies (family Sarcophagidae) pupariate over a period spanning more than two weeks (Denlinger et al. 1988a). Development prior to the onset of diapause is frequently accelerated or retarded, as discussed in Section 6.1, a feature further reflecting plasticity of developmental timing and its close association with diapause. One can easily imagine a scenario where natural selection, under certain circumstances, favors a delay in developmental timing, a delay that in many ways defines the core element of the diapause response. Continued selection for a slow-developing phenotype could generate discontinuity where fast and slow development progressively become more discrete states. The slow-developing phenotype, with likely a lower metabolic rate and other attributes of quiescence, may eventually be canonized as diapause. Perhaps the quiescent state we see in adults of *Drosophila melanogaster* (Section 1.1.3) represents an evolutionary transition toward a *bona fide* diapause?

11.1.2   Diapause Evolved Numerous Times

The fact that diapause intercedes at so many different stages of development implies there is no single stage that is consistently more suitable for diapause. And, the fact that some closely related species rely on different stages for diapause underscores the idea that diapause has originated multiple times within the insect lineage and, as discussed in Chapter 9, can also be found in different instars within the same species and is subject to selection. The peculiarities of a host's phenology or other seasonal attributes can thus allow natural selection to favor a range of time points for insertion of diapause into the insect life cycle. A less likely scenario, in my view, is that the capacity for diapause evolved once and then jumped around to different stages in different species. If there were a single gene or single hormonal basis for diapause such a scenario might be plausible, but that is clearly not the case.

11.1.3   Need for a Timekeeping Mechanism

Halting development must be linked firmly to a timekeeping mechanism. Which came first – the developmental halt or timekeeping? If a stage-specific slow down as an extension of quiescence came first, one then has to struggle with how the timekeeping mechanism governing its onset and termination was added. Since clocks are present in all plants and animals I argue that an intact clock was already present in insects from the beginning. The challenge still remains how to link this clock to diapause. But, we know that hormonal cycles are driven by circadian clock mechanisms, thus it is not a huge step to connect the clock to patterns of hormonal release that govern insect development. If a daily time-specific (gated) response such as hormone release is delayed even further the result would be a halt in development. Granted, the seasonal cycle of photoperiodism differs from the daily circadian cycle, but by adding some sort of mechanism for counting the number of short or long days it should be possible for the circadian clock to regulate photoperiodic responses. Cyclic changes in environmental factors other than photoperiod (*e.g.*, temperature, stress, food availability) also offer timekeeping mechanisms that can give rise to specific patterns of hormone release. Thus, photoperiod is not the only conduit for connecting environmental cycles to patterns of development.

It is also interesting to note that the daily **sleep cycle** in insects (*e.g.,* Dubowy and Sehgal 2017, Guo et al. 2018) already has the features of a clock-driven period of rest. Is the daily sleep rhythm a possible point of origin for diapause? Diapause differs from sleep, of course, in the timescale of the cycle. But, could evolution capitalize on this daily recurring developmental pause by extending its duration? Might it be possible to select for longer periods of sleep that extend from hours to days or months? Both sleep and diapause share a timekeeping element as well as a rhythmic temporal pause, key attributes needed for both types of arrested activity.

11.1.4   Where Did Diapause Arise?

We often think of diapause as a response operating primarily in temperate latitudes, but there are no fundamental reasons to argue for an origin where winter is the dominant seasonal threat. Diapause is common in the tropics (Denlinger 1986), a region that may be more similar to the environment of the early world inhabited by insects. What was that **Devonian** environment like? It was clearly a seasonal world, but the Earth was apparently warm and rather dry, thus surviving a cold winter may not have been a challenge confronting those first insects. Surviving periods of drought may have been a more significant challenge. Dry season and winter survival rely on many common physiological mechanisms (*e.g.,* desiccation tolerance) capable of providing protection against either form of seasonal stress, thus the evolution of such universal protective mechanisms sets the stage for surviving diverse seasons of stress. And, selection for seasonally synchronized life cycles may have been an even more important driver for the evolution of diapause than the inimical seasons themselves. Environments we currently see in tropical environments may most closely reflect environments encountered by early insects. Clearly, dormancy strategies abound among the current tropical insect fauna and serve to both mitigate the challenges of unfavorable seasons as well as offer a powerful means for synchronizing seasonal cycles.

Some taxa such as the flesh fly family Sarcophagidae (Denlinger 1979), *Drosophila melanogaster* (Zonato et al. 2017, Erickson et al. 2020), black scavenger flies in the genus *Sepsis* (Zeender et al. 2019), and the Colorado potato beetle *Leptinotarsa decemlineata* (Lehmann et al. 2014b) are thought to have a tropical or subtropical origin and subsequently dispersed into temperate latitudes with a diapause or other form of dormancy already in place. Thus, there appear to be no good reasons to assume that diapause had its origin in cool temperate latitudes.

11.1.5   Is There a Diapause Toolbox?

If diapause had a common origin, one might expect to see a genetic toolbox reflective of this common origin. As discussed earlier (Chapter 7), there are a few genes expressed during diapause in a wide range of species and stages of diapause, but there are numerous exceptions at the transcript level. What is common to most diapauses is not the transcriptional signature but the phenotypic response. Thus, a cell cycle arrest may be common to all diapauses, but which component of the cell cycle is targeted to bring about the arrest varies among species. Likewise, many insects upregulate heat shock proteins during diapause, but not all do, and even among those that do upregulate their heat shock proteins, different members of the family may be upregulated. The gluconeogenic enzyme PEPCK is commonly upregulated during diapause but not in all species. Likewise, insulin signaling is shut down in many diapauses, but different components of the insulin signaling pathway appear to be targeted in different systems. It is thus hard to argue for a common toolbox when the tools used to generate the phenotype differ.

The transcription factor FoxO (Section 9.3) may be a good candidate for inclusion in a common toolbox, but it remains too early to know how widespread its role may be. MicroRNAs (Section 9.10), with their abundant targets, may also be good candidates for inclusion in a toolbox regulating diapause, however there appear to be no specific microRNAs consistently associated with diapause in multiple species. But, their complex regulatory webs direct numerous critical metabolic processes offering links to clock functions, and with their recognized responsiveness to environmental perturbations they have some of the attributes of an ideal heritable driver for the evolution of diapause.

The concept of a common diapause toolbox is likely quite valid for closely related species and is still a useful concept if one substitutes general regulatory schemes for specific transcripts. Yes, tools for shutting down the cell cycle, suppressing metabolism, boosting defense responses, etc. exist for each diapause, but the types of tools employed to perform these tasks differ slightly between species, resulting in toolboxes with slightly different contents. For the origin of diapause, these differences mean that there are many ways to achieve the same end point, and the diverse manners in which these end points are attained reflects the diverse pathways that can lead to the evolution of diapause.

In summary, the evolution of diapause requires a halt in development to somehow be linked to a timekeeping mechanism capable of monitoring seasonal anticipatory cues. Common responses evoking a developmental shutdown or slowing of activity can be seen in transcriptional responses of quiescence, food restriction, sleep, and diapause, but only sleep and diapause share a timekeeping mechanism. The hormonal patterns operating to drive development provide a mechanism for starting and stopping development at precise developmental stages, and these pulses of hormone release are frequently gated responses (occurring at a specific time of day), implying a link to circadian clocks. Circadian clocks have well-established synaptic connections with numerous neurons within regulatory centers in the brain. For the evolutionary story of diapause to be completed simply requires a connection between a circadian clock and the specific genes and neurons regulating development (and, of course, nontrivial mechanisms for counting days and storing this information for later usage!).

11.2 Evolutionary Changes in the Diapause Response

Once diapause has been inserted into the life cycle (the hard part) its dimensions can be rather quickly altered, as evidenced by evolutionary changes in critical daylength,

length of the photosensitive period and its placement during development, temperature responses, diapause intensity, and other variable attributes discussed in Chapters 3, 5, and 10. How quickly the responses can be modified is of paramount importance as a species invades new territory or adapts to climate change, or, as discussed in Chapter 8, adapts to a new host with a different seasonal profile.

11.2.1   Responses along a Latitudinal Cline

The response most widely noted across a latitudinal cline is a systematic change in **critical daylength** (CDL), as noted in Chapter 5. Numerous examples of latitudinal clines are discussed in that chapter, thus the emphasis here is an exploration of underlying genetic and physiological mechanisms shaped by evolution to generate such clines.

The introduction of diapausing eggs of *Aedes albopictus* into North America in a shipment of used tires from Japan at a precise time and place (1985 in Houston, Texas, 32° N) provides a rich template for monitoring changes in the CDL and other diapause traits as the mosquito expanded its range over the next 35 years into regions as far south as southern Florida (26° N) and as far north as 42° N in eastern North America (Armbruster 2016). During those early years (1985–1988) the correlation between CDL and latitude was rather weak for North American populations in comparison to the well-established cline noted for populations native to Japan (Figure 11.1A), but by 2008, a much more robust relationship was apparent (Figure 11.1B), implying a rapid adaptive evolutionary response in North American populations (Urbanski et al. 2012). Differences in other characteristics such as body size, egg size, wing length, development time, adult survivorship, and reproductive output are not apparent, but several diapause-related features such as cold tolerance, diapause duration, and post-diapause

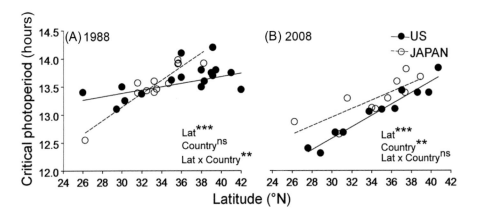

Figure 11.1 Effect of latitude and region (USA, solid circle; Japan, open circle) in (A) 1988 and (B) 2008 on critical daylength for diapause induction in embryos of the Asian tiger mosquito *Aedes albopictus*.
From Armbruster (2016), with permission from John Wiley & Sons.

starvation tolerance also show genetic divergence among the different North American populations (Batz et al. 2020). Migration of *A. albopictus* east and south from Houston into warm, southern Florida was accompanied by a drop in the incidence of diapause (Lounibos et al. 2003). *A. albopictus* was also introduced into Brazil at nearly the same time it arrived in North America, but founder populations on the two American continents appear to be from different sources because most Brazilian populations lack the capacity for diapause. Diapause has been noted in the southern region of Brazil (>26° S), but levels of diapause are quite low (Lounibos et al. 2003).

A multivoltine species such as *A. albopictus* must "know" whether sufficient time remains in the season to complete another generation. If not, they must enter diapause rather than risk death. What information best predicts this decision point? For *A. albopictus*, a degree day model most accurately predicts the deadline for deciding between developing or entering diapause (Mushegian et al. 2021). As noted above, following its invasion into North America from Japan in 1985 *A. albopictus* has spread over a range of approximately 16° in latitude. CDL latitudinal clines across the mosquito's native and invasive range after a 10-year period are parallel, but the CDL is longer in Japan, implying an earlier onset of diapause in Japan. This difference is best explained by a model of degree days that delineates a location-specific deadline after which it is not possible to complete another generation, rather than a deadline predicted by some other feature such as the date of first frost.

Another **invasive species** with a well-defined time and place of introduction is the fall webworm, *Hyphantria cunea*. This species was introduced into Japan from North America in 1945, and populations in Japan, like their North American counterparts, are bivoltine and require approximately 800 degree days above 0°C to complete one generation, but beginning in the 1970s there was a shift to trivoltinism in southwestern Japan (Gomi and Takeda 1996). Both the CDL and developmental time decreased with the shift from bivoltinism to trivoltinism. In the newly evolved trivoltine populations only 700 degree days are required to complete a generation. In this example, geographic variation in the CDL is not a simple cline, but a step-wise shift that occurs around 36° N. Acceleration of development associated with the shift to trivoltinism seen in *H. cunea* is also noted for the European corn borer *Ostrinia nubilalis* as it shifted from having one to more than three generations in North America following its introduction from Europe (Showers 1979). Duration of immature stages decreased as the number of generations increased. In the late 1940s, the bruchid *Acanthoscelides pallidipennis* was introduced into Japan from North America, and since that time a latitudinal cline in CDL has become well established between Kobe (34° 75′ N) and Owani (40° 51′ N), with a longer CDL in the northern population and a switch from being bivoltine in the south to being univoltine in the north, adaptations that have evolved within the past 70 years (Sadakiyo and Ishihara 2011).

Introductions that can be linked to a specific time offer a particularly valuable resource for monitoring the rapidity of change, but clines are also well defined among native species and consistently demonstrate the fine-tuning of CDL and other diapause traits along **latitudinal and altitudinal gradients** as discussed in Chapter 5. Among the best-studied examples of a latitudinal cline in photoperiodism in North America is the work from Bradshaw and Holzapfel (2001a, 2017) and their students on larval diapause induction in the pitcher plant mosquito *Wyeomyia smithii*, a species that ranges from its ancestral home in the Gulf Coast (CDL of approximately 13 hours) to derived populations in Canada (CDL of 17 hours). Their work shows that average heterozygosity at protein-coding loci decreases with latitude, with genetic variation for the photoperiodic response increasing with latitude. Hybridization experiments point to epistasis as the basis for the genetic differences in CDL. They attribute increased variation in the response to increased additive variance during successive founder events during the northward dispersal. T experiments (Section 5.6.1), in which a fixed daylength is coupled with night lengths of varying duration, consistently show reduced rhythmicity in northern versus southern populations (Figure 11.2) (Bradshaw and Holzapfel 2001b), a feature noted for a number of other species as well. Bradshaw and Holzapfel propose that circadian rhythms are more strongly coupled to southern population responses, thus reducing the role of circadian rhythmicity in photoperiodic responses among northern populations.

Experiments with *W. smithii* also reveal intriguing latitudinal differences in expression of the clock gene *timeless* (Mathias et al. 2005). Expression levels during the

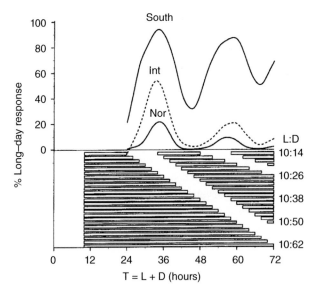

Figure 11.2 Differences in intensity of a circadian rhythm in northern and southern populations of the pitcher plant mosquito *Wyeomyia smithii*, as demonstrated in T-experiments (See Section 5.6.1). The experiments couple a 10-hour photophase with scotophases of varying lengths to create T = 24–72 hours. The rhythmic long-day response (no diapause) indicates an underlying rhythmic sensitivity to light and implicates circadian involvement in photoperiodic time measurement. The southern population from the Gulf Coast shows a more robust circadian rhythm than the population from northern population from Maine.
From Bradshaw and Holzapfel (2001b), with permission from Elsevier.

photosensitive larval stage are approximately 20% higher in southern (38–40° N) than in northern (50° N) populations, suggesting a possible relationship between expression of this clock gene and latitudinal change. A latitudinal link between *timeless* and diapause expression is also apparent in European populations of *Drosophila melanogaster*, as discussed in more detail in Section 5.6.2. In *D. melanogaster*, two forms of *timeless* are involved: a short form of the mRNA (*s-tim*) and a long form (*ls-tim*). The *ls-tim* variant correlates with higher levels of diapause (Tauber et al. 2007). This predicts that *ls-tim* should be more abundant in more northern populations, but the opposite is true. Latitudinal clines for diapause are evident in North American and Australian populations of *Drosophila melanogaster*, but only a shallow cline is noted in Europe (Pegoraro et al. 2017), possibly a consequence of *ls-tim*, which appeared as a mutation in southeastern Italy a few thousand years ago. The incidence of this allele is approximately 80% in southern Italy but only 20% in Scandinavia; very low levels of this allele are present in North America and Australia. The molecular basis for the enhanced diapause response appears to be a reduced physical interaction of L-TIM with the circadian photoreceptor CRYPTOCHROME1, leading to a more light-stable TIM in *ls-tim* flies. As a consequence, the circadian clock in *ls-tim* flies is less light-sensitive, possibly resulting in a fly that is more prone to enter diapause under long daylengths, thus blurring the expected diapause cline due to the high incidence of this diapause-enhancing allele in southern Europe.

The clock genes *timeless* and **period** also feature prominently in latitudinal divergence in the speckled wood butterfly *Pararge aegeria* (Pruisscher et al. 2017). Five populations across a latitudinal cline in Sweden have a range of CDL values for the programming of pupal diapause, from 15.8 hours in the south to 19.1 hours in the north. A genome-wide scan for divergence between the two extreme populations reveals significant SNP associations in four genes present in two chromosomal regions, *per* on the sex chromosome and *kinesin, carnitine O-acetyltransferase* and *tim* on an autosome. There is especially strong selection against heterozygotes of *tim* and *per*, as evidenced by the fact that numerous heterozygotes are present in the cline, but *tim* and *per* appear mainly as homozygotes of either the southern or northern alleles. It is thus interesting that *tim* emerges as a key player in latitudinal variation not only from targeted studies but also from a genome-wide scan.

The Colorado potato beetle *Leptinotarsa decemlineata*, a native of North America, invaded Europe approximately 150 years ago and has proceeded to expand its range at an annual speed of 80–130 km. During the past 30 years, expansion in Europe has slowed at the northern edge of its range, possibly due to the low amount of heritable variability in traits relevant to winter survival (Lehmann et al. 2015b). A comparison of populations from Italy and Russia shows little change in CDL, but latitudinal clines are evident for diapause incidence and the age at which adult beetles burrow. Resurfacing behavior in beetles that had already burrowed increases with daylength, a response that seems counterintuitive because such behavior could lead to higher overwintering mortality. This odd linkage between resurfacing behavior and long daylength could be the constraint on using a longer daylength for northern expansion.

The challenge of adapting to a new environment can be seen in species that are recent immigrants to new areas. From the early 1960s to 2000 the stink bug *Nezara viridula* expanded its range northward in Japan and reached Osaka (Musolin and Numata 2003b). Although some females in Osaka continue to lay eggs until late November, progeny from any eggs laid after late September are destined to die during the winter, thus exerting strong selective pressure for a longer CDL, and consequently earlier entry into diapause and avoidance of late autumn reproduction. But, the impact of a cold winter in Osaka may be mitigated in the future by continued warming that would decrease winter and spring mortality and advance post-diapause reproduction (Takeda et al. 2010). A Kyoto (35° N) population of the bean bug *Riptortus clavatus* transferred to conditions of the Russian forest-steppe zone (50° 38′ N) enters diapause in September, long after the eggs it produces could successfully hatch and develop, indicating that this population is maladapted to these new environmental conditions (Musolin et al. 2001).

Though latitudinal clines are well documented, few experiments have probed the physiological basis for such differences. What changes occur to elicit a change in the photoperiodic response? An elegant set of experiments by Yamaguchi and Goto (2019) address this question by examining latitudinal and sexual differences in photoperiodic induction of pupal diapause in the flesh fly *Sarcophaga similis*. In Japan, this fly shows the expected variation in CDL across at latitudinal cline from Nagasaki (32.7° N) to Otaru (43.2° N), with CDL increasing in populations from more northern areas. This fly nicely conforms to an "external coincidence model" (Section 5.6.1 and Figure 5.10) in which measuring daylength depends on whether the **photo-inducible phase** (φ_i), determined by the circadian clock, is exposed to light. The position of φ_i can be determined by light interruption experiments, and φ_i shows a latitudinal cline, a result that supports the hypothesis that the latitudinal cline in CDL is a consequence of shifts in timing of φ_i. There is also a sexual difference for diapause induction in this species (Tagaya et al. 2010), with females being more sensitive to photoperiodic cues for diapause induction. This sexual difference also reflects differences in CDL, but there is no sexual difference in the position of φ_i, showing that males are simply less sensitive than females to a light pulse given at φ_i. Thus, in this species, distinct mechanisms underlie latitudinal and sexual differences in CDL used for the induction of diapause.

By their very nature, some diapause traits are binary, thus latitudinal clines may be disjunct for some traits, rather than showing a gradual transition. This is noted both for stages of diapause as well as the type of diapause (obligate vs. facultative). For example, the eastern tiger swallowtail *Papilio glaucus* and the Canadian tiger swallowtail *P. canadensis*, species that diverged 0.5–0.6 million years ago, share many attributes reflecting common ancestry, but *P. glaucus* has a facultative pupal diapause, while the diapause of *P. canadensis* is obligate (Ryan et al. 2016). A narrow (<50 km) hybrid zone in the northeastern United States separates the two species, a zone thought to be maintained by a climate-mediated developmental threshold that dictates whether one or more generations can be produced. The genetic divergence is distributed across the genome, but the Z chromosome and four autosomes show a disproportionate amount of the genetic differentiation. Numerous traits differ in these two butterflies, but it appears that the distinction between a facultative and obligate diapause is not the consequence of one or just a few mutations. It is not apparent which type of diapause is ancestral for these two tiger swallowtails, but the transition for a trait such as this cannot be gradual. Presumably, examples in earlier stages of divergence will reflect differences in the diapause program that are less complex.

Invasion of new geographic regions is frequently the consequence of one or few individuals, potentially limiting genetic diversity in the founding population. As a consequence, one might anticipate **genetic bottlenecks** that may influence diapause decisions in a new area. A case in point is the introduction of the walnut husk fly *Rhagoletis completa* from its native American Midwest into California. There appear to have been multiple introductions, but only two out of six populations show evidence of genetic bottlenecks (Chen et al. 2006b), and all of the introduced populations are fully capable of entering pupal diapause at a time appropriate for the new location. A shorter diapause, a two-week advance in the timing of diapause termination, has evolved in the warmer conditions prevailing in California, thus sufficient genetic diversity was present in the invading populations to enable a rapid response to selection of a population with a different diapause trait.

Photoperiodism and circadian rhythmicity appear to have evolved separately, as pointed out by Bradshaw and Holzapfel (2007, 2017). Good evidence for this is noted in several insects including *Sarcophaga similis* (Goto 2009), as discussed further in Chapter 5. A mutant fly that fails to respond to short daylength by entering pupal diapause metamorphoses into an adult with a perfectly intact circadian clock. But, there are other examples in which photoperiodism and circadian rhythmicity are affected by the same mutation, a mutation that presumably is centered on the clock mechanism. A mutant of another flesh fly, *S. bullata*, fails to respond to photoperiodic cues for pupal diapause induction and is arrhythmic as an adult (Goto et al. 2006), a response also seen for a variant of the linden bug *Pyrrhocoris apterus* (Hodková et al. 2003) and the fly *Chymomyza costata* (Pavelka et al. 2003). If photoperiodism and circadian rhythmicity are independent it should be possible to alter the photoperiodic response without interfering with the circadian clock. In such cases selection would likely be acting primarily on genes downstream of the clock to modify the diapause response. Presumably, most evolutionary responses noted across latitudinal gradients are the result of selection acting on genes downstream of the clock mechanism. It is perhaps a "safer" option to alter downstream genes than to alter something as fundamental as the clock.

Overall, the amount of phenotypic and genetic variation evident within a cline differs with species, as evident from a comparison of latitudinal variation observed across 19° latitude for *Drosophila melanogaster* and *D. simulans* (Machado et al. 2016). Consistent with phenotypic studies, less variation is seen in *D. simulans* (2.1% latitudinally clinal variants on autosomes versus 4.3% in *D. melanogaster*), and the cline in *D. simulans* is less stable across time, in spite of the fact that these two closely related species share a similar range, ecology, and evolutionary history. One distinct difference, however, is that *D. simulans* is unable to overwinter in northern portions of its range, perhaps resulting in a stronger winter bottleneck that may account for a less stable cline from year to year. A comparison between these same two species for populations ranging from Panama to Maine also highlights differences in temperature responses and reveals that genome × environment interactions are more common within populations of *D. melanogaster* than in *D. simulans* (Zhao et al. 2015). Though neither of these experiments focused strictly on diapause-related traits, the distinctions reported suggest that we can expect considerable clinal variation in diapause traits between even closely related species. Variation in diapause traits is nicely demonstrated in an investigation of seven clinal populations of *D. montana* in Finland, results that reveal considerable variation in the CDL used for programming adult diapause: This variation increases with latitude, reaching a maximum of 3 hours in the most northern population (Lankinen et al. 2013). Steepness of the photoperiodic response curve also differs significantly among populations of *D. montana*; in this case, the difference does not show clinal variation but instead is a response based on how long the population has remained in laboratory culture. With longer time in culture, the response curve becomes steeper, showing that some of the variability present in wild populations is lost with time, perhaps due to inbreeding, inadvertent direct selection, or correlated response to that selection (pleiotropy).

Commonly, more than a single diapause trait displays geographic variation. For example, clinal variation may be evident for both diapause induction and diapause duration, as seen in the cotton bollworm *Helicoverpa armigera* (Chen et al. 2013). Five geographic populations of this moth examined in China show variation in CDL for pupal diapause induction, differences in responsiveness to photoperiod and temperature, as well as differences in diapause duration. Across a latitudinal cline stretching from Corsica to Finland, populations of the parasitoid *Nasonia vitripennis* differ in several diapause attributes: proportion of larval diapause in the female's progeny, maternal age at which the female switches to production of diapausing progeny, and the CDL for diapause induction (Paolucci et al. 2013). Northern populations make an earlier switch to production of diapausing progeny, a higher proportion of the population enters diapause and the CDL is longer. It will be interesting to know whether a common genetic basis underlies variation in these different traits or if they are each under separate selection.

11.2.2   Responses to Climate Change

Just as invasion into new territory offers a method for examining the rapidity of evolutionary changes across space, a comparison of diapause responses across time provides valuable insight into naturally occurring evolutionary responses prompted by climate change. A striking example of this approach is offered by Bradshaw and Holzapfel (2001a) who observed **shifts in CDL** within eastern North America populations of the pitcher plant mosquito *Wyeomyia smithii* over a nearly 30-year interval, but changes could be detected in as short a time span as 5 years. CDLs in seven populations examined were shorter in 1996 than in 1972, reflecting a shift toward a more southern phenotype with a longer growing season. The shift toward a shorter

CDL was more pronounced in northern populations, likely shaped by the greater rate of increase in length of the growing season in the north. Similarly, during a seven-year interval (1995–2002) in Fukui, Japan, the CDL of the fall webworm *Hyphantria cunea* shortened by 14 minutes and rather than being exclusively bivoltine, a portion of the population completed three generations per year, shifts likely caused by the documented warming of this region (Gomi et al. 2007). During the 1991–2002 interval in Kochi, Japan, critical daylength for adult diapause induction in the water strider *Aquarius paludum* shortened by approximately 30 minutes, a response thought to be the consequence of warming during this short interval (Harada et al. 2005).

Though there are still relatively few examples documenting changes in the diapause response across time, the surge of interest in climate change has fostered thoughtful reviews on how insect life cycles, including diapause responses, are and will likely continue to be impacted by our changing environment (Bale and Hayward 2010, Bradshaw and Holzapfel 2010d, Sgrò et al. 2016). One of the most obvious effects is that a warmer environment lengthens the favorable growing season, thus leading to selection on voltinism, the number of generations completed each year. The challenge is that temperature is changing, but photoperiod is not, thus successful adjustments to a warmer environment require a shift in the underlying photoperiodic response if an insect is to remain in synchrony with its host plant or other seasonal features. But, the impact of climate change goes beyond a simple extension of the growing season. As demonstrated for temperate latitude trees, lengthening the growing season does not simply extend tree productivity but instead drives earlier autumn leaf senescence (Zani et al. 2020). Such results are likely to affect a number of biotic interactions, including the life cycles of insects that rely on these trees as hosts. Added to temperature changes are distortions of rainfall patterns, and in tropical environments, the evaporation of clouds in the cloud forest, subjecting insects to hotter and longer dry seasons (Janzen and Hallwachs 2021), all of which can be expected to alter seasonal cycles that have evolved to cope with an earlier climate.

Since high temperatures frequently reduce the photoperiodic induction of diapause, we can expect lower incidences of diapause, and since high temperatures also frequently advance the timing of diapause termination and shorten the duration of post-diapause quiescence we can expect both later entry into diapause and earlier spring emergence, observations that are already being realized for butterflies in England (Roy and Sparks 2000), Sweden (Karlsson 2014), in the Pacific Northwest of the United States (Maurer et al. 2018), as well as for European forest caterpillars (van Asch and and Visser 2007), and dragonflies in the Netherlands (Dingemanse and Kalkman 2008), among others. Warmer winters are likely to reduce cold-stress-induced mortality for some species, but others may actually be more prone to cold injury if warm winters result in less snow cover and hence less buffering of the winter habitat. A warm winter can also be expected to take a toll on depletion of energy reserves. The cold of winter works in concert with diapause to reduce metabolic rate, thus conserving energy reserves. A warmer environment thus has multiple implications for how many generations are produced, the timing of diapause entry and exit, as well as how reserves are managed during diapause. In addition, an obvious impact of a warming climate is range expansion, which in turn, necessitates selective changes in response to photoperiod, even if range expansion results in a congruent thermal environment. Range shifts also result in latitudinal shifts in hybrid zones (Larson et al. 2019), rich zones for examining evolutionary processes demarking species boundaries and interactions that impact diapause.

The Bradshaw and Holzapfel laboratory evaluated the impact of shifts in both temperature and photoperiod on fitness in *W. smithii* in a laboratory experiment simulating rapid (180–200 years) climate warming (Bradshaw et al. 2004). In brief, northern populations achieve greater fitness in a "southern" year than they do in their native northern thermal year when they remain under their northern photoperiod, but when challenged with a southern photoperiod, they lose 88% of their fitness, a consequence of the mismatch between the southern photoperiod and the timing of diapause entry and termination. Exposure to incorrect daylength, not higher temperatures, is the cause of the fitness decline.

Complexity in the relationship between temperature and photoperiod in determining **voltinism** is demonstrated in a model developed for predicting the impact of climate change on the grape berry moth *Paralobesia viteana* (Tobin et al. 2008). It's not just the quantity of degree days that accumulate throughout the year that determine voltinism but also when those degree days are received. The critical photoperiod for populations of the grape berry moth in the grape-growing region along the Great Lakes in North America occurs around August 3. If the moth completes a second generation prior to that date, a third generation can be produced. A third generation appears to now be more common, possibly driven by an increase in the number of degree days received prior to the August 3 switch date. A warming climate can also be expected to favor a switch from **univoltinism to bivoltinism** or multivoltism in the European spruce bark beetle *Ips typographus*. Both facultative and obligate diapause populations of the bark beetle currently exist in the mountains of Central Europe. A warming climate is likely to favor populations with facultative diapause, thus increasing the number of generations produced each year, which in turn increases the consequent damage to the spruce forests (Schebeck et al. 2017). Selection for bivoltinism would appear to be more challenging for the spruce beetle *Dendroctonus rufipennis*, a univoltine species in North America, because this species has an obligate adult diapause, and switching to bivoltinism would require a novel switch from an obligate to a facultative diapause. Projections for the codling moth *Cydia pomonella* in Switzerland predict that the risk of adding a second generation will increase from <20% currently to 70–100% by 2045–2074, and the risk of a third generation will increase from 0–2 to 100%, assuming a mean temperature rise of 2.0–2.7°C (Stoeckli et al. 2012). The accuracy of this projection is even greater if the codling moth is able to make photoperiodic adjustments to the programming of its facultative diapause.

The historic record indicates that increases in voltinism are already underway, as documented for butterfly and moth species in Central Europe (Altermatt 2009). Using a dataset extending back to the mid-nineteenth century for an area within a 30 km radius of Basel, Switzerland, a significant proportion of the 263 species examined show increases in the number of annual generations, and 44 species have increased generation number since 1980, an increase most likely caused by earlier spring emergence and more rapid development in the warmer environment. These features allow more generations to develop before the onset of diapause in the autumn. For polyphagous species, an increase in voltinism may also shift host specialization, as predicted for the butterfly *Polygonia c-album* in Sweden (Audusseau et al. 2013). The current univoltine population feeds on a range of herbaceous and woody plants, but late in the season performance is always better on herbaceous hosts. Rising temperatures and a shift to bivoltinism may favor specialization on the herbaceous hosts.

Another possible consequence of a warming environment is that some species may shift from having a winter diapause to **skipping diapause** entirely. Such a scenario is most likely for a species that relies heavily on temperature for programming diapause. This is the possible fate of a eusocial bee species *Plebeia droryana* from southern Brazil (Dos Santos et al. 2015). This bee enters adult diapause when autumn temperatures drop to 8–10°C. The expectation is that by 2080, median winter temperature will rise to 13.4°C, thus likely causing a large portion of the population to avert diapause.

How much potential there is for responding to climate change depends in a large part on how much genetic variability resides within a population. This feature has been evaluated in isofemale lines of the spider mite *Tetranychus pueraricola* (Ito 2014). Although there is little variability in the diapause response at low temperatures (nearly all enter diapause), diapause response curves generated at higher temperatures display highly variable diapause incidences, indicating that significant genetic variation for diapause induction is maintained. This genetic variation, in turn, suggests that spider mite populations have the capacity to respond quickly to climate shifts. This inherent genetic variation provides the possibility for rapid evolutionary change.

Climate change is likely to remain one of the primary drivers of evolutionary change in the years ahead. Though much of this emphasis has focused on lengthening the growing season, climate change is also rapidly modifying winter conditions including winter temperatures, variability of winter conditions, and snow cover (Williams et al. 2015). All of these features can be expected to continuously impact the timing of diapause entry and exit, as well as overwinter energy expenditure and survival. As argued by Bradshaw and Holzapfel (2008), most contemporary genetic as well as phenotypic changes in response to climate change are reflected in changes in phenology, not thermal physiology.

11.3   Diapause as a Step toward Evolution of Sociality

Adult diapause may have been a critical evolutionary step fostering evolution of sociality in Hymenoptera. This idea, proposed by Hunt and colleagues (Hunt and Amdam 2005, Hunt et al. 2007, 2010), gains credence from observations on paper wasps in the genus *Polistes*. Unlike the more evolutionarily advanced social species, castes of *Polistes* are morphologically indistinguishable. Distinctions are strictly behavioral, with physiological underpinnings, suggesting an ancestral form of sociality. As shown in Figure 11.3, distinctions between the **G1** (workers, capable of reproducing immediately after adult eclosion if the queen dies) and **G2** (gynes, diapausing stage that will become queens in the next generation) developmental pathways begin during larval development, with those entering the G1 pathway having slow larval development time, short pupation time (from cocoon spinning to adult eclosion), and low levels of storage protein. In contrast, those in the G2 trajectory are characterized by fast larval development, long pupation time, and greater stockpiling of storage protein. Females in the G2 pathway are destined for diapause: they delay reproduction, their adult lifespan is long, they fail to forage for protein, they are involved in neither brood care nor nest construction, more energy reserves are sequestered, and they are more cold-tolerant than G1 individuals. Those in the G1 trajectory are fully prepared to replace the queen if she dies or take on worker duties of caring for the queen's progeny as long as the queen survives. G2 individuals, by contrast, show no maternal behavior during their first year, but eventually mate and, after a period of diapause, go on to become queens that will found a new nest the following spring. This is a bivoltine life cycle with two pathways having markedly different seasonal attributes and distinct developmental outcomes. The evolution of this peculiar bivoltine cycle required the capacity to insert an adult diapause, the G2 trajectory, into the life cycle. In this scenario, the diapause (gyne) and nondiapause (worker) phenotypes are the precedents for more highly differentiated behavioral and morphologically specialized castes seen in the highly eusocial species (Hunt et al. 2007, Piekarski et al. 2018).

The storage protein **hexamerin** is thought to be the causal basis for this developmental transition (Hunt et al. 2007). Diapause-destined wasps (G2) synthesize and sequester more hexamerin during larval life, and an abundance of hexamerin is hypothesized to silence JH signaling, a prerequisite for adult diapause and a link consistent with the well-documented role for JH in caste determination in other social insects. Thus, larval provisioning, in addition to daylength (Yoshimura and Yamada 2018), is central to the diapause decision and consequently to the formation of developmental phenotypes and castes of the progeny. It is not the general state of nutrition that is important, nor are all storage proteins of equal importance. Four hexamerins are present in *Polistes metricus*, but only one, Hex-1, differs in abundance between larvae that will mature into the two types of wasps. High abundance of Hex-1 in gynes is thought to suppress the JH level, resulting in a caste that does not engage in reproductive behavior until the following year. A broad comparison of gene expression and protein profiles of the two developmental tracks in *P. metricus* with castes of honey bees reveals numerous similarities for pathways associated with developmental trajectories and nutritional differences, similarities that support a common foundation for castes in the primitively eusocial paper wasps and more advanced eusocial Hymenoptera (Hunt et al. 2010). Nutritional regulation of diapause is also evident in the solitary bee *Megachile rotundata* (Fischman et al. 2017), suggesting that this form of plasticity may be ancestral, a condition facilitating evolution of castes in social bees.

A role for hexamerins in caste determination goes beyond the Hymenoptera. Hexamerins regulate caste determination in termites as well (Zhou et al. 2006).

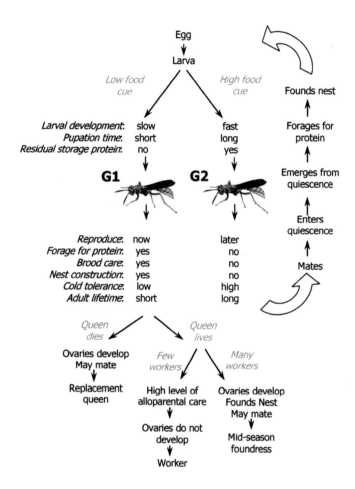

Figure 11.3 The *Polistes* life cycle incorporates a bivoltine ground plan. Larvae respond during development to food cues and diverge into one of two trajectories. Scant provisioning leads to the G1 pathway that results in reproduction during the first year, while abundant food during larval development results in the G2 pathway, in which the adult females delay reproduction, enter diapause and emerge the following year as reproductive queens.
From Hunt and Amdam (2005), with permission from AAAS.

In a lower termite, *Reticulitermes flavipes*, two hexamerins, Hex-1 and Hex-2, partici-pate in differentiation of totipotent workers into either soldiers or reproductive caste phenotypes. The presence of these two proteins suppresses differentiation of workers into soldiers, resulting in colony build-up of the worker phenotype. But, in the termite example, JH drives the synthesis of the hexamerins, rather than the opposite scenario proposed for *Polistes* in which hexamerins suppress JH synthesis.

A transcriptomic analysis of diapause in the bumble bee *Bombus terrestris* demon-strates substantial overlap between genes related to caste determination and diapause (Amsalem et al. 2015), a result further supporting the idea that diapause was co-opted as an evolutionary path toward sociality. In *B. terrestris*, as in *Polistes* wasps, the role of JH has also been expanded to encompass a role in caste differentiation and division of labor.

A link between adult diapause and evolution of sociality gains further support from a broad phylogenetic analysis comparing stages of diapause and presence of sociality in 150 species of bees (Santos et al. 2019). Most bee species diapause in the last larval instar, but interestingly, none of the evolutionarily derived social bee species do so. Most social bees diapause as adults, setting the stage for the evolution of sociality. Or, is the opposite true? Did sociality set the stage for adult diapause?

Many of the evolutionary questions posed in this chapter can likely never be fully answered, but the rich insect literature offers a fertile field for testing hypotheses about life cycle evolution and seasonality. Insects are at the forefront of research probing evolutionary responses associated with invasion of new geographic regions and responses to climate change. Diapause is central to both challenges. The ubiquity of diapause across taxa, developmental stages, and in all corners of Earth underscores the centrality of this trait to the evolutionary history of insects and their future success in new territories and new times.

12 Wider Implications

Diapause research offers more than answering how insects bridge unfavorable seasons and how they exploit this mechanism for synchronizing their life cycles. Understanding diapause is critical for developing effective pest management strategies, managing domesticated species, and fostering storage methods for biological control agents. Peculiarities of diapause physiology also offer fascinating models for examining issues in human health (Denlinger 2008). This chapter provides some of these broader implications of diapause research, connections that place diapause in the context of its rich potential for probing diverse biological questions as well as developing new tools for pest management.

12.1   Population Modeling

Predictive models that can accurately define an insect's destructive period require information about the insect's seasonal distribution. This, in turn, requires knowledge of environmental cues that initiate diapause as well as factors required to terminate diapause and elicit resumption of development.

For example, successful modeling of spring emergence of the gypsy moth *Lymantria dispar* demands detailed knowledge of the chilling requirement needed to capacitate development, as well as information on the thermal units required to complete development prior to egg hatch (Shaub et al. 1995, Gray et al. 2001). By knowing the timing of diapause termination and post-diapause development, a reliable model predicting phenology of the codling moth *Cydia pomonella* in Swiss apple orchards is not only accurate currently but should be able to predict future changes in phenology of this pest species as temperatures rise (Graf et al. 2018). Termination of adult diapause in the apple blossom weevil *Anthonomus pomorum* and dispersal into Swiss apple orchards can be predicted accurately by the accumulation of 161 degree days calculated from January 1 (Toepfer et al. 2002). Diapausing eggs of the European red mite *Panonychus ulmi* hatch in northern Greece after receiving 129 degree days above a threshold of 7.4°C (Broufas and Koveos 2000). Such information on temperature requirements for diapause termination and post-diapause development have been recorded for many species and have proven enormously useful for predicting the timing of outbreaks of pest species. A model developed for mosquitoes reveals differences in population growth based on whether the mosquito has an embryonic diapause (*Aedes albopictus*) or adult diapause (*Culex pipiens*) and suggests that killing mosquitoes during diapause will lower population levels, but it is not as effective as killing mosquitoes during the nondiapause phase of their life cycle (Lou et al. 2019).

12.2   Exploiting Diapause Traits for Pest Management

Distinct physiological and behavioral attributes of diapausing insects reveal features that can be exploited for pest management, as nicely discussed by Tauber et al. (1986). **Cultural practices** that include timing of crop planting, harvesting, and crop rotation can all be used to disrupt diapause, but of course this is an evolutionary cat and mouse game where the pest species, in time, will likely make the needed adjustments to match a new cropping system.

A range of diapause features offer useful tips for pest management. For example, adult females of the codling moth *Cydia pomonella* that have been through larval diapause retain mobility following gamma radiation much better than those that have not experienced diapause (Bloem et al. 2006), a discovery with important implications for the application of the sterile insect technique (SIT) for this species. An understanding of diapause biology also enhances pest management in the apple blossom weevil *Anthonomus pomorum*. When overwintering beetles terminate diapause in the spring, they are attracted to warm sites during their first few days of activity (Hausmann et al. 2004b, 2005). Thus, by providing a warm shelter in the apple orchard it is possible to attract a large number of beetles during the first two to four days after diapause has been broken. Because this thermal attraction is lost after six days, shelters are especially useful for monitoring the appearance of new adults and calculating the correct time for intervention (Hausmann et al. 2004a).

Knowing when diapause is initiated and terminated is critical for planning the correct timing of control efforts. In Australia, the red-legged earth mite *Halotydeus destructor* enters an embryonic summer diapause in response to daylength. The model that best controls this important plant pest (99% fewer mites in the autumn) is based on a single chemical application applied in the spring, two weeks before production of 90% diapause eggs (Ridsdill-Smith et al. 2005). In the far north of Norway, Atlantic cod are dried in the sun as they have been for at least the past 1,000 years. Most fish are caught in March and dried outdoors until June. Although the early phase of drying is free of flies, the blow fly *Calliphora vicina* causes yearly losses of €1–2 million, thus knowing when the fly will terminate diapause in the spring is essential for developing hygienic standards for this ancient industry (Aak et al. 2011).

As told to me by Dr. Rostislav Zemek from the Czech Republic, early cultural practices in Czech apple orchards involved removing tips of apple trees during the winter pruning season. But, this was counterproductive from an insect control perspective because it is precisely in this location that diapausing predatory mites, critical for biological control, are located during the winter months. With this information in hand, it is possible to conserve these valuable predators. By knowing overwintering preferences, attractive artificial shelters have been developed for adults of green lacewings, genus *Chrysoperla*, to protect against predators and disease (Canard 2005).

Locating overwintering sites of pest species can be extremely useful in planning control strategies. In the autumn, adult females of the soybean aphid *Aphis glycines*, a major pest of soybeans across the Midwest region of the United States, fly to their overwintering host, buckthorn, where the embryos overwinter. Interestingly, diapausing soybean aphids appear to successfully overwinter only within a narrow geographic area around 41° N (Crossley and Hogg 2015), and from that location adults disperse over hundreds of kilometers to establish new populations. This restricted overwintering location offers a limited region that can be targeted for control. Adults of the annual bluegrass weevil *Listronotus maculicollis* leave manicured golf courses in the autumn and fly to overwintering sites near the tree line (Diaz and Peck 2007), making it possible to deploy control measures both in time and space. The fact that diapausing adults of the Colorado potato beetle *Leptinotarsa decemlineata* preferentially overwinter at field edges and from those sites immigrate into potato fields suggests an effective control strategy of using border treatments laced with a plant systemic or transgenic insecticide (Blom et al. 2004). Before rice is harvested, larvae of the rice yellow stem borer, *Scirpophaga incertulas*, retreat to the base of the rice stem where they overwinter in larval diapause. If fields of stubble are flooded for several days, up to 90% of the diapausing larvae are killed (Luo et al. 2021). Flooding, however, is not effective in killing diapausing larvae of the rice striped stem borer *Chilo suppressalis*, a species that coexists in the rice fields with *S. incertulas*. Diapausing larvae of *C. suppressalis* are highly mobile and are fully capable of escaping a water-logged environment, resulting in little mortality (10%) when subjected to the same flooding regime.

In some cases, pest species form aggregations in their overwintering sites, making it possible to target large groups of the pest at one time. For example, the brown marmorated stink bug *Halyomorpha halys* is well known for selecting human structures as well as dry crevices in dead standing trees for overwintering, thus their numbers are concentrated in select sites that can be identified by humans or canine detectors (Lee et al. 2014).

One factor to consider in **till versus no-till agriculture** is the impact of tillage on overwintering insect pests. Post-season tillage of cotton fields in Mississippi destroys most diapausing pupae of the tobacco budworm *Heliothis virscens* and the corn earworm *Helicoverpa zea* (Schneider 2003), but the overall impact may not be so great because many pupae overwinter in sites beyond the cotton fields.

Diapause traits are sometimes altered in concert with changes in seemingly unrelated traits. Though one strain of *Helicoverpa armigera* resistant to *Bacillus thuringiensis* (Bt) is just as likely to enter diapause as a Bt-susceptible strain (Bird and Akhurst 2005), another Bt-resistant strain of *H. armigera* is more prone to enter diapause (Liang et al. 2007), thus the evolution of Bt resistance may or may not alter the diapause response. A Bt-resistant strain of the pink bollworm *Pectinophora gossypiella* enters and exits pupal diapause at the same time as a susceptible strain, but successful emergence from diapause in the spring is 71% lower in three highly resistant strains (Carrière et al. 2001). The exact cause of death during the winter is unknown, but results of this sort have clear implications for resistance management.

12.3   Diapause Disruption

The diapause literature reports a number of physical, chemical, and genetic manipulations that have the potential to be diapause disruptors, tools that may be useful in pest management or for manipulating the diapause status of insects in culture (Denlinger 2021). Some such tools are routinely used to artificially bring diapause to an end for experimental purposes, but there is also an urgent need to develop more such tools that can be used on insects for which diapause is an obstacle to overcome. Subverting the diapause response is often an attractive goal for generating the huge numbers needed for mass rearing in sterile insect release programs or for food and commercial production. Thus far, few such manipulations have been developed for pest management, but the prospect of developing agents of **ecological suicide** (breaking or entering diapause at the wrong time of year) holds promise. This section briefly highlights some of the many manipulations known to be effective in preventing, shortening, and terminating diapause or for artificially inducing a diapause-like state.

12.3.1   Physical Manipulations

The most prominent and perhaps most promising physical manipulation of diapause is altering the natural scotophase with pulses of light. The utility of this approach under field conditions has been demonstrated both by artificially **extending daylength** beyond the insect's critical photoperiod and by administering **pulses of light** at select times during the night. Daylength extensions with artificial light have proven effective in disrupting the diapause response in a few species of Lepidoptera, including the leafroller *Adoxophyes reticulana* (Ankersmit 1968, Berlinger and Ankersmit 1976), the European corn borer *Ostrinia nubilalis*, and the codling moth *Cydia pomonella* (Hayes et al. 1970, 1974, Schechter et al. 1971). Extending daylength to 17 hours in September through the use of 1,000-watt mercury arc lamps prevented diapause in 70–96% of corn borer larvae compared to no prevention of diapause under natural conditions; a similar response was seen for the codling moth (Schechter et al. 1971). Brief light pulses are effective in some species (Hayes et al. 1974, Shah et al. 2011) but not in others (Hayes et al. 1970, Berlinger and Ankersmit 1976). Although there was a flurry of activity in this field in the early 1970s few recent attempts have tested the viability of this approach. If successful, photoperiodic manipulations would be extremely attractive due to their non-reliance on chemicals. But, costs of energy and light installation likely restrict economic feasibility, yet it may have potential in some situations, especially with the recent development of more energy-efficient light technologies.

   Simple physical manipulations are sometimes also effective in preventing diapause. For example, rearing flesh fly larvae on a shaking platform that generates

160 oscillations per minute dramatically reduces the incidence of pupal diapause (Denlinger et al. 1988b), although such treatment would likely have limited practical application. Anoxia is highly effective in breaking pupal diapause in the flesh fly *Sarcophaga crassipalpis* (Kukal et al. 1991) and shortening embryonic diapause in the false melon beetle *Atrachya menetriesi* (Kidokoro and Ando 2006). Other physical manipulations including electric shock, pricking, singeing, and wet treatment are also effective in preventing or breaking diapause in certain species (Denlinger 2021).

12.3.2   Chemical Manipulations

The most obvious chemicals that can be used to avert or break diapause are the hormones discussed in Chapter 9. Ecdysteroids readily break most larval and pupal diapauses, but the drawback to their use for anything but experimental purposes is that they cannot be used topically because, as steroids, they do not readily penetrate the insect cuticle. That is not an issue, however, with juvenile hormones and their many analogs. These isoprenoids do penetrate the cuticle, thus they can be applied topically. They are highly effective in terminating most adult diapauses, and they can elevate metabolic rate in some diapausing pupae, thus reducing diapause duration. Diapause hormone, a neuropeptide, can be used to induce embryonic diapause in the commercial silk moth and, paradoxically, is an effective pupal diapause terminator for moths in the *Heliothis/Helicoverpa* complex of agricultural pests. In spite of their potential as diapause disruptors, thus far hormones have found limited application in disrupting diapause for control purposes, but two notable examples have shown promise, including disruption of adult diapause in the apple blossom weevil *Anthonomus pomorum* in the Czech Republic (Žďárek et al. 2000) and the pear psylla *Cacopsylla pyricola* in North America (Krysan 1990). In both cases, juvenile hormone analogs were effective in disrupting diapause under field conditions.

In addition to the potential for hormones and hormone analogs to disrupt diapause, numerous other chemical agents are capable of such action. Examples of a wide array of chemicals that can either **prevent or terminate diapause** are listed in Table 12.1. In some cases it is known how the effect is elicited, but in other cases it remains unknown. Xylol terminates embryonic diapause of the grasshopper *Melanoplus differentialis* by removing the wax coating of the egg, thus giving the embryo access to external water, a feature that results in diapause termination. The action of HCl in preventing or breaking embryonic diapause in *Bombyx mori* is possibly linked to the destruction of two small proteins in the chorion that help form an oxygen barrier essential for the establishment and maintenance of diapause (Tsurumaru et al. 2010). Hexane's potent action in terminating pupal diapause in flesh flies (*Sarcophaga*) is possibly similar to the effect of an electric shock stimulating release of the neuropeptide PTTH, leading to ecdysteroid release, but its actual mode of action remains undefined. Action of the imidazole derivative KK-42 in breaking embryonic and pupal diapause is possibly related to its purported ecdysteroid activation. Several psychodysleptics, such as mescaline sulfate and lysergic acid diethylamide (LSD), elicit their effect only during the photosensitive period, thus suggesting the effect is

Table 12.1 Nonhormonal chemical agents capable of breaking or preventing diapause

Diapause stage	Species	Chemical agent	Reference
Embryo	*Bombyx mori*	Hydrochloric, sulfuric acids	Henneguy (1904)
		"	Kogure (1933)
	"	Ouabain	Takeda and Hasegawa (1975)
	"	DMSO	Yamamoto et al. (2013)
	Teleogryllus commodus	Urea	Hogan (1962)
	"	Gaseous ammonia	Hogan (1964)
	Melanoplus differentialis	Xylol	Slifer (1946)
	Atrachya menetriesi	Carbon dioxide	Kidokoro and Ando (2006)
	"	Nitrogen	"
	"	Mercury compounds	Kurihara and Ando (1969)
	Lymantria dispar	KK-42	Suzuki et al. (1993)
	Antheraea yamamai	"	Suzuki et al. (1990)
	"	Numerous imidazoles	Kuwano et al. (1991)
	Artemia franciscana	Hydrogen peroxide	Robbins et al. (2010)
	"	Nitric oxide	"
Larva	*Loxostege sticticalcis*	Xylol	Pepper (1973)
	"	Carbon tetrachloride	"
	Ostrinia nubilalis	Ammonium acetate	Beck et al. (1969)
Pupa	*Sarcophaga crassipalpis*	Hexane and other alkanes	Denlinger et al. (1980)
	"	Diethyl ether	"
	"	Hexene	"
	"	Deuterium oxide	Rockey and Denlinger (1983)
	"	Cholera toxin	Denlinger (1976)
	"	Imidazole, pilocarpine	"
	Manduca sexta	Hexane, acetone	"
	Pieris brassicae	Mescaline sulfate/LSD	Vuillaume and Berkaloff (1974)
	Helicoverpa armigera	KK-42	Wang et al. (1999)
	"	Pyruvate	Wang et al. (2018)
Adult	*Bombus terrestris*	Carbon dioxide	Roseler (1985)
	Osmia bicornis	Hexane	Hayward (unpublished)

Source: Adapted from Denlinger (2021).

achieved by altering the transduction of the photoperiodic signal. Likewise, heavy water (deuterium oxide) possibly sabotages the timekeeping mechanism by slowing the clock that measures daylength. Ouabain's effect on diapause is likely related to its action as an inhibitor of active transport of certain inorganic cations. The action of cholera toxin in breaking diapause is presumably linked to adenylate cyclase activation and the subsequent boost in cyclic AMP production, a response associated with cyclic AMP's well-known role as a second messenger for PTTH and other neuropeptides.

Many of these chemical tricks for breaking diapause are species-specific. Though hexane vapor or topical application is extremely effective in breaking pupal diapause in flesh flies, hexane does not work in other insects, including flies such as *Rhagoletis* that also have a pupal diapause. Hexane can break pupal diapause in the tobacco hornworm *Manduca sexta*, but only when injected. Carbon dioxide, which has proven extremely useful for preventing adult diapause in bumble bees, is also effective in preventing embryonic diapause in the beetle *Atrachya menetriesi*, but fails to alter the diapause response in most other species.

Though mechanisms of action may be unknown many of these agents are proving useful for terminating diapause and allowing precise monitoring of developmental sequences associated with diapause termination, precision that can seldom be attained if one has to rely on environmental termination of diapause. Having tools that can avert diapause or break diapause at will makes it possible to accelerate experiments without waiting for diapause to end naturally. Work on many species could profit from being able to artificially prevent or break diapause. The wide range of agents that are effective in this regard (Table 12.1) offer promise that certain chemicals can be found that will have similar effects on other diapausing species.

In addition to agents that prevent or break diapause, a few chemicals are, interestingly, capable of **inducing diapause** or a diapause-like state. The neuropeptide diapause hormone induces embryonic diapause of the silk moth *Bombyx mori*, and it appears to do so by elevating levels of sorbitol within the egg. It is thus possible to force an embryo not environmentally programmed for diapause to enter diapause by simply culturing the embryo with sorbitol (Horie et al. 2000). Yamamarin, a palmitoyl conjugate of an insect pentapeptide (DILRGa) isolated from diapausing embryos of the silk moth *Antheraea yamamai*, will also cause females of *B. mori* to lay diapausing eggs (Yang et al. 2007b). This peptide has been tested extensively on insect and human cell cultures and is quite effective in arresting cell proliferation and respiration (Sato et al. 2010), attributes common to diapause. Feeding Dopa to the last-instar larvae of the cabbage armyworm *Mamestra brassicae* induces pupal diapause (Uryu et al. 2003). Dopa-fed larvae have higher concentrations of dopamine in their hemolymph and central nervous system, an effect that possibly inhibits synthesis or release of PTTH, the neuropeptide that triggers adult development.

Adults of the Colorado potato beetle *Leptinotarsa decemlineata* enter a diapause-like state when exposed to **precocene**, a dimethylchromene extracted from seeds of the common garden flower *Ageratum houstonianum* (Bowers 1976). Precocene acts by chemically destroying the corpora allata, thus removing the source of JH and simulating the natural shutdown in JH production seen in adult diapause. Likewise, suppression of insulin-like peptide-1, an upstream regulator of JH synthesis, results in a diapause-like state in adult females of the northern house mosquito *Culex pipiens* (Sim and Denlinger 2009b), and suppression of genes encoding Methoprene-tolerant (Met) and Taiman (Tai), components of the JH signaling pathway, mimic the effect of adult diapause in the linden bug *Pyrrhocoris apterus* (Smykal et al. 2014).

Agents capable of suppressing synthesis or secretion of PTTH or ecdysteroids could also generate a diapause-likes state in larvae or pupae. Artificial reduction of an insect's ecdysteroid titer is not easy to achieve, but one interesting agent capable of doing so is ecdysteroid-22-oxidase (E220) (Kamimura et al. 2012). This compound, extracted from the fungus *Nomuraea rileyi*, is capable of suppressing the ecdysteroid titer in several species. Larvae of the crambid moth *Haritalodes basipunctalis* injected with E220 enter a diapause-like state that, like a normal diapause, can be broken by chilling.

The imidazole derivative KK-42 does not force a nondiapause-programmed insect to enter diapause, but topical application or larval feeding can significantly boost the incidence of pupal diapause in the moth *Helicoverpa zea* and flesh fly *Sarcophaga crassipalpis* (Liu et al. 2015). The polyamine putrescine elicits a similar effect in *H. armigera*: Diet supplementation elevates diapause incidence at short daylengths (Wu 2010).

Chemical tools that **prolong diapause** would delay spring emergence and thus be as effective as premature emergence in desynchronizing a pest with its host plant. A few chemicals exert such an effect. Pupal diapause in the corn earworm *Helicoverpa zea* can be broken with diapause hormone (DH) as well as several agonists, but in addition to these diapause terminators, a DH antagonist has been developed that blocks termination of diapause (Zhang et al. 2011a). The antagonist was developed by incorporating a dihydroimidazole trans-Proline mimetic motif into one of the DH agonists, thereby converting an agonist into an antagonist. Injection of luzindone, a melatonin receptor antagonist, into pupae of the Chinese silk moth *Antheraea pernyi* delays diapause termination, presumably by inhibiting release of PTTH (Wang et al. 2013). The imidazole derivative KK-42 elicits a similar effect in *A. pernyi* and *H. zea* (Liu et al. 2015): When injected into diapausing pupae, KK-42 delays diapause termination by one to two months.

12.3.3   Genetic Manipulations

Genetic approaches to vector control are becoming increasingly more common, and as pointed out by Armbruster (2016), diapause-disrupting constructs could be spread by releasing male mosquitoes early in the season when there is no need for diapause, with lethal effects setting in with the arrival of winter. Similar strategies could also be employed for agricultural pests, but the technology is currently more highly developed for mosquitoes. Numerous physiological traits are linked to a successful diapause, thus a wide range of potential targets could be identified for genetic manipulation, for example, desiccation resistance, cold hardiness, metabolic depression, sequestration, and utilization of energy stores. The potential for genetic manipulations of diapause offers a promise that is yet to be realized.

12.4   Matching Seasonal Cycles of Biological Control Agents and Their Hosts

For a biological control program to be successful, the introduced control agent must be present at the same time as the insect or plant being targeted. If seasonal timing of the two is not aligned, there is little chance of success. This implies that the phenologies of the two must be in **synchrony**, a feature that has a higher chance of being realized if the biological control agent originates from a similar latitude, altitude, and climatic zone as the target.

The challenge of finding a biological control agent with the desired features for a new environment is exemplified in the introduction of the tamarisk leaf beetle *Diorhabda elongata* from China (44° N) into the western United States to control the invasive weed saltcedar (*Tamarisk* sp.). Though the chrysomelid is becoming established at northern sites, little success is noted south of 38° N, presumably because the introduced beetle is univoltine and enters a premature diapause due to prevailing daylengths at lower latitudes (Bean et al. 2007a,b). Ecotypes of *D. elongata* collected recently from more southern latitudes in China (36, 38, 43° N) produce up to four generations a year and enter diapause at a more suitable time (Dalin et al. 2010), thus offering a better match for biological control of *Tamarisk* at lower latitudes. The ideal match for release in central Texas and points further south is a population of *D. elongata* native to Crete, a population that produces four to five summer generations and continues to lay eggs until mid-October (Milbrath et al. 2007).

Flea beetles, chrysomelids in the genus *Aphthona*, were introduced to the Great Plains of North America from Europe for biological control of the weed leafy spurge and have proven effective except in warm winters when diapause is broken prematurely (Joshi et al. 2009). Two biological control agents from Eurasia (tortricid moth *Agapeta zoegana* and weevil *Cyphocleonus achates*) introduced for control of spotted knapweed in the western United States are fairly successful in their new habitat although features of their overwintering differ from that observed in Eurasia (Corn et al. 2009). Larvae of both potential biocontrol species enter diapause at any earlier instar than in their native range, and slower development in their new, colder habitat results in only one generation per year in Montana versus two or three generations in Europe.

Invasion of the horse chestnut leafminer, *Cameraria ohridella* into Europe has been devasting to horse chestnut trees. Though a number of native European parasitoids attack the leafminer, the problem is one of synchrony (Grabenweger 2004). Large numbers of native parasitoids successfully overwinter, but they emerge too late in the spring to have an impact on first-generation leafminers. Success with a biological control agent will require a parasitoid that emerges from diapause earlier in the year.

Predacious flower bugs, genus *Orius*, are effective predators of thrips and other small arthropods. A comparison of 12 strains representing 5 species in Japan reveals, as one might expect, a wide range of critical daylengths, diapause capacity, and winter survival, thus offering strains that are potentially useful for biological control under diverse conditions (Shimizu and Kawasaki 2001). Of special utility are strains that do not enter adult diapause in greenhouses during the winter months. A nondiapausing strain from Okinawa is thus especially promising (Nakashima and Hirose 1997), as are other southern multivoltine species that do not diapause as long as food is available (Saulich and Musolin 2019). Other approaches to maintaining active predator populations in greenhouses are to extend daylength during the winter months (Chambers et al. 1993) or to elevate temperature, thereby preventing diapause (Musolin and Ito 2008). Another predatory mirid, *Pilophorus gallicus*, plays an important role in suppressing populations of the pear psyllid in pear orchards in the Mediterranean region, but the fact that populations of this natural predator include a mixture of univoltine and multivoltine individuals reduces its potential as a biological control agent (Ramírez-Soria et al. 2019).

Diapause is a factor to consider in release programs of *Oobius agrili*, a hymenopteran egg parasitoid of the emerald ash borer, *Agrilus planipennis* (Hoban et al. 2016). The parasitoid kills the host before the ash borer can cause damage, thus making it a promising biological control agent. This species has a maternally induced diapause, and adults that have gone through diapause are more effective in early summer field releases because they have higher reproductive output than those that have not been through diapause, and they are less likely to produce diapausing progeny. Thus, the most effective parasitoids are those that have been through diapause. Fecundity in the chrysomelid beetle *Zygogramma bicolorata*, a biological control agent imported to India from Mexico for weed control, is likewise enhanced by the adult having gone through diapause (Hasan et al. 2018). This feature, plus the fact that adults terminate diapause at the onset of the monsoons, just as the target weed is beginning to proliferate, is resulting in a promising release program. The fact that diapause capacity can be lost readily in laboratory cultures (Chapter 10) foretells a potential problem for biological control agents that require a diapause for successful establishment in their new environment. Loss of diapause capacity in a Chinese strain of the parasitoid *Binodoxys communis* appears to be why the parasitoid failed to establish and provide control of the soybean aphid *Aphis glycines* in North America (Gariepy et al. 2015).

Responses of biological control agents can be expected to mirror the types of adaptive responses observed with invasive species, and from that literature we know that life-history attributes of many invasive species rapidly evolve to match parameters of their new environment. Though not all biological control agents have evolved rapidly enough to succeed in their new habitat, for example, the parasitoid *Aphidius ervi* released to control pea aphids (Hufbauer 2002), other releases do show rapid evolution that nicely matches the new environment. For example, the ragwort flea beetle, *Longitarsus jacobaeae*, introduced from Italy to Oregon for control of tansy ragwort, quickly adapted, in approximately 30 generations, to a new, cooler high altitude site near Mt. Hood by losing summer diapause and accelerating the onset of egg-laying (Szücs et al. 2012). Rapid evolution, together with the inherent variation available from populations collected from diverse locales, offer promise for successfully matching biological control agents with their targeted hosts.

12.5   Increasing Shelf Life of Biological Control Agents

A challenge for the biological control industry is being able to store parasitoids, predatory mites, and other biological control agents without incurring a significant loss of vitality. Diapause is a great way to increase shelf life, but for this to work it must be possible to break diapause on demand and be certain that diapause does not exact a toll on post-diapause fitness. Several features of diapause such as cold hardiness also make diapausing insects ideal for long-distance, commercial transport.

Among insects most widely used for biological control are the lacewings, members of the family Chrysopidae. Diapausing adults of *Chrysoperla carnea* can be stored at a low temperature (5°C) for up to 31 weeks without impairing survival or jeopardizing reproductive potential (Tauber et al. 1993, Chang et al. 1996), thus offering flexibility and efficiency for mass production. Diverse populations and biotypes of *C. carnea* display differing capacities for long-term storage (ranging from 18 to 39 weeks) and subsequent survival and reproduction (Chang et al. 2000), underscoring choices that can be made in selecting the best options for mass production and release. Nondiapausing adults can also be stored for a limited period, but diapause greatly increases shelf life. A species from Honduras that does not enter diapause, *C. externa*, can be stored successfully as adults at 10°C, but the maximum shelf life in this case is 16 weeks, considerably shorter than the shelf life of diapausing adults of *C. carnea* (Tauber et al. 1997). *Chrysopa formosa*, a green lacewing widely used for biological control in Asia, overwinters in larval diapause, and inducing diapause makes it possible to store these lacewings at 5°C for up to 300 days with high survival (Li et al. 2018b).

Egg parasitoids of several species in the genus *Trichogramma* are commonly deployed for biological control, and again, diapause in the late larval stage can be used to increase shelf life. Diapausing larvae of *T. cordubensis* can be stored up to 6 months at 3°C, if they are first held 30–40 days at 10°C (Ventura Garcia et al. 2002). Larvae of *T. cacoeciae* that are not in diapause but are quiescent can be stored for one to two months, but if larvae are in diapause the storage period can be extended up to one year (Pizzol and Pintureau 2008). A tropical species *T. chilonis* can be stored in diapause for up to 125 days (35 days at 10°C, followed by 90 days at 5°C) (Ghosh and Ballal 2018). Good adult emergence in *T. dendrolimi* is retained for up to 100 days of storage (30 days at 12°C, followed by 70 days at 3°C) (Zhang et al. 2018b). Though some strains of *Trichogramma* are infected with the endosymbiotic bacteria *Wolbachia*, this infection does not appear to either harm or boost the diapausing parasitoid's success in emerging as an adult or its fecundity (Pintureau et al. 2003). Egg parasitoids *Trissolcus basalis* and *Telenomus podisi* diapause as adults, and again, storage can be enhanced by diapause: At 15°C *T. basalis* can be stored for 13 months, and *T. podisi* for 10 months, but fecundity drops if storage exceeds 180 days (Foerster and Doetzer 2006). These parasitoids, used for stink bug control, can thus be adequately stored between the two soybean crop seasons in Brazil. Storage successes as demonstrated in these examples benefit from diapause and allow mass rearing and well-timed releases during the field season.

Populations of the predatory gall midge *Aphidoletes aphidimyza*, used widely for aphid control in greenhouses, differ in levels of metabolic depression during larval diapause and have different diapause durations, rates of glycogen depletion, and cold hardiness (Koštál et al. 2001). Midges with higher diapause intensity (greater metabolic depression and diapause duration) can be stored more effectively and are thus more desirable for commercial use.

The predatory phytoseiid mite *Euseius finlandicus*, an important biological control agent of the red spider mite, has a facultative adult diapause. The diapausing stage is the best option for long-term storage (Broufas et al. 2006). Cold storage at 5°C for 10–13 weeks allows synchronous termination of diapause within 2 weeks, an attribute that may prove valuable for mass releases.

Coccinelids, the ladybird beetles, are also widely used in the biological control industry, and they too can be effectively stored at low temperature as diapausing adults. Diapause-destined adults of the multicolored Asian ladybird beetle *Harmonia axyridis*, collected in the autumn, can be stored for up to 150 days at 3–6°C with minimal loss of survival or reduction in post-storage reproduction (Ruan et al. 2012).

12.6   Managing Domesticated Species and Experimental Lines

Methods for long-term insect storage and management of diapause are useful not only for biological control agents but also for other commercially important insects, including pollinators, butterflies used in butterfly gardens, silk producers, insects used for organic decomposition, edible insects, insects that are mass-reared for sterile insect release programs, and laboratory lines that have desirable traits worthy of preservation. In some cases it is desirable to use diapause for storage, but in other cases the need is to break diapause.

Bumble bees and solitary bees are important **pollinators** in greenhouses and field crops, respectively. Commercial pollination services rely on domesticated strains of bees, and because both bumble bees and solitary bees have a diapause (adult queens of bumble bees and larvae of solitary bees), commercial producers must understand how this diapause is environmentally regulated and be able to produce active bees at the correct time of year. Pollination is frequently required at times when native pollinators are absent or inadequate to meet demand, and in these situations, stockpiled pollinators provide a vital service. For example, overwinter storage of diapausing larvae of the solitary bee *Osmia cornifrons* at 1–2°C makes it possible to have cool-season pollinators early in the spring before other pollinators are available (Wilson and Abel 1996). Specific treatments that improve storage survival, such as fluctuating thermal regimes (Colinet et al. 2018), can be used to boost storage duration of pollinators as well as other species.

The commercial silk moth *Bombyx mori* and wild species of silk moths used for **silk production** have diapauses that growers must understand and be able to manipulate. Tools such as hydrochloric acid have been used in the silk moth industry for a long time to prevent or break diapause, thus facilitating continuous culturing of silk moths without the delay of a long diapause (Tsurumaru et al. 2010), but there are also times when it is desirable to use diapause for storing eggs. Maintenance of multivoltine silk moths without diapause requires rearing five generations per year, an undertaking that requires considerable cost, labor, and infrastructure, as well as exposure to disease, parasitism, and predation, thus an attractive alternative is to put embryos into diapause and store them at low temperatures until needed, an approach that also maintains genetic stability (Saravanakumar et al. 2010).

To **reduce costs** of maintaining an ongoing colony of the sugarcane borer *Diatraea evanesens*, larvae can be stored in diapause for up to five months at 10°C, a procedure that saves both rearing supplies and labor costs when the insects are not needed (White 2008). Similar cost-savings can be achieved for many insects that are not needed continuously for research. The proliferation of new genetic lines poises a particularly costly challenge for culture maintenance. For some species, diapause offers an attractive method for reducing costs of continuous propagation.

Storage of the black soldier fly *Hermetia illucens*, a stratiomyid used widely for organic waste decomposition, is a valuable method for making flies available on demand. Huge numbers of flies may be needed, but the need may be sporadic. Larvae can be stored in diapause for 114 days at 28°C, and presumably much longer at lower temperatures, provided humidity remains high (Samayoa and Hwang 2018).

Defining the optimal microclimate is essential for successful storage. Finding the right moisture and temperature conditions (moist sand at 4–6°C) makes the difference between being able to store the sugarbeet root maggot *Tetanops myopaeformis* for up to five years rather than seven days (Yocum et al. 2012). Optimal conditions for long-term storage of the maggot are for the larvae to be in diapause or post-diapause quiescence, to be maintained in a moist microclimate, and to have acquired ample metabolic reserves.

Tools for breaking diapause are urgently needed for some insect species consumed as **human food**. According to Professor Sansan Da from Burkina Faso, the large (6.4 g) shea nut caterpillar *Cirina butyrospermum* is a highly desirable food in regions of West Africa (Figure 12.1), but this species has a 10-month pupal diapause, making it available as a food source for only a brief time each year. Year round production of this edible insect would offer opportunities for producers as well as offer a continuous nutritious food source, but the obstacle is lack of a mechanism for artificially breaking diapause to generate continuous development. Similar demand for breaking diapause is urgently needed for a number of species that are being considered for **sterile insect release programs**. Current efforts at the International Atomic Energy Agency in Vienna to mass rear several species for sterile release are hampered by diapause, for example, apple maggot *Rhagoletis pomonella*, European cherry fruit fly *R. cerasi*, Chinese citrus fly *Bactrocera minax*, and Russian melon fly *Myiopardalis pardalina*. Tools to short-circuit the obligate pupal diapauses in these species would facilitate mass rearing.

The ultimate mechanism for long-term storage is **cryopreservation**, a technique that is especially useful for preserving valuable mutants and genetically transformed lines (Leopold 2007, Leopold and Rinehart 2010, Štetina et al. 2018). Though we are at the threshold of a technology that will make it possible to vitrify and thus cryopreserve insects indefinitely, the fine details must be worked out for each species. Tools for inducing this chemically induced dormancy may benefit from some of the chemical adjustments (*e.g.*, sugars such as trehalose) used by cold-hardy or freeze-

Figure 12.1 Larvae of the edible shea nut caterpillar *Cirina butyrospermum* at a market in Burkina Faso. Due to the 10-month pupal diapause in this species it is only available as a food source for a limited time each year.
Photo courtesy of Sansan Da (INERA, Burkina Faso).

tolerant diapausing insects. Success in cryopreserving ovaries from the ladybird beetle *Harmonia axyridis* (Kawaguchi and Niimi 2018) offers hope for developing effective methods for cryopreserving biological control agents as well as other non-model species.

12.7   Issues of Light Pollution

Artificial light at night (ALAN) is well recognized as a disruptor of physiological function and behavior in birds and mammals, including humans (Stevenson et al. 2015). Attraction of insects to street lights and possible impacts of artificial light on bioluminescent signaling underscore the effect that night light may exert on insects as well (Rich and Longcore 2006). Direct emission of light is one issue, but the second aspect of nighttime light is **nightglow**, the brightening of the night sky caused mainly by upwardly emitted and reflected light that is scattered in the atmosphere by water, dust, and gas molecules (Gaston 2018). The growing use of light-emitting diodes (LEDs) has also shifted the dominant spectra of artificial light, with more emission of blue wavelengths, a portion of the spectra to which insects are highly sensitive.

As discussed in earlier chapters, insect diapause is particularly responsive to daylength, and artificial extension of daylength and night interruption with pulses of light are quite effective in preventing diapause in most species. The question is whether the abundance of light in the night environment is actually having an impact on insect diapause. The question has not been addressed extensively, but several papers hint that night light can be expected to have such an impact. Low levels of artificial light used to extend daylength for larvae of the cabbage moth *Mamestra brassicae* prevent the onset of pupal diapause (van Geffen et al. 2014), as demonstrated for numerous other species as well. Low levels of artificial light alter clock gene expression in the mosquito *Culex pipiens* f. *molestus* (Honnen et al. 2019), a result also expected from many previous experiments. In the mosquito *Aedes albopictus* simulated ALAN reduces the incidence of embryonic diapause by 40% (Westby and Medley 2020). Although light has been used to alter the diapause response for control purposes (see Section 12.3.1), as far as I know, it has not yet been demonstrated that photoperiodic responses in natural populations are altered by light at night. Such a scenario, however, is quite plausible. The most likely scenario would be abortion of the diapause program due to artificial light extending the perceived daylength.

12.8   Insect Conservation

Protecting **overwintering sites** is an important dimension of insect conservation, a measure that is especially critical for species overwintering in large aggregations at restricted locations. Amassing huge numbers of individuals in a specific site makes a species vulnerable to major calamities, both natural and human-driven.

Protection of the montane sites used by the monarch butterfly *Danaus plexippus* in Mexico for overwintering is of utmost importance for conservation of this charismatic species (Oberhauser et al. 2017). Since millions of adults from eastern North America become concentrated at a small site for overwintering, the monarch is particularly vulnerable during diapause. An epic winter ice storm in 2002 is estimated to have killed 200–272 million monarchs, representing 74 and 81% of the total populations in the two main overwintering sites (Brower et al. 2004), and even more problematic is the continued logging efforts that have severely reduced (at a rate of approximately 0.89 ha/year since 1993) the area suitable for overwintering (Oberhauser et al. 2017). Political agreements between Mexico, the United States, and Canada are designed to protect these overwintering sites, and numerous nongovernment agencies also support monarch conservation, efforts that hopefully can be sustained over time to protect this vulnerable species.

Like the monarch, the ladybird beetle *Hippodamia undecimnotata* overwinters in spectacular aggregations at the same sites each winter in southern and eastern Europe, yet little has been done to protect these sites (Susset et al. 2017b). Habitat destruction and fragmentation of aggregations appear to be the most threatening challenges the species faces. Protection of *H. undecimnotata* aggregation sites would also have a cascading effect since the overwintering sites are shared with several other highly beneficial aphidophagous species. Up to 80% of *H. undecimnotata* may die in the overwintering aggregations, and over 90% mortality was noted in aggregations of another ladybird beetle *Coccinella septempunctata*, mainly a consequence of entomopathogenic fungal infection (Guven et al. 2015). Catastrophic events at such sites have significant consequences for successful employment of these beetles for natural biological control of agricultural pests.

The stink bug *Encosternum delegorguei*, a local food source in Southern Africa, is currently in danger, possibly due to overharvesting by humans who capture huge numbers at the bug's dry-season aggregation site (Dzerefos et al. 2015). Overexploitation is a common problem with edible insects (van Huis et al. 2013). Protecting overwintering sites and limiting the number of insects that can be harvested may be required to sustain long-term productivity.

The large copper butterfly *Lycaena dispar batavus* has not been seen in the United Kingdom since the middle of the nineteenth century. Attempts to re-introduce *L. dispar* to the United Kingdom from Europe have failed to establish self-sustaining populations. Although this is a wetland species that feeds on water dock, failure of the species to become reestablished is likely due to high mortality caused by the excessively wet winter habitat into which it was introduced (Nicholls and Pullin 2003). Overwintering larvae can survive remarkably long under water (28 days), but for longer survival it must be able to access plant refuges that protrude above the waterline.

Removal of fence rows, hedges, and other favored insect overwintering sites have likely been factors contributing to the demise of insect diversity across much of the agricultural landscape in North America and Europe. Until producers recognize the benefits of biodiversity and sustainability, overwintering sites used by insects are likely to continue to disappear.

12.9   Role in Disease Transmission

Insect vectors of human and animal diseases may harbor their pathogens during diapause, thus facilitating propagation of the pathogen the following season. Two such examples include the mosquitoes *Culex pipiens*, a major North American vector of **West Nile virus** (WNV), and *Ochlerotatus triseriatus,* the vector of **La Crosse virus** (LAC).

Diapausing adults of *C. pipiens* harboring WNV have been collected in midwinter in New York, New Jersey, and Pennsylvania (Nasci et al. 2001, Bugbee and Forte 2004, Farajollahi et al. 2005, Andreadis et al. 2010). The fact that females do not normally take a blood meal prior to entering diapause suggests a vertical route of transmission. **Vertical transmission** of WNV by *C. pipiens* is further supported by experiments examining progeny of infected females: Following diapause, such progeny, without being reinfected, were capable of transmitting WNV to a hamster in the laboratory (Anderson and Main (2016). The virus does not replicate during diapause, but replication resumes as soon as diapause is broken in the spring. The situation in the more mild Sacramento Valley of California is less clear. Both *C. pipiens* and *C. tarsalis* vector WNV in this region, and some samples collected during winter test positive for WNV (Nelms et al. 2013). Although most individuals overwinter in diapause, especially *C. tarsalis*, the overwintering population of *Culex* mosquitoes is a mixture of diapausing, nondiapausing, and quiescent individuals, suggesting the possibility that both vertically and horizontally infected *Culex* females survive winter and possibly transmit WNV in the spring.

The eastern treehole mosquito *O. triseriatus*, a species that overwinters as a diapausing embryo, is the principal vector of LAC. The rodent hosts of LAC do not sustain the virus through the winter. Instead, the virus is transovarially transmitted and resides overwinter in diapausing embryos of the host mosquito (Blitvich et al. 2001). At least three QTL act collectively to permit **transovarial transmission** of the virus into the diapausing embryo (Graham et al. 2003).

In both of these examples, the diapausing stage enables the virus to persist in the mosquito population from one year to the next. What is particularly intriguing is that the insect's entry into diapause halts viral replication, thus suggesting a mechanism within diapausing insects to shut down the replication process. The virus is thus finely tuned to the developmental status of its mosquito host, but how this is achieved remains unknown.

Another intriguing question relevant to disease transmission is whether the experience of diapause has an impact on **vector competence**. The answer appears to be "no," at least for a population of *C. pipiens* in the Netherlands. Females emerging from winter diapause have similar WNV infection rates (30.4%) as summer females that have not been through diapause (35.3%) (Koenraadt et al. 2019).

12.10   Models for Human Health

Our understanding of human biology has benefited considerably from insights gained from insect genetics, neurobiology, endocrinology, and other aspects of insect physiology and behavior, but are there also features of insect diapause that could serve as models for human health? I would argue that select features of diapause could indeed serve as models for exploring human health issues such as aging, obesity, ischemia, and possibly others.

12.10.1   Links to Aging

Diapause is an impressive extension of life. Insects that live for only a month or so without diapause are, when the diapause program is invoked, suddenly capable of living 10 months or more, the entomological equivalent of extending human life to 800 years! The dauer stage (nematode equivalent of diapause) of *Caenorhabditis elegans* has been at the forefront of aging research for quite a few years, demonstrating parallels between **insulin signaling** and aging (*e.g.*, Kimura et al. 1997, Kim 2007). The slow aging process during insect diapause has many similarities to the nematode dauer state. In the laboratory, mortality rates of post-diapause adults of *Drosophila melanogaster* are similar to rates of newly eclosed, young flies, suggesting that senescence during diapause is slow or negligible (Tatar and Yin 2001, Tatar et al.

2001). Like other adult diapauses, the dormancy in *D. melanogaster* is controlled by downregulation of juvenile hormone (JH), which in turn has links with insulin signaling. Mutations in the insulin receptor (*InR*) result in reduced JH synthesis and consequently longer life. Application of a JH analog reverses the effect, generating a short-lived fly (Tatar et al. 2001, Tu et al. 2005). Many gene networks implicated in adult senescence in *D. melanogaster* (*e.g.*, insulin, TOR, and germ-line signaling pathways) are similar to those that underlie life span extension during adult diapause (Kučerová et al. 2016).

Reactive oxygen species (ROS) are thought to promote aging and are associated with diverse medical disorders, but experiments with yeast, *C. elegans* and *D. melanogaster* have shown that increased ROS generated by chemical inhibition or mutation can also lengthen life span. Diapausing insects have contributed to this question. Naturally occurring, physiological levels of ROS increase the life span of the cotton bollworm *Helicoverpa armigera* by channeling pupae into diapause (Zhang et al. 2017b). ROS appears to exert this effect by acting through components of the insulin signaling pathway. These results suggest that high levels of ROS accelerate aging but more modest, physiological levels of ROS, paradoxically, extend life span.

Insects also offer great models for probing trade-offs between life span and reproduction. Such trade-offs are evident in some species but not all: Production of eggs does not affect female longevity in the linden bug *Pyrrhocoris apterus* (Hodková 2008). Eliminating egg production in *P. apterus* by ablation of the ovary does not extend life, but removal of the corpora allata does so, indicating that the production of eggs is not what takes a toll on life span but instead life span is dictated by an upstream regulatory mechanism located within the brain's pars intercerebralis, the control center that presides over the corpora allata. The greatest longevity in *P. apterus* is generated by diapause, a naturally occurring state governed by the brain-corpora allata axis.

The fact that so many insects have the capacity to vary their longevity through diapause suggests that this group of organisms offers a rich repository of valuable models to probe essential elements of aging.

12.10.2   Model for Obesity Research?

The pending obesity crisis has prompted a new emphasis on obesity research. Do insects offer useful insights? Perhaps. What is especially interesting is that many insects "decide" to sequester more fat reserves if they are programmed for diapause (Hahn and Denlinger 2007, Section 6.2). This suggests some sort of developmental switch that tells the insect to garner additional reserves and/or to convert its reserves into fat. The net result is that diapausing insects may contain more than twice the amount of fat found in their nondiapausing counterparts. Packing on additional fat requires metabolic adjustments, but diapausing insects appear able to dodge the health problems that plague obese humans. Diapausing insects must also carefully parse out their fat reserves before diapause comes to an end. In some cases, the diapausing insect will switch from reliance on fat to other reserves at specific points during diapause. Little is known about these decision points related to fat sequestration and utilization. Interactions between fat storage, aging, and insulin signaling have been effectively probed in *C. elegans* (Wang et al. 2008), suggesting that insect diapause models could also offer valuable insights into this important issue of human health.

12.10.3 Model for Responding to Ischemia?

Restricted oxygen availability, a major problem in cardiac arrest as well as organ transplantation, is a challenge routinely confronted by diapausing insects. In insects there is frequently a self-imposed reduction in oxygen brought about by reduced rates of oxygen consumption during diapause. Oxygen consumption may be halted for days at a time. Oxygen may also be unavailable as a consequence of the anoxic environments (*e.g.,* wet or frozen soils) where some diapausing insects spend the winter. Diapausing insects have mechanisms in place to survive bouts of anoxia (Section 7.13.5) and can perhaps offer templates to enhance survival of human tissues under similar circumstances.

12.11   Pharmacological Prospecting

Traditionally, the pharmaceutical industry has relied on medicinal plant compounds that can easily be harvested in high quantities, but recent progress in molecular biology makes it possible to produce compounds of medical interest that are normally produced in small amounts. Diapausing insects may be an especially valuable resource for pharmacological prospecting because they mount impressive defenses against bacteria, viruses, and fungi during the long months of diapause. Human, animal, and plant health can all expect to benefit from defense agents that can be tapped from such insect sources.

Cecropin, the antibacterial peptide first described from diapausing pupae of the silk moth, *Hyalophora cecropia*, displays cytotoxicity against several mammalian tumor cell lines (Moore et al. 1994), and insertion of *cecropin* into human bladder carcinoma cell lines results in a loss of tumorigenicity (Winder et al. 1998). Low mammalian toxicity and high antimicrobial activity against pathogenic bacteria make cecropins and their analogs promising pharmaceutical agents (Brady et al. 2019). Diapausing larvae of the blow fly, *Calliphora vicina*, produce novel defense compounds, **alloferons**, that stimulate cytotoxicity of human peripheral blood lymphocytes, induce interferon synthesis in mice and human models, and enhance antiviral and antitumor resistance in mice (Chernysh et al. 2002). One of the alloferons, alloferon-1, is currently used to treat persistent viral infections and appears to enhance anti-tumor activity when combined with cytotoxic drugs (Chernysh et al. 2012). Several alloferon analogs have been developed with even higher antiviral effects than the native peptide (Kuczer et al. 2016). The pentapeptide **yamamarin**, first described as an inhibitor of development during pupal diapause in the Japanese oak silk moth *Antheraea yamamai*, also has the capacity to suppress proliferation of rat hepatoma cells (Yang et al. 2004, 2007b), while **sericin** extracted from cocoons of diapausing pupae of *A.*

yamamai promotes proliferation of both insect and mammalian cells (Cui et al. 2009). **Diapausin**, a 41 amino acid diapause-specific peptide isolated from diapausing adults of the leaf beetle *Gastrophyusa atrocyanea*, exhibits strong anti-fungal activity and blocks Ca^{2+} channels (Sato et al. 2002, Tanaka et al. 2003). The amino acid sequence has little homology with other known Ca^{2+} blockers or antifungal peptides, thus diapausin appears to represent a new class of biopharmacological agents (Kouno et al. 2007). Transgenic expression of diapausin in *Arabidopsis* enables the plant to resist phytopathogenic fungal attack (Wu et al. 2013). Discovery of more such interesting compounds from diapausing insects can be anticipated in the years ahead.

Manipulating diapause, probing mechanisms that insects themselves use for regulating diapause, and unraveling tactics used by diapausing insects to protect against environmental injury and to ward off attack by microbes combine to offer rich opportunities for translating the secrets of diapause into tools with wide potential application.

References

Aak, A., T. Birkemoe, and H.P. Leinaas. 2011. Phenology and life history of the blowfly *Calliphora vicina* in stockfish production areas. *Ent. Exp. Appl.* 139:35–46.

Abarca, M. 2019. Herbivore seasonality responds to conflicting cues: untangling the effects of host, temperature, and photoperiod. *PLoS One* 14:e0222227.

Abdelrahman, H., J.P. Rinehart, G.D. Yocum, K.J. Greenlee, B.R. Helm, W.P. Kemp, C.H. Schulz, and J.H. Bowsher. 2014. Extended hypoxia in the alfalfa leafcutting bee, *Megachile rotundata*, increases survival but causes sub-lethal effects. *J. Insect Physiol.* 64:81–89.

Abrieux, A., Y. Xue, Y. Cali, K.M. Lewald, H.N. Nguyen, Y. Zhang, and J.C. Chiu. 2020. EYES ABSENT and TIMELESS integrate photoperiodic and temperature cues to regulate seasonal physiology in *Drosophila*. *Proc. Nat'l. Acad. Sci., USA* 117:15293–15304.

Adedokun, T.A. and D.L. Denlinger. 1985. Metabolic reserves associated with pupal diapause in the flesh fly, *Sarcophaga crassipalpis*. *J. Insect Physiol.* 31:229–233.

Adkisson, P.L. 1964. Action of the photoperiod in controlling insect diapause. *Am. Nat.* 98:357–374.

Adkisson, P.L. 1966. Internal clocks and insect diapause. *Science* 154:234–241.

Agui, N. 1975. Activation of prothoracic glands of brains *in vitro*. *J. Insect Physiol.* 21:903–913.

Agui, N., N.A. Granger, L.I. Gilbert, and W.E. Bollenbacher. 1979. Cellular localization of insect prothoracicotropic hormone. *Proc. Nat'l. Acad. Sci., USA* 76:5684–5690.

Ahmadi, F., S. Moharramipour, and A. Mikani. 2018. The effect of temperature and photoperiod on diapause induction in pupae of *Scrobipalpa ocellatella* (Lepidoptera: Gelechiidae). *Environ. Ent.* 47:1314–1322.

Akitomo, S., Y. Egi, Y. Nakamura, Y. Suetsugu, K. Oishi, and K. Sakamoto. 2017. Genome-wide microarray screening of *Bombyx mori* genes related to transmitting the determination outcome of whether to produce diapause or nondiapause eggs. *Insect Sci.* 24:187–193.

Akkawi, M.M. and D.R. Scott. 1984. The effect of age of parents on the progeny of diapaused and nondiapaused *Heliothis zea*. *Ent. Exp. Appl.* 35:235–239.

Ala-Honkola, Q., H. Kauranen, V. Tyukmaeva, F.A. Boetzl, A. Hoikkala, and T. Schmitt. 2018. Diapause affects cuticular hydrocarbon composition and mating behavior of both sexes in *Drosophila montana*. *Insect Sci.* 27:304–316.

Al-Anzi, B., V. Sapin, C. Waters, K. Zinn, R.J. Wyman, and S. Benzer. 2009. Obesity-blocking neurons in *Drosophila*. *Neuron* 63:329–341.

Allen, M.J. 2007. What makes a fly enter diapause? *Fly* 1:307–310.

Altermatt, F. 2009. Climatic warming increases voltinism in European butterflies and moths. *Proc. R. Soc. B* 277:1281–1287.

Amevoin, K., I.A. Glitho, J.P. Monge, and J. Huignard. 2005. Why *Callosobruchus rhodesianus* causes limited damage during storage of cowpea sees in a tropical humid zone in Togo. *Ent. Exp. Appl.* 116:175–182.

Amouroux, P., F. Normand, H. Delatte, A. Roques, and S. Nibouche. 2014. Diapause incidence and duration in the pest mango blossom gall midge, *Procontrinia mangiferae* (Felt), on Reunion Island. *Bull. Ent. Res.* 104:661–670.

Amsalem, E., D.A. Galbraith, J. Cnaani, P.E.A. Teal, and C.M. Grozinger. 2015. Conservation and modification of genetic and physiological toolkits underpinning diapause in bumble bee queens. *Mol. Ecol.* 24:5596–5615.

An, Y., T. Nakajima, and K. Suzuki. 1998. Immunohistochemical demonstration of mammalian- and FMRFamide-like peptides in the gut innervation and endocrine cells of the wild silkmoth, *Antheraea yamamai* (Lepidoptera: Saturniidae) during diapause and post-diapause of pharate first-instar larvae. *Eur. J. Ent.* 95:185–196.

Anchordoguy, T.J. and S.C. Hand. 1994. Acute blockage of the ubiquitin-mediated proteolytic pathway during invertebrate quiescence. *Am. J. Physiol.* 267: R895–R900.

Anderson, J.F. and A.J. Main. 2006. Importance of vertical and horizontal transmission of West Nile virus by *Culex pipiens* in the Northeastern United States. *J. Infect. Diseases* 194:1577–1579.

Andrade, T.O., M. Herve, Y. Outreman, L. Krespi, and J. van Baaren. 2013. Winter host exploitation influences fitness traits in a parasitoid. *Ent. Exp. Appl.* 147:167–174.

Andreadis, T.G., P.A. Armstrong, and W.J. Bajwa. 2010. Studies on hibernating populations of *Culex pipiens* (Diptera: Culicidae) from a West Nile virus endemic focus in New York City: parity rates and isolation of West Nile virus. *Am. J. Mosq. Cont. Assoc.* 26:257–264.

Andreatta, G., C.P. Kyriacou, T. Flatt, and R. Costa. 2018. Aminergic signaling controls ovarian dormancy in *Drosophila*. *Sci. Rep.* 8:2030.

Andreatta, G., C. Broyart, C. Borghgraef, K. Vadiwala, V. Kozin, A. Polo, A. Bileck, I. Beets, L. Schoofs, C. Gerner, and F. Raible. 2020. Corazonin signaling integrates energy homeostasis and lunar phase to regulate aspects of growth and sexual maturation in *Platynereis*. *Proc. Nat'l. Acad. Sci., USA* 117:1097–1106.

Andrewartha, H.G. 1952. Diapause in relation to the ecology of insects. *Biol. Rev. Camb. Philos. Soc.* 27:50–107.

Andrews, M.T. 2019. Molecular interactions underpinning the phenotype of hibernation in mammals. *J. Exp. Biol.* 222: jeb160606.

Anduaga, A.M., D. Nagy, R. Costa, and C.P. Kyriaou. 2018. Diapause in *Drosophila melanogaster* – photoperiodicity, cold tolerance and metabolites. *J. Insect Physiol.* 105:46–53.

Ankersmit, G.W. 1968. The photoperiod as a control agent against *Adoxophyes reticulana* (Lepidoptera; Tortricidae). *Ent. Exp. Appl.* 11:213–240.

Appenroth, D., V.J. Melum, A.C. West, H. Dardente, D.G. Hazlerigg, and G.C. Wagner. 2020. Photoperiodic induction without light-mediated circadian entrainment in a high Arctic resident bird. *J. Exp. Biol.* 223:jeb220699.

Appleby, J.H. and P.F. Credland. 2007. The role of temperature and larval crowding in morph determination in a tropical beetle, *Callosobruchus subinnotatus*. *J. Insect Physiol.* 53:983–993.

Arbaciauskas, K. 2004. Daphnia. *J. Limnol.* 63 (Suppl. 1):7–15.

Arfat, Y., H. Chang, and Y. Ga. 2018. Stress-responsive miRNAs are involved in reprogramming of metabolic functions in hibernators. *J. Cell. Physiol.* 233:2695–2704.

Arias-Cordero, E., L. Ping, K. Reichwald, H. Delb, M. Platzer, and W. Boland. 2012. Comparative evaluation of the gut microbiota associated with the below- and above-ground life stages (larvae and beetles) of the forest cockchafer, *Melolontha hippocastani*. *PLoS One* 7: e51557.

Armbruster, P.A. 2016. Photoperiodic diapause and the establishment of *Aedes albopictus* (Diptera: Culicidae) in North America. *J. Med. Ent.* 53:1013–1023.

Arpagaus, M. 1987. Vertebrate insulin induces diapause termination in *Pieris brassicae* pupae. *Roux's Arch. Dev. Biol.* 196:527–530.

Arrese, E.L. and J.L. Soulages. 2010. Insect fat body: energy, metabolism, and regulation. *Ann. Rev. Ent.* 55:207–225.

Asano, W., F.N. Munyiri, Y. Shintani, and Y. Ishikawa. 2004. Interactive effects of photoperiod and temperature on diapause induction and termination in the yellow-spotted longicorn beetle, *Psacothea hilaris*. *Physiol. Ent.* 29:458–463.

van Asch, M. and M.E. Visser. 2007. Phenology of forest caterpillars and their host trees: the importance of synchrony. *Ann. Rev. Ent.* 52:37–55.

Audusseau, H., S. Nylin, and N. Janz. 2013. Implications of a temperature increase for host plant range: predictions for a butterfly. *Ecol. Evol.* 3:3021–3029.

Austin, A.B.M., G.M. Tatchell, R. Harrington, and J.S. Bale. 1996. Adaptive significance of changes in morph production during the transition from parthenogenetic to sexual reproduction in the aphid *Rhopalosiphum padi* (Homoptera: Aphididae). *Bull. Ent. Res.* 86:93–99.

Baik, L.S., K.J. Fogle, L. Roberts, A.M. Galschiodt, J.A. Chevez, Y. Recinos, V. Nguy, and T.C. Holmes. 2017. CRYPTOCHROME mediates behavioral executive choice in response to UV light. *Proc. Nat'l. Acad. Sci., USA* 114:776–781.

Bajgar, A., D. Doležel, and M. Hodková. 2013a. Endocrine regulation of non-circadian behavior of circadian genes in insect gut. *J. Insect Physiol.* 59:881–886.

Bajgar, A., M. Jindra, and D. Doležel. 2013b. Autonomous regulation of the insect gut by circadian genes acting downstream of juvenile hormone signaling. *Proc. Nat'l Acad. Sci., USA* 110:4415–4421.

Baker, D.A. and S. Russell. 2009. Gene expression during *Drosophila melanogaster* egg development before and after reproductive diapause. *BMC Genomics* 10:242.

Baker, K.D. and C.S. Thummel. 2007. Diabetic larvae and obese flies – emerging studies of metabolism in *Drosophila*. *Cell. Metab.* 6:257–266.

Bale, J.S. and S.A.L. Hayward. 2010. Insect overwintering in a changing climate. *J. Exp. Biol.* 213:980–994.

Bao, B. and W.-H. Xu. 2011. Identification of gene expression changes associated with the initiation of diapause in the brain of the cotton bollworm, *Helicoverpa armigera*. *BMC Genomics* 12:224.

Bao, B., B. Hong, Q.-L. Feng, and W.-H. Xu. 2011. Transcription factor fork head regulates the promoter of diapause hormone gene in the cotton bollworm, *Helicoverpa armigera*, and the modification of SUMOylation. *Insect Biochem. Mol. Biol.* 41:670–679.

Barat, M., P. Vernon, M. Tarayre, and A. Atlan. 2010. Overwintering strategy of two weevils infesting three gorse species: when cold hardiness meets plant–insect interactions. *J. Insect Physiol.* 56:170–177.

Barberà, M., B. Mengual, J.M. Collantes-Alegre, T. Cortés, A. González, and D. Martínez-Torres. 2013. Identification, characterization and analysis of expression of genes encoding arylalkylamine N-acetyltransferase in the pea aphid *Acyrthosiphon pisum*. *Insect Mol. Biol.* 22:623–634.

Barberà, M., J.M. Collantes-Alegre, and D. Martínez-Torres. 2017. Characterization, analysis of expression and localization of circadian clock genes from the perspective of photoperiodism in the aphid

Acyrthosiphon pisum. Insect Biochem. Mol. Biol. 83:54–67.

Barberà, M., L. Escrivá, J.M. Collantes-Alegre, G. Meca, E. Rosato, and D. Martínez-Torres. 2018. Melatonin in the seasonal response of the aphid *Acyrthosiphon pisum. Insect Sci.*27:224–238.

Barberà, M., R. Cañas-Cañas, and D. Martínez-Torres. 2019. Insulin-like peptides involved in photoperiodism in the aphid *Acyrthosiphon pisum. Insect Biochem. Mol. Biol.* 112:103185.

Barbosa, M. and G.W. Fernandes. 2019. Moisture stimulates galler emergence from dry season dormancy in the leaf litter. *Phytoparasitica* 47:629–635.

Batz, Z.A., A.C. Goff, and P.A. Armbruster. 2017. MicroRNAs are differentially abundant during *Aedes albopictus* diapause maintenance but not diapause induction. *Insect Mol. Biol.* 26:721–733.

Batz, Z.A. and P.A. Armbruster. 2018. Diapause-associated changes in the lipid and metabolite profiles of the Asian tiger mosquito, *Aedes albopictus. J. Exp. Biol.* 221:189480.

Batz, Z.A., C.S. Brent, M.R. Marias, J. Sugijanto, and P.A. Armbruster. 2019. Juvenile hormone III but not 20-hydroxyecdysone regulates the embryonic diapause of *Aedes albopictus. Front. Physiol.* 10:1352.

Batz, Z.A., A.J. Clemento,J. Fitzenwanker, T.J. Ring, J.C. Garza, and P.A. Armbruster. 2020. Rapid adaptive evolution of the diapause program during range expansion of an invasive mosquito. *Evolution* 74:1451–1465.

Baumgartner, M.F. and A.M. Tarrant. 2017. The physiology and ecology of diapause in marine copepods. *Ann. Rev. Mar. Sci.* 9:387–411.

Beach, R. 1978. The required day number and timely induction of diapause in geographic strains of the mosquito *Aedes atropalpus. J. Insect Physiol.* 24:449–455.

Bean, D.W., T.L. Dudley, and J.C. Keller. 2007a. Seasonal timing of diapause induction limits the effective range of *Diorhabda elongate deserticola* (Coleoptera: Chrysomelidae) as a biological control agent for tamarisk (*Tamarisk* spp.). *Environ. Ent.* 36:15–25.

Bean, D.W., T. Wang, R.J. Bartlet, and B.W. Zilkowski. 2007b. Diapause in the leaf beetle *Diorhabda elongata* (Coleoptera: Chrysomelidae) a biological control agent for tamarisk (*Tamarix* spp.). *Environ. Ent.* 36:531–540.

Beards, G.W. and F.E. Strong. 1966. Photoperiod in relation to diapause in *Lygus hesperus* Knight. *Hilgardia* 37:345–362.

Bebas, P., B. Cymborowski, M. Kazek, and M.A. Polanska. 2018. Effects of larval diapause and the juvenile hormone analog, fenoxycarb, on testis development and spermatogenesis in the wax moth, *Galleria mellonella* (Lepidoptera: Pyralidae). *Eur. J. Ent.* 115:400–417.

Beck, S.D. 1968. *Insect Photoperiodism.* New York: Academic Press.

Beck, S.D. 1980. *Insect Photoperiodism,* second ed. New York: Academic Press.

Beck, S.D. 1982. Thermoperiodic induction of larval diapause in the European corn borer, *Ostrinia nubilalis. J. Insect Physiol.* 28:273–277.

Beck, S.D. 1985. Effects of thermoperiod on photoperiodic determination of diapause in *Ostrinia nubilalis. J. Insect Physiol.* 31:41–46.

Beck, S.D. 1987. Thermoperiod–photoperiod interactions in the determination of diapause in *Ostrinia nubilalis. J. Insect Physiol.* 33:707–712.

Beck, S.D. 1991. Thermoperiodism. In *Insects at Low Temperature,* ed. R.E. Lee, Jr. and D.L. Denlinger, New York: Chapman & Hall, pp. 199–228.

Beck, S.D. and W. Hanec. 1960. Diapause in the European corn borer, *Pyrausta nubilalis* (Hbn.). *J. Insect Physiol.* 4:304–318.

Beck, S.D., J.L. Shane, and J.A. Garland. 1969. Ammonium-induced termination of diapause in the European corn borer, *Ostrinia nubilalis*. *J. Insect Physiol.* 15:945–951.

Beer, K. and C. Helfrich-Förster. 2020. Model and non-model insects in chronobiology. *Front. Behav. Neurosci.* 14:601676.

Belozerov, V.N. 2008. Diapause and quiescence as two main kinds of dormancy and their significance in life cycles of mites and ticks (Chelicerata: Arachnida: Acari). Part 1. Acariformes. *Acarina* 16:79–130.

Bel-Venner, M.-C., N. Mondy, F. Arthaud, J. Marandet, D. Giron, S. Venner, and F. Menu. 2009. Ecophysiological attributes of adult overwintering in insects: insights from a field study of the nut weevil, *Curculio nucum*. *Physiol. Ent.* 34:61–70.

Benetta, E.D., L. W. Beukeboom, and L. van de Zande. 2019. Adaptive differences in circadian clock gene expression patterns and photoperiodic diapause induction in *Nasonia vitripennis*. *Am. Nat.* 193:881–896.

Bennett, M.M., J.P. Rinehart, G.D. Yocum, C. Doetkott, and K.J. Greenlee. 2018. Cues for cavity nesters: investigating relevant Zeitgebers for emerging leafcutting bees, *Megachile rotundata*. *J. Exp. Biol.* 221:jeb175406.

Bennett, V.A., O. Kukal, and R.E. Lee, Jr. 1999. Metabolic opportunists: feeding and temperature influence the rate and pattern of respiration in the high arctic woollybear caterpillar *Gynaephora groenlandica* (Lymantriidae). *J. Exp. Biol.* 202:47–53.

Benoit, J.B. and D.L. Denlinger. 2007. Suppression of water loss during adult diapause in the northern house mosquito, *Culex pipiens*. *J. Exp. Biol.* 210:217–226.

Benoit, J.B., G. Lopez-Martinez, Z.P. Phillips, K.R. Patrick, and D.L. Denlinger. 2010a. Heat shock proteins contribute to mosquito dehydration tolerance. *J. Insect Physiol.* 56:151–156.

Benoit, J.B., P.K. Morton, S.E. Cambron, K.R. Patrick, and B.J. Schemerhorn. 2010b. Aestivation and diapause syndromes reduce the water balance requirements for pupae of the Hessian fly, *Mayetiola destructor*. *Ent. Exp. Appl.* 136:89–96.

Benoit, J.B., Q. Zhang, E.C. Jennings, A.J. Rosendale, and D.L. Denlinger. 2015. Suppression of net transpiration by multiple mechanisms conserves water resources during pupal diapause in the corn earworm, *Helicoverpa zea*. *Physiol. Ent.* 40:336–342.

Benton, F.P. 2015. Insect diapause of very variable duration: the case of the palm nut bruchid *Pachymerus nucleorum* (Coleoptera: Chrysomelidae). *Int'l. J. Ent.* 51:10–13.

Bentz, B.J. and E.M. Hansen. 2017. Evidence for a prepupal diapause in the mountain pine beetle (*Dendroctonus ponderosae*). *Environ. Ent.* 47:175–183.

Bergland, A.O., M. Agotsch, D. Mathias, W.E. Bradshaw, and C.M. Holzapfel. 2005. Factors influencing the seasonal life history of the pitcher-plant mosquito, *Wyeomyia smithii*. *Ecol. Ent.* 30:129–137.

Bergonzi, S., M.C. Albani, E.V.L. van Themaat, K.J.V. Nordström, R. Wang, K. Schneeberger, P.D. Moerland, and G. Coupland. 2013. Mechanisms of age-dependent response to winter temperature in perennial flowering of *Arabis alpina*. *Science* 340:1094–1097.

Berlinger, M.J. and G.W. Ankersmit. 1976. Manipulation with photoperiod as a method of control of *Adoxophyes orana* (Lepidoptera, Tortricidae). *Ent. Exp. Appl.* 19:96–107.

Bertolini, E., F.K. Schubert, D. Zanini, H. Sehadova, C. Helfrich-Forster, and P. Menegazzi. 2019. Life at high latitudes does not require circadian behavioral rhythmicity under constant darkness. *Curr. Biol.* 29:1–9.

Bertossa, R.C., L.W. van de Zande, and D.G.M. Beersma. 2014. Phylogeny and oscillating expression of period and cryptochrome in short and long photoperiods suggest a conserved function in *Nasonia vitripennis*. *Chronobiol. Inter.* 31:749–760.

Biggar, K.K. and K.B. Storey. 2018. Functional impact of microRNA regulation in models of extreme stress adaptation. *J. Mol. Cell Biol.* 10:93–101.

Bird, L.J. and R.J. Akhurst. 2005. Fitness of Cry1A-resistant and -susceptible *Helicoverpa armigera* (Lepidoptera: Nocutidae) on transgenic cotton with reduced levels of Cry1Ac. *J. Econ. Ent.* 98:1311–1319.

Biron, D.G., E. Brunel, K. Tiilikkala, S. Hellqvist, and P.L. Dixon. 2003. Expression of a bimodal emergence pattern in diapausing and nondiapausing *Delia floralis*: a phenological survival strategy. *Ent. Exp. Appl.* 107:163–166.

Blake, G.M. 1958. Diapause and regulation of development in *Anthrenus verbasci* (L.) (Col., Dermestidae). *Bull. Ent. Res.* 49:751–775.

Blanckenhorn, W.U. 1998a. Altitudinal differentiation in the diapause response of two species of dung flies. *Ecol. Ent.* 23:1–8.

Blanckenhorn, W.U. 1998b. Adaptive phenotypic plasticity in growth, development, and body size in the yellow dung fly. *Evolution* 52:1394–1407.

Blanckenhorn, W.U., C. Henseler, D.U. Burkhard, and H. Briegel. 2001. Summer decline in populations of the yellow dung fly: diapause or quiescence? *Physiol. Ent.* 26:260–265.

Blau, W.S. 1981. Life history variation in the black swallowtail butterfly. *Oecologia* 48:116–122.

Blitvich, B.J., A. Rayms-Keller, C.D. Blair, and B.J. Beaty. 2001. Identification and sequence determination of mRNAs detected in dormant (diapausing) *Aedes triseriatus* mosquito embryos. *DNA Sequence* 12:197–202.

de Block, M., M.A. McPeek, and R. Stoks. 2007. Winter compensatory growth under field conditions partly offsets low energy reserves before winter in a damselfly. *Oikos* 116:1975–1982.

Bloem, S., J.E. Carpenter, and S. Dorn. 2006. Mobility of mass-reared diapaused and nondiapaused *Cydia pomonella* (Lepidoptera: Tortricidae): effects of mating status and treatment with gamma radiation. *J. Econ. Ent.* 99:699–706.

Blom, P.E., S.J. Fleischer, and C.L. Harding. 2004. Modeling colonization of overwintering immigrant *Leptinotarsa decemlineata* (Coleoptera: Chrysomelidae). *Environ. Ent.* 33:267–274.

Bodnaryk, R.P. 1977. Stages of diapause development in the pupa of *Mamestra configurata* based on β-ecdysone sensitivity index. *J. Insect Physiol.* 23:537–542.

Bodnaryk, R.P. 1985. Ecdysteroid levels during postdiapause development and 20-hydroxyecdysone induced development in male pupae of *Mamestra configurata* Wlk. *J. Insect Physiol.* 31:53–58.

Boivin, T., J.-C. Bouvier, D. Beslay, and B. Sauphanor. 2004. Variability in diapause propensity within populations of a temperate insect species: interactions between insecticide resistance genes and photoperiodism. *Biol. J.* 83:341–351.

Bonnett, T.R., J.A. Robert, C. Pitt, J.D. Fraser, C.I. Keeling, J. Bohlmann, and D.P.W. Huber. 2012. Global and comparative proteomic profiling of overwintering and developing mountain pine beetle, *Dendroctonus ponderosae* (Coleoptera: Curculionidae), larvae. *Insect Biochem. Mol. Biol.* 42:890–901.

Borah, B.K., Z. Renthlei, and A.K. Trivedi. 2020. Hypothalamus but not liver retains daily expression of clock genes during hibernation in terai tree frog (*Polypedates teraiensis*). *Chronobiol. Int'l.* 2020:1726373.

Bosch, J. and W.P. Kemp. 2003. Effect of wintering duration and temperature on survival and emergence time in males of the orchard pollinator *Osmia lignaria* (Hymenoptera: Megachilidae). *Environ. Ent.* 32:711–716.

Bosch, J., F. Sgolastra, and W.P. Kemp. 2010. Timing of eclosion affects diapause development, fat body consumption and longevity in *Osmia lignaria*, a univoltine, adult-wintering solitary bee. *J. Insect Physiol.* 56:1949–1957.

Bosse, T.C. and A. Veerman. 1996. Involvement of vitamin A in the photoperiodic induction of diapause in the spider mite *Tetranychus urticae* is demonstrated by rearing an albino mutant on a semi-synthetic diet and without β-carotene or vitamin A. *Physiol. Ent.* 21:188–192.

Boulay, J., C. Aubernon, G.D. Ruxton, V. Hedouin, J.-L. Deneubourg, and D. Charabidze. 2019. Mixed-species aggregations in arthropods. *Insect Sci.* 26:2–19.

Bova, J., J. Soghigian, and S. Paulson. 2019. The prediapause stage of *Aedes japonicas japonicas* and the evolution of embryonic diapause in Aedini. *Insects* 10:222.

Bowen, M.F. 1992. Patterns of sugar feeding in diapausing and nondiapausing *Culex pipiens* (Diptera: Culicidae) females. *J. Med. Ent.* 29:843–849.

Bowen, M.F., W.E. Bollenbacher, and L.I. Gilbert. 1984. *In vitro* studies on the role of the brain and prothoracic glands in the pupal diapause of *Manduca sexta. J. Exp. Biol.* 108:9–24.

Bowers, W.S. 1976. Discovery of insect anti-allatotropins. In *The Juvenile Hormones*, ed. L.I. Gilbert, New York: Plenum, pp. 384–408.

Boychuk, E.C., J.T. Smiley, E.P. Dahlhoff, M.A. Bernards, N.E. Rank, and B.J. Sinclair. 2015. Cold tolerance of the montane Sierra leaf beetle, *Chrysomela aeneicollis. J. Insect Physiol.* 81:157–166.

Boyle, J.H., P.M.A. Rastas, X. Huang, A.G. Garner, I. Vythilingam, and P.A. Armbruster. 2021. A linkage-based genome assembly for the mosquito *Aedes albopictus* and identification of chromosomal regions affecting diapause. *Insects* 12:167.

Bradfield, J.Y. and D.L. Denlinger. 1980. Diapause development in the tobacco hornworm: a role for ecdysone or juvenile hormone? *Gen. Comp. Endocrin.* 41:101–107.

Bradshaw, W.E. 1969. Major environmental factors inducing the termination of larval diapause in *Chaoborus americanus* Johannsen (Diptera: Culicidae). *Biol. Bull.* 136:2–8.

Bradshaw, W.E. 1973. Homeostasis and polymorphism in vernal development of *Chaoborus americanus. Ecology* 54:1247–1259.

Bradshaw, W.E. 1974. Photoperiodic control of development in *Chaborus americanus* with special reference to photoperiodic action spectra. *Biol. Bull.* 146:11–19.

Bradshaw, W.E. 1976. Geography of photoperiodic response in a diapausing mosquito. *Nature* 262:384–386.

Bradshaw, W.E. 1980. Thermoperiodism and the thermal environment of the pitcher-plant mosquito, *Wyeomyia smithii. Oecologia* 46:13–17.

Bradshaw, W.E. and L.P. Lounibos. 1972. Photoperiodic control of development in the pitcher-plant mosquito, *Wyeomyia smithii. Can. J. Zool.* 50:713–719.

Bradshaw, W.E. and L.P. Lounibos. 1977. Evolution of dormancy and its photoperiodic control in pitcher-plant mosquitoes. *Evolution* 31:546–567.

Bradshaw, W.E., P.A. Armbruster, and C.M. Holzapfel. 1998. Fitness consequences of hibernal diapause in the pitcher-plant mosquito, *Wyeomyia smithii. Ecology* 79:1458–1462.

Bradshaw, W.E. and C.M. Holzapfel. 2001a. Genetic shift in photoperiodic response correlated with global warming. *Proc. Nat'l. Acad. Sci., USA* 98:14509–14511.

Bradshaw, W.E. and C.M. Holzapfel. 2001b. Phenotypic evolution and the genetic architecture underlying photoperiodic time measurement. *J. Insect Physiol.* 47:809–820.

Bradshaw, W.E., M.C. Quebodeaux, and C.M. Holzapfel. 2003a. Circadian rhythmicity and photoperiodism in the pitcher-plant mosquito: adaptive response to photic environment or correlated response to the seasonal environment? *Am. Nat.* 161:735–748.

Bradshaw, W.E., M.C. Quebodeaux, and C.M. Holzapfel. 2003b. The contribution of an hourglass timer to the evolution of photoperiodic response in the pitcher-plant mosquito, *Wyeomyia smithii. Evolution* 57:2342–2349.

Bradshaw, W.E., P.A. Zani, and C.M. Holzapfel. 2004. Adaptation to temperate climates. *Evolution* 58:1748–1762.

Bradshaw, W.E. and C.M. Holzapfel. 2007. Evolution of animal photoperiodism. *Ann. Rev. Ecol. Evol. Syst.* 38:1–25.

Bradshaw, W.E. and C.M. Holzapfel. 2008. Genetic response to rapid climate change: it's seasonal timing that matters. *Mol. Ecol.* 17:157–166.

Bradshaw, W.E. and C.M. Holzapfel. 2010a. Light, time, and the physiology of biotic response to rapid climate change in animals. *Ann. Rev. Physiol.* 72:147–166.

Bradshaw, W.E. and C.M. Holzapfel. 2010b. Circadian clock genes, ovarian development and diapause. *BMC Biol.* 8:115.

Bradshaw, W.E. and C.M. Holzapfel. 2010c. What season is it anyway? Circadian tracking vs. photoperiodic anticipation in insects. *J. Biol. Rhyth.* 25:155–165.

Bradshaw, W.E. and C.M. Holzapfel. 2010d. Insects at not so low temperature: climate change in the temperate zone and its biotic consequences. In *Low Temperature Biology of Insects*, ed. D.L. Denlinger and R.E. Lee, Jr., Cambridge: Cambridge University Press, pp. 242–275.

Bradshaw, W.E., K.J. Emerson, J.M. Catchen, W.A. Cresko, and C.M. Holzapfel. 2012. Footprints in time: comparative quantitative trait loci mapping of the pitcher-plant mosquito, *Wyeomyia smithii. Proc. R. Soc. Lond., Ser. B* 279:4551–4558.

Bradshaw, W.E. and C.M. Holzapfel. 2017. Natural variation and genetics of

photoperiodism in *Wyeomyia smithii. Adv. Gen.* 99:39–71.

Brady, D., A. Grapputo, O. Romoli, and F. Sandrelli. 2019. Insect cecropins, antimicrobial peptides with potential therapeutic applications. *Int. J. Mol. Sci.* 20:5862.

Brent, C.S. 2012. Classification of diapause status by color phenotype in *Lygus hesperus. J. Insect Sci.* 12:136.

Bresnahan, S.T., M.A. Döke, T. Giray, and C.M. Grozinger. 2021. Tissue-specific transcriptional patterns underlie seasonal phenotypes in honey bees (*Apis mellifera*). *Mol. Ecol.* (in press).

Briers, T. and A. de Loof. 1981. Moulting hormone activity in the adult Colorado potato beetle, *Leptinotarsa decemlineata* Say in relation to reproduction and diapause. *Int'l. J. Invert. Reprod.* 3:145–155.

Briers, T., M. Peferoen, and A. de Loof. 1982. Ecdysteroids and adult diapause in the Colorado potato beetle, *Leptinotarsa decemlineata. Physiol. Ent.* 7:379–386.

Broufas, G. 2002. Diapause induction and termination in the predatory mite *Euseius finlandicus* in peach orchards in northern Greece. *Exp. App. Acarol.* 25:921–932.

Broufas, G.D. and D.S. Koveos. 2000. Threshold temperature for post-diapause development and degree-days to hatching of winter eggs of the European red mite (Acari: Tetranychidae) in northern Greece. *Environ. Ent.* 29:710–713.

Broufas, G.D., M.L. Pappas, and D.S. Koveos. 2006. Effect of cold exposure and photoperiod on diapause termination of the predatory mite *Euseius finlandicus* (Acari: Phytoseiidae). *Environ. Ent.* 35:1216–1221.

Brower, L.P., W.H. Calvert, L.E. Hedrick, and J. Christian. 1977. Biological observations on an overwintering colony of monarch butterflies (*Danaus plexippus,* Danaidae) in Mexico. *J. Lep. Soc.* 31:232–242.

Brower, L.P., D.R. Kust, E. Rendon Salinas, E.G. Serrano, K.R. Kust, J. Miller, C.

Fernandez del Rey, and K. Pape. 2004. Catastropic winter storm mortality of monarch butterflies in Mexico during January 2002. In *The Monarch Butterfly: Biology and Conservation*, ed. K.M. Oberhauser and M. Solensky, Ithaca: Cornell Univesity Press, pp. 151–166.

Brower, L.P., E.H. Williams, L.S. Fink, D.A. Slayback, M.I. Ramirez, M. Van Limon Garcia, R.R. Zubieta, S.B. Weiss, W.H. Calvert, and W. Zuchowski. 2011. Overwintering clusters of the monarch butterfly coincide with the least hazardous vertical temperatures in the oyamel forest. *J. Lep. Soc.* 65:27–46.

Brown, J.C.L., D.J. Chung, K.R. Belgrave, and J.F. Staples. 2011. Mitochondrial metabolic suppression and reactive oxygen species production in liver and skeletal muscle of hibernating thirteen-lined ground squirrels. *Am. J. Physiol. Regul. Integr. Comp. Physiol.* 302:R15–R28.

Brown, J.J. 1980. Haemolymph protein reserves of diapausing and nondiapausing codling moth, *Cydia pomonella. J. Insect Physiol.* 26:487–492.

Brown, J.J. and G.M. Chippendale. 1978. Juvenile hormone and a protein associated with the larval diapause of the southwestern corn borer, *Diatraea grandiosella. Insect Biochem.* 8:359–367.

Brown, M.W. and M.E. Norris. 2004. Survivorship advantage of conspecific necrophay in overwintering boxelder bugs (Heteroptera: Rhopalidae). *Ann. Ent. Soc. Am.* 97:500–503.

Brown, V.K. 1983. Developmental strategies in British Dictyoptera: seasonal variation. In *Diapause and Life-Cycle Strategies in Insects*, ed. V.K. Brown and I. Hodek, the Hague: Dr. W. Junk, pp. 111–125.

Browning, T.O. 1981. Ecdysteroids and diapause in pupae of Heliothis punctiger. *J. Insect Physiol.* 27:715–719.

Bryon, A., N. Wybouw, W. Dermauw, L. Tirry, and T. van Leeuwen. 2013. Genome wide gene-expression analysis of facultative reproductive diapause in the two-spotted spider mite *Tetranychus urticae. BMC Genomics* 14:815.

Bryon, A., A.H. Kurlovs, T. van Leeuwen and R.M. Clark. 2017a. A molecular-genetic understanding of diapause in spider mites: current knowledge and future directions. *Physiol. Ent.* 42:211–224.

Bryon, A., A.H. Kurlovs, W. Dermauw, R. Greenhalgh, M. Riga, M. Grbic, L. Tirry, M. Osakabe, J. Vontas, R.M. Clark, and T. van Leeuwen. 2017b. Disruption of a horizontally transferred phytoene desaturase abolishes carotenoid accumulation and diapause in *Tetranychus urticae. Proc. Nat'l. Acad. Sci., USA* 114:E5871–E5880.

Buckner, J.S., M.C. Mardaus, and D.R. Nelson. 1996. Cuticular lipid composition of *Heliothis virescens* and *Helicoverpa zea* pupae. *Comp. Biochem. Physiol.* 114:207–216.

Bugbee, L.M. and L.R. Forte. 2004. The discovery of West Nile virus in overwintering *Culex pipiens* (Diptera: Culicidae) mosquitoes in Lehigh County, Pennsylvania. *J. Am. Mosq. Cont. Assoc.* 20:326–327.

Bünning, E. 1936. Die endogene Tagesrhythmik als Grundlage der Photoperiodioschen Reaktion. *Ber. Deut. Bot. Ges.* 54:590–607.

Bünsow, R.C. 1953. Uber Tages- und Jahresrhythmische Anderugen der Photoperiodischen Lichteropfindlischkeit bie *Kalanchoe blossfeldiana* und ihre Bezieghugen zuer endogogen Tagesrhythmik. *Z. Bot.* 41:257–276.

Burges, H.D. 1960. Studies on the dermestid beetle *Trogoderma granarium* Everts-IV. Feeding, growth, and respiration with particular reference to diapause larvae. *J. Insect Physiol.* 5:317–334.

Burke, S., A.S. Pullin, R.J. Wilson, and C.D. Thomas. 2005. Selection for discontinuous life-history tratits along a continuous thermal gradient in the butterfly *Aricia agestis. Ecol. Ent.* 30:613–619.

Burkett, B.N. and H.A. Schneiderman. 1974. Roles of oxygen and carbon dioxide in the control of spiracular function in cecropia pupae. *Biol. Bull.* 147:274–293.

Burmester, T. 1999. Evolution and function of the insect hexamerins. *Eur. J. Ent.* 96:213–225.

Bush, G.L. 1969. Sympatric host race formation and speciation in frugivorous flies of the genus *Rhagoletis* (Diptera: Tephritidae). *Evolution* 23:237–251.

Cabrera, B.J. and S.T. Kamble. 2001. Effects of decreasing thermophotoperiod on the eastern subterranean termite (Isoptera: Rhinotermitidae). *Environ. Ent.* 30:166–171.

Calle, Z., B.O. Schlumpberger, L. Piedrahita, A. Leftin, S.A. Hammer, A. Tye, and R. Borchert. 2010. Seasonal variationin daily insolation induces synchronous bud break and flowering in the tropics. *Trees: Struc. Funct.* 24:865–877.

Canard, M. 2005. Seasonal adaptations of green lacewings (Neuroptera: Chrysopidae). *Eur. J. Ent.* 102:317–324.

Canzano, A.A., R.E. Jones, and J.E. Seymour. 2003. Diapause termination in two species of tropical butterfly, *Euploea core* (Cramer) and *Euploea sylvester* (Fabricius) (Lepidoptera: Nymphalidae). *Aust. J. Ent.* 42:352–356.

Canzano, A.A., A.A. Krockenberger, R.E. Jones, and J.E. Seymour. 2006. Rates of metabolism in diapausing and reproductively active tropical butterflies, *Euploea core* and *Euploea sylvester* (Lepidoptera: Nymphalidae). *Physiol. Ent.* 31:184–189.

Cao, L.-J., W. Song, L. Yue, S.-K. Guo, J.-C. Chen, Y.-J. Gong, A.A. Hoffmann, and S.-J. Wei. 2020. Chromosome-level genome of the peach fruit moth *Carposina sasakii* (Lepidoptera: Carposinidae) provides a resource for evolutionary studies in moths. *Mol. Ecol. Resources* 21:834–848. doi: 10.1111/1755-0998.13288.

Caporale, A., H.P. Romanowski, and N.O. Mega. 2017. Winter is coming: diapause in the subtropical swallowtail butterfly *Euryades corethrus* (Lepidoptera, Papilionidae) is triggered by the shortening of day length and reinforced by low temperature. *J. Exp. Zool.* 327:182–188.

Carrière, Y., D.A. Roff, and J.-P. Deland. 1995. The joint evolution of diapause and insecticide resistance: a test of an optimality model. *Ecology* 76:1497–1505.

Carrière, Y., C. Ellers-Kirk, A.L. Patin, M.A. Sims, S. Meyer, Y.-B. Liu, T.J. Dennehy, and B.E. Tabashnik. 2001. Overwintering cost associated with resistance to transgenic cotton in the pink bollworm (Lepidoptera: Gelechiidae). *J. Econ. Ent.* 94:935–941.

Carton, Y. and J. Claret. 1982. Adaptive significance of a temperature induced diapause in a cosmopolitan parasitoid of Drosophila. *Ecol. Ent.* 7:239–247.

Černecká, L., M. Dorková, B. Jarčuška, and P. Kanuch. 2021. Elevational variation in voltinism demonstrates climatic adaptation in the dark bush-cricket. *Ecol. Ent.* 46: 360–367.

Ceryngier, P. 2000. Overwintering of *Coccinella septempuncata* (Coleoptera: Coccinellidae) at different altitudes in the Karkonosze Mts, SW Poland. *Eur. J. Ent.* 97:323–328.

Chambers, R.J., S. Long, and N.L. Heyer. 1993. Effectiveness of *Orius laevigatus* (Hem.: Anthocoridae) for the control of *Frankliniella occidentalis* on cucumber and pepper in the UK. *Biocontrol Sci. Tech.* 3:295–307.

Champlin, D.T. and J.W. Truman. 1998. Ecdysteroid control of cell proliferation during optic lobe neurogenesis in the moth *Manduca sexta*. *Development* 125:269–277.

Chang, J., J. Singh, S. Kim, W.C. Hockaday, C. Sim, and S.J. Kim. 2016. Solid-state NMR reveals differential carbohydrate utilization in diapausing Culex pipiens. *Sci. Rep.* 6:37350.

Chang, V. and M.E. Meuti. 2020. Circadian transcription factors differentially regulate features of the adult overwintering diapause in the Northern house mosquito, *Culex pipiens. Insect Biochem. Mol. Biol.* 121:103365.

Chang, Y.-F., M.J. Tauber, and C.A. Tauber. 1996. Reproduction and quality of F_1 offspring in *Chrysoperla carnea*: differential influence of quiescence artificially-induced diapause, and natural diapause. *J. Insect Physiol.* 42:521–528.

Chang, Y.-F., M.J. Tauber, C.A. Tauber, and J.P. Nyrop. 2000. Interpopulation variation in *Chrysoperla carnea* reproduction: implications for mass-rearing and storage. *Ent. Exp. Appl.* 95:293–302.

Chaplin, S.B. and P.H. Wells. 1982. Energy reserves and metabolic expenditures of monarch butterflies overwintering in Southern California. *Ecol. Ent.* 7:249–256.

Charlesworth, B., J.A. Coyne, and N.H. Barton. 1987. The relative rates of evolution of sex chromosomes and autosomes. *Am. Nat.* 130:113–146.

Cheesman, D.F., W.L. Lee, and P.F. Zagalsky. 1967. Carotenoproteins in invertebrates. *Biol. Rev.* 42:131–160.

Chen, B., T. Kayukawa, H. Jiang, A. Monteiro, S. Hoshizaki, and Y. Ishikawa. 2005a. *DaTrypsin,* a novel clip-domain serine proteinase gene up-regulated during winter and summer diapauses of the onion maggot, *Delia antiqua. Gene* 347:115–123.

Chen, B., T. Kayukawa, A. Monteiro, and Y. Ishikawa. 2005b. The expression of the *HSP90* gene in response to winter and summer diapauses and thermal-stress in the onion maggot, *Delia antiqua. Insect Mol. Biol.* 14:697–702.

Chen, B., T. Kayukawa, A. Monteiro, and Y. Ishikawa. 2006a. Cloning and characterization of the HSP70 gene, and its expression in response to diapauses and thermal stress in the onion maggot, *Delia antiqua. J. Biochem. Mol. Biol.* 39:749–758.

Chen, C., Q.-W. Xia, Y.-S. Chen, H.-J. Xiao, and F.-S. Xue. 2012. Inheritance of photoperiodic control of pupal diapause in the cotton bollworm, *Helicoverpa armigera* (Hübner). *J. Insect Physiol.* 58:1582–1588.

Chen, C., X.-T. Wei, H.J. Xiao, H.-M. He, Q.-W. Xia, and F.-S. Xue. 2014a. Diapause induction and termination in *Hyphantria cunea* (Drury) (Lepidoptera: Arctiinae). *PLoS One* 9:e98145.

Chen, C., R. Mahar, M.E. Merritt, D.L. Denlinger, and D.A. Hahn. 2021. ROS and hypoxia signaling regulate arousal during insect dormancy to coordinate glucose, amino acid, and lipid metabolism. *Proc. Nat'l. Acad. Sci., USA* 118: e2017603118.

Chen, C.-P. and D.L. Denlinger. 1990. Activation of phosphorylase in response to cold and heat stress in the flesh fly, *Sarcophaga crassipalpis. J. Insect Physiol.* 36:549–553.

Chen, J., D.-N. Cui, H. Ullah, S. Li, F. Pan, C.-M. Xu, X.-B. Tu, and Z.-H Zhang. 2020. The function of *LmPrx6* in diapause regulation in *Locusta migratoria* through the insulin signaling pathway. *Insects* 11:0763.

Chen, J.-H. and O. Yamashita. 1989. Activity changes in guanylate cyclase and cyclic GMP phosphodiesterase related to the accumulation of cyclic GMP in developing ovaries of the silkmoth, *Bombyx mori. Comp. Biochem. Physiol. B* 93:385–390.

Chen, L., W. Ma, X. Wang, C. Niu, and C. Lei. 2010. Analysis of pupal head proteome and its alteration in diapausing pupae of *Helicoverpa armigera. J. Insect Physiol.* 56:247–252.

Chen, L., R.E. Barnett, M. Horstmann, V. Bamberger, L. Heberle, N. Krebs, J.K. Colbourne, R.Gomez, and L.C. Weiss. 2018. Mitotic activity patterns and cytoskeletal changes throughout the progression of diapause developmental program in *Daphnia. BMC Cell Biol.* 19:30.

Chen, M., D.R. MacGregor, A. Dave, H. Florance, K. Moon, K. Paszkiewicz, N. Smirnoff, I.A. Graham, and S. Penfield. 2014b. Maternal temperature history activates Flowering Locus T in fruits to control progeny dormancy according to time of year. *Proc. Nat'l Acad. Sci., USA* 111:18787–18792.

Chen, W. and W.-H. Xu. 2014. Wnt/β-catenin signaling regulates *Helicoverpa armigera* pupal development by up-regulating c-Myc and AP-4. *Insect Biochem. Mol. Biol.* 53:44–53.

Chen, W., Z. Liu, T. Li, R. Zhang, Y. Xue, Y. Zhong, W. Bai, D. Zhou, and Z. Zhao. 2014c. Regulation of *Drosophila* circadian rhythms by miRNA let-7 is mediated by a regulatory cycle. *Nat. Comm.* 5:5549.

Chen, X. and M. Rosbash. 2016. MicroRNA-92a is a circadian modulator of neuronal excitability in *Drosophila*. *Nat. Comm.* 8:14707.

Chen, Y.H., S.B. Opp, S.H. Berlocher, and G.K. Roderick. 2006b. Are bottlenecks associated with colonization? Genetic diversity and diapause variation of native and introduced *Rhagoletis completa* populations. *Oecologia* 149:656–667.

Chen, Y.-R., T.Jiang, J. Zhu, Y.-C. Xie, Z.-C. Tan,Y.-H. Chen, S.-M. Tang, B.-F. Hao, S.-P. Wang, J.-S. Huang, and X.-J. Shen. 2017. Transcriptome sequencing reveals potential mechanisms of diapause preparation in biovoltine silkworm *Bombyx mori* (Lepidoptera: Bombycidae). *Comp. Biochem. Physiol. D* 24:68–78.

Chen, Y.-S., C. Chen, H.-M. He, Q.-W. Xia, and F.-S. Xue. 2013. Geographic variation in diapause induction and termination of the cotton bollworm, *Helicoverpa armigera* Hübner (Lepidoptera: Noctuidae). *J. Insect Physiol.* 59:855–862.

Cheng, W., D. Li, Y. Wang, Y. Liu, and K. Zhu-Salzman. 2016. Cloning of heat shock protein genes (*hsp70, hsc70* and *hsp90*) and their expression in response to larval diapause and thermal stress in the wheat blossom midge, *Sitodiplosis mosellana*. *J. Insect Physiol.* 95:66–77.

Cheng, W., Z. Long, Y. Zhang, T. Liang, and K. Zhu-Salzman. 2017. Effects of temperature, soil moisture and photoperiod on diapause termination and post-diapause

development of the wheat blossom midge, *Sitodiplosis mosellana* (Gehen)(Diptera: Cecidomyiidae). *J. Insect Physiol.* 103:78–85.

Cheng, W.-N., X.-L. Li, F. Yu, Y.-P. Li, J.-J. Li, and J.-X. Wu. 2009. Proteomic analysis of pre-diapause, diapause and post-diapause larvae of the wheat blossom midge, *Sitodiplosis mosellana* (Diptera: Cecidomyiidae). *Eur. J. Ent.* 106:29–35.

Cheng, X., A.A. Hoffmann, J.L. Maino, and P.A. Umina. 2018. A cryptic diapause strategy in *Halotydeus destructor* (Tucker) (Trombidiformes: Penthaleidae) induced by multiple cues. *Pest. Manag. Sci.* 74:2618–2625.

Cheng, X., A.A. Hoffmann, J.L. Maino, and P.A. Umina. 2019. Summer diapause intensity influenced by parental and offspring environmental conditions in the pest mite, *Halotydeus destructor*. *J. Insect Physiol.* 114:92–99.

Cherkasova, V., S. Ayyadevara, N. Egilmez, and R.S. Reis. 2000. Diverse *Caenorhabditis elegans* genes that are upregulated in dauer larvae also show elevated transcript levels in long-lived, aged, or starved adults. *J. Mol. Biol.* 300:433–448.

Chernysh, S., S.I. Kim, G. Bekker, V.A. Pleskach, N.A. Filatova, V.B. Anikin, V.G. Platonov, and P. Bulet. 2002. Antiviral and antitumor peptides from insects. *Proc. Nat'l. Acad. Sci., USA* 99:12628–12632.

Chernysh, S., K. Irina, and A. Irina. 2012. Anti-tumor activity of immunomodulatory peptide alloferon-1 in mouse tumor transplantation model. *Int. Immunopharmacol.* 12:312–314.

Chippendale, G.M. 1977. Hormonal regulation of larval diapause. *Ann. Rev. Ent.* 22:121–138.

Chippendale, G.M. and A.S. Reddy. 1972. Diapause of the southwestern corn borer, *Diatraea grandiosella*: transition from spotted to immaculate mature larvae. *Ann. Ent. Soc. Am.* 65:882–887.

Chippendale, G.M., A.S. Reddy, and C.L. Catt. 1976. Photoperiodic and thermoperiodic interaction in the regulation of the larval diapause of *Diatraea grandiosella*. *J. Insect Physiol.* 22:823–828.

Chirumamilla, A., G.D. Yocum, M.A. Boetel, and R.J. Dregseth. 2008. Multi-year survival of sugarbeet root maggot (*Tetanops myopaeformis*) larvae in cold storage. *J. Insect Physiol.* 54:691–699.

Cho, J.R., M. Lee, H.S. Kim, and K.S. Boo. 2007. Effect of the juvenile hormone analog, fenoxycarb on termination of reproductive diapause in *Scotinophara lurida* (Burmeister) (Heteroptera: Pentatomidae). *J. Asia Pac. Entomol.* 10:145–150.

Christie, A.E., V. Roncalli, and P.H. Lenz. 2016. Diversity of insulin-like peptide signaling system proteins in *Calanus finmarchicus* (Crustacea; Copepoda) – possible contributors to seasonal pre-adult diapause. *Gen. Comp. Endocrinol.* 236:157–173.

Cira, T.M., R.L. Koch, E.C. Burkness, W.D. Hutchison, and R.C. Venette. 2018. Effects of diapause on *Halyomorpha halys* (Hemiptera: Pentatomidae) cold tolerance. *Environ. Ent.* 47:997–1004.

Claret, J. 1966. Mise en évidence du role photorécepteur du cerveau dans l'induction de la diapause, chez *Pieris brassicae* (Lepido.). *Ann. Endocr.* 27:311–320.

Claret, J. and N. Volkoff. 1992. Vitamin A is essential for two processes involved in the photoperiodic reaction in *Pieris brassicae*. *J. Insect Physiol.* 38:569–574.

Clark, M.S., M.A.S. Thorne, J. Purac, G. Grubor-Lasjic, M. Kube, R. Reinhardt, and M.R. Worland. 2007. Surviving extreme polar winters by desiccation: clues from Arctic springtail (*Onychiurus arcticus*) EST libraries. *BMC Genomics* 8:475.

Clegg, J.S., L.E. Drinkwater, and P. Sorgeloos. 1996. The metabolic status of diapause embryos of *Artemia franciscana* (SFB). *Physiol. Zool.* 69:49–66.

Clemmensen, S.F. and D.A. Hahn. 2015. Dormancy cues alter insect temperature-size relationships. *Oecologia* 177:113–121.

Cobben, R.H. 1968. Evolutionary trends in Heteroptera. Part I. Eggs, architecture of

the shell, gross embryology and eclosion. *Meded. Landb. Hogesh. Wageningen* 151:1–475.

Cogni, R., C. Kuczynski, S. Koury, E. Lavington, E.L. Behrman, K.R. O'Brien, P.S. Schmidt, and W.F. Eanes. 2013. The intensity of selection acting on the *couch potato* gene – spatial-temporal variation in a diapause cline. *Evolution* 68:538–548.

Colgan, T.J., S. Finlay, M.J.F. Brown, and J.C. Carolan. 2019. Mating precedes selective immune priming which is maintained throughout bumblebee queen diapause. *BMC Genomics* 20:959.

Colgan, T.J., J.C. Carolan, S. Sumner, M.L. Blaxter, and M.J.F. Brown. 2020. Infection by the castrating parasitic nematode *Sphaerularia bombi* changes gene expression in *Bombus terrestris* bumblebee queens. *Insect Mol. Biol.* 29:170–182.

Colinet, H., F. Muratori, and T. Hance. 2010. Cold-induced expression of diapause in *Praon volucre*: fitness cost and morpho-physiological characterization. *Physiol. Ent.* 35:301–307.

Colinet, H., D. Renault, B. Charoy-Guével, and E. Com. 2012. Metabolic and proteomic profiling of diapause in the aphid parasitoid *Praon volucre*. *PLoS One* 7:e32606.

Colinet, H., C. Pineau, and E. Com. 2017. Large scale phosphoprotein profiling to explore *Drosophila* cold acclimation regulatory mechanisms. *Sci. Rep.* 7:1713.

Colinet, H., J.P. Rinehart, G.D. Yocum, and K.J. Greenlee. 2018. Mechanisms underpinning the beneficial effects of fluctuating thermal regimes in insect cold tolerance. *J. Exp. Biol.* 221:164806.

Collantes-Alegre, J.M., F. Mattenberger, M. Barbera, and D. Martinez-Torres. 2018. Characterisation, analysis of expression and localization of the opsin gene repertoire from the perspective of photoperiodism in the aphid *Acyrthosiphon pisum*. *J. Insect Physiol.* 104:48–59.

Colombani, J., S. Raisin, S. Pantalacci, T. Radimerski, L. Montagne, and P. Leopold.

2003. A nutrient sensor mechanism controls *Drosophila* growth. *Cell* 114:739–749.

Convey, P. 2010. Life-history adaptations to polar and alpine environments. In *Low Temperature Biology of Insects*, ed. D.L. Denlinger and R.E. Lee, Jr., Cambridge: Cambridge University Press, pp. 297–321.

Cook, S.C., M.D. Eubanks, R.E. Gold, and S.T. Behmer. 2016. Summer and fall ants have different physiological responses to food macronutrient content. *J. Insect Physiol.* 87:35–44.

Cook, S.J., K. Balmanno, A. Garner, T. Millar, C. Taverner, and D. Todd. 2000. Regulation of cell cycle re-entry by growth, survival and stress signaling pathways. *Biochem. Soc. Trans.* 28:233–240.

Copeland, R.S. and G.B. Craig, Jr. 1989. Winter cold influences the spatial and age distributions of the North American treehole mosquito *Anopheles barberi*. *Oecologia* 79:287–292.

Corn, J.G., J.M. Story, and L.J. White. 2009. Comparison of larval development and overwintering stages of the spotted knapweed biological control agent *Agapeta zoegana* (Lepidoptera: Tortricidae) and *Cyphocleonus achates* (Coleoptera: Curculionidae) in Montana versus eastern Europe. *Environ. Ent.* 38:971–976.

Cornette, R. and T. Kikawada. 2011. The induction of anhydrobiosis in the sleeping chironomid: current status of our knowledge. *Int'l. Union Biochem. Mol. Biol. Life* 63:419–429.

Costa, C.P., M.A. Duennes, K. Fisher, J.P. Der, K.M. Watrous, N. Okamoto, N. Yamanaka, and S.H. Woodward. 2020. Transcriptome analysis reveals nutrition- and age-related patterns of gene expression in the fat body of pre-overwintering bumble bee queens. *Mol. Ecol.* 29:720–737.

Courteau, L.A., K.B. Storey, and P.J. Morin. 2012. Differential expression of microRNA species in a freeze tolerant insect, *Eurosta solidaginis*. Cryobiology 65:210-214.

Cox, G.K. and T.E. Gillis. 2020. Surviving anoxia: the maintenance of energy production and tissue integrity during anoxia and reoxygenation. *J. Exp. Biol.* 223:207613.

Crossley, M.S. and D.B. Hogg. 2015. Potential overwintering locations of soybean aphid (Hemiptera: Aphididae) colonizing soybean in Ohio and Wisconsin. *Environ. Ent.* 44:210–222.

Crosthwaite, J.C., S. Sobek, D.B. Lyons, M.A. Bernards, and B.J. Sinclair. 2011. The overwintering physiology of the emerald ash borer, *Agrilus planipennis* Fairmaire (Coleoptera: Buprestidae). *J. Insect Physiol.* 57:166–173.

Crozier, A.J.G. 1979. Supradian and infradian cycles of oxygen uptake in diapausing pupae of *Pieris brassicae*. *J. Insect Physiol.* 25:575–582.

Cruz-Bustos, J., P. Montoya, G. Pérez-Lachaud, J. Valle-Mora, and P. Liedo. 2020. Biological attributes of diapausing and non-diapausing *Doryctobracon areolatus* (Hymenoptera, Braconidae), a parasitoid of *Anastrepha* spp. (Diptera, Tephritidae) fruit flies. *J. Hymen. Res.* 78:41–56.

Cui, D.-N., X.-B. Tu, K. Hao, A. Raza, J. Chen, M. McNeill, and Z.-H. Zhang. 2019. Identification of diapause-associated proteins in migratory locust, *Locusta migratoria* L. (Orthoptera: Acridoidea) by label-free quantification analysis. *J. Integrat. Agric.* 18:2579–2588.

Cui, X., S. Urita, S. Imanishi, T. Nagasawa, and K. Suzuki. 2009. Isolation and characterization of a 41 kDa sericin from the wild silkmoth *Antheraea yamamai*. *J. Insect Biotech. Ser.* 78:11–16.

Czypionka, T., P.D. Fields, J. Routtu, E. van den Berg, D. Ebert, and L. De Meester. 2018. The genetic architecture underlying diapause termination in a planktonic crustacean. *Mol. Ecol.* 28:998–1008.

Dai, L., S. Ye, H.-W. Li, D.-F. Chen, H.-L. Want, S.-N. Jia, C. Lin, J.-S. Yang, F. Yang, H. Ngasawa, and W.-J. Yang. 2017.

SETD4 regulates cell quiescence and catalyzes the trimethylation of H4K20 during diapause formation in *Artemia*. *Mol. Cell. Biol.* 37:e00453.

Dalin, P., D.W. Bean, T.L. Dudley, V.A. Carney, D. Eberts, K.T. Gardner, E. Hebertson, E.N. Jones, D.J. Kazmer, G.J. Michels, Jr., S.A. O'Meara, and D.C. Thompson. 2010. Seasonal adaptations to day length in ecotypes of *Diorhabda* spp. (Coleoptera: Chrysomelidae) inform selection of agents against saltcedars (*Tamarix* spp.). *Environ. Ent.* 39:1666–1675.

Dalin, P. and S. Nylin. 2012. Host-plant quality adaptively affects the diapause threshold: evidence from leaf beetles in willow plantations. *Ecol. Ent.* 37:490–499.

Damos, P.T. and M. Savopoulou-Soultani. 2010. Synchronized diapause termination of the peach twig borer *Anarsia lineatella* (Lepidoptera: Gelechiidae): Brownian motion with drift? *Physiol. Ent.* 35:64–75.

Dancau, T., P.G. Mason, and N. Cappuccino. 2018. Elusively overwintering: a review of diamondback moth (Lepidoptera: Plutellidae) cold tolerance and overwintering strategy. *Can. Ent.* 150:156–173.

Danforth, B.N. 1999. Emergence dynamics and bet hedging in a desert bee, *Perdita portalis*. *Proc. R. Soc. Lond. B* 266:1985–1994.

Danforth, B.N., R.L. Minckley, and J.L. Neff. 2019. *The Solitary Bees, Biology, Evolution and Conservation*. Princeton: Princeton University Press.

Danilevskii, A.S. 1965. *Photoperiodism and Seasonal Development of Insects*. London: Oliver & Boyd (English translation).

Danks, H.V. 1987. *Insect Dormancy: An Ecological Perspective*. Ottawa: Biological Survey of Canada.

Danks, H.V. 2000. Dehydration in dormant insects. *J. Insect Physiol.* 46:837–852.

Danks, H.V. 2002. The range of insect dormancy responses. *Eur. J. Ent.* 99:127–142.

Danks, H.V. 2004a. Seasonal adaptations in Arctic insects. *Integr. Comp. Biol.* 44:85–94.

Danks, H.V. 2004b. The roles of insect cocoons in cold conditions. *Eur. J. Ent.* 101:433–437.

Danks, H.V. 2005. How similar are daily and seasonal biological clocks? *J. Insect Physiol.* 51:609–619.

Danneels, E.L., E.M. Formesyn, D.A. Hahn, D.L. Denlinger, D. Cardoen, P. Verleyen, T. Wenseleers, L. Schoofs, and D.C. de Graaf. 2013. Early changes in the pupal transcriptome of the flesh fly *Sarcophaga crassipalpis* to parasitization by the ectoparasitic wasp, *Nasonia vitripennis*. *Insect Biochem. Mol. Biol.* 43:1189–1200.

Dao, A., A.S. Yaro, M. Diallo, S. Timbine, D.L. Huestis, Y. Kassogue, A.I. Traore, Z.L. Sanogo, D. Samake, and T. Lehmann. 2014. Signatures of aestivation and migration in Sahelian malaria mosquito populations. *Nature* 516:387–390.

Datta, M.S., A.A. Almada, M.F. Baumgartner, T.J. Mincer, A.M. Tarrant, and M.F. Polz. 2018. Inter-individual variability in copepod microbiomes reveals bacterial networks linked to host physiology. *ISME J.* 12:2103–2113.

Davies, A.B., P. Eggleton, B.J. van Rensburg and C.L. Parr. 2015. Seasonal activity patterns of African savanna termites vary across a rainfall gradient. *Insect Soc.* 62:157–165.

Davis, D.E. 1945. The annual cycle of plants, mosquitoes, birds, and mammals in two Brazilian forests. *Ecol. Monogr.* 15:243–295.

Dean, C.A.E., N.M. Teets, V. Koštál, P. Šimek, and D.L. Denlinger. 2016. Enhanced stress responses and metabolic adjustments linked to diapause and onset of migration in the large milkweed bug *Oncopeltus fasciatus*. *Physiol. Ent.* 41:152–161.

Deans, C. and K.A. Maggert. 2015. What do you mean, "Epigenetic"? *Genetics* 199:887–896.

Deichmann, U. 2016. Epigenetics: the origins and evolution of a fashionable topic. *Dev. Biol.* 416:249–254.

Delanoue, R., E. Meschi, N. Agrawal, A. Mauri, Y. Tsatskis, H. McNeill, and P. Léopold. 2016. *Drosophila* insulin release is triggered by adipose Stunted ligand to brain Methuselah receptor. *Science* 353:1553–1556.

Delany, C.E., A.T. Chen, J.V. Graniel, K.D. Dumas, and P.J. Hu. 2017. A histone H4 lysine 20 methyltransferase couples environmental cues to sensory neuron control of developmental plasticity. *Development* 144:1273–1282.

Delisle, J., L. Royer, M. Bemier-Cardou, E. Bauce, and A. Labrecque. 2009. The combined effect of photoperiod and temperature on egg dormancy in an island and a mainland population of the hemlock looper, *Lambdina fiscellaria*. *Ent. Exp. Appl.* 133:232–243.

Demont, M. and W.U. Blanckenhorn. 2008. Genetic differentiation in diapause response along a latitudinal cline in European yellow dung fly populations. *Ecol. Ent.* 33:197–201.

Denlinger, D.L. 1971. Embryonic determination of pupal diapause induction in the flesh fly *Sarcophaga crassipalpis* Macquart. *J. Insect Physiol.* 17:1815–1822.

Denlinger, D.L. 1972a. Seasonal phenology of diapause in the flesh fly *Sarcophaga bullata*. *Ann. Ent. Soc. Am.* 65:410–414.

Denlinger, D.L. 1972b. Induction and termination of pupal diapause in *Sarcophaga* (Diptera: Sarcophagidae). *Biol. Bull.* 142:11–24.

Denlinger, D.L. 1974. Diapause potential in tropical flesh flies. *Nature* 252:223–224.

Denlinger, D.L. 1976. Preventing insect diapause with hormones and cholera toxin. *Life Sci.* 19:1485–1490.

Denlinger, D.L. 1978. The developmental response of flesh flies (Diptera: Sarcophagidae) to tropical seasons: variation in generation time and diapause in East Africa. *Oecologia* 35:105–107.

Denlinger, D.L. 1979. Pupal diapause in tropical flesh flies: environmental and endocrine regulation, metabolic rate and genetic selection. *Biol. Bull.* 156:31–46.

Denlinger, D.L. 1980. Seasonal and annual variation of insect abundance in the Nairobi National Park, Kenya. *Biotropica* 12:100–106.

Denlinger, D.L. 1981. Basis for a skewed sex ratio in diapause-destined flesh flies. *Evolution* 35:1247–1248.

Denlinger, D.L. 1985. Hormonal control of diapause. In *Comprehensive Insect Physiology, Biochemistry and Pharmacology*, ed. G.A. Kerkut and L.I. Gilbert, Vol. 8, Oxford: Pergamon Press, pp. 353–412.

Denlinger, D.L. 1986. Dormancy in tropical insects. *Ann. Rev. Ent.* 31:239–264.

Denlinger, D.L. 1991. Relationship between cold-hardiness and diapause. In *Insects at Low Temperature*, ed. R.E. Lee, Jr. and D.L. Denlinger, New York: Chapman & Hall, pp. 174–198.

Denlinger, D.L. 1994. The beetle tree. *Am. Ent. Fall* 1994:168–171.

Denlinger, D.L. 1998. Maternal control of fly diapause. In *Maternal Effects as Adaptations*, ed. T.A. Mousseau and C.W. Fox, Oxford: Oxford University Press, pp. 275–287.

Denlinger, D.L. 2001. Interrupted development: the impact of temperature on insect diapause. In *Environment of Animal Development: Genes, Life Histories and Plasticity*, ed. D. Atkinson and M. Thorndyke, Oxford: BIOS Scientific Publishers, pp. 235–250.

Denlinger, D.L. 2002. Regulation of diapause. *Ann. Rev. Ent.* 47:93–122.

Denlinger, D.L. 2008. Why study diapause? *Ent. Res.* 38:1–9.

Denlinger, D.L. 2021. Exploiting tools for manipulating insect diapause. *Bull. Ent. Res.* (in press).

Denlinger, D.L., J.H. Willis, and G. Fraenkel. 1972. Rates and cycles of oxygen consumption during pupal diapause in *Sarcophaga* flesh flies. *J. Insect Physiol.* 18:871–882.

Denlinger, D.L. and P. Wingard. 1978. Cyclic GMP breaks pupal diapause in the flesh fly

Sarcophaga crassipalpis. J. Insect Physiol. 24:715–719.

Denlinger, D.L., J.J. Campbell, and J.Y. Bradfield. 1980. Stimulatory effect of organic solvents on initiating development in diapausing pupae of the flesh fly *Sarcophaga crassipalpis* and the tobacco hornworm Manduca sexta. *Physiol. Ent.* 5:7–15.

Denlinger, D.L. and J.Y. Bradfield. 1981. Duration of pupal diapause in the tobacco hornworm is determined by number of short days received by the larva. *J. Exp. Biol.* 91:331–337.

Denlinger, D.L., M. Shukla, and D.L. Faustini. 1984. Juvenile hormone involvement in pupal diapause of the flesh fly *Sarcophaga crassipalpis*: regulation of infradian cycles of O_2 consumption. *J. Exp. Biol.* 109:191–199.

Denlinger, D.L., C.-P. Chen, and S. Tanaka. 1988a. The impact of diapause on the evolution of other life history traits in flesh flies. *Oecologia* 77:350–356.

Denlinger, D.L., J. Giebultowicz, and T. Adedokun. 1988b. Insect diapause: dynamics of hormone sensitivity and vulnerability to environmental stress. In *Endocrinological Frontiers in Physiological Insect Ecology*, ed. F. Sehnal, A. Zabża, and D.L. Denlinger, Wroclaw: Wroclaw Technical University Press, pp. 309–324.

Denlinger, D.L. and S. Tanaka. 1989. Cycles of juvenile hormone esterase activity during the juvenile hormone-driven cycles of oxygen consumption in pupal diapause of flesh flies. *Experientia* 45:474–476.

Denlinger, D.L., S. Tanaka, W.L. Downes, H. Wolda, and M. Guardia. 1991. Does lack of diapause result in less insect seasonality? *Oecologia* 87:152–154.

Denlinger, D.L., J.P. Rinehart, and G.D. Yocum. 2001. Stress proteins: a role in insect diapause? In *Insect Timing: Circadian Rhythmicity to Seasonality*, ed. D.L. Denlinger, J.M. Giebultowicz, and

D.S. Saunders, Amsterdam: Elsevier, pp. 155–171.

Denlinger, D.L., G.D. Yocum, and J.P. Rinehart. 2005. Hormonal control of diapause. In *Comprehensive Molecular Insect Science*, ed. L.I. Gilbert, K. Iatrou, and S.S. Gill, Vol. 3, Oxford: Elsevier, pp. 615–650.

Denlinger, D.L. and R.E. Lee, Jr., eds. 2010. *Low Temperature Biology of Insects*, Cambridge: Cambridge University Press.

Denlinger, D.L., G.D. Yocum, and J.P. Rinehart. 2012. Hormonal control of diapause. In *Insect Endocrinology*, ed. L.I. Gilbert, San Diego: Academic Press, pp. 430–463.

Denlinger, D.L. and P.A. Armbruster. 2014. Mosquito diapause. *Ann. Rev. Ent.* 59:73–93.

Denlinger, D.L. and P.A. Armbruster. 2016. Molecular physiology of mosquito diapause. In *Advances in Insect Physiology*, ed. A.S. Raikhel, Vol. 51, San Diego: Elsevier, pp. 329–361.

Denlinger, D.L., D.A. Hahn, C. Merlin, C.M. Holzapfel, and W.E. Bradshaw. 2017. Keeping time without a spine: what can the insect clock teach us about seasonal adaptation? *Phil. Trans. R. Soc. B* 371:20160257.

De Souza, K., J. Holt, and J. Colvin. 1995. Diapause, migration and pyrethroid-resistance dynamics in the cotton bollworm, *Helicoverpa armigera* (Lepidoptera: Noctuidae). *Ecol. Ent.* 20:333–342.

Des Marteaux, L.E. and R.H. Hallett. 2019. Swede midge (Diptera: Cecidomyiidae) diapause initiation under stable conditions: not a family affair. *Can. Ent.* 151:465–474.

Deveson, E.D. and J.D. Woodman. 2014. Embryonic diapause in the Australian plague locust relative to parental experience of cumulative photophase decline. *J. Insect Physiol.* 70:1–7.

Dewhurst, C.F., W.W. Page, and D.J.W. Rose. 2001. The relationship between outbreaks, rainfall and low density populations of the African armyworm, *Spodoptera exempta*, in Kenya. *Ent. Exp. Appl.* 98:285–294.

Dhillon, M.K., F. Hasan, A.K. Tanwar, J. Jaba, N. Singh, and H.C. Sharma. 2020. Genetic regulation of diapause and associated traits in *Chilo partellus* (Swinhoe). *Sci. Rep.* 10:1793.

Diaz, M.D.C. and D.C. Peck. 2007. Overwintering of annual bluegrass weevils, *Listronotus maculicollis*, in the golf course landscape. *Ent. Exp. Appl.* 125:259–268.

Dickson, R.C. 1949. Factors governing the induction of diapause in the oriental fruit moth. *Ann. Ent. Soc. Am.* 42:511–537.

Dillon, M.E. and J.D. Lozier. 2019. Adaptation to the abiotic environment in insects: the influence of variability on ecophysiology and evolutionary genomics. *Curr. Opin. Insect Sci.* 36:131–139.

Dillwith, J.W., C. L. Goodman, and G.M. Chippendale. 1985. An immunological study of the diapause-associated protein of the southwestern corn borer, *Diatraea grandiosella*. *Insect Biochem.* 15:711–722.

Ding, L., Y. Li, and M. Goto. 2003. Physiological and biochemical changes in summer and winter diapause and non-diapause pupae of the cabbage armyworm, *Mamestra brassicae* L. during long-term cold acclimation. *J. Insect Physiol.* 49:1153–1159.

Dingemanse, N.J. and V.J. Kalkman. 2008. Changing temperature regimes have advanced the phenology of Odonata in the Netherlands. *Ecol. Ent.* 33:394–402.

Dingle, H. 1974. Diapause in a migrant species, the milkweed bug *Oncopeltus fasciatus* (Dallas)(Hemiptera: Lygaeidae). *Oecologia* 1:1–10.

Dingle, H., M.P. Zalucki, and W.A. Rochester. 1999. Season-specific directional movement in migratory Australian butterflies. *Aust. J. Ent.* 38:323–329.

Diniz, D.F.A., C.M.R. de Albuquerque, L.O. Oliva, M.A.V. de Melo-Santos, and C.F.J. Ayres. 2017. Diapause and quiescence:

dormancy mechanisms that contribute to the geographic expansion of mosquitoes and their evolutionary success. *Parasit. Vect.* 10:310.

Ditrich, T. and V. Koštál. 2011. Comparative analysis of overwintering physiology in nine species of semi-aquatic bugs (Heteroptera: Gerromorpha). *Physiol. Ent.* 36:261–270.

Ditrich, T., V. Janda, H. Vanečková, and D. Doležel. 2018. Climatic variation of supercooling point in the linden bug *Pyrrhocoris apterus* (Heteroptera: Pyrrhocoridae). *Insects* 9:144.

Dockx, C. 2007. Directional and stabilizing selection on wing size and shape in migrant and resident monarch butterflies, *Danaus plexippus* (L.), in Cuba. *Biol. J. Linn. Soc.* 92:605–616.

Doellman, M.M., G.J. Ragland, G.R. Hood, P.J. Meyers, S.P. Egan, T.H.Q. Powell, P. Lazorchak, M.M. Glover, C. Tait, H. Schuler, D.A. Hahn, S.H. Berlocher, J.J. Smith, P. Nosil, and J.L. Feder. 2018. Genomic differentiation during speciation-with-gene-flow: comparing geographic and host-related variation in divergent life history adaptation in *Rhagoletis pomonella*. *Genes* 9:262.

Dogan,C., S. Hänniger, D.G. Heckel, C. Coutu, D.D. Hegedus, L. Crubaugh, R.L. Groves, S. Bayram, and U. Toprak. 2020. Two calcium-binding chaperones from the fat body of the Colorado potato beetle, *Leptinotarsa decemlineata* (Coleoptera: Chrysomelidae) involved in diapause. *Arch. Insect Biochem. Physiol.* 106: e21755.

Doherty, J.-F., J.-F. Guay, and C. Cloutier. 2018. Embryonic stage obligatory diapause and effects of abiotic conditions on egg hatching in the balsam twig aphid, *Mindarus abietinus*. *Ent. Exp. Appl.* 166:628–637.

Doležel, D., H. Vanečková, I. Šauman, and M. Hodková. 2005. Is *period* gene causally involved in the photoperiodic regulation of reproductive diapause in the linden bug, *Pyrrhocoris apterus*? *J. Insect Physiol.* 51:655–659.

Doležal, P., O. Habuštová, and F. Sehnal. 2007. Effects of photoperiod and temperature on the rate of larval development, food conversion efficiency, and imaginal diapause in *Leptinotarsa decemlineata*. *J. Insect Physiol.* 53:849–857.

Doležel, D., L. Zdechovanova, I. Šauman, and M. Hodková. 2008. Endocrine-dependent expression of circadian clock genes in insects. *Cell. Mol. Life Sci.* 65:964–969.

Dolfi, L., R. Ripa, A. Antebi, D.R. Valenzano, and A. Cellerino. 2019. Cell cycle dynamics during diapause entry and exit in an annual killifish revealed by FUCCI technology. *EvoDevo* 10:29.

Dong, L., H. Udaka, H. Numata, and C. Ito. 2021. Regulation of *Krüppel homologue 1* expression by photoperiod in the bean bug, *Riptortus pedestris*. *Physiol. Ent.* 46:87–93.

Dong, Y.-C., Z.-Z. Chen, A.R. Clarke, and C.-Y. Niu. 2019. Changes in energy metabolism trigger pupal diapause transition of *Bactrocera minax* after 20-hydroxyecysone application. *Front. Physiol.* 10:1288.

Dopman, E.B., L. Pérez, S.M. Bogdanowicz, and R.G. Harrison. 2005. Consequences of reproductive barriers for genealogical discordance in the European corn borer. *Proc. Nat'l. Acad. Sci., USA* 102:14706–14711.

Dopman, E.B., P.S. Robbins, and A. Seaman. 2010. Components of reproductive isolation between North American pheromone strains of the European corn borer. *Evolution* 64:881–902.

Dos Santos, C.F., A.L. Acosta, P. Nunes-Silva, A.M. Saraiva, and B. Blochtein. 2015. Climate warming may threaten reproductive diapause of a highly eusocial bee. *Environ. Ent.* 44:1172–1181.

Dostálková, S., P. Dobeš, M. Kunc, J. Hurychová, M. Škrabišová, M. Petrivalsky, D. Titera, J. Havlík, P. Hyršl, and J. Danihlík. 2021. Winter honeybee (*Apis mellifera*) populations show greater potential to induce immune responses than summer populations after immune stimuli. *J. Exp. Biol.* 224:232595.

Douglas, A.E. 2000. Reproductive diapause and the bacterial symbiosis in the sycamore aphid *Drepanosiphum platanoidis*. *Ecol. Ent.* 25:256–261.

Dowle, E.J., T.H.Q. Powell, M.M. Doellman, P.J. Meyers, M.B. Calvert, K.K.O. Walden, H.M. Robertson, S.H. Berlocher, J.L. Feder, D.A. Hahn, and G.J. Ragland. 2020. Genome-wide variation and transcriptional changes in diverse developmental processes underlying the rapid evolution of seasonal adaptation. *Proc. Nat'l. Acad. Sci., USA* 117:23960–23969.

Dowling, D.K. and L.W. Simmons. 2009. Reactive oxygen species as universal constraints in life-history evolution. *Proc. Roy. Soc. B* 276:1737–1745.

Dreyer, D., B. el Jundi, D. Kishkinev, C. Suchentrunk, L. Campostrini, B.J. Frost, T. Zechmeister, and E.J. Warrant. 2018. Evidence for a southward autumn migration of nocturnal noctuid moths in Central Europe. *J. Exp. Biol.* 221:179218.

Duan, T.-F., L. Li, Y. Tan, Y.-Y. Li, and B.-P. Pang. 2021. Identification and functional analysis of microRNAs in the regulation of summer diapause in *Galeruca daurica*. *Comp. Biochem. Physiol. D* 37:100786.

Dubowy, C. and A. Sehgal. 2017. Circadian rhythms and sleep in *Drosophila melanogaster*. *Genetics* 205:1373–1397.

Dulcis, D., N.T. Davis, and J.G. Hildebrand. 2001. Neuronal control of heart reversal in the hawkmoth *Manduca sexta*. *J. Comp. Physiol. A* 187:837–849.

Du Merle, P. 1999. Egg development and diapause: ecophysiological and genetic basis of phenological polymorphism and adaptation to varied hosts in the green oak tortrix, *Tortrix viridana* L. (Lepidoptera: Tortricidae). *J. Insect Physiol.* 45:599–611.

Duman, J.G., K.R. Walters, T. Sforo, M.A. Carrasco, P.K. Nickell, X. Lin, and B.M. Barnes. 2010. Antifreeze and ice-nucleator proteins. In *Low Temperature Biology of Insects*, ed. D.L. Denlinger and R.E. Lee, Jr., Cambridge: Cambridge University Press, pp. 59–90.

Duplouy, A., G. Minard, M. Lahteenaro, S. Rytteri, and M. Saastamoinen. 2018. Silk properties and overwinter survival in gregarious butterfly larvae. *Ecol. Evol.* 8:12443–12455.

Dupuis, J.R., B.A. Mori, and F.A.H. Sperling. 2016. *Trogus* parasitoids of *Papilio* butterflies undergo extended diapause in western Canada (Hymenoptera, Ichneumonidae). *J. Hymenop. Res.* 50:179–190.

Durak, R., J. Dampc, J. Dampc, S. Bartoszewski, and A. Michalik. 2020. Uninterrupted development of two aphid species belonging to *Cinara* genus during winter diapause. *Insects* 11:150.

Durant, D.R., A.J. Berens, A.L. Toth, and S.M. Rehan. 2016. Transcriptional profiling of overwintering gene expression in the small carpenter bee, *Ceratina calcarata*. *Apidologie* 47:572–582.

van Dyck, H. and C. Wiklund. 2002. Seasonal butterfly design: morphological plasticity among three developmental pathways related to sex, flight and thermoregulation. *J. Evol. Biol.* 15:216–225.

Dzerefos, C.M., B.F.N. Erasmus, E.T.F. Witkowski, and D. Guo. 2015. Modelling the current and future dry-season distribution of the edible stinkbug *Encosternum delegorguei* in sub-Saharan Africa. *Ent. Exp. Appl.* 156:1–13.

Eddy, S.F., J.D. McNally, and K.B. Storey. 2005. Up-regulation of a thioredoxin peroxidase-like protein, proliferation-associated gene, in hibernating bats. *Arch. Biochem. Biophys.* 435:103–111.

Egan, S.P., G.J. Ragland, L. Assour, T.H.Q. Powell, G.R. Hood, S. Emrich, P. Nosil, and J.L. Feder. 2015. Experimental evidence of genome-wide impact of ecological selection during early stages of speciation-with-gene-flow. *Ecol. Letters* 18:817–825.

Egi, Y., H. Tsubouchi, and K. Sakamoto. 2016. Does DNA methylation play a role in photoperiodic diapause of moths? *J. Ent. Stud.* 4:458–460.

Eizaguirre, M., C. López, L. Asin, and R. Albajes. 1994. Thermoperiodism, photoperiodism and sensitive stage in the diapause induction of *Sesamia nonagrioides* (Lepidoptera: Noctuidae). *J. Insect Physiol.* 40:113–119.

Eizaguirre, M., C. Schafellner, C. López, and F. Sehnal. 2005. Relationship between an increase in juvenile hormone titer in early instars and the induction of diapause in fully grown larvae of *Sesamia nonagrioides*. *J. Insect Physiol.* 51:1127–1134.

Eizaguirre, M., C. López, and R. Albajes. 2008. Factors affecting the natural duration of diapause and post-diapause development in the Mediterranean corn borer *Sesamia nonagrioides* (Lepidoptera: Noctuidae). *J. Insect Physiol.* 54:1057–1063.

Ellers, J. and J.J.M. van Alphen. 2002. A trade-off between diapause duration and fitness in female parasitoids. *Ecol. Ent.* 27:279–284.

Emerson, K.J., A.D. Letaw, W.E. Bradshaw, and C.M. Holzapfel. 2008. Extrinsic light: dark cycles, rather than endogenous circadian cycles, affect the photoperiodic counter in the pitcher-plant mosquito, *Wyeomyia smithii*. *J. Comp. Physiol. A* 194:611–615.

Emerson, K.J., S.J. Dake, W.E. Bradshaw, and C.M. Holzapfel. 2009a. Evolution of photoperiodic time measurement is independent of the circadian clock in the pitcher-plant mosquito, *Wyeomyia smithii*. *J. Comp. Physiol. A* 195:385–391.

Emerson, K.J., W.E. Bradshaw, and C.M. Holzapfel. 2009b. Complications of complexity: integrating environmental, genetic and hormonal control of insect diapause. *Trends Gen.* 25:217–225.

Emerson, K.J., A.M. Uyemura, K.L. McDaniel, P.S. Schmidt, W.E. Bradshaw, and C.M. Holzapfel. 2009c. Environmental control of ovarian dormancy in natural populations of *Drosophila melanogaster*. *J. Comp. Physiol. A* 195:825–829.

Emerson, K.J., W.E. Bradshaw, and C.M. Holzapfel. 2010. Microarrays reveal early transcriptional events during the termination of larval diapause in natural populations of the mosquito, *Wyeomyia smithii*. *PLoS One* 5:e9574.

Endo, K., Y. Fujimoto, M. Kondo, A. Yamanaka, M. Watanabe, K. Weihua, and K. Kumagai. 1997. Stage-dependent changes of the prothoracicotropic hormone (PTTH) activity of brain extracts and of the PTTH sensitivity of the prothoracic glands in the cabbage armyworm, *Mamestra brassicae*, before and during winter and aestival pupal diapause. *Zool. Sci.* 14:127–133.

Epstein, N.R., K. Saez, A. Polat, S.R. Davis, and M. L. Aardema. 2021. The urban-adapted underground mosquito, *Culex molestus*, maintains exogenously influenced circadian rhythms despite an absence of photoperiodically induced dormancy. *J. Exp. Biol.* 224:jeb242231.

Erickson, P.A., C.A. Weller, D.Y. Song, A.S. Bangerter, P. Schmidt, and A.O. Bergland. 2020. Unique genetic signatures of local adaptation over space and time for diapause, an ecologically relevant complex trait, in *Drosophila melanogaster*. *PLoS Genet.* 16:e1009110.

Eriksson, M., N. Janz, S. Nylin, and M.A. Carlsson. 2020. Structural plasticity of olfactory neuropils in relation to insect diapause. *Ecol. Evol.* 10:13323–14434.

Erion, R. and A. Sehgal. 2013. Regulation of insect behavior via the insulin-signaling pathway. *Front. Physiol.* 4:353.

Esperk, T., C. Stefanescu, T. Teder, C. Wiklund, A. Kaasik, and T. Tammaru. 2013. Distinguishing between anticipatory and responsive plasticity in a seasonally polyphenic butterfly. *Evol. Ecol.* 27:315–332.

Evenden, M.L., G. Armitage, and R. Lau. 2007. Effects of nutrition and methoprene treatment upon reproductive diapause in *Caloptilia fraxinella* (Lepidoptera: Gracillariidae). *Physiol. Ent.* 32:275–282.

Everman, E.R., P.J. Freda, M. Brown, A.J. Schieferecke, G.J. Ragland, and T.J. Morgan. 2018. Ovary development and cold tolerance of the invasive pest *Drosophila suzukii* (Matsumura) in the central plains of Kansas, United States. *Environ. Ent.* 47:1013–1023.

Evison, S.E.F., J.D. Gallagher, J.J.W. Thompson, M.T. Siva-Jothy, and S.A.O. Armitage. 2017. Cuticular colour reflects underlying architecture and is affected by a limiting resource. *J. Insect Physiol.* 98:7–13.

Fan, W., Y. Zhong, B. Qin, F. Lin, H. Chen, W. Li, and J. Lin. 2017. Differentially expressed miRNAs in diapausing versus HCl-treated *Bombyx* embryos. *PLoS One* 12:e0180085.

Fantinou, A.A. and E.A. Kagkou. 2000. Effect of thermoperiod on diapause induction of *Sesamia nonagrioides* (Lepidoptera-Nocutuidae). *Environ. Ent.* 29:489–494.

Fantinou, A.A., C.S. Chatzoglou, and E.A. Kagkou. 2002. Thermoperiodic effects on diapause of *Sesamia nonagrioides* (Lepidoptera: Noctuidae). *Eur. J. Ent.* 99:421–425.

Fantinou, A.A., A.T. Kourti, and C.J. Saitanis. 2003. Photoperiodic and temperature effects on the intensity of larval diapause in Sesamia nonagrioides. *Physiol. Ent.* 28:82–87.

Fantinou, A.A., D.C. Perdikis, and K. Zota. 2004. Reproductive responses to photoperiod and temperature by diapausing and nondiapausing populations of *Sesamia nonagrioides* Lef. (Lepidoptera-Noctuidae). *Physiol. Ent.* 29:169–173.

Faraji, A. and R. Gaugler. 2015. Experimental host preference of diapause and non-diapause induced *Culex pipiens pipiens* (Diptera: Culicidae). *Parasit. Vect.* 8:389.

Farajollahi, A., W.J. Crans, P. Bryant, B. Wolf, K.L. Burkhalter, M.S. Godsey, S.E. Aspen, and R.S. Nasci. 2005. Detection of West Nile viral RNA from an overwintering pool of *Culex pipiens pipiens* (Diptera:

Culicidae) in New Jersey, 2003. *J. Med. Ent.* 42:490–494.

Farooqui, T. and A.A. Farooqui (eds.). 2012. *Oxidative Stress in Vertebrates and Invertebrates.* Hobokon: Wiley-Blackwell.

Fauser, A., C. Sandrock, P. Neumann, and B.M. Sadd. 2017. Neonicotinoids override a parasite exposure impact on hibernation success of a key bumblebee pollinator. *Ecol. Ent.* 42:306–314.

Feder, J.L., X.F. Xie, J. Rull, S. Velez, A. Forbes, B. Leung, H. Dambrowski, K.E. Filchak, and M. Aluia. 2005. Mayr, Dobzhansky, and Bush and the complexities of sympatric speciation in *Rhagoletis. Proc. Nat'l. Acad. Sci., USA* 102:6573–6580.

Feder, J.L., T.H.Q. Powell, K. Filchak, and B. Leung. 2010. The diapause response of *Rhagoletis pomonella* to varying environmental conditions and its significance for geographic and host plant-related adaptation. *Ent. Exp. Appl.* 136:31–44.

Feder, M.E. and G.E. Hofmann. 1999. Heat-shock proteins, molecular chaperones, and the stress response: evolutionary and ecological physiology. *Ann. Rev. Physiol.* 61:243–282.

Ferenz, H.J. 1975. Photoperiodic and hormonal control of reproduction in male beetles, *Pterostichus nigrita. J. Insect Physiol.* 21:331–341.

Ferguson, L.V., R. Kortet, and B.J. Sinclair. 2018a. Eco-immunology in the cold: the role of immunity in shaping the overwintering survival of ectotherms. *J. Exp. Biol.* 221:jeb163873.

Ferguson, L.V., P. Dhakal, J.E. Lebenzon, D.E. Heinrichs, C. Bucking, and B.J. Sinclair. 2018b. Seasonal shifts in the insect gut microbiome are concurrent with changes in cold tolerance and immunity. *Funct. Ecol.* 32:2357–2368.

Fielenbach, N. and A. Antebi. 2008. *C. elegans* dauer formation and the molecular basis of plasticity. *Genes Dev.* 22:2149–2165.

Fischer, K. and K. Fiedler. 2001. Sexual differences in life-history traits in the butterfly *Lycaena tityrus*: a comparison between direct and diapause development. *Ent. Exp. Appl.* 100:325–330.

Fischer, K. and K. Fiedler. 2002. Life-history plasticity in the butterfly *Lycaena hippothoe*: local adaptations and trade-offs. *Biol. J. Linn. Soc.* 75:173–185.

Fischer, S., M.S. De Majo, C.M. Di Battista, P. Montini, V. Loetti, and R.E. Campos. 2019. Adaptation to temperate climates: evidence of photoperiodic-induced embryonic dormancy in *Aedes aegypti* in South America. *J. Insect Physiol.* 117:103887.

Fischman, B.J., T.L. Pitts-Singer, and G.E. Robinson. 2017. Nutritional regulation of phenotypic plasticity in a solitary bee (Hymenoptera: Megachilidae). *Environ. Ent.* 46:1070–1079.

Flannagan, R.D., S.P. Tammariello, K.H. Joplin, R.A. Cikra-Ireland, G.D. Yocum, and D.L. Denlinger. 1998. Diapause specific gene expression in pupae of the flesh fly, *Sarcophaga crassipalpis. Proc. Nat'l. Acad. Sci., USA* 95:5616–5620.

Foerster, L.A. and A.K. Doetzer. 2006. Cold storage of the egg parasitoids *Trissolcus basalis* (Wollaston) and *Telenomus podisi* Ashmead (Hymenoptera: Scelionidae). *Biol. Cont.* 36:232–237.

Forbes, A.A., K.S. Pelz-Stelinski, and R. Isaacs. 2010. Transfer of life-history phenology from mothers to progeny in a solitary univoltine parasitoid. *Physiol. Ent.* 35:192–195.

Förster, T.D. and S.K. Hetz. 2010. Spiracle activity in moth pupae – the role of oxygen and carbon dioxide revisited. *J. Insect Physiol.* 56:492–501.

Fraenkel, G. and C. Hsiao. 1968. Morphological and endocrinological aspects of pupal diapause in a flesh fly, *Sarcophaga argyrostoma* (Robineau-Desvoidy). *J. Insect Physiol.* 14:707–718.

Frago, E., M. Guara, J. Pujade-Villar, and J. Selfa. 2010. Winter feeding leads to a shifted phenology in the browntail moth *Euproctis chrysorrhoea* on the evergreen strawberry tree *Arbutus unedo*. *Agric. Forest Ent.* 12:381–388.

Franc, A. and M.H. Luong-Skovmand. 2009. Life cycle, reproductive maturation, and wing color changes in *Nomadacris septemfasciata* (Orthoptera: Acrididae) in Madagascar. *Environ. Ent.* 38:569–576.

Frank, J.H. 1967. The effect of pupal predators on a population of winter moth, *Operophtera brumata* (L.) (Hydriomenidae). *J. Animal Ecol.* 36:611–621.

Fraser, J.D., T.R. Bonnett, C.I. Keeling, and D.P.W. Huber. 2017. Seasonal shifts in accumulation of glycerol biosynthetic gene transcripts in mountain pine beetle, *Dendroctonus ponderosae* Hopkins (Coleoptera: Curculionidae), larvae. *PeerJ* 5:e3284.

Freedman, M.G., H. Dingle, C.A. Tabuloc, J.C. Chiu, L.H. Yang, and M.P. Zalucki. 2018. Non-migratory monarch butterflies, *Danaus plexippus* (L.), retain developmental plasticity and a navigational mechanism associated with migration. *Biol. J. Linn. Soc.* 123:265–278.

Frentiu, F.D., F. Yuan, W.K. Savage, G.D. Bernard, S.P. Mullen, and A.D. Briscoe. 2015. Opsin clines in butterflies suggest novel roles for insect photopigments. *Mol. Biol. Evol.* 32:368–379.

Frey, D.F. and K.L.H. Leong. 1993. Can microhabitat selection or differences in catchability explain male-biased sex ratios in overwintering populations of monarch butterflies. *Animal Behav.* 45:1025–1027.

Friberg, M., J. Dahlerus, and C. Wiklund. 2012. Strategic larval decision-making in a bivoltine butterfly. *Oecologia* 169:623–635.

Friedlander, M. and S. Reynolds. 1992. Intratesticular ecdysteroid titres and the arrest of sperm production during pupal diapause in the tobacco hornworm, *Manduca sexta*. *J. Insect Physiol.* 38:693–703.

Froy, O., A.L. Gotter, A.L. Casselman, and S.M. Reppert. 2003. Illuminating the circadian clock in monarch butterfly migration. *Science* 300:1303–1305.

Fu, D., L. Dai, H. Gao, Y. Sun, B. Liu, and H. Chen. 2019. Identification, expression patterns and RNA interference of aquaporins in *Dendroctonus aramandi* (Coleoptera: Scolytinae) larvae during overwintering. *Front. Physiol.* 10:967.

Fujita, K., K. Shimomura, K.-I. Yamamoto, T. Yamashita, and K. Suzuki. 2006. A chitinase structurally related to the glycoside hydrolase family 48 is indispensable for the hormonally induced diapause termination in a beetled. *Biochem. Biophys. Res. Com.* 345:502–507.

Fujiwara, H., M. Jindra, R. Newirtt, S.R. Palli, K. Hiruma, and L.M. Riddiford. 1995. Cloning of an ecdysone receptor homologue from *Manduca sexta* and the developmental profile of its mRNA in wings. *Insect Biochem. Mol. Biol.* 25:845–856.

Fujiwara, Y. and K. Shiomi. 2006. Distinct effects of different temperatures on diapause termination, yolk morphology and MAPK phosphorylation in the silkworm, *Bombyx mori*. *J. Insect Physiol.* 52:1194–1201.

Fujiwara, Y., C. Shindome, M. Takeda, and K. Shiomi. 2006a. The roles of ERK and p38 MAPK signaling cascades on embryonic diapause initiation and termination of the silkworm, *Bombyx mori*. *Insect Biochem. Mol. Biol.* 36:47–53.

Fujiwara, Y., Y. Tanaka, K.-I. Iwata, R.O. Rubio, T. Yaginuma, O. Yamashita, and K. Shiomi. 2006b. ERK/MAPK regulates ecdysteroid and sorbitol metabolism for embryonic diapause termination in the silkworm, *Bombyx mori*. *J. Insect Physiol.* 52:569–575.

Fujiwara, Y. and D.L. Denlinger. 2007. High temperature and hexane break pupal diapause in the flesh fly, *Sarcophaga crassipalpis*, by activating ERK/MAPK. *J. Insect Physiol.* 53:1276–1282.

Fukuda, S. 1951. Factors determining the production of non-diapausing eggs in the silkworm. *Proc. Jap. Acad.* 27:582–586.

Fukumoto, E., H. Numata, and S. Shiga. 2006. Effects of temperature of adults and eggs on the induction of embryonic diapause in the band-legged ground cricket, *Dianemobius nigrofasciatus*. *Physiol. Ent.* 31:211–217.

Gäde, G., P. Šimek, and H.G. Marco. 2020. The adipokinetic peptides in Diptera: structure, function, and evolutionary trends. *Front. Physiol.* 11:153.

Gadenne, C., L. Varjas, and B. Mauchamp. 1990. Effects of the non-steroidal ecdysone mimic, RH-5849, on diapause and non-diapause larvae of the European corn borer, *Ostrinia nubilalis*. *J. Insect Physiol.* 36:555–559.

Gadenne, C., M.-C. Dufour, F. Rossignol, J.-M. Bécard, and F. Couillaud. 1997. Occurrence of non-stationary larval moults during diapause in the corn-stalk borer, *Sesamia nonagrioides* (Lepidoptera: Noctuidae). *J. Insect Physiol.* 43:425–431.

Gallot, A., C. Rispe, N. Leterme, J.-P. Gauthier, S. Jaubert-Possamai, and D. Tagu. 2010. Cuticular proteins and seasonal photoperiodism in aphids. *Insect Biochem. Mol. Biol.* 40:235–240.

Gao, N., M. von Schantz, R. Foster, and J. Hardie. 1999. The putative brain photoperiodic photoreceptors in the vetch aphid, *Megoura viciae*. *J. Insect Physiol.* 45:1011–1019.

Gao, Q., B.-X. Wei, W. Liu, J.-L. Wang, X.-M. Zhou, and X.-P. Wang. 2019. Differences in the development of internal reproductive organs, feeding amount and nutrient storage between pre-diapause and pre-reproductive *Harmonia axyridis* adults. *Insects* 10:243.

Garcia-Rogers, E.M., E. Lubzens, D. Fontaneto, and M. Serra. 2019. Facing adversity: dormant embryos in rotifers. *Biol. Bull.* 237:119–144.

Gariepy, V., G. Boivin, and J. Brodeur. 2015. Why two species of parasitoids showed promise in the laboratory but failed to control the soybean aphid under field conditions. *Biol. Cont.* 80:1–7.

Gaston, K.J. 2018. Lighting up the nighttime. *Science* 362:744–746.

Ge, F., F.-J. Chen, M.N. Parajulee, and E.N. Yardim. 2005. Quantification of diapausing fourth generation and suicidal fifth generation cotton bollworm, *Helicoverpa armigera*, in cotton and corn in northern China. *Ent. Exp. Appl.* 116:1–7.

van Geffen, K.G., R.H.A. van Grunsven, J. van Ruijven, F. Berendse, and E.M. Veenendaal. 2014. Artificial light at night causes diapause inhibition and sex-specific life history changes in a moth. *Ecol. Evol.* 4:2082–2089.

Gehrken, U. and Y.O. Doumbia. 1996. Diapause and quiescence in eggs of a tropical grasshopper *Oedaleus senegalensis* (Krauss). *J. Insect Physiol.* 42:483–491.

Gelman, D.B. and C.W. Woods. 1983. Haemolymph ecdysteroid titers of diapause- and nondiapause-bound fifth instars and pupae of the European corn borer, *Ostrinia nubilalis* (Hubner). *Comp. Biochem. Physiol. A* 76:367–375.

Gelman, D.B., B.S. Thyagaraja, T.J. Kelly, E.P. Masler, R.A. Bell, and A.B. Borkovec. 1992. Prothoracicotropic hormone levels in brains of the European corn borer, *Ostrinia nubilalis*: diapause vs. the non-diapause state. *J. Insect Physiol.* 38:383–375.

Gerling, D., E. Erel, M. Guershon, and M. Inbar. 2009. Bionomics of *Encarsia scapeata* Rivnay (Hymenoptera: Aphelinidae), tritrophic relationships and host-induced diapause. *Biol. Cont.* 49:201–206.

Ghosh, E. and C.R. Ballal. 2018. Maternal influence on diapause induction: an approach to improve long-term storage of *Trichogramma chilonis*. *Phytoparasitica* 46:383–389.

Gibbens, Y.Y., J.T. Warren, L.I. Gilbert, and M.B. O'Connor. 2011. Neuroendocrine regulation of *Drosophila* metamorphosis requires TGFbeta/activin signaling. *Development* 138:2693–2703.

Gibbs, D. 1975. Reversal of pupal diapause in *Sarcophaga argyrostoma* by temperature shifts after puparium formation. *J. Insect Physiol.* 21:1179–1186.

Giebultowicz, J.M. 2001. Peripheral clocks and their role in circadian timing: insights from insects. *Phil. Trans. R. Soc. Lond. B* 356:1791–1799.

Giebultowicz, J.M. and D.L. Denlinger. 1986. Role of the brain and ring gland in relation to pupal diapause in the flesh fly, *Sarcophaga crassipalpis*. *J. Insect Physiol.* 32:161–166.

Gillespie, M., I.D. Hodkinson, E.J. Cooper, J.M. Bird, and I.S. Jónsdóttir. 2007. Life history and host–plant relationships of the rare endemic Arctic aphid *Acyrthosiphon calvulus* in a changing environment. *Ent. Exp. Appl.* 123:229–237.

Gkouvitsas, T., D. Kontogiannatos, and A. Kourti. 2008. Differential expression of two small Hsps during diapause in the corn stalk borer *Sesamia nonagrioides* (Lef.). *J. Insect Physiol.* 54:1503–1510.

Gkouvitsas, T., D. Kontogiannatos, and A. Kourti. 2009a. Cognate *Hsp70* gene is induced during deep diapause in the moth *Sesamia nonagrioides*. *Insect Mol. Biol.* 14:697–702.

Gkouvitsas, T., D. Kontogiannatos, and A. Kourti. 2009b. Expression of the *Hsp83* gene in response to diapause and thermal stress in the moth *Sesamia nonagrioides*. *Insect Mol. Biol.* 18:759–768.

Glitho, I.A., A. Lenga, D. Pierre, and J. Huignard. 1996. Changes in the responsiveness during two phases of diapause termination in *Bruchidius atrolineatus* Pic (Coleoptera: Bruchidae). *J. Insect Physiol.* 42:953–960.

Glockner, M.O., A. Madi, S. Mikkat, C. Koy, B. Ringel, and H.J. Thiesen. 2008. Mass spectrometric proteome analysis suggests anaerobic shift in metabolism of dauer larvae of *Caenorhabditis elegans*. *Biochim. Biophys. Acta Proteins Proteom.* 1784:1763–1770.

Gnagey, A.L. and D.L. Denlinger. 1984. Photoperiodic induction of pupal diapause in the flesh fly, *Sarcophaga crassipalpis*: embryonic sensitivity. *J. Comp. Physiol. B* 154:91–96.

Godlewski, J., B. Kłudkiewicz, K. Grzelak, and B. Cymborowski. 2001. Expression of larval hemolymph proteins (*Lhp*) genes and protein synthesis in the fat body of greater wax moth (*Galleria mellonella*) larvae during diapause. *J. Insect Physiol.* 47:759–766.

Goehring, L. and K.S. Oberhauser. 2002. Effects of photoperiod, temperature, and host plant age on induction of reproductive diapause and development time in *Danaus plexippus*. *Ecol. Ent.* 27:674–685.

Goldson, S.L., M.R. McNeill, and J.R. Proffitt. 1993. Effect of host condition and photoperiod on the development of *Microctonus hyperodae* Loan, a parasitoid of the Argentine stem weevil (*Listronotus bonariensis* (Kuschel)). *N. Zealand J. Zool.* 20:89–94.

Golec, J.R. and X.P. Hu. 2015. Preoverwintering copulation and female ratio bias: life history characteristics contributing to the invasiveness and rapid spread of *Megacopta cribraria* (Heteroptera: Plataspidae). *Environ. Ent.* 44:411–417.

Gomi, T. and M. Takeda. 1996. Changes in life-history traits in the fall webworm within half a century of introduction to Japan. *Funct. Ecol.* 10:384–389.

Gomi, T., M. Nagasaka, T. Fukuda, and H. Hagihara. 2007. Shifting of the life cycle and life-history traits of the fall webworm in relation to climate change. *Ent. Exp. Appl.* 125:179–184.

Gong, J., X. Zheng, S. Zhao, L. Yang, Z. Xue, Z. Fan, and M. Tang. 2020. Early molecular events during onset of diapause in silkworm eggs revealed by transcriptome analysis. *Int'l. J. Mol. Sci.* 21:6180.

Goto, M., Y.-P. Li, S. Kayaba, S. Outani, and K. Suzuki. 2001. Cold hardiness in summer and winter diapause and post-diapause pupae of the cabbage armyworm, *Mamestra brassicae* L. under temperature acclimation. *J. Insect Physiol.* 47:709–714.

Goto, S.G. 2009. Genetic analysis of diapause capacity and association between larval and pupal photoperiodic responses in the flesh fly *Sarcophaga similis*. *Physiol. Ent.* 34:46–51.

Goto, S.G. 2013. Roles of circadian clock genes in insect photoperiodism. *Ent. Sci.* 16:1–16.

Goto, S.G., K.M. Yoshida, and M.T. Kimura. 1998. Accumulation of *Hsp70* mRNA under environmental stresses in diapausing and nondiapausing adults of *Drosophila triauraria*. *J. Insect Physiol.* 44:1009–1015.

Goto, S.G. and D.L. Denlinger. 2002. Short-day and long-day expression patterns of genes involved in the flesh fly clock mechanism: *period, timeless, cycle* and cryptochrome. *J. Insect Physiol.* 48:803–816.

Goto, S.G., B. Han, and D.L. Denlinger. 2006. A nondiapausing variant of the flesh fly, *Sarcophaga bullata*, that shows arrhythmic adult eclosion and elevated expression of two circadian clock genes, *period* and timeless. *J. Insect Physiol.* 52:1213–1218.

Goto, S.G., K. Doi, S. Nakayama, and H. Numata. 2008. Maternal control of cold and desiccation tolerance in eggs of the band-legged ground cricket *Dianemobius nigrofasciatus* in relation to embryonic diapause. *Ent. Res.* 38:17–23.

Goto, S.G. and H. Numata. 2009. Possible involvement of distinct photoreceptors in the photoperiodic induction of diapause in the flesh fly *Sarcophaga similis*. *J. Insect Physiol.* 55:401–407.

Goto, S.G., S. Shiga, and H. Numata. 2010. Photoperiodism in insects: perception of light and the role of clock genes. In *Photoperiodism, the Biological Calendar*, ed. R.J. Nelson, D.L. Denlinger, and D.E. Somers, Oxford: Oxford University Press, pp. 258–286.

Gotthard, K., S. Nylin, and C. Wiklund. 1999. Seasonal plasticity in two satyrine butterflies: state-dependent decision making in relation to daylength. *Oikos* 84:453–462.

Gotthard, K. and D. Berger. 2010. The diapause decision as a cascade switch for adaptive developmental plasticity in body mass in a butterfly. *J. Evol. Biol.* 23:1129–1137.

Gotthard, K. and C.W. Wheat. 2019. Diapause: circadian clock genes are at it again. *Curr. Biol.* 29:R1245.

Goymann, W., B. Helm, W. Jensen, I. Schwabl, and I.T. Moore. 2012. A tropical bird can use the equatorial change in sunrise and sunset times to synchronize its circannual clock. *Proc. Roy. Soc. B* 279:3527–3534.

Grabenweger, G. 2004. Poor control of the horse chestnut leafminer, *Cameraria ohridella* (Lepidoptera: Gracillariidae), by native European parasitoids: a synchronization problem. *Eur. J. Ent.* 101:189–192.

Graf, B., H. Höhn, H.U. Höpli, and S. Kuske. 2018. Predicting the phenology of codling moth, *Cydia pomonella*, for sustainable pest management in Swiss apple orchards. *Ent. Exp. Appl.* 166:L618–L627.

Graham, D.H., J.L. Holmes, B.J. Beaty, and W.C. Black IV. 2003. Quantitative trait loci conditioning transovarial transmission of La Crosse virus in the eastern treehole mosquito, *Ochlerotatus triseriatus*. *Insect Mol. Biol.* 12:307–318.

Gray, D.R., F.W. Ravlin, and J.A. Braine. 2001. Diapause in the gypsy moth: a model of inhibition and development. *J. Insect Physiol.* 47:173–184.

Green, D.A. and M.R. Kronforst. 2019. Monarch butterflies use an environmentally sensitive, internal timer to control overwintering dynamics. *Mol. Ecol.* 28:3642–3655.

Greenberg, S.M., T.W. Sappington, M. Setamou, J.S. Armstrong, R.J. Coleman, and T.-X. Liu. 2007. Reproductive potential of overwintering, F_1, and F_2 female boll weevils (Coleoptera: Curculionidae) in the lower Rio Grande Valley of Texas. *Environ. Ent.* 36:256–262.

Greenfield, M.D. and M.P. Pener. 1992. Alternative schedules of male reproductive diapause in the grasshopper *Anacridium aegyptium* (L.): effects of the corpora allata on sexual behavior (Orthoptera: Acrididae). *J. Insect Behav.* 5:245–261.

Gregg, P.C., B. Roberts, and S.L. Wentworth. 1987. Levels of ecdysteroids in diapause and non-diapause eggs of the Australian plague locust, *Chortoicetes terminifera* (Walker). *J. Insect Physiol.* 33:237–242.

Grewal, S.S. 2009. Insulin/TOR signaling in growth and homeostasis: a view from the fly world. *J. Biochem. Cell Biol.* 41:1006–1010.

Grüner, C. and S. Masaki. 1994. Summer diapause in the polymorphic life cycle of the noctuid moth *Mamestra brassicae*. In *Insect Life-cycle Polymorphism*, ed. by H.V. Danks, pp. 191–204, Amsterdam: Kluwer Academic.

Gu, S.-H., H.-Y. Hsieh and P.-L. Lin. 2017. Regulation of protein phosphatase 2A during embryonic diapause process in the silkworm, *Bombyx mori. J. Insect Physiol.* 103:117–124.

Gu, S.-H., P.-L. Lin, and H.-Y. Hsieh. 2019. Bombyxin/Akt signaling in relation to the embryonic diapause process of the silkworm, *Bombyx mori. J. Insect Physiol.* 116:32–40.

Gu, S.-H., C.-H. Chen, H.-Y. Hsieh, and P.-L. Lin. 2020. Expression of protein kinase C in relation to the embryonic diapause process in the silkworm, *Bombyx mori. J. Insect Physiol.* 121:104010.

Gu, S.-H., C.-H. Chen, and P.-L. Lin. 2021. Expression of protein tyrosine phosphatases and *Bombyx* embryonic development. *J. Insect Physiol.* 130:104198.

Guidetti, R., T. Altiero, and L. Rebecchi. 2011. On dormancy strategies in tardigrades. *J. Insect Physiol.* 57:567–576.

Güney, G., U. Toprak, D.D. Hegedus, S. Bayram, C. Coutu, D. Bekkaoui, D. Baldwin, D.G. Heckel, D. Hänniger, D. Cedden, D.A. Mutlu, and Z. Suludere. 2020. A look into Colorado potato beetle lipid metabolism through the lens of lipid storage droplet proteins. *Insect Biochem. Mol. Biol.* 133:103473. doi: 10.1016/j.ibmb.2020.103473

Guo, F., M. Holla, M.M. Diaz, and M. Rosbash. 2018. A circadian output circuit controls sleep-wake arousal in *Drosophila. Neuron* 100:624–635.

Guo, S., D. Sun, Z. tian, W. Liu, F. Zhu, and X.-P. Wang. 2019. The limited regulatory roles of juvenile hormone degradation pathways in reproductive diapause preparation of the cabbage beetle, *Colaphellus bowringi. J. Insect Physiol.* 119:103967.

Guo, S., Z. Tian, Q.-W. Wu, K. King-Jones, W. Liu, F. Zhu, and X.-P. Wang. 2021. Steroid hormone ecdysone deficiency stimulates preparation for photoperiodic reproductive diapause. *PLoS Genet.* 17:e1009352.

Guven, O., H. Golluoglu, and P. Ceryngier. 2015. Aestivo-hibernation of *Coccinella septempunctata* (Coleoptera: Coccinellidae) in a mountainous area in southern Turkey: is dormancy at high altitudes adaptive? *Eur. J. Ent.* 112:41–48.

Guz, N., U. Toprak, A. Dageri, M.O. Gurkan, and D.L. Denlinger. 2014. Identification of a putative antifreeze protein gene that is highly expressed during preparation for winter in the sunn pest, *Eurygaster maura. J. Insect Physiol.* 68:30–35.

Hadley, N.F. 1994. *Water Relationships of Terrestrial Arthropods.* New York: Academic Press.

Hafker, N.S., M. Teschke, L. Huppe, and B. Meyer. 2018. *Calanus finmarchicus* diel and seasonal rhythmicity in relation to endogenous timing under extreme polar photoperiods. *Mar. Ecol. Prog. Ser.* 603:79–92.

Hagen, K.S. 1962. Biology and ecology of predaceous Coccinellidae. *Ann. Rev. Ent.* 7:289–326.

Hahn, D.A. and D.L. Denlinger. 2007. Meeting the energetic demands of insect diapause: nutrient storage and utilization. *J. Insect Physiol.* 53:760–773.

Hahn, D.A. and D.L. Denlinger. 2011. Energetics of insect diapause. *Ann. Rev. Ent.* 56:103–121.

Hairston, N.G.J. 1987. Diapause as a predator-avoidance adaptation. In *Predation: Direct and Indirect Impacts on Aquatic Communities*, ed. W.C. Kerfoot and A. Sih, Hanover: University Press of New England, pp. 281–290.

Halbritter, D.A. 2020. Exposed *Neophasia terlooii* (Lepidoptera: Pieridae) eggs are resistant to desiccation during quiescence. *Environ. Ent.* 49:918–923.

Hall, J.C. 2003. Genetics and molecular biology of rhythms in *Drosophila* and other insects. *Adv. Genet.* 48:1–280.

Hallworth, M.T., P.P. Marra, K.P. McFarland, S. Zahendra, and C.E. Studds. 2018. Tracking dragons: stable isotopes reveal the annual cycle of a long-distance migratory insect. *Biol. Lett.* 14:20180741.

Hamedi, N., S. Moharramipour, and M. Barzegar. 2013. Temperature-dependent chemical components accumulation in *Hippodamia variegata* (Coleoptera: Coccinellidae) during overwintering. *Environ. Ent.* 42:375–380.

Hamel, M., C. Geri, and A. Auger-Rozenberg. 1998. The effects of 20-hydroxyecdysone on breaking diapause of *Diprion pini* L. (Hym., Diprionidae). *Physiol. Ent.* 23:337–346.

Hammell, C.M., X. Karp, and V. Ambros. 2009. A feedback circuit involving let-7-family miRNAs and DAF-12 integrates environmental signals and developmental timing in *Caenorhabditis elegans*. *Proc. Nat'l. Acad. Sci., USA* 106:18668–18673.

Han, B. and D.L. Denlinger. 2009a. Length variation in a specific region of the *period* gene correlates with differences in pupal diapause incidence in the flesh fly, *Sarcophaga bullata. J. Insect Physiol.* 55:415–418.

Han, B. and D.L. Denlinger. 2009b. Mendelian inheritance of pupal diapause in the flesh fly, *Sarcophaga bullata. J. Heredity* 100:251–255.

Hand, S.C., M.A. Menze, A. Borcar, Y. Patil, J.A. Covi, J.A. Reynolds, and M. Toner. 2011. Metabolic restructuring during energy-limited states: insights from *Artemia franciscana* embryos and other animals. *J. Insect Physiol.* 57:584–594.

Hand, S.C., D.L. Denlinger, J.E. Podrabsky, and R. Roy. 2016. Mechanisms of animal diapause: recent developments from nematodes, crustaceans, insects, and fish. *Am. J. Physiol. Regul. Integr. Comp. Physiol.* 310:R1193–R1211.

Hand, S.C., D.S. Moore, and Y. Patil. 2018. Challenges during diapause and anhydrobiosis: mitochondrial bioenergetics and desiccation tolerance. *Int'l. Union Biochem. Mol. Biol.* 9999:1–9.

Hanski, I. 1988. Four kinds of extra long diapause in insects: a review of theory and observations. *Ann. Zool. Fennici* 25:37–53.

Hao, K., X. Tu, H. Ullah, M.R. McNeill, and Z. Zhang. 2019. Novel Lom-dh genes play potential role in promoting egg diapause of *Locusta migratoria* L. *Front. Physiol.* 10:767.

Hao, Y., Y. Zhang, F. Si, D. Fu, Z. He, and B. Chen. 2016. Insight into the possible mechanism of the summer diapause of *Delia antiqua* (Diptera: Anthomyiidae) through digital gene expression analysis. *Insect Sci.* 23:438–451.

Hao, Y.-J., W.-S. Li, Z.-B. He, F.-L. Si, YU. Ishikawa, and B. Chen. 2012. Differential gene expression between summer and winter diapause pupae of the onion maggot *Delia antiqua*, detected by suppressive subtractive hybridization. *J. Insect Physiol.* 58:1444–1449.

Harada, T., S. Nitta, and K. Ito. 2005. Photoperiodic changes according to global warming in wing-form determination and diapause induction of a water strider, *Aquarius paludum* (Heteroptera: Gerridae). *Appl. Ent. Zool.* 40:461–466.

Hardie, J. 2010. Photoperiodism in insects: aphid polyphenism. In *Photoperiodism, the Biological Calendar*, ed. R.J. Nelson, D.L. Denlinger, and D.E. Somers, Oxford: Oxford University Press, pp. 342–363.

Hardin, P.E. 2005. The circadian timekeeping system of *Drosophila*. *Curr. Biol.* 15: R714–R722.

Härkönen, L. and A. Kaitala. 2013. Months of asynchrony in offspring production but synchronous adult emergence: the role of diapause in an ectoparasite's life cycle. *Environ. Ent.* 42:1408–1414.

Hartman, M.J. and C.D. Hynes. 1980. Embryonic diapause in *Tipula simplex* and the action of photoperiod in its termination (Diptera: Tipulidae). *Pan-Pacific Entomol.* 56:207–212.

Harvey, G.T. 1961. Second diapause in spruce budworm from eastern Canada. *Can. Ent.* 93:594–602.

Harvey, W.R. and C.M. Williams. 1961. The injury metabolism of the Cecropia silkworm – -I. Biological amplification of the effects of localized injury. *J. Insect Physiol.* 7:81–99.

Hasan, F., M.S. Ansari, M.K. Dhillon, M. Muslim, A.S. Bhadauriya, A.K. Tanwar, and S. Ahmad. 2018. Diapause regulation in *Zygogramma bicolorata* (Coleoptera: Chrysomelidae), a biocontrol agent of *Parthenium hysterophorus*. *Int'l. J. Trop. Insect Sci.* 38:145–158.

Hasanvand, H., H. Izadi, and M. Mohammadzadeh. 2020. Overwintering physiology and cold tolerance of the sunn pest, *Eurygaster integriceps*, an emphasis on the role of cryoprotectants. *Front. Physiol.* 11:321.

Hasebe, M. and S. Shiga. 2021. Oviposition-promoting pars intercerbralis neurons show *period*-dependent photoperiodic changes in their firing activity in the bean bug. *Proc. Nat'l. Acad. Sci., USA* 118:e2018823118.

Hasegawa, K. 1951. Studies in voltinism in the silkworm, *Bombyx mori* L., with special reference to the organs concerning determination of voltinism (a preliminary note). *Proc. Jap. Acad.* 27:667–671.

Hasegawa, K. and I. Shimizu. 1987. *In vivo* and *in vitro* photoperiodic induction of diapause using isolated brain-subesophageal ganglion complexes of the silkworm, *Bombyx mori*. *J. Insect Physiol.* 33:959–966.

Hausmann, C., J. Smietz, and S. Dorn. 2004a. Monitoring the dynamics of orchard colonization by *Anthonomus pomorum* in spring. *Ent. Exp. Appl.* 110:207–216.

Hausmann, C., J. Samietz, and S. Dorn. 2004b. Significance of shelter traps for spring monitoring of *Anthonomus pomorum* in apple orchards. *Ent. Exp. Appl.* 112:29–36.

Hausmann, C., J. Samietz, and S. Dorn. 2005. Thermal orientation of *Anthonomus pomorum* (Coleoptera: Curculionidae) in early spring. *Physiol. Ent.* 30:48–53.

Hawley, W.A., P. Reiter, R.S. Copeland, C.B. Pumpuni, and G.B. Craig, Jr. 1987. *Aedes albopictus* in North America: probable introduction in used tires from northern Asia. *Science* 236:1114–1116.

Hayes, D.K., W.N. Sullivan, M.Z. Oliver, and M.S. Schechter. 1970. Photoperiod manipulation of insect diapause: a method of pest control? *Science* 169:382–383.

Hayes, D.K., B.M. Cawley, W.N. Sullivan, V.E. Adler, and M.S. Schechter. 1974. The effect of added light pulses on overwintering and diapause, under natural lights and temperature conditions, of four species of Lepidoptera. *Environ. Ent.* 3:863–865.

Hayward, S.A.L., S.C. Pavlides, S.P. Tammariello, J.P. Rinehart, and D.L. Denlinger. 2005. Temporal expression patterns of diapause-associated genes in flesh fly pupae from the onset of diapause through post-diapause quiescence. *J. Insect Physiol.* 51:631–640.

Hayward, S.A.L., J.P. Rinehart, L.H. Sandro, R.E. Lee, Jr., and D.L. Denlinger. 2007. Slow dehydration promotes desiccation and freeze tolerance in the Antarctic midge,

Belgica antarctica. J. Exp. Biol. 210:836–844.

He, H.-M., Z.-H. Xian, F. Huang, X.-P. Liu, and F.-S. Xue. 2009. Photoperiodism of diapause induction in *Thyrassia penangae* (Lepidoptera: Zygaenidae). *J. Insect Physiol.* 55:1003–1008.

Heestand, B., M. Simon, S. Frenk, D. Titov, and S. Ahmed. 2018. Transgenerational sterility of Piwi mutants represents a dynamic form of adult reproductive diapause. *Cell Rep.* 23:156–171.

Hejnikova, M., M. Paroulek, and M. Hodková. 2016. Decrease in *Methoprene tolerant* and *Taiman* expression reduces juvenile hormone effects and enhances the levels of juvenile hormone circulating in males of the linden bug *Pyrrhocoris apterus. J. Insect Physiol.* 93–94:72–80.

Held, C. and H. Spieth. 1999. First evidence of pupal summer diapause in *Pieris brassicae* L.: the evolution of local adaptedness. *J. Insect Physiol.* 45:587–598.

Helfrich-Förster, C. 2003. The neuroarchitecture of the circadian clock in the brain of *Drosophila melanogaster. Microsc. Res. Tech.* 62:94–102.

Helfrich-Förster, C. 2006. The neural basis of Drosophila's circadian clock. *Sleep Biol. Rhythms* 4:224–234.

Helfrich-Förster, C., T. Edwards, K. Yasuyama, B. Wisotzki, S. Schneuwly, R. Stanewsky, I.A. Meinertzhagen, and A. Hofbauer. 2002. The extraretinal eyelet of *Drosophila*: development, ultrastructure, and putative circadian function. *J. Neurosci.* 22:9255–9266.

Hemmati, C., S. Moharramipour, and A.A. Talebi. 2017. Diapause induced by temperature and photoperiod affects fatty acid compositions and cold tolerance of *Phthorimaea operculella* (Lepidoptera: Gelechiidae). *Environ. Ent.* 46:1456–1463.

Henneguy, L.F. 1904. *Les Insectes. Morphologie, Reproduction, Embryogenie.* Paris: Masson.

Henrich, V.C. and D.L. Denlinger. 1982a. A maternal effect that eliminates pupal diapause in progeny of the flesh fly, *Sarcophaga bullata. J. Insect Physiol.* 28:881–884.

Henrich, V.C. and D.L. Denlinger. 1982b. Selection for late pupariation affects diapause incidence and duration in the flesh fly, *Sarcophaga bullata. Physiol. Ent.* 7:407–411.

Henrich, V.C. and D.L. Denlinger. 1983. Genetic differences in pupal diapause incidence between two selected strains of the flesh fly. *J. Heredity* 74:371–374.

Hetz, S.K. and T.J. Bradley. 2005. Insects breathe discontinuously to avoid oxygen toxicity. *Nature* 433:516–519.

Hickner, P.V., A. Mori, E. Zeng, J.C. Tan, and D.W. Severson. 2015. Whole transcriptome responses among females of the filariasis and arbovirus vector mosquito *Culex pipiens* implicates TGF-β signaling and chromatin modification as key drivers of diapause induction. *Funct. Integr. Genomics* 15:439–447.

Higaki, M. 2006. Repeated cycles of chilling and warming effectively terminate prolonged larval diapause in the chestnut weevil, *Curculio sikkimensis. J. Insect Physiol.* 52:514–519.

Higaki, M. 2016. Prolonged diapause and seed predation by the acorn weevil, *Curculio robustus*, in relation to masting of the deciduous oak *Quercus acutissima. Ent. Exp. Appl.* 159:338–346.

Higaki, M. and Y. Ando. 2005. Effects of temperature during chilling and pre-chilling periods on diapause and post-diapause development in a katydid, *Eobiana engelhardti subtropica. J. Insect Physiol.* 51:709–716.

Higaki, M., F. Ihara, M. Toyama, and K. Mishiro. 2010. Thermal response and reversibility of prolonged larval diapause in the chestnut weevil, *Curculio sikkimensis. J. Insect Physiol.* 56:616–621.

Higaki, M. and M. Toyama. 2012. Evidence for reversible change in intensity of prolonged diapause in the chestnut weevil *Curculio sikkimensis*. *J. Insect Physiol.* 58:56–60.

Hiiesaar, K., T. Kaart, I.H. Williams, A. Luik, L. Metspalu, A. Ploomi, E. Kruss, K. Jogar, and M. Mand. 2018. Dynamics of supercooling ability and cold tolerance of the alder beetle (Coleoptera: Chrysomelidae). *Environ. Ent.* 47:1024–1029.

Hinton, H.E. 1951. A new chironomid from Africa, the larva of which can be dehydrated without injury. *Proc. Zool. Soc. Lond.* 121:371–280.

Hiraga, S. 2006. Interactions of environmental factors influencing pupal coloration in swallowtail butterfly *Papilio xuthus*. *J. Insect Physiol.* 52:826–838.

Hirai, T., N. Shibayama, S. Akashi, and S.Y. Park. 2008. Crystal structures of the clock protein EA4 from the silkworm *Bombyx mori*. *J. Mol. Biol.* 377:630–635.

Hiroyoshi, S., G.V.P. Reddy, and J. Mitsuhashi. 2017. Effects of juvenile hormone analogue (methoprene) and 20-hydroxyecdysone on reproduction in *Polygonia c-aureum* (Lepidoptera: Nymphalidae) in relation to adult diapause. *J. Comp. Physiol. A* 203:635–647.

Ho, D.H. and W.W. Burggren. 2010. Epigenetics and transgenerational transfer: a physiological perspective. *J. Exp. Biol.* 213:3–16.

Hoback, W.W. and D.W. Stanley. 2001. Insects in hypoxia. *J. Insect Physiol.* 47:533–542.

Hoban, J., J.J. Duan, and J. Hough-Goldstein. 2016. Effects of temperature and photoperiod on the reproductive biology and diapause of *Oobius agrili* (Hymenoptera: Encyrtidae), an egg parasitoid of emerald ash borer (Coleoptera: Buprestidae). *Environ. Ent.* 45:726–731.

Hockham, L.R., J.A. Graves, and M.G. Ritchie. 2001. Variable maternal control of facultative egg diapause in the bushcricket *Ephippiger ephippiger*. *Ecol. Ent.* 26:143–147.

Hodek, I. 1968. Diapause in females of *Pyrrhocoris apterus* L. (Heteroptera). *Acta Ent. Bohemoslov.* 65:422–435.

Hodek, I. 1971a. Sensitivity to photoperiod in *Aelia acuminate* (L.) after adult diapause. *Oecologia* 6:152–155.

Hodek, I. 1971b. Termination of adult diapause in *Pyrrhocoris apterus* (Heteroptera: Pyrrhocoridae) in the field. *Ent. Exp. Appl.* 14:212–222.

Hodek, I. 2002. Controversial aspects of diapause development. *Eur. J. Ent.* 99:163–173.

Hodek, I. 2003. Role of water and moisture in diapause development (a review). *Eur. J. Ent.* 100:223–232.

Hodek, I. 2012. Diapause/Dormancy. In *Ecology and Behaviour of the Ladybird Beetles (Coccinellidae)*, ed. I. Hodek, H.F. van Emden, and A. Honek. Oxford: Blackwell, pp. 275–342.

Hodek, I. and M. Hodková. 1986. Diapause development and photoperiodic activation in starving females of *Pyrrhocoris apterus* (Heteroptera). *J. Insect Physiol.* 32:615–621.

Hodek, I. and M. Hodková. 1988. Multiple role of temperature during insect diapause: a review. *Ent. Exp. Appl.* 49:153–165.

Hodek, I. and P. Ceryngier. 2000. Sexual activity in Coccinelidae (Coleoptera): a review. *Eur. J. Ent.* 97:449–456.

Hodkinson, I.D. 2005. Terrestrial insects along elevation gradients: species and community responses to altitude. *Biol. Rev.* 80:489–513.

Hodková, M. 1976. Nervous inhibition of corpora allata by photoperiod in *Pyrrhocoris apterus*. *Nature* 263:521–523.

Hodková, M. 1992. Storage of the photoperiodic information within the implanted neuroendocrine complexes in females of the linden bug, *Pyrrhocoris apterus* (L.) (Heteroptera). *J. Insect Physiol.* 38:357–363.

Hodková, M. 1994. Photoperiodic regulation of mating behavior in the linden bug *Pyrrhocoris apterus*, is mediated by a brain inhibitory factor. *Experientia* 50:742–744.

Hodková, M. 2008. Tissue signaling pathways in the regulation of life-span and reproduction in females of the linden bug, *Pyrrhocoris apterus*. *J. Insect Physiol.* 54:508–517.

Hodková, M. 2015. Why is the number of days required for induction of adult diapause in the linden bug *Pyrrhocoris apterus* fewer in the larval than in the adult stage? *J. Insect Physiol.* 77:39–44.

Hodková, M. and I. Hodek. 1987. Photoperiodic summation is temperature-dependent in *Pyrrhocoris apterus* (L.) (Heteroptera). *Experientia* 43:454–456.

Hodková, M., T. Okuda, and R. Wagner. 2001. Regulation of corpora allata in females of *Pyrrhocoris apterus* (Heteroptera). *Cell. Dev. Biol.* 37:560–563.

Hodková, M., Z. Syrova, D. Doležel, and I. Sauman. 2003. *Period* gene expression in relation to seasonality and circadian rhythms in the linden bug, *Pyrrhocoris apterus* (Heteroptera). *Eur. J. Ent.* 100:267–273.

Hodková, M. and I. Hodek. 2004. Photoperiod, diapause and cold-hardiness. *Eur. J. Ent.* 101:445–458.

Hodková, M. and R. Socha. 2006. Endocrine regulation of the reproductive arrest in the long-winged females of a flightless bug, *Pyrrhocoris apterus* (Heteroptera: Pyrrhocoridae). *Eur. J. Ent.* 103:523–529.

Hodková, M. and T. Okuda. 2019. Three kinds of regulatory signals for production of juvenile hormone in females of the linden bug, *Pyrrhocoris apterus*. *J. Insect Physiol.* 113:17–23.

Hoffmann, E.J., J. Vanderjagt, and M.E. Whalon. 2007. Pyriproxyfen activates reproduction in pre-diapause northern strain plum curculio (*Conotrachelus nenuphar* Herbst). *Pest Manag. Sci.* 63:835–840.

Hogan, T.W. 1962. The effect of ammonia on the rate of termination of diapause in eggs of *Acheta commodus* (Walk.) (Orthoptera: Gryllidae). *Aust. J. Biol. Sci.* 15:538–542.

Hogan, T.W. 1964. Further data on the effect of ammonia on the termination of diapause in eggs of *Teleogryllus commodus* (Walk.) (Orthoptera: Gryllidae). *Aust. J. Biol. Sci.* 17:752–757.

Holmstrup, M., M. Bayley, S.A. Pedersen, and K.E. Zachariassen. 2010. Interactions between cold, desiccation and environmental toxins. In *Low Temperature Biology of Insects*, ed. D.L. Denlinger and R.E. Lee, Jr., Cambridge: Cambridge University Press, pp. 166–187.

Holzapfel, C.M. and W.E. Bradshaw. 1981. Geography of larval dormancy in the tree-hole mosquito, *Aedes triseriatus* (Say). *Can. J. Zool.* 59:1014–1021.

Homma, T., K. Watanabe, S. Tsurumaru, H. Kataoka, K. Imai, M. Kamba, T. Niimi, O. Yamashita, and T. Yaginuma. 2006. G protein-coupled receptor for diapause hormone, an inducer of *Bombyx* embryonic diapause. *Biochem. Biophy. Res. Comm.* 344:386–393.

Hondelmann, P. and H.-M. Poehling. 2007. Diapause and overwintering of the hoverfly *Episyrphus balteatus*. *Ent. Ent. Appl.* 124:189–200.

Honnen, A.-C., J.L. Kypke, F. Hölker, and M.T. Monaghan. 2019. Artificial light at night influences clock-gene expression, activity, and fecundity in the mosquito *Culex pipiens* f. *molestus*. *Sustainability* 11:6220.

Hood, G.R., A.A. Forbes, T.H.Q. Powell, S.P. Egan, G. Hamerlinck, J.J. Smith, and J.L. Feder. 2015. Sequential divergence and the multiplicative origin of community diversity. *Proc. Nat'l. Acad. Sci., USA* 112: e5980–e5989.

Horie, Y., T. Kanda, and Y. Mochida. 2000. Sorbitol as an arrester of embryonic development in diapausing eggs of the silkworm, *Bombyx mori*. *J. Insect Physiol.* 46:1009–1016.

van Houten, Y.M. 1989. Photoperiodic control of adult diapause in the predacious mite,

Amblyseius potentillae: repeated diapause induction and termination. *Physiol. Ent.* 14:341–348.

Hoy, M.A. 1977. Rapid response to selection for a non-diapausing gypsy moth. *Science* 196:1462–1463.

Hu, C.-K., W. Wang, J. Brind'Amour, P.P. Singh, G.A. Reeves, M.C. Lorincz, A.S. Alvarado, and A. Brunet. 2020. Vertebrate diapause preserves organisms long term through Polycomb complex members. *Science* 367:870–874.

Hu, G., K.S. Lim, N. Horvitz, S.J. Clark, D.R. Reynolds, N. Sapir, and J.W. Chapman. 2016. Mass seasonal bioflows of high-flying insect migrants. *Science* 354:1584–1587.

Hu, G., C. Stefanescu, T.H. Oliver, D.B. Roy, T. Brereton, C. Van Swaay, D.R. Reynolds, and J.W. Chapman. 2021. Environmental drivers of annual population fluctuations in a trans-Saharan insect migrant. *Proc. Nat'l. Acad. Sci., USA* 118:e2102762118.

Hua, A., D. Yang, S. Wu, and F.-S. Xue. 2005a. Photoperiodic control of diapause in *Pseudopidorus fasciata* (Lepidoptera: Zygaenidae) based on a qualitative time measurement. *J. Insect Physiol.* 51:1261–1267.

Hua, A., F.-S. Xue, H.-J. Xiao, and X.-F. Zhu. 2005b. Photoperiodic counter of diapause induction in *Pseudopidorus fasciata* (Lepidoptera: Zygaenidae). *J. Insect Physiol.* 51:1287–1294.

Huang, H., C. Chen, F. Yao, X. Li, Y. Wang, Y. Shao, X. Wang, X. Zhang, T. Jiang, and L. Hou. 2019. Investigation of the possible role of RAD9 in post-diapaused embryonic development of the brine shrimp *Artemia sinica*. *Genes* 10:768.

Huang, L., F. Xue, G. Wang, R. Han, and F. Ge. 2005. Photoperiodic response of diapause induction in the pine caterpillar, *Dendrolimus punctatus*. *Ent. Exp. Appl.* 117:127–133.

Huang, L.-L., C. Chen, L. Xiao, Q. Xia, L.-T. Hu, and F. Xue. 2013. Geographic variation and inheritance of the photoperiodic response controlling larval diapause in two distinct voltine ecotypes of the Asian cornborer Ostrinia furnacalis. *Physiol. Ent.* 38:126–132.

Huang, X., M.F. Poelchau, and P.A. Armbruster. 2015. Global transcriptional dynamics of diapause induction in non-blood-fed and blood-fed *Aedes albopictus*. *PLoS Negl. Trop. Dis.* 9:e0003724.

Huestis, D.L. and J.L. Marshall. 2006. Interaction between maternal effects and temperature affects diapause occurrence in the cricket *Allonemobius socius*. *Oecologia* 146:513–520.

Huey, R.B., L. Ma, O. Levy, and M.R. Kearney. 2021. Three questions about the eco-physiology of overwintering underground. *Ecol. Lett.* 24:170–185.

Hufbauer, R.A. 2002. Evidence for nonadaptive evolution in parasitoid virulence following a biological control introduction. *Ecol. Appl.* 12:66–78.

van Huis, A., J. Van Itterbeeck, H. Klunder, E. Mertens, A. Halloran, G. Muir, and P. Vantomme. 2013. Edible insects: future prospects for food and feed security. FAO Forestry Paper 171, Rome.

Hunt, J.H. and G.V. Amdam. 2005. Bivoltinism as an antecedent to eusociality in the paper wasp genus *Polistes*. *Science* 308:264–267.

Hunt, J.H., B.J. Kensinger, J.A. Kossuth, M.T. Henshaw, K. Norberg, F. Wolschin, and G.V. Amdam. 2007. A diapause pathway underlies the gyne phenotype in *Polistes* wasps, revealing an evolutionary route to caste-containing insect societies. *Proc. Nat'l. Acad. Sci., USA* 104:14020–14025.

Hunt, J.H., F. Wolschin, M.T. Henshaw, T.C. Newman, A.L. Toth, and G.V. Amdam. 2010. Differential gene expression and protein abundance evince ontogenetic bias toward castes in a primitively eusocial wasp. *PLoS One* 5:e10674.

Hunter, M.D. and J.N. McNeil. 1997. Host-plant quality influences diapause and voltinism in a polyphagous insect herbivore. *Ecology* 78:977–986.

Hut, R.A. 2011. Photoperiodism: shall EYA compare thee to a summer's day? *Curr. Biol.* 21:R22–R25.

Hut, R.A., S. Paolucci, R. Dor, C.P. Kyriacou, and S. Daan. 2013. Latitudinal clines: an evolutionary view of biological rhythms. *Proc. Roy. Soc. B.* 280:20130433.

Ichikawa, T., K. Hasegawa, I. Shimizu, K. Katsuno, H. Kataoka, and A. Suzuki. 1995. Structure of neurosecretory cells with immunoreactive diapause hormone and pheromone biosynthesis activating neuropeptide in the silkworm, *Bombyx mori. Zool. Sci.* 12:703–712.

Iga, M., T. Nakaoka, Y. Suzuki, and H. Kataoka. 2014. Pigment dispersing factor regulates ecdysone biosynthesis via *Bombyx* neuropeptide G protein coupled receptor-B2 in the prothoracic glands of *Bombyx mori. PLoS One* 9:e103239.

Iiams, S.E., A.B. Lugena, Y. Zhang, A.N. Hayden, and C. Merlin. 2019. Photoperiodic and clock regulation of the vitamin A pathway in the brain mediates seasonal responsiveness in the monarch butterfly. *Proc. Nat'l. Acad. Sci., USA* 116:25214–25221.

Ikeda, K., H. Numata, and S. Shiga. 2005. Roles of the mushroom bodies in olfactory learning and photoperiodism in the blow fly *Protophormia terraenovae. J. Insect Physiol.* 51:669–680.

Ikeda, M., Z.-H. Su, H. Saito, K. Imai, Y. Sato, M. Isobe, and O. Yamashita. 1993. Induction of embryonic diapause and stimulation of ovary trehalase activity in the silkworm, *Bombyx mori*, by synthetic diapause hormone. *J. Insect Physiol.* 39:889–895.

Ikeda-Kikue, K. and H. Numata. 1994. Effect of low temperature on the termination of photoperiodic and food-meidated diapause in the cabbage bug, *Eurydema rugosa* Motshulsky (Heteroptera: Pentatomidae). *Appl. Ent. Zool.* 29:229–236.

Ikeda-Kikue, K. and H. Numata. 2001. Timing of diapause induction in the cabbage bug *Eurydema rugosum* (Heteroptera: Pentatomidae) on different host plants. *Acta Soc. Zool. Bohem.* 65:197–205.

Ikeno, T., H. Numata, and S. Goto. 2008. Molecular characterization of the circadian clock genes in the bean beetle, *Riptortus pedestris*, and their expression patterns under long- and short-day conditions. *Gene* 419:56–61.

Ikeno, T., S. Tanaka, H. Numata, and S. Goto. 2010. Photoperiodic diapause under the control of circadian clock genes in an insect. *BMC Biol.* 8:116.

Ikeno, T., C. Katagiri, H. Numata, and S.G. Goto. 2011a. Causal involvement of *mammalian-type cryptochrome* in the circadian cuticle deposition rhythm in the bean bug *Riptortus pedestris. Insect Mol. Biol.* 20:409–415.

Ikeno, T., H. Numata, and S.G. Goto. 2011b. Photoperiodic response requires *mammalian-type cryptochrome* in the bean bug *Riptortus pedestris. Biochem. Biophys. Res. Com.* 410:394–397.

Ikeno, T., H. Numata, and S.G. Goto. 2011c. Circadian clock genes *period* and *cycle* regulate photoperiodic diapause in the bean bug *Riptortus pedestris* males. *J. Insect Physiol.* 57:935–938.

Ikeno, T., K. Ishikawa, H. Numata, and S.G. Goto. 2013. Circadian clock gene *Clock* is involved in the photoperiodic response of the bean bug *Riptortus pedestris. Physiol. Ent.* 38:157–162.

Ikeno, T., H. Numata, S.G. Goto, and S. Shiga. 2014. Involvement of the brain region containing pigment-dispersing factor-immunoreactive neurons in the photoperiodic response of the bean beetle, *Riptortus pedestris. J. Exp. Biol.* 217:453–462.

Imai, C. 2004. Photoperiodic induction and termination of summer diapause in adult *Epilachna admirabilis* (Coleoptera: Coccinellidae) from a warm temperate region. *Eur. J. Ent.* 101:523–529.

Imai, K., T. Konno, Y. Nakazawa, T. Komiya, M. Isobe, K. Koga, T. Goto, T. Yaginuma, K. Sakakibara, K. Hasegawa, and O. Yamashita. 1991. Isolation and structure of diapause hormone of the silkworm, *Bombyx mori. Proc. Jap. Acad. B.* 67:98–101.

Imhof, M.O., S. Rusconi, and M. Lezzi. 1993. Cloning of the *Chironomus tentans* cDNA encoding a protein (cEcRH) homologous to the *Drosophila melanogaster* ecdysteroid receptor (dEcR). *Insect Biochem. Mol. Biol.* 23:115–124.

Ingrisch, S. 1986. The plurennial life cycles of the European Tettigoniidae (Insecta: Orthoptera) 1. The effect of temperature on embryonic development and hatching. *Oecologia* 70:606–616.

Ingrisch, S. 1996. Evidence of an embryonic diapause in a tropical Phaneropterinae (Insecta Ensifera Tettigonioidae). *Trop. Zool.* 9:431–439.

Inoue, T. and J.H. Thomas. 2000. Targets of TGF-β signaling in *Caenorhabditis elegans* dauer formation. *Dev. Biol.* 217:192–204.

Irwin, J.T. and R.E. Lee, Jr. 2000. Mild winter temperatures reduce survival and potential fecundity of the goldenrod gall fly, *Eurosta solidaginis* (Diptera: Tephritidae). *J. Insect Physiol.* 46:655–661.

Irwin, J.T. and R.E. Lee, Jr. 2003. Cold winter microenvironments conserve energy and improve overwintering survival and potential fecundity of the goldenrod gall fly, *Eurosta solidaginis. Oikos* 100:71–78.

Isabel, G., L. Gordoux, and R. Moreau. 2001. Changes of biogenic amine levels in haemolymph during diapausing and non-diapausing status in *Pieris brassicae* L. *Comp. Biochem. Physiol. A* 128:117–127.

Ishihara, M. and M. Shimada. 1995. Trade-off in allocation of metabolic reserves: effects of diapause on egg production and adult longevity in a multivoltine bruchid, *Kytorhinus sharpianus. Funct. Ecol.* 9:618–624.

Ishihara, M. and T. Ohgushi. 2006. Reproductive inactivity and prolonged developmental time induced by seasonal decline in host plant quality in the willow leaf beetle *Plagiodera versicolora* (Coleoptera: Chrysomelidae). *Environ. Ent.* 35:524–530.

Ishikawa, Y., T. Yamashita, and M. Nomura. 2000. Characteristics of summer diapause in the onion maggot, *Delia antiqua* (Diptera: Anthomyiidae). *J. Insect Physiol.* 46:161–167.

Ishimaru, Y., S. Tomonari, Y. Matsuoka, T. Watanabe, K. Miyawaki, T. Bando, K. Tomioka, H. Ohuchi, S. Noji, and T. Mito. 2016. TGF-β signaling in insects regulates metamorphosis via juvenile hormone biosynthesis. *Proc. Nat'l. Acad. Sci., USA* 113:5634–5639.

Isobe, M., H. Kai, T.Kurahashi, S. Suwan, S. Pitchayawasin-Thapphasaraphong, T. Franz, N. Tani, K. Higashi, and H. Nishida. 2006. The molecular mechanism of the termination of insect diapause, part 1: A timer protein, TIME-EA4, in the diapause eggs of the silkworm *Bombyx mori* is a metallo-glycoprotein. *Chembiochem.* 7:1590–1598.

Ito, K. 2004. Deteriorating effects of diapause duration on postdiapause life history traits in the Kanzawa spider mite. *Physiol. Ent.* 29:453–457.

Ito, K. 2007. Negative genetic correlation between diapause duration and fecundity after diapause in a spider mite. *Ecol. Ent.* 32:643–650.

Ito, K. 2010. Effect of host plants on diapause induction in immature and adult *Tetranychus kanzawai* (Acari: Tetranychidae). *Exp. Appl. Acarol.* 52:11–17.

Ito, K., N. Yokoyama, Y. Kumekawa, H. Hayakawa, Y. Minamiya, K. Nakaishi, T. Fukuda, R. Arakawa, and Y. Saito. 2012. Effects of inbreeding on variation in diapause duration and early fecundity in the Kanzawa spider mite. *Ent. Exp. Appl.* 144:202–208.

Ito, K. 2014. Intra-population genetic variation in diapause incidence of adult-diapausing *Tetranychus pueraricola* (Acari: Tetranychidae). *Ecol. Ent.* 39:186–194.

Iwai, S., Y. Fukui, Y. Fujiwara, and M. Takada. 2006. Structure and expression of two circadian clock genes, *period* and *timeless* in the commercial silkmoth, *Bombyx mori. J. Insect Physiol.* 52:625–637.

Iwasaki, H., M. Takahashi, T. Niimi, O. Yamashita, and T. Yaginuma. 1997. Cloning of cDNAs encoding *Bombyx* homologues of Cdc2 and Cdc2-related kinase from eggs. *Insect Mol. Biol.* 6:131–141.

Iwata, K.-I., C. Shindome, Y. Kobayashi, M. Takeda, O. Yamashita, K. Shiomi, and Y. Fujiwara. 2005a. Temperature-dependent activation of ERK/MAPK in yolk cells and its role in embryonic diapause termination in the silkworm *Bombyx mori. J. Insect Physiol.* 51:1306–1312.

Iwata, K.-I., Y. Fujiwara, and M. Takeda. 2005b. Effects of temperature, sorbitol, alanine and diapause hormone on the embryonic development in *Bombyx mori*: in vitro tests of old hypotheses. *Physiol. Ent.* 30:317–323.

Izumi, Y., K. Anniwaer, H. Yoshieda, S. Sonoda, K. Fujisaki, and H. Tsumuki. 2005. Comparison of cold hardiness and sugar content between diapausing and non-diapausing pupae of the cotton bollworm, *Helicoverpa armigera* (Lepidoptera: Noctuidae). *Physiol. Ent.* 30:36–41.

Izumi, Y., S. Sonoda, and H.Tsumuki. 2007. Effects of diapause and cold-acclimation on the avoidance of freezing injury in fat body tissue of the rice stem borer, *Chilo suppressalis* Walker. *J. Insect Physiol.* 53:685–690.

Izzo, V.M., J. Armstrong, D. Hawthorne, and Y. Chen. 2014. Time of the season: the effect of host photoperiodism on diapause induction in an insect herbivore, *Leptinotarsa decemlineata. Ecol. Ent.* 39:75–82.

James, D.G. and T.A. James. 2019. Migration and overwintering in Australian monarch butterflies (*Danaus plexippus* (L.) (Lepidoptera: Nymphalidae): a review with new observations and research needs. *J. Lepidop. Soc.* 73:177–190.

Jansson, R.K., A.E. Zitzman, Jr., and J.H. Lashcomb. 1989. Effects of food plant and diapause on adult survival and fecundity of Colorado potato beetle (Coleoptera: Chrysomelidae). *Environ. Ent.* 18:291–297.

Janzen, D.H. 1984. Weather-related color polymorphism of *Rothschildia lebeau* (Saturniidae). *Bull. Ent. Soc. Am.* 30:16–20.

Janzen, D.H. and T.W. Schoener. 1968. Differences in insect abundance and diversity between wetter and drier sites during a tropical dry season. *Ecology* 49:96–110.

Janzen, D.H. and W. Hallwachs. 2021. To us insectometers, it is clear that insect decline in our Costa Rican tropics is real, so let's be kind to the survivors. *Proc. Nat'l. Acad. Sci., USA.* 118:e2002546117.

Jaquiery, J., S. Stoeckel, C. Larose, P. Nouhaud, C. Rispe, L. Mieuzet, J. Bonhomme, F. Maheo, F. Legeai, J.-P. Gauthier, N. Prunier-Leterme, D. Tagu, and J.-C. Simon. 2014. Genetic control of contagious asexuality in the pea aphid. *PLoS Genet.* 10:e10044838.

Jarwar, A.R., K. Hao, E.V. Bitume, H. Ullah, D. Cui, X. Nong, G. Wang, X. Tu, and Z. Zhang. 2019. Comparative transcriptomic analysis reveals molecular profiles of central nervous system in maternal diapause induction of *Locusta migratoria. G3* 119:400475.

Jeong, P.-Y., M.-S. Kwon, H.-J. Joo, and Y.-K. Paik. 2009. Molecular time-course and the metabolic basis of entry into dauer in *Caenorhabditis elegans. PLoS One* 5: e4162.

Jiang, X.F., S.H. Huang, L.Z. Luo, Y. Liu, and L. Zhang. 2010. Diapause termination, post-diapause development and reproduction in the beet webworm, *Loxostege sticticalis* (Lepidoptera: Pyralidae). *J. Insect Physiol.* 56:1325–1331.

Jiang, X.F., S.H. Huang, and L.Z. Luo. 2011. Juvenile hormone changes associated with diapause induction, maintenance, and termination in the beet webworm, *Loxostege sticticalis* (Lepidoptera: Pyralidae). *Arch. Insect Biochem. Physiol.* 77:134–144.

Jindra, M., S.R. Palli, and L.M. Riddiford. 2013. The juvenile hormone signaling pathway in insect development. *Ann. Rev. Ent.* 58:181–204.

Joanisse, D.R. and K.B. Storey. 1994. Mitochondrial enzymes during overwintering in two species of cold-hardy gall insects. *Insect Biochem. Mol. Biol.* 24:145–150.

Joanisse, D.R. and K.B. Storey. 1996. Fatty acid content and enzymes of fatty acid metabolism in overwintering cold-hardy gall insects. *Physiol. Zool.* 69:1079–1095.

Johansen, A.I., A. Exnerová, K.H. Svádová, P. Stys, G. Gamberale-Stille, and B.S. Tullberg. 2010. Adaptive change in protective coloration in adult striated shieldbugs *Graphosoma lineatum* (Heteroptera: Pentatomidae): test of detectability of two colour forms by avian predators. *Ecol. Ent.* 35:602–610.

Joplin, K.H. and D.L. Denlinger. 1989. Cycles of protein synthesis during pupal diapause in the flesh fly, *Sarcophaga crassipalpis*. *Arch. Insect Biochem. Physiol.* 12:111–122.

Joplin, K.H., G.D. Yocum, and D.L. Denlinger. 1990. Diapause specific proteins expressed by the brain during pupal diapause of the flesh fly, *Sarcophaga crassipalpis*. *J. Insect Physiol.* 36:775–783.

Joplin, K.H., D.L. Stetson, J.G. Diaz, and D.L. Denlinger. 1993. Cellular differences in ring glands of flesh fly pupae as a consequence of diapause programming. *Tissue Cell* 25:245–257.

Jordan, R.G. 1980. Embryonic diapause in three popluations of the western tree hole mosquito, *Aedes sierrensis*. *Ann. Ent. Soc. Am.* 73:357–359.

Joschinski, J. and D. Bonte. 2021. Diapause and bet-hedging in insects: a meta-analysis of reaction norm shapes. *Oikos* 130:1240–1250.

Joshi, A., D.L. Olson, and D.R. Carey. 2009. Overwintering survival of *Aphtona* beetles (Coleoptera: Chrysomelidae): a biological control agent of leafy spurge released in North Dakota. *Environ. Ent.* 38:1539–1545.

Jovanović-Galović, A., D.P. Blagojević, G. Grubor-Lajšić, M.R. Worland, and M.B. Spasić. 2007. Antioxidant defense in mitochondria during diapause and postdiapause development of European corn borer (*Ostrinia nubilalis*, Hubn.). *Arch. Insect Biochem. Physiol.* 64:111–119.

Jurenka, R.A., D. Holland, and E.S. Krafsur. 1998. Hydrocarbon profiles of diapausing and nondiapausing adult face flies (*Musca autumnalis*). *Arch. Insect Biochem. Physiol.* 37:206–214.

Kadener, S., J.S. Menet, K. Sugino, M.D. Horwich, U. Weissbein, P. Nawthean, V.V. Vagin, P.D. Zamore, S.B. Nelson, and M. Rosbash. 2009. A role for microRNAs in the *Drosophila* circadian clock. *Genes Devel.* 23:2179–2191.

Kageyama, T. and E. Ohnishi. 1971. Carbohydrate metabolism in the eggs of the silkworm *Bombyx mori* II. Anaerobiosis and polyol formation. *Dev. Growth Differ.* 15:47–55.

Kai, H., T. Kawai, and Y. Kawai. 1987. A time-interval activation of esterase A4 by cold. *Insect Biochem.* 17:367–372.

Kai, H., Y. Kotani, Y. Miao, and M. Azuma. 1995. Time interval measuring enzyme for resumption of embryonic development in the silkworm, *Bombyx mori. J. Insect Physiol.* 41:905–910.

Kalberer, N.M., T.C.J. Turlings, and M. Rahier. 2005. An alternative hibernation strategy involving sun-exposed "hotspots," dispersal by flight, and host plant finding by olfaction in an alpine leaf beetle. *Ent. Exp. Appl.* 114:189–196.

Kaltenpoth, M., W. Göttler, G. Herzner, and E. Strohm. 2005. Symbiotic bacteria protect wasp larvae from fungal infestations. *Curr. Biol.* 15:475–479.

Kalushkov, P., M. Hodkova, O. Nedvêd, and I. Hodek. 2001. Effect of diapause intensity in *Pyrrhocoris apterus* (Heteroptera Pyrrhocoridae). *J. Insect Physiol.* 47:55–61.

Kamei, Y., Y. Hasegawa, T. Niimi, O. Yamashita, and T. Yaginuma. 2011. *trehalase-2* protein contributes to trehalase activity enhanced by diapause hormone in developing ovaries of the silkworm, *Bombyx mori. J. Insect Physiol.* 57:6C–613.

Kamimura, M., H. Saito, R. Niwa, T. Niimi, K. Toyoda, C. Ueno, Y. Kanamori, S. Shimura, and M. Kiuchi. 2012. Fungal ecdysteroid-22-oxidase, a new tool for manipulating ecdysteroid signaling and insect development. *J. Biol. Chem.* 287:16488–16498.

Kandel, E.R. 2001. The molecular biology of memory storage: a dialogue between genes and synapses. *Science* 294:1030–1038.

Kandel, E.R., Y. Dudai, and M.R. Mayford. 2014. The molecular and systems biology of memory. *Cell* 157:163–185.

Kang, D.S., D.L. Denlinger, and C. Sim. 2014. Suppression of allatotropin simulates reproductive diapause in the mosquito *Culex pipiens. J. Insect Physiol.* 64:48–53.

Kang, D.S., M.A. Cotton, D.L. Denlinger, and C. Sim. 2016. Comparative transcriptomics reveals key gene expression differences between diapausing and non-diapausing adults of *Culex pipiens. PLoS One* 11: e0154892.

Kankare, M., T. Salminen, A. Laiho, L. Vesala, and A. Hoikkala. 2010. Changes in gene expression linked with adult reproductive diapause in a northern malt fly species: a candidate gene microarray study. *BMC Ecology* 10:3.

Kankare, M., T.S. Salminen, H. Lampinen, and A. Hoikkala. 2012. Sequence variation in *couch potato* and its effects on life-history traits in a northern malt fly,

Drosophila montana. J. Insect Physiol. 58:256–264.

Karl, I., K. Hoffmann, and K. Fischer. 2010. Cuticular melanisation and immune response in a butterfly: local adaptation and lack of correlation. *Ecol. Ent.* 35:523–528.

Karlsson, B. 2014. Exended season for northern butterflies. *Int. J. Biometeorol.* 58:691–701.

Karlsson, B. and A. Johansson. 2008. Seasonal polyphenism and developmental trade-offs between flight ability and egg laying in a pierid butterfly. *Proc. R. Soc. B* 275:2131–2136.

Kashima, T., T. Nakamura, and S.J. Tojo. 2006. Uric acid recycling in the shield bug, *Parastrachia japonensis* (Hemiptera: Parastrachiidae), during diapause. *J. Insect Physiol.* 52:816–825.

Katsoyannos, P., D.C. Kontodimas, and G. Stathas. 2005. Summer diapause and winter quiescence of *Hippodamia* (*Semiadalia*) *undecimnotata* (Coleoptera: Coccinellidae) in central Greece. *Eur. J. Ent.* 102:453–457.

Katz, D.J., T.M. Edwards, V. Reinke, and W.G. Kelly. 2009. A *C. elegans* LSD1 demethylase contributes to germline immortality by reprogramming epigenetic memory. *Cell* 137:308–320.

Kauranen, H., V. Tyukmaeva, and A. Hoikkala. 2013. Involvement of circadian oscillation(s) in the photoperiodic time measurement and the induction of reproductive diapause in a northern *Drosophila* species. *J. Insect Physiol.* 59:662–666.

Kauranen, H., J. Kinnunen, D. Hopkins, and A. Hoikkala. 2019a. Direct and correlated responses to bi-directional selection on pre-adult development time in *Drosophila montana. J. Insect Physiol.* 116:77–89.

Kauranen, H., J. Kinnunen, A.-L. Hiilos, P. Lankinen, D. Hopkins, R.A.W. Wiberg, M.G. Ritchie, and A. Hoikkala. 2019b. Selection for reproduction under short photoperiods changes diapause-associated traits and induces widespread genomic divergence. *J. Exp. Biol.* 222:205831.

Kawaguchi, H. and T. Niimi. 2018. A method for cryopreservation of ovaries of the ladybird beetle, *Harmonia axyridis*. *J. Insect Biotech. Ser.* 87:35–44.

Kawaguchi, S., Y. Manabe, T. Sugawara, and M. Osakabe. 2016. Imaginal feeding for progression of diapause phenotype in the two-spotted spider mite (Acari: Tetranychidae). *Environ. Ent.* 45:1568–1573.

Kawakami, Y., K. Yamazaki, and K. Ohashi. 2017. Protogyny after hibernation and aestivation in *Cheilomenes sexmaculata* (Coleoptera: Coccinellidae) in central Japan. *Eur. J. Ent.* 114:275–278.

Kefuss, J.A. 1978. Influence of photoperiod on the behavior and brood-rearing activities of honeybees in a flight room. *J. Apicult. Res.* 17:137–151.

Kelty, J.D. and R.E. Lee, Jr. 2000. Diapausing pupae of the flesh fly *Sarcophaga crassipalpis* (Diptera: Sarcophagidae) are more resistant to inoculative freezing than non-diapausing pupae. *Physiol. Ent.* 25:120–126.

Kemp, D.J. and R.E. Jones. 2001. Phenotypic plasticity in field populations of the tropical butterfly *Hypolimnas bolina* (L.) (Nymphalidae). *Biol. J. Linnean Soc.* 72:33–45.

Kfir, R. 1991. Effect of diapause on development and reproduction of the stem borers *Busseola fusca* (Lepidoptera: Noctuidae) and *Chilo partellus* (Lepidoptera: Pyralidae). *J. Econ. Ent.* 84:1677–1680.

Khan, M.A. 1988. Brain-controlled synthesis of juvenile hormone in adult insects. *Ent. Exp. Appl.* 46:3–17.

Kidokoro, K. and Y. Ando. 2006. Effect of anoxia on diapause termination in eggs of the false melon beetle, *Atrachya menetriesi*. *J. Insect Physiol.* 52:87–93.

Kidokoro, K., K. Iwata, Y. Fujiwara, and M. Takeda. 2006a. Effects of juvenile hormone analogs and 20-hydroxyecdysone on diapause termination in eggs of *Locusta*

migratoria and *Oxya yezoensis*. *J. Insect Physiol.* 52:473–479.

Kidokoro, K., K. Iwata, M. Takeda, and Y. Fujiwara. 2006b. Involvement of ERK/MAPK in regulation of diapause intensity in the false melon beetle, *Atrachya menetriesi*. *J. Insect Physiol.* 52:1189–1193.

Kikawada, T., Y. Nakahara, Y. Kanmori, K.I. Iwata, M. Watanabe, B. McGee, A. Tunnacliffe, and T. Okuda. 2006. Dehydration-induced expression of LEA proteins in an anhydrobiotic chironomid. *Biochem. Biophy. Res. Com.* 348:56–61.

Kikukawa, S. and K. Ohde. 2007. The role of the main photophase on dark-time measurement used for diapause determination in the Indian meal moth, *Plodia interpunctella*. *Physiol. Ent.* 32:351–356.

Kikukawa, S., T. Minamizuka, and W. Matoba. 2008. Responses to stepwise photoperiodic changes for the larval diapause of the Indian meal moth *Plodia interpunctella*. *Physiol. Ent.* 33:360–364.

Kikukawa, S., Y. Arakawa, K. Hayakawa, M. Hayashi, K. Katou, J. Kaneshige, M. Kimura, T. Nakamura, Y. Nakamura, and H. Watanabe. 2009. Effects of skeleton photoperiods on the induction of larval diapause in the Indian meal moth *Plodia interpunctella*. *Physiol. Ent.* 34:180–184.

Kim, K.S., B.W. French, D.V. Sumerford, and T.W. Sappington. 2007. Genetic diversity in laboratory colonies of western corn rootworm (Coleoptera: Chrysomelidae), including a nondiapause colony. *Environ. Ent.* 36:637–645.

Kim, M., R.M. Robich, J.P. Rinehart, and D.L. Denlinger. 2006. Upregulation of two actin genes and redistribution of actin during diapause and cold stress in the northern house mosquito, *Culex pipiens*. *J. Insect Physiol.* 52:1226–1233.

Kim, M. and D.L. Denlinger. 2009. Decrease in expression of *beta-tubulin* and microtubule abundance in flight muscles during diapause in adults of *Culex pipiens*. *Insect Mol. Biol.* 18:295–302.

Kim, M. and D.L. Denlinger. 2010. A potential role for ribosomal protein S2 in the gene network regulating reproductive diapause in the mosquito *Culex pipiens*. *J. Comp. Physiol. B* 180:171–178.

Kim, M., C. Sim, and D.L. Denlinger. 2010. RNA interference directed against ribosomal protein S3a suggests a link between this gene and arrested ovarian development during adult diapause in *Culex pipiens*. *Insect Mol. Biol.* 19:27–33.

Kim, S.K. 2007. Proteins that promote long life. *Science* 317:603–605.

Kimura, K.D., H.A. Tissenbaum, Y. Liu, and G. Ruvkun. 1997. *daf-2,* an insulin receptor-like gene that regulates longevity and diapause in *Caenorhabditis elegans*. *Science* 277:942–946.

King, A.M. and T.H. MacRae. 2015. Insect heat shock proteins during stress and diapause. *Ann. Rev. Ent.* 60:59–75.

Kipyatkov, V.E. 1995. Role of endogenous rhythms in regulation of annual cycles of development in ants (Hymenoptera, Formicidae). *Ent. Rev.* 74:1–15.

Kipyatkov, V.E. 2006. The evolution of seasonal cycles in cold-temperate and boreal ants: patterns and constraints. In *Life Cycles in Social Insects: Behaviour, Ecology and Evolution*, ed. V.E. Kipyatkov, St. Petersburg: St. Petersburg University Press, pp. 63–84.

Kipyatkov, V.E. and E.B. Lopatina. 1999. Social regulation of larval diapause induction and termination by the worker ants of three species of the genus *Myrmica* Latreille (Hymenoptera, Formicidae). *Ent. Rev.* 79:1138–1144.

Kistenpfennig, C., M. Nakayama, R. Nihara, K. Tomioka, C. Helfrich-Förster, and T. Yoshii. 2018. A tug-of-war between cryptochrome and the visual system allows the adaptation of evening activity to long photoperiods in *Drosophila melanogaster*. *J. Biol. Rhy.* 33:24–34.

Kitching, R.L. and M.P. Zalucki. 1981. Observations on the ecology of *Euploea core corinna* (Nymphalidae) with special reference to an overwintering population. *J. Lep. Soc.* 35:106–119.

Knapp, M. and M. Rericha. 2020. Effects of the winter temperature regime on survival, body mass loss and post-winter starvation resistance in laboratory-reared and field-collected ladybirds. *Sci. Rep.* 10:4970.

Kobelkova, A., S.G. Goto, J.T. Peyton, T. Ikeno, R.E. Lee, Jr., and D.L. Denlinger. 2015. Continuous activity and no cycling of clock genes in the Antarctic midge during the polar summer. *J. Insect Physiol.* 81:90–96.

Koch, P.B. and D. Bückmann. 1987. Hormonal control of seasonal morphs by the timing of ecdysteroid release in *Araschnia levana* L. (Nymphalidae: Lepidoptera). *J. Insect Physiol.* 33:823–829.

Koch, R.L. 2003. The multicolored Asian lady beetle, *Harmonia axyridis*: a review of its biology, uses in biological control, and non-target impacts. *J. Insect Sci.* 3:32.

Koenraadt, C.J.M., T.W.R. Mohlmann, N.O. Verhulst, J. Spitzen, and C.B.F. Vogels. 2019. Effect of overwintering on survival and vector competence of the West Nile virus vector *Culex pipiens*. *Parasit. Vect.* 12:147.

Koga, M., H. Ushirogawa, and K. Tomioka. 2005. Photoperiodic modulation of circadian rhythms in the cricket *Gryllus bimaculatus*. *J. Insect Physiol.* 51:681–690.

Kogure, M. 1933. The influence of light and temperature on certain characteristics os the silkworm, *Bombyx mori*. *J. Dept. Agric., Kyushu Univ.* 4:1–93.

Komata, S. and T. Sota. 2017. Seasonal polyphenism in body size and juvenile development of the swallowtail butterfly *Papilio xuthus* (Lepidoptera: Papilionidae). *Eur. J. Ent.* 114:365–371.

Kontogiannatos, D., T. Gkouvitsas, and A. Kourti. 2016. The expression patterns of the clock genes *period* and *timeless* are affected by photoperiod in the

Mediterranean corn stalk borer, *Sesamia nonagrioides*. *Arch. Insect Biochem. Physiol.* 94:e21366.

Kontogiannatos, D., T. Gkouvitsas, and A. Kourti. 2017. The expression of clock gene *cycle* has rhythmic pattern and is affected by photoperiod in the moth *Sesamia nonagrioides*. *Comp. Biochem. Physiol. B* 208–209:1–6.

Koopmanschap, A.B., J.H.M. Lammers, and C.A.D. de Kort. 1995. The structure of the gene encoding diapause hormone protein 1 of the Colorado potato beetle (*Leptinotarsa decemlineata*). *J. Insect Physiol.* 41:509–518.

de Kort, C.A.D. 1969. Hormones and the structural and biochemical properties of the flight muscles in the Colorado beetle. *Meded. Landb. Hogesch. Wageningen* 69:1–63.

de Kort, C.A.D. 1990. Thirty-five years of diapause research with the Colorado potato beetle. *Ent. Exp. Appl.* 56:1–13.

de Kort, C.A.D. and A.B. Koopmanschap. 1994. Nucleotide and deduced amino acid sequence of a cDNA clone encoding Diapause Protein-1, an arylphorin-type storage hexamer of the Colorado potato beetle. *J. Insect Physiol.* 40:527–535.

Koštál, V. 2006. Eco-physiological phases of insect diapause. *J. Insect Physiol.* 52:113–127.

Koštál, V. 2010. Cell structural modifications in insects at low temperatures. In *Low Temperature Biology of Insects*, ed. D.L. Denlinger and R.E. Lee, Jr., Cambridge: Cambridge University Press, pp. 116–140.

Koštál, V. 2011. Insect photoperiodic calendar and circadian clock: independence, cooperation, or unity? *J. Insect Physiol.* 57:538–556.

Koštál, V. and I. Hodek. 1997. Photoperiodism and control of summer diapause in the Mediterranean tiger moth *Cymbalophora pudica*. *J. Insect Physiol.* 43:767–777.

Koštál, V., H. Noguchi, K. Shimada, and Y. Hayakawa. 1998. Developmental changes in dopamine levels in larvae of the fly *Chymomyza costata:* comparison between widl-type and mutant-nondiapause strains. *J. Insect Physiol.* 44:605–614.

Koštál, V., K. Shimada, and Y. Hayakawa. 2000a. Induction and development of winter larval diapause in a drosophilid fly, *Chymomyza costata*. *J. Insect Physiol.* 46:417–428.

Koštál, V., H. Noguchi, K. Shimada, and Y. Hayakawa. 2000b. Circadian component influences the photoperiodic induction of diapause in a drosophilid fly, *Chymomyza costata*. *J. Insect Physiol.* 46:887–896.

Koštál, V. and K. Shimada. 2001. Malfunction of circadian clock in the *non-photoperiodic-diap*ause mutants of the drosophilid fly, *Chymomyza costata*. *J. Insect Physiol.* 47:1269–1274.

Koštál, V., J. Havelka, and P. Šimek. 2001. Low-temperature storage and cold hardiness in two populations of the predatory midge *Aphidoletes aphidimyza*, differing in diapause intensity. *Physiol. Ent.* 26:320–328.

Koštál, V., P. Berkova, and P. Šimek. 2003. Remodeling of membrane phospholipids during the transition to diapause and cold-acclimation in the larvae of *Chymomyza costata* (Drosophilidae). *Comp. Biochem. Physiol. B* 135:407–419.

Koštál, V., M. Tamura, M. Tollarová, and H. Zahradníčková. 2004. Enzymatic capacity for accumulation of polyol cryoprotectants changes during diapause in the adult red firebug, *Pyrrhocoris apterus*. *Physiol. Ent.* 29:344–355.

Koštál, V., M. Tollarová, and D. Doležel. 2008. Dynamism in physiology and gene transcription during reproductive diapause in a heteropteran bug, *Pyrrhocoris apterus*. *J. Insect Physiol.* 54:77–88.

Koštál, V., R. Závodská, and D.L. Denlinger. 2009. Clock genes *period* and *timeless* are rhythmically expressed in brains of newly hatched, photosensitive larvae of the fly, *Sarcophaga crassipalpis*. *J. Insect Physiol.* 55:408 414.

Koštál, V., H. Zahradníčková, and P. Šimek. 2011. Hyperprolinemic larvae of the drosophilid fly, *Chymomyza costata*, survive cryopreservation in liquid nitrogen. *Proc. Nat'l Acad. Sci., USA* 108:13041–13046.

Koštál, V., P. Šimek, H. Zahradníčková, J. Cimlová, and T. Štetina. 2012. Conversion of the chill susceptible fruit fly larva (*Drosophila melanogaster*) to a freeze tolerant organism. *Proc. Nat'l. Acad. Sci., USA* 109:3270–3274.

Koštál, V., T. Urban, L. Rimnáčová, P. Berková, and P. Šimek. 2013. Seasonal changes in minor membrane phospholipid classes, sterols and tocopherols in overwintering insect, *Pyrrhocoris apterus*. *J. Insect Physiol.* 59:934–941.

Koštál, V., B. Miklas, P. Dolezal, J. Rozsypal, and H. Zahradníčková. 2014. Physiology of cold tolerance in the bark beetle, *Pityogenes chalcographus* and its overwintering in spruce stands. *J. Insect Physiol.* 63:62–70.

Koštál, V., J. Korbelová, R. Poupardin, M. Moos, and P. Šimek. 2016a. Arginine and proline applied as food additives stimulate high freeze tolerance in larvae of *Drosophila melanogaster*. *J. Exp. Biol.* 219:2358–2367.

Koštál, V., M. Mallaei, and K. Schottner. 2016b. Diapause induction as an interplay between seasonal token stimuli, and modifying and directly limiting factors: hibernation in *Chymomyza costata*. *Physiol. Ent.* 41:344–357.

Koštál, V., T. Stetina, R. Poupardin, J. Korbelová, and A.W. Bruce. 2017. Conceptual framework of the ecophysiological phases of insect diapause development justified by transcriptomic profiling. *Proc. Nat'l. Acad. Sci., USA* 114:8532–8537.

Kotaki, T., T. Shinada, K. Kaihara, Y. Ohfune, and H. Numata. 2011. Biological activities of juvenile hormone III skipped bisepoxide in last instar nymphs and adults of a stink bug, *Plautia stali*. *J. Insect Physiol.* 57:147–152.

Kotwica-Rolinska, J., L. Pivarciova, H. Vaneckova, and D. Doležel. 2017. The role of circadian clock genes in the photoperiodic timer of the linden bug *Pyrrhocoris apterus* during the nymphal stage. *Physiol. Ent.* 42:266–273.

Kouno, T., M. Mizuguchi, H. Tanaka, P. Yang, Y. Mori, H. Shinoda, K. Unoki, T. Aizawa, M. Demura, K. Suzuki, and K. Kawano. 2007. The structure of a novel insect peptide explains it Ca^{2+} channel blocking and antifungal activities. *Biochemistry* 46:13733–13741.

Kourti, A. and D. Kontogiannatos. 2018. The *cryptochrome1* (*cry1*) gene has oscillating expression under short and long photoperiods in *Sesamia nonagrioides*. *Int'l. J. Mol. Theor. Physics* 2:1–9.

Kovacs, J.L. and M.A.D. Goodisman. 2012. Effects of size, shape, genotype, and mating status on queen overwintering survival in the social wasp *Vespula maculifrons*. *Environ. Ent.* 41:1612–1620.

Kozak, G.M., C.B. Wadsworth, S.C. Kahne, S.M. Bogdanowicz, R.G. Harrison, B.S. Coates, and E.B. Dopman. 2019. Genomic basis of circannual rhythm in the European corn borer moth. *Curr. Biol.* 29:2019.08.053.

Kozuki, Y. and M. Takeda. 2004. Split life cycle and differentiations in diapause characteristics in three host-habitat strains of *Atrophaneura alcinous* (Lepidoptera: Papilionidae). *Jpn. J. Environ. Ent. Zool.* 15:157–168.

Krajacich, B.J., M. Sullivan, R. Faiman, L. Veru, L. Graber, and T. Lehmann. 2020. Induction of long-lived potential aestivation states in laboratory *An. gambiae* mosquitoes. *Parasit. Vect.* 13:412.

Krishnan, N., D. Kodrik, F. Turanli, and F. Sehnal. 2007. Stage-specific distribution of oxidative radicals and antioxidant enzymes in the midgut of *Leptinotarsa decemlineata*. *J. Insect Physiol.* 53:67–74.

Kroon, A., R.L. Veenendaal, J. Bruin, M. Egas, and M.W. Sabelis. 2008. "Sleeping with the enemy" – predator-induced diapause in a mite. *Naturwissenschaften* 95:1195–1198.

Krysan, J.L. 1990. Fenoxycarb and diapause: a possible method of control for pear psylla (Homoptera; Psyllidae). *J. Econ. Ent.* 83:293–299.

Krysan, J.L., D.E. Foster, T.F. Branson, K.R. Ostlie, and W.S. Cranshaw. 1986. Two years before the hatch: rootworms adapt to crop rotation. *Bull. Ent. Soc. Am.* 32:250–253.

Kuang, X.-J., J. Xu, Q. Xia, H.-M. He, and F. Xue. 2011. Inheritance of the photoperiodic response controlling imaginal summer diapause in the cabbage beetle, Colaphellus bowringi. *J. Insect Physiol.* 57:614–619.

Kubrak, O.I., L. Kučerová, U. Theopold, and D.R. Nässel. 2014. The sleeping beauty: how reproductive diapause affects hormone signaling, metabolism, immune response and somatic maintenance in *Drosophila melanogaster*. *PLoS One* 9:e113051.

Kubrak, O.I., L. Kučerová, U. Theopold, S. Nylin, and D.R. Nässel. 2016. Characterization of reproductive dormancy in male *Drosophila melanogaster*. *Front. Physiol.* 7:572.

Kučerová, L., O.I. Kubrak, J.M. Bengtsson, H. Strnad, S. Nylin, U. Theopold, and D.R. Nässel. 2016. Slowed aging during reproductive dormancy is reflected in genome-wide transcriptome changes in *Drosophila melanogaster*. *BMC Genomics* 17:50.

Kuczer, M., E. Czarniewska, A. Majewska, M. Różanowska, G. Rosiński, and M. Lisowski. 2016. Novel analogs of alloferon: synthesis, conformational studies, pro-apoptotic and antiviral activity. *Bioorg. Chem.* 66:12–20.

Kukal, O., D.L. Denlinger, and R.E. Lee, Jr. 1991. Developmental and metabolic changes induced by anoxia in diapausing and non-diapausing flesh fly pupae. *J. Comp. Physiol. B* 160:683–689.

Kurihara, M. and Y. Ando. 1969. The effect of mercury compounds on breaking of

diapause in the eggs of the false melon beetle, *Atrachya menestriesi* Falderman (Coleoptera: Chrysomelidae). *Appl. Ent. Zool.* 4:149–151.

Kurota, H. and M. Shimada. 2001. Photoperiod- and temperature-dependent induction of larval diapause in a multivoltine bruchid, *Bruchidius dorsalis*. *Ent. Exp. Appl.* 107:11–18.

Kurota, H. and M. Shimada. 2003a. Geographic variation in photoperiodic induction of larval diapause in the bruchid beetle, *Bruchidius dorsalis*: polymorphism in overwintering stages. *Ent. Exp. Appl.* 107:11–18.

Kurota, H. and M. Shimada. 2003b. Photoperiod-dependent adult diapause within a geographical cline in the multivoltine bruchid *Bruchidius dorsalis*. *Ent. Exp. Appl.* 107:177–185.

Kurota, H. and M. Shimada. 2007. Overwintering stage polymorphism of a bruchine beetle: geographic variation in optimal diapause strategy. *Ecol. Ent.* 32:722–731.

Kutaragi, Y., A. Tokuoka, Y. Tomiyama, M. Nose, T. Watanabe, T. Bando, Y. Moriyama, and K. Tomioka. 2018. A novel photic entrainment mechanism for the circadian clock in an insect: involvement of *c-fos* and cryptochromes. *Zool. Letters* 4:26.

Kutcherov, D., A. Saulich, E. Lopatina, and M. Ryzhkova. 2015. Stable and variable life-history responses to temperature and photoperiod in the beet webworm, *Loxostege sticticalis*. *Ent. Exp. Appl.* 154:228–241.

Kutcherov, D., E.B. Lopatina, and S. Balashov. 2018. Convergent photoperiodic plasticity of developmental rate in two species of insects insect with widely different thermal phenotypes. *Eur. J. Ent.* 115:624–631.

Kuwano, E., T.Fujisawa, K. Suzuki, and M. Eto. 1991. Termination of egg diapause by imidazoles in the silkmoth,

Antheraea yamamai. Agric. Biol. Chem. 55:1185–1186.

Kyriacou, C.P., A.A. Peixoto, F. Sandrelli, R. Costa, and E. Tauber. 2008. Clines in clock genes: fine-tuning circadian rhythms to the environment. *Trends Genet.* 24:124–132.

Labrie, G., D. Coderre, and E. Lucas. 2008. Overwintering strategy of multicolored Asian lady beetle (Coleoptera: Coccinellidae): cold-free space as a factor of invasive success. *Ann. Ent. Soc. Am.* 101:860–866.

Lacour, G., F. Vernichon, N. Cadilhac, S. Boyer, C. Lagneau, and T. Hance. 2014. When mothers anticipate: effects of the pre-diapause stage on embryo development and of maternal photoperiod on eggs of a temperate and a tropical strains of *Ades albopictus* (Diptera: Culicidae). *J. Insect Physiol.* 71:87–96.

Lagueux, M., P. Harry, and J.A. Hoffman. 1981. Ecdysteroids are bound to vitellin in newly laid eggs of *Locusta. Mol. Cell. Endocrin.* 24:325–338.

Lai, X.-T., D. Yang, S.-H. Wu, X.-F. Zhu, and F.-S. Xue. 2008. Diapause incidence of progeny in relation to parental geographic origin, host plant and rearing density in the cabbage beetle, *Colaphellus bowringi. Ent. Exp. Appl.* 129:117–123.

Lam, W.-K. and L.P. Pedigo. 2000a. Cold tolerance of overwintering bean leaf beetles (Coleoptera: Chrysomelidae). *Environ. Ent.* 29:157–163.

Lam, W.-K. and L.P. Pedigo. 2000b. A predictive model for survival of overwintering bean leaf beetles (Coleoptera: Chrysomelidae). *Environ. Ent.* 29:800–806.

Lambhod, C., A. Pathak, A.K. Munjai, and R. Parkash. 2017. Tropical *Drosophila ananassae* of wet-dry seasons shows cross resistance to heat, drought and starvation. *Biol. Open* 6:1698–1706.

Lankinen, P. 1986. Geographical variation in circadian eclosion rhythm and photoperiodic adult diapause in *Drosophila littoralis. J. Comp. Physiol. A* 159:123–142.

Lankinen, P. and P. Forsman. 2006. Independence of genetic geographical variation between photoperiodic diapause, circadian eclosion rhythm, and the Thr-GLY repeat region of the *period* gene in *Drosophila littoralis. J. Biol. Rhyth.* 21:3–12.

Lankinen, P., V.I. Tyukmaeva, and A. Hoikkala. 2013. Northern *Drosophila montana* flies show variation both within and between cline populations in the critical day length evoking reproductive diapause. *J. Insect Physiol.* 59:745–751.

Larson, E.L., R.M. Tinghitella, and S.A. Taylor. 2019. Insect hybridization and climate change. *Front. Ecol. Evol.* 7:348.

Le Berre, J.-R. 1953. Contribution a l'etude biologique du criquet migrateur des lands (*Locusta migratoria gallica* Remaudiere). *Bull. Biol.* 87:227–273.

Le Hesran, S., T. Groot, M. Knapp, J.E. Nugroho, G. Beretta, L.F. Salomé-Abarca, Y.H. Choi, M. Vancová, A.M. Moreno-Rodenas, and M. Dicke. 2019. Proximate mechanisms of drought resistance in *Phytoseiulus persimilis* eggs. *Exp. Appl. Acar.* 79:279–298.

Le Trionnaire, G., J. Hardie, S. Jaubert, J.-C. Simon, and D. Tagu. 2008. Shifting from clonal to sexual reproduction in aphids: physiological and developmental aspects. *Biol. Cell* 100:441–451.

Le Trionnaire, G., F. Francis, S. Jaubert, J. Bonhomme, E. De Pauw, J.-P. Gauthier, E. Haubruge, F. Legeai, N. Prunier-Leterme, J.-C. Simon, S. Tanguy, and D. Tagu. 2009. Transcriptomic and proteomic analyses of seasonal photoperiodism in the pea aphid. *BMC Genomics* 10:1–14.

Le Trionnaire, G., S. Jaubert-Possamai, J. Bonhomme, J.-P. Gauthier, G. Guernec, A. Le Cam, F. Legeai, J. Monfort, and D. Tagu. 2012. Transcriptomic profiling of the reproductive mode switch in the pea aphid in response to natural autumnal photoperiod. *J. Insect Physiol.* 58:1517–1524.

Leal, L., V. Talla, T. Kallman, M. Friberg, C. Wiklund, V. Dinca, R. Vila, and

N. Backstrom. 2018. Gene expression profiling across ontogenetic stages in the wood white (*Leptidea sinapis*) reveals pathways linked to butterfly diapause regulation. *Mol. Ecol.* 2018:1–14.

Lebenzon, J.E., L. Mohammad, K.E. Mathers, K.F. Turnbull, J.F. Staples, and B.J. Sinclair. 2020. Burning down the powerhouse: does mitophagy drive metabolic suppression during diapause in the Colorado potato beetle (*Leptinotarsa decemlineata*)? *Integ. Comp. Biol.* 59: E357.

Lee, D.-H., J.P. Cullum, J.L. Anderson, J.L. Daugherty, L.M. Beckett, and T.C. Leskey. 2014. Characterization of overwintering sites of the invasive brown marmorated stink bug in natural landscapes using human surveyors and detector canines. *PLoS One* 9:e91575.

Lee, K.-Y. and D.L. Denlinger. 1996. Diapause-regulated proteins in the gut of pharate first instar larvae of the gypsy moth, *Lymantria dispar*, and the effects of KK-42 and neck ligation on expression. *J. Insect Physiol.* 42:423–431.

Lee, K.-Y. and D.L. Denlinger. 1997. A role for ecdysteroids in the induction and maintenance of the pharate first instar diapause of the gypsy moth, *Lymantria dispar. J. Insect Physiol.* 43:289–296.

Lee, K.-Y., A.P. Valaitis, and D.L. Denlinger. 1997. Further evidence that a diapause in the gypsy moth, *Lymantria dispar*, is regulated by ecdysteroids: a comparison of diapause and nondiapause strains. *J. Insect Physiol.* 43:897–903.

Lee, K.-Y., S. Hiremath, and D.L. Denlinger. 1998. Expression of actin in the central nervous system is switched off during diapause in the gypsy moth, *Lymantria dispar. J. Insect Physiol.* 44:221–226.

Lee, K.-Y., F.M. Horodyski, A.P. Valaitis, and D.L. Denlinger. 2002. Molecular characterization of the insect immune protein hemolin and its high induction during embryonic diapause in the gypsy moth, *Lymantria dispar. Insect Biochem. Mol. Biol.* 32:1457–1467.

Lee, R.E., Jr. 2010. A primer on insect cold-tolerance. In *Low Temperature Biology of Insects*, ed. D.L. Denlinger and R.E. Lee, Jr., Cambridge: Cambridge University Press, pp. 3–34.

Lee, R.E., Jr., C.-P. Chen, and D.L. Denlinger. 1987a. A rapid cold-hardening process in insects. *Science* 238:1415–1417.

Lee, R.E., Jr., C.-P. Chen, M.H. Meacham, and D.L. Denlinger. 1987b. Ontogenetic patterns of cold-hardiness and glycerol production in *Sarcophaga crassipalpis. J. Insect Physiol.* 33:587–592.

Lee, R.E., Jr. and D.L. Denlinger. 2010. Rapid cold hardening: ecological significance and underpinning mechanisms. In *Low Temperature Biology of Insects*, ed. D.L. Denlinger and R.E. Lee, Jr., Cambridge: Cambridge University Press, pp. 35–58.

Lee, R.E., Jr. and D.L. Denlinger. 2015. Stress tolerance in a polyextremophile: the southernmost insect. *Can. J. Zool.* 93:679–685.

Lee, S.F., C.M. Sgro, J. Shirriffs, C.W. Wee, L. Rako, B. van Heerwaarden, and A.A. Hoffmann. 2011. Polymorphism in the *couch potato* gene clines in eastern Australia but is not associated with ovarian dormancy in *Drosophila melanogaster. Mol. Ecol.* 20:2973–2984.

Lees, A.D. 1955. *The Physiology of Diapause in Arthropods*. Cambridge: Cambridge University Press.

Lees, A.D. 1964. The location of the photoperiodic receptors in the aphid *Megoura viciae* Buckton. *J. Exp. Biol.* 41:119–133.

Lees, A.D. 1966. The control of polymorphism in aphids. *Adv. Insect Physiol.* 3:207–277.

Lees, A.D. 1973. Photoperiodic time measurement in the aphid *Megoura viciae. J. Insect Physiol.* 19:2279–2316.

Lees, A.D. 1981. Action spectrum for the photoperiodic control of polymorphism in the aphid *Megoura viviae*. *J. Insect Physiol.* 27:761–771.

Lefevere, K.S. 1989. Endocrine control of diapause termination in the adult female Colorado potato beetle, *Leptinotarsa decemlineata*. *J. Insect Physiol.* 35:197–203.

Lefevere, K.S. and C.A.D. de Kort. 1989. Adult diapause in the Colorado potato beetle, *Leptinotarsa decemlineata*: effects of external factors on maintenance, termination and post-diapause development. *Physiol. Ent.* 14:299–308.

Lefevere, K.S., A.B. Koopmanschap, and C.A.D. de Kort. 1989. Changes in the concentrations of metabolites in hemolymph during and after diapause in female Colorado potato beetle *Leptinotarsa decemlineata*. *J. Insect Physiol.* 35:121–128.

Legett, H.D., V.A. Baranov, and X.E. Bernal. 2018. Seasonal variation in abundance and diversity of eavesdropping frog-biting midges (Diptera, Corethrellidae) in a neotropical rainforest. *Ecol. Ent.* 43:226–233.

Lehmann, P., S. Piiroinen, M. Kankare, A. Lyytinen, M. Paljakka, and L. Lindstrom. 2014a. Photoperiodic effects on diapause-associated gene expression trajectories in European *Leptinotarsa decemlineata* populations. *Insect Mol. Biol.* 23:566–578.

Lehmann, P., A. Lyytinen, S. Piiroinen, and L. Lindström. 2014b. Northward range expansion requires synchronization of both overwintering behavior and physiology with photoperiod in the invasive Colorado potato beetle (*Leptinotarsa decemlineata*). *Oecologia* 176:57–68.

Lehmann, P., S. Piiroinen, A. Lyytinen, and L. Lindström. 2015a. Response in metabolic rate to changes in temperature in diapausing Colorado potato beetle *Leptinotarsa decemlineata* from three European populations. *Physiol. Ent.* 40:123–130.

Lehmann, P., A. Lyytinen, S. Piiroinn, and L. Lindström. 2015b. Latitudinal differences in diapause related photoperiodic responses of European Colorado potato beetles (*Leptinotarsa decemlineata*). *Evol. Ecol.* 29:269–282.

Lehmann, P., P. Pruisscher, D. Posledovich, M. Carlsson, R. Kakela, P. Tang, S. Nylin, C.W. Wheat, C. Wiklund, and K. Gotthard. 2016a. Energy and lipid metabolism during direct and diapause development in a pierid butterfly. *J. Exp. Biol.* 219:3049–3060.

Lehmann, P., A. Margus, and L. Lindström. 2016b. Inheritance patterns of photoperiodic diapause induction in *Leptinotarsa decemlineata*. *Physiol. Ent.* 41:218–223.

Lehmann, P., S. Nylin, K. Gotthard, and M.A. Carlsson. 2017a. Idiosyncratic development of sensory structures in brains of diapausing butterfly pupae: implications for information processing. *Proc. R. Soc. B* 284:20170897.

Lehmann, P., W. van der Bijl, S. Nylin, C.W. Wheat, and K. Gotthard. 2017b. Timing of diapause termination in relation to variation in winter climate. *Physiol. Ent.* 42:232–238.

Lehmann, P., P. Pruisscher, V. Koštál, M. Moos, P. Šimek, S. Nylin, R. Agren, L. Varemo, C. Wiklund, C.W. Wheat, and K. Gotthard. 2018. Metabolome dynamics of diapause in the butterfly *Pieris napi*: distinguishing maintenance, termination and post-diapause phases. *J. Exp. Biol.* 221: jeb.169508.

Lehmann, P., M. Westberg, P. Tang, L. Lindström, and R. Käkelä. 2020. The diapause lipidomes of three closely related beetle species reveal mechanisms for tolerating energetic and cold stress in high-latitude seasonal environments. *Front. Physiol.* 11:576617.

Leonard, D.E. 1968. Diapause in the gypsy moth. *J. Econ. Ent.* 61:596–598.

Leopold, R.A. 2007. Cryopreservation of nonmammalian metazoan systems. In *Advances in Biopreservation*, ed. J.G. Baust and J.M. Baust, Boca Raton: CRC Press, pp. 271–289.

Leopold, R.A. and J.P. Rinehart. 2010. A template for insect cryopreservation. In *Insects at Low Temperature*, ed. D.L. Denlinger and R.E. Lee, Jr., Cambridge: Cambridge University Press, pp. 325–341.

Lepage, M.P., G. Boivin, J. Broudeur, and G. Bourgeois. 2014. Oviposition pattern of early and late-emerging genotypes of *Delia radicum* (Diptera: Anthomyiidae) at different temperatures. *Environ. Ent.* 43:178–186.

Lester, J.D. and J.T. Irwin. 2012. Metabolism and cold tolerance of overwintering adult mountain pine beetles (*Dendroctonus ponderosae*): evidence of facultative diapause. *J. Insect Physiol.* 58:808–815.

Levin, D.B., H.V. Danks, and S.A. Barber. 2003. Variations in mitochondrial DNA and gene transcription in freezing-tolerant larvae of *Eurosta solidaginis* (Diptera: Tephritidae) and *Gynaephora groenlandica* (Lepidoptera: Lymantriidae). *Insect Mol. Biol.* 12:281–289.

Levine, E., H. Oloui-Sadeghi, and J.R. Fisher. 1992. Discovery of multiyear diapause in Illinois and South Dakota northern corn rootworm (Coleoptera: Chrysomelidae) eggs and incidence of prolonged diapause trait in Illinois. *J. Econ Ent.* 85:262–267.

Levy, R.C., G.M. Kozak, and E.B. Dopman. 2018. Non-pleiotropic coupling of daily and seasonal temporal isolation in the European corn borer. *Genes* 9:180.

Levy, R.I. and H.A. Schneiderman. 1966. Discontinuous respiration in insects II: the direct measurement and significance of changes in tracheal gas composition during the respiratory cycles of silkworm pupae. *J. Insect Physiol.* 12:83–104.

Lewis, D.K., D. Spurgeon, T.W. Sappington, and L.L. Keeley. 2002. A hexamerin protein, AgSP-1, is associated with diapause in the boll weevil. *J. Insect Physiol.* 48:887–901.

Li, A. and D.L. Denlinger. 2009. Pupal cuticle protein is abundant during early adult diapause in the mosquito *Culex pipiens*. *J. Med. Ent.* 46:1382–1386.

Li, A., F. Xue, A. Hua, and J. Tang. 2003. Photoperiodic clock of diapause termination in *Pseudopidorus fasciata* (Lepidoptera: Zygaenidae). *Eur. J. Ent.* 100:287–293.

Li, A.Q., A. Popova-Butler, D.H. Dean, and D.L. Denlinger. 2007. Proteomics of the flesh fly brain reveals an abundance of upregulated heat shock proteins during pupal diapause. *J. Insect Physiol.* 53:385–391.

Li, A., J.P. Rinehart, and D.L. Denlinger. 2009a. Neuropeptide-like precursor 4 is uniquely expressed during pupal diapause in the flesh fly. *Peptides* 30:518–521.

Li, A., M.R. Michaud, and D.L. Denlinger. 2009b. Rapid elevation of Inos and decreases in abundance of other proteins at pupal diapause termination in the flesh fly *Sarcophaga crassipalpis*. *Biochim. Biophys. Acta Proteins Proteom.* 1794:663–668.

Li, A., L. Wang, M. Liang, Q. Zhou, G. Wang, and F. Hou. 2012. Proteomic approach to understanding the maternal effect in the flesh fly, *Sarcophaga bullata* (Diptera: Sarcophagidae). *Eur. J. Ent.* 109:7–14.

Li, F.-Q., N.-N. Fu, C. Qu, R. Wang, Y.-H. Xu, and C. Luo. 2017a. Understanding the mechanisms of dormancy in an invasive alien sycamore lace bug, *Corythucha ciliata* through transcript and metabolite profiling. *Sci. Rep.* 7:2631.

Li, H.-Y., T. Wang, Y.-P. Yang, S.-L. Geng, and W.-H. Xu. 2017b. TGF-β signaling regulates p-Akt levels via PP2A during diapause entry in the cotton bollworm, *Helicoverpa armigera*. *Insect Biochem. Mol. Biol.* 87:165–173.

Li, H.-Y., X.-W. Lin, S.-L. Geng, and W.-H. Xu. 2018a. TGF-β and BMP signals regulate insect diapause through Smad1-POU-TFAM pathway. *BBA-Mol. Cell Res.* 1865:1239–1249.

Li, Y., L. Zhang, Q. Zhang, H. Chen, and D.L. Denlinger. 2014. Host diapause status and host diets augmented with cryoprotectants enhance cold hardiness in the parasitoid *Nasonia vitripennis*. *J. Insect Physiol.* 70:8–14.

Li, Y., L. Zhang, H. Chen, V. Koštál, P. Šimek, M. Moos, and D.L. Denlinger. 2015. Shifts in metabolomics profiles of the parasitoid *Nasonia vitripennis* associated with elevated cold tolerance induced by the parasitoid's diapause, host diapause and host diet augmented with proline. *Insect Biochem. Mol. Biol.* 63:34–46.

Li, Y., M. Wang, H. Zhang, H. Chen, M. Wang, C. Liu, and L. Zhang. 2018b. Exploiting diapause and cold tolerance to enhance the use of the green lacewing *Chrysopa formosa* for biological control. *Biol. Cont.* 127:116–126.

Li, Y.-P., M. Goto, S. Ito, Y. Sato, K. Sasaki, and N. Goto. 2001. Physiology of diapause and cold hardiness in the overwintering pupae of the fall webworm *Hyphantria cunea* (Lepidoptera: Arctiidae) in Japan. *J. Insect Physiol.* 47:1181–1187.

Li, Y.-P., S. Oguchi, and M. Goto. 2002a. Physiology of diapause and cold hardiness in overwintering pupae of the apple leaf miner *Phyllonorycter ringoneiella* in Japan. *Physiol. Ent.* 27:92–96.

Li, Y.-P., M. Goto, L. Ding, and H. Tsumuki. 2002b. Diapause development and acclimation regulating enzymes associated with glycerol synthesis in the Shonai ecotype of the rice stem borer larva, *Chilo suppressalis* Walker. *J. Insect Physiol.* 48:303–310.

Liang, G.-M., K.-M. Wu, B. Rector, and Y.Y. Guo. 2007. Diapause, cold hardiness and flight ability of Cry1Ac-resistant and -susceptible strains of *Helicoverpa armigera* (Lepidoptera: Noctuidae). *Eur. J. Ent.* 104:699–704.

Lighton, J.R.B. 1996. Discontinuous gas exchange in insects. *Ann. Rev. Ent.* 41:309–324.

Lin, J.L., P.L. Lin, and S.H. Gu. 2009. Phosphorylation of glycogen synthase kinase-3 beta in relation to diapause processing in the silkworm, *Bombyx mori*. *J. Insect Physiol.* 55:593–598.

Lin, X.-W. and W.-H. Xu. 2016. Hexokinase is a key regulator of energy metabolism and ROS activity in insect lifespan extension. *Aging* 8:245–258.

Lirakis, M., M. Dolezal, and C. Schlötterer. 2018. Redefining reproductive dormancy in *Drosophila* as a general stress response to cold temperatures. *J. Insect Physiol.* 107:175–185.

Liu, M., T.Y. Zhang, and W.-H. Xu. 2005. A cDNA encoding diazepam-binding inhibitor/acyl-CoA-binding protein in *Helicoverpa armigera*: molecular characterization and expression analysis associated with pupal diapause. *Comp. Biochem. Physiol. C* 141:168–176.

Liu, W., Y. Li, L. Zhu, F. Zhu, C.-L. Lei, and X.-P. Wang. 2016a. Juvenile hormone facilitates the antagonism between adult reproduction and diapause through the *methoprene-tolerant* gene in the female *Colaphellus bowringi*. *Insect Biochem. Mol. Biol.* 74:50–60.

Liu, W., Y. Li, S. Guo, H. Yin, C.-L. Lei, and X.-P. Wang. 2016b. Association between gut microbiota and diapause preparation in the cabbage beetle: a new perspective for studying insect diapause. *Sci. Rep.* 6:38900.

Liu, W., Q.Q. Tan, L. Zhu, Y. Li, F. Zhu, C.-L. Lei, and X.-P. Wang. 2017. Absence of juvenile hormone signaling regulates the dynamic expression profiles of nutritional metabolism genes during diapause preparation in the cabbage beetle *Colaphellus bowrinigi*. *Insect Mol. Biol.* 26:530–542.

Liu, Y., Q. Zhang, and D.L. Denlinger. 2015. Imidazole derivative KK-42 boosts pupal diapause incidence and delays diapause termination in several insect species. *J. Insect Physiol.* 74:38–44.

Liu, Y., S. Liao, J.A. Veenstra, and D.R. Nässel. 2016c. *Drosophila* insulin-like peptide 1 (DILP1) is transiently expressed during non-feeding stages and reproductive dormancy. *Sci. Rep.* 6:26620.

Liu, Z., P. Gong, K. Wu, J. Sun, and D. Li. 2006. A true summer diapause induced by high temperatures in the cotton bollworm, *Helicoverpa armigera* (Lepidoptera: Noctuidae). *J. Insect Physiol.* 52:1012–1020.

Liu, Z., P. Gong, K. Wu, W. Wei, J. Sun, and D. Li. 2007. Effects of larval host plants on over-wintering preparedness and survival of the cotton bollworm, *Helicoverpa armigera* (Hübner) (Lepidoptera: Noctuidae). *J. Insect Physiol.* 53:1016–1026.

Liu, Z., P. Gong, D.G. Heckel, W. Wei, J. Sun, and D. Li. 2009. Effects of larval host plants on over-wintering physiological dynamics and survival of the cotton bollworm, *Helicoverpa armigera* (Hübner) (Lepidoptera: Noctuidae). *J. Insect Physiol.* 55:1–9.

Liu, Z., P. Gong, D. Li, and W. Wei. 2010. Pupal diapause of *Helicoverpa armigera* (Hübner)(Lepidoptera: Noctuidae) mediated by larval host plants: pupal weight is important. *J. Insect Physiol.* 56:1863–1870.

Liu, Z., Y. Xin, Y. Zhang, J. Fan, and J. Sun. 2016d. Summer diapause induced by high temperatures in the oriental tobacco budworm: ecological adaptation to hot summers. *Sci. Rep.* 6:27443.

Lohmeyer, K.H., J.M. Pound, and J.E. George. 2009. Effects of photoperiod on reproduction, nymphal developmental timing, and diapause in *Amblyomma maculatum* (Acari: Ixodidae). *J. Med. Ent.* 46:1299–1302.

Lopatina, E.B., S.V. Balashov, and V.E. Kipyatkov. 2007. First demonstration of the influence of photoperiod on the thermal requirements for development in insects and in particular the linden-bug, *Pyrrhocoris apterus* (Heteroptera: Pyrrhocoridae). *Eur. J. Ent.* 104:23–31.

López-Martinez, G. and D.A. Hahn. 2012. Short-term anoxic conditioning hormesis boosts antioxidant defenses, lowers oxidative damage following irradiation, and enhances male sexual performance in the Caribbean fruit fly, *Anastrepha suspensa*. *J. Exp. Biol.* 215:2150–2161.

Lorenz, M.W. and G. Gäde. 2009. Hormonal regulation of energy metabolism in insects as a driving force for performance. *Integr. Comp. Biol.* 49:380–392.

Lou, Y., K. Liu, D. He, D. Gao, and S. Ruan. 2019. Modelling diapause in mosquito population growth. *J. Math. Biol.* 78:2259–2288.

Lounibos, L.P. and W.E. Bradshaw. 1975. A second diapause in *Wyeomyia smithii*: seasonal incidence and maintenance by photoperiod. *Can. J. Zool.* 53:215–221.

Lounibos, L.P., R.L. Escher, and R. Lourenço De Oliveira. 2003. Asymmetric evolution of photoperiodic diapause in temperate and tropical invasive populations of *Aedes albopictus* (Diptera: Culicidae). *Ann. Ent. Soc. Am.* 96:512–518.

Lounibos, L.P., R.L. Escher, and N. Nishimura. 2011. Retention and adaptiveness of photoperiodic egg diapause in Florida populations of invasive *Aedes albopictus*. *J. Am. Mosq. Cont. Assoc.* 27:433–436.

Lu, M.-X., S.-S. Cao, Y.-Z. Du, Z.-X. Liu, P. Liu, and J. Li. 2013a. Diapause, signal and molecular characteristics of overwintering *Chilo suppressalis* (Insecta: Lepidoptera: Pyralidae). *Sci. Rep.* 3:3211.

Lu, W., Q.-J. Meng, N.J.C. Tyler, K.-A. Stokkan, and A.S.I. Loudon. 2010. A circadian clock is not required in an Arctic mammal. *Curr. Biol.* 20:533–537.

Lu, Y.-X. and W.-H. Xu. 2010. Proteomic and phosphoproteomic analysis at diapause initiation in the cotton bollworm, *Helicoverpa armigera*. *J. Proteome Res.* 9:5053–5064.

Lu, Y.-X., D.L. Denlinger, and W.-H. Xu. 2013b. Polycomb complex 2 (PRC2) protein ESC regulates insect developmental timing by mediating H3K27me3 and activating prothoracicotropic hormone gene expression. *J. Biol. Chem.* 288:23554–23564.

Lubawy, J., A. Urbański, H. Colinet, H.-J. Pflüger, and P. Marciniak. 2020. Role of the insect neuroendocrine system in the response to cold stress. *Front. Physiol.* 11:376.

Lumme, J. 1978. Phenology and photoperiodic diapause in northern populations of *Drosophila*. In *Evolution of Insect Migration and Diapause*, ed. H. Dingle, New York: Springer Verlag, pp. 145–170.

Lumme, J. and A. Oikarinen. 1977. The genetic basis of the geographically variable photoperiodic diapause in *Drosophila littoralis*. *Hereditas* 86:129–142.

Luna-Cozar, J., O. Martinez-Madera, and R.W. Jones. 2020. Ball moss, *Tillandsia recurvata* L., as a refuge site for arthropods in a seasonally dry tropical forest of Central Mexico. *Southwest. Ent.* 45:445–460.

Luo, G.-H., Z.-X. Luo, Z.-L, Zhang, Y. Sun, M.-H. Lu, Z.-L. Shu, Z.-H. Tian, A.A. Hoffmann, and J.-C. Fang. 2021. The response to flooding of two overwintering rice stem borers likely accounts for their changing impacts. *J. Pest Sci.* 94:451–461.

Lyons, P.J., L. Govaere, N. Crapoulet, K.B. Storey, and P.J. Morin. 2016. Characterization of cold-associated microRNAs in the freeze-tolearnt gall fly *Eurosta solidaginis* using high-throughput sequencing. *Comp. Biochem. Physiol. D, Genomics Proteomics* 20:95–100.

Lyons-Sobaski, S. and S. Berlocher. 2009. Life history phenology differences between southern and northern populations of the apple maggot fly, *Rhagoletis pomonella*. *Ent. Exp. Appl.* 130:149–159.

Ma, H.-Y., X.-R. Zhou, Y. Tan, and B.-P. Pang. 2019. Proteomic analysis of adult *Galeruca daurica* (Coleoptera: Chrysomelidae) at different stages during summer diapause. *Comp. Biochem. Physiol. D* 29:351–357.

Ma, H.-Y., Y.-Y. Li, L. Li, Y. Tan, and B.-P. Pang. 2021. Regulation of juvenile hormone on summer diapause of *Galeruca daurica* and its pathway analysis. *Insects* 12:237.

Ma, W.-M., H.-W. Li, Z.-M. Dai, J.-S. Yang, F. Yang, and W.-J. Yang. 2013. Chitin-

binding proteins of *Artemia* diapause cysts participate in formation of the embryonic cuticle layer of cyst shells. *Biochem. J.* 449:285–294.

Machado, H., A.O. Bergland, K.R. O'Brien, E.L. Behrman, P.S. Schmidt, and D.A. Petrov. 2016. Comparative population genomics of latitudinal variation in *Drosophila simulans* and *Drosophila melanogaster*. *Mol. Ecol.* 25:723–740.

MacLeod, E.G. 1967. Experimental induction and elimination of diapause and autumnal coloration in *Chrysopa carnea* (Neuroptera). *J. Insect Physiol.* 13:1343–1349.

MacMillan, H.A., J.M. Knee, A.B. Dennis, H. Udaka, K.E. Marshall, T.J.S. Merritt, and B.J. Sinclair. 2016. Cold acclimation wholly reorganizes the *Drosophila melanogaster* transcriptome and metabolome. *Sci. Rep.* 6:28999.

MacRae, T.H. 2016. Stress tolerance during diapause and quiescence of the brine shrimp, *Artemia*. *Cell Stress Chap.* 21:9–18.

Maeto, K. and K. Ozaki. 2003. Prolonged diapause of specialist seed-feeders makes predator satiation unstable in masting of *Quercus crispula*. *Oecologia* 137:392–398.

Majewska, A.A. and S. Altizer. 2019. Exposure to non-native tropical milkweed promotes reproductive development in migratory monarch butterflies. *Insects* 10:253.

Malod, K., C.R. Archer, M. Karsten, R. Cruywagen, A. Howard, S.W. Nicolson, and C.W. Weldon. 2020. Exploring the role of host specialization and oxidative stress in interspecific lifespan variation in subtropical tephritid flies. *Sci. Rep.* 10:5601.

Mansingh, A. 1971. Physiological classification of dormancies in insects. *Can. Ent.* 103:983–1009.

Marchal, E., H.P. Vandersmissen, L. Badisco, S. van de Velde, H. Verlinden, M. Iga, P. van Wielendaele, R. Huybrechts, G. Simonet, G. Smagghe, and J. vanden

Broeck. 2010. Control of ecdysteroidogenesis in prothoracic glands of insects: a review. *Peptides* 31:506–519.

Marcus, N.H. and L.P. Scheef. 2010. Photoperiodism in copepods. In *Photoperiodism, the Biological Calendar.* ed. R.J. Nelson, D.L. Denlinger, and D.E. Somers, Oxford: Oxford University Press, pp. 193–217.

Margaritopoulos, J.T. and M.E. Tzanakakis. 2006. Diapause completion in the almond seed wasp, *Eurytoma amygdali* (Hymenoptera: Eurytomidae) following early low temperature treatment. *Eur. J. Ent.* 103:733–742.

Margus, A. and L. Lindström. 2020. Prolonged diapause has sex-specific fertility and fitness costs. *Evol. Ecol.* 34:41–57.

Marshall, K.E. and B.J. Sinclair. 2015. The relative importance of number, duration, and intensity of cold stress events in determining survival and energetics of an overwintering insect. *Funct. Ecol.* 29:357–366.

Martin, K.L.M. and J.E. Podrabsky. 2017. Hit pause: developmental arrest in annual killifishes and their close relatives. *Devel. Dynamics* 246:858–866.

Masaki, S. 1967. Geographic variation and climatic adaptation in a field cricket (Orthoptera: Gryllidae). *Evolution* 21:725–741.

Masaki, S. 1978. Seasonal and latitudinal adaptations in the life cycles of crickets. In *Evolution of Insect Migration and Diapause*, ed. H. Dingle, New York: Springer-Verlag, pp. 72–100.

Masaki, S. 1980. Summer diapause. *Ann. Rev. Ent.* 25:1–25.

Masaki, S. 2002. Ecophysiological consequence of variability in diapause intensity. *Eur. J. Ent.* 99:143–154.

Masaki, S. and S. Kikukawa. 1981. The diapause clock in a moth: response to temperature signals. In *Biological Clocks in Seasonal Reproductive Cycles*, ed. B.K. Follet and D.E. Follett, Bristol: Wright, pp. 101–112.

Masaki, S. and T.J. Walker. 1987. Cricket life cycles. *Evol. Biol.* 21:349–423.

Maslennikova, V.A. 1958. On the conditions determining the diapause in the parasitic Hymenoptera, *Apanteles glomeratus* L. (Braconidae) and *Pteromalus puparum* L. (Chalcididae). *Ent. Rev.* 37:466–472.

Massaqué, J. 2012. TGFβ signaling in control. *Nature Rev. Mol. Cell Biol.* 13:616–630.

Mathias, D., L. Jacky, W.E. Bradshaw, and C.M. Holzapfel. 2005. Geographic and developmental variation in expression of the circadian rhythm gene, *timeless*, in the pitcher-plant mosquito, *Wyeomyia smithii*. *J. Insect Physiol.* 51:661–667.

Mathias, D., L. Jacky, W.E. Bradshaw, and C.M. Holzapfel. 2007. Quantitative trait loci associated with photoperiodic response and stage of diapause in the pitcher-plant mosquito, *Wyeomyia smithii. Genetics* 176:391–402.

Matsuda, N., T. Kanbe, S.-I. Akimoto, and H. Numata. 2017. Transgenerational seasonal timer for suppression of sexual morph production in the pea aphid, *Acyrthosiphon pisum. J. Insect Phyiol.* 101:1–6.

Matsumoto, M. and M. Takeda. 2002. Changes in brain monoamine contents in diapausing pupae of *Antheraea pernyi* when activated under long-day conditions. *J. Insect Physiol.* 48:765–771.

Matsumoto, M., H. Numata, and S. Shiga. 2013. Role of the brain in photoperiodic regulation of juvenile hormone biosynthesis in the brown-winged green bug *Plautia stali. J. Insect Physiol.* 59:387–393.

Matsunaga, K., H. Takahashi, T. Yoshida, and M.T. Kimura. 1995. Feeding, reproductive and locomotor activities in diapausing and non-diapausing adults of *Drosophila. Ecol. Res.* 10:87–93.

Matsuo, J., S. Nakayama, and H. Numata. 1997. Role of the corpus allatum in control of adult diapause in the blow fly, *Protophormia terraenovae. J. Insect Physiol.* 43:211–216.

Matsuo, Y. 2006. Cost of prolonged diapause and its relationship to body size in a seed predator. *Funct. Ecol.* 20:300–306.

Maurer, J.A., J.H. Shephard, L.G. Crabo, P.C. Hammond, R.S. Zack, and M.A. Peterson. 2018. Phenological responses of 215 moth species to interannual climate variation in the Pacific Northwest from 1895 to 2013. *PLoS One* 13:e0202850.

Mayekar, H.V. and U. Kodandaramaiah. 2021. Partially coupled reaction norms of pupal colour and spot size in a butterfly. *Evol. Ecol.* 35:201–216. doi: 10.1007/s10682–020-10090-w.

Mayr, E. 1963. *Animal Species and Evolution.* Cambridge: Harvard University Press

McBrayer, Z., H. Ono, M.J. Shimell, J.-P. Parvy, R.B. Beckstead, J.T. Warren, C.S. Thummel, C. Dauphin-Villemanr, L.I. Gilbert, and M.B. O'Connor. 2007. Prothoracicotropic hormone regulates developmental timing and body size. *Dev. Cell* 13:857–871.

McDaniel, C.N. and S.J. Berry. 1967. Activation of the prothoracic glands of *Antheraea polyphemus. Nature* 214:1032–1034.

McEvoy, P.B., K.M. Higgs, E.M. Coombs, E. Karaçetin, and L.A. Starcevich. 2012. Evolving while invading: rapid adaptive evolution in juvenile development time for a biological control organism colonizing a high-elevation environment. *Evol. Appl.* 5:524–536.

McKee, F.R. and B.H. Aukema. 2015. Successful reproduction by the eastern larch beetle (Coleoptera: Curculionidae) in the absence of an overwintering period. *Can. Ent.* 147:602–610.

McMullen, D.C. and K.B. Storey. 2008a. Mitochondria of cold hardy insects: responses to cold and hypoxia assessed at enzymatic , mRNA, and DNA levels. *Insect Biochem. Mol. Biol.* 38:367–373.

McMullen, D.C. and K.B. Storey. 2008b. Suppression of Na$^+$K$^+$-ATPase activity by reversible phosphorylation over winter in a freeze-tolerant insect. *J. Insect Physiol.* 54:1023–1027.

McWatters, H.G. and D.S. Saunders. 1996. The influence of each parent and geographic origin on larval diapause in the blow fly, *Calliphora vicina. J. Insect Physiol.* 42:721–726.

McWatters, H.G. and D.S. Saunders. 1997. Inheritance of the photoperiodic response controlling larval diapause in the blow fly, *Calliphora vicina. J. Insect Physiol.* 43:709–717.

McWatters, H.G. and D.S. Saunders. 1998. Maternal temperature has different effects on the photoperiodic response and duration of larval diapause in blow fly (*Calliphora vicina*) strains collected at two latitudes. *Physiol. Ent.* 23:369–375.

Medina, V., P.M. Sardoy, M. Soria, C.A. Vay, G.O. Gutkind, and J.A. Zavala. 2018. Characterized non-transient microbiota from stinkbug (*Nezara viridula*) midgut deactivates soybean chemical defenses. *PLoS One* 13:e0200161.

Mehrnejad, M.R. and M.J.W. Copland. 2005. Diapause strategy in the parasitoid *Psyllaephagus pistaciae. Ent. Exp. Appl.* 116:109–114.

Meisel, R.P. and T. Connallon. 2013. The faster-X effect: integrating theory and data. *Trends Gen.* 29:537–544.

Meister, P., S. Schoot, C. Bedet, Y. Xiao, S. Rohner, S. Bodennec, B. Hudry, L. Molin, F. Solari, S.M. Gasser, and F. Palladino. 2011. *Caenorhabditis elegans* heterochromatin protein 1 (HPL-2) links developmental plasticity, longevity and lipid metabolism. *Genome Biol.* 12: R123.

Mellanby, K. 1938. Diapause and metamorphosis of the blowfly *Lucilia sericata* Mg. *Parasitology* 30:392–402.

Mellström, H.L., M. Friberg, A.-K. Borg-Karlson, R. Murtazina, M. Palm, and C. Wiklund. 2010. Seasonal polyphenism in life history traits: time costs of direct development in a butterfly. *Behav. Ecol. Sociobiol.* 64:1377–1383.

Menu, F. and D. Debouzie. 1993. Coin-flipping plasticity and prolonged diapause in insects: example of the chestnut weevil *Curculio elephas* (Coleoptera: Curculionidae). *Oecologia* 93:367–373.

Meola, R.W. and P.L. Adkisson. 1977. Release of prothoracicotropic hormone and potentiation of developmental ability during diapause in the bollworm, *Heliothis zea. J. Insect Physiol.* 23:683–688.

Meuti, M.E. and D.L. Denlinger. 2013. Evolutionary links between circadian clocks and photoperiodic diapause in insects. *Integ. Comp. Biol.* 53:131–143.

Meuti, M.E., M. Stone, T. Ikeno, and D.L. Denlinger. 2015. Functional circadian clock genes are essential for the overwintering diapause of the Northern house mosquito, *Culex pipiens. J. Exp. Biol.* 218:412–422.

Meuti, M.E., R. Bautista-Jimenez, and J.A. Reynolds. 2018. Evidence that microRNAs are part of the molecular toolkit regulating adult reproductive diapause in the mosquito, *Culex pipiens. PLoS One* 13:e0203015.

Meyers, P.J., T.H.Q. Powell, K.K.O. Walden, A.J. Schieferecke, J.L. Feder, D.A. Hahn, H.M. Robertson, S.H. Berlocher, and G.J. Ragland. 2016. Divergence of the diapause transcriptome in apple maggot flies: winter regulation and post-winter transcriptional repression. *J. Exp. Biol.* 219:2613–2622.

Michaud, J.P. and J.A. Qureshi. 2005. Induction of reproductive diapause in *Hippodamia convergens* (Coleoptera: Coccinellidae) hinges on prey quality and availability. *Eur. J. Ent.* 102:483–487.

Michaud, M.R. and D.L. Denlinger. 2006. Oleic acid is elevated in cell membranes during rapid cold hardening and pupal diapause in the flesh fly, *Sarcophaga crassipalpis. J. Insect Physiol.* 52:1073–1082.

Michaud, M.R. and D.L. Denlinger. 2007. Shifts in the carbohydrate, polyol, and amino acid pools during rapid cold-hardening and diapause-associated cold-hardening in flesh flies (*Sarcophaga crassipalpis*): a metabolomic comparison. *J. Comp. Physiol. B* 177:753–763.

Michaud, M.R. and D.L. Denlinger. 2010. Genomics, proteomics and metabolomics:

finding the other players in insect cold hardiness. In *Low Temperature Biology of Insects*, ed. D.L. Denlinger and R.E. Lee, Jr., Cambridge: Cambridge University Press, pp. 91–115.

Michaud, M.R., N.M. Teets, J.T. Peyton, B.M. Blobner, and D.L. Denlinger. 2011. Heat shock response to hypoxia and its attenuation during recovery in the flesh fly, *Sarcophaga crassipalpis. J. Insect Physiol.* 57:203–210.

Miki, T., T. Shinohara, S. Chafino, S. Noji, and K. Tomioka. 2020. Photoperiod and temperature separately regulate nymphal development through JH and insulin/TOR signaling pathways in an insect. *Proc. Nat'l. Acad. Sci., USA* 117:5525–5531.

Milbrath, L.R., C.J. Deloach, and J.L. Tracy. 2007. Overwintering survival, phenology, voltinism, and reproduction among different populations of the leaf beetle *Diorhabda elongata* (Coleoptera: Chrysomelidae). *Environ. Ent.* 36:1356–1364.

Milonas, P.G. and M. Savopoulou-Soultani. 2004. Diapause termination in overwintering larvae of a Greek strain of *Adoxophyes orana* (Lepidoptera: Tortricidae). *Environ. Ent.* 33:513–519.

Minder, I.F. 1976. Hibernation conditions and survival rate of the Colorado potato beetle in different types of soil. In *Ecology and Physiology of Diapause in the Colorado beetle*, ed. K.V. Arnoldi, New Delhi: Indian National Scientific Documentation Centre (translation from Russian), pp. 28–58.

Minder, I.F. and D.V. Petrova. 1976. Repeated diapause in the Colorado beetle under Transcarpathian conditions. In *Ecology and Physiology of Diapause in the Colorado beetle*, ed. K.V. Arnoldi, New Delhi: Indian National Scientific Documentation Centre (translation from Russian), pp. 279–296.

Mirth, C.K., H.Y. Tang, S.C. Makohon-Moore, S. Salhadar, R.H. Gokhale, R.D. Warner, T. Koyama, L.M. Riddiford, and

A.W. Shingleton. 2014. Juvenile hormone regulates body size and perturbs insulin signaling in *Drosophila*. *Proc. Nat'l. Acad. Sci., USA* 111:7018–7023.

Mishra, P.K., S.K. Sharan, D. Kumar, B. Singh, M.K. Subrahmanyam, and N. Suryanarayana. 2008. Effects of ecdysone on termination of pupal diapause and egg production in *Antheraea mylita* Drury. *J. Adv. Zool.* 29:128–136.

Mitchell, C.J. and H. Briegel. 1989. Inability of diapausing *Culex pipiens* (Diptera: Culicidae) to use blood for producing lipid reserves for overwintering survival. *J. Med. Entomol.* 26:318–326.

Miyawaki, R., S.I. Tanaka, and H. Numata. 2003. Photoperiodic receptor in the nymph of *Poecilocoris lewisi* (Heteroptera: Scutelleridae). *Eur. J. Entomol.* 100:301–303.

Miyazaki, Y., T. Nisimura, and H. Numata. 2009. A circadian system is involved in photoperiodic entrainment of the circannual rhythm of *Anthrenus verbasci*. *J. Insect Physiol.* 55:494–498.

Miyazaki, Y., T. Nisimura, and H. Numata. 2014. Circannual rhythms in insects. In *Annual, Lunar and Tidal Clocks*, ed. H. Numata and B. Helms, Tokyo: Springer Japan, pp. 333–352.

Miyazaki, Y., Y. Watari, and H. Numata. 2016. Resetting of the circannual rhythm of the varied carpet beetle *Anthrenus verbasci* by low-temperature pulses. *Physiol. Ent.* 41:390–399.

Mizoguchi, A., S. Ohsumi, K. Kobayashi, N. Okamoto, N. Yamada, K. Tateishi, Y. Fujimoto, and H. Kataoka. 2013. Prothoracicotropic hormone acts as a neuroendocrine switch between pupal diapause and adult development. *PLoS One* 8:e60824.

Mizoguchi, A., M. Kamimura, M. Kiuchi, and H. Kataoka. 2015. Positive feedback regulation of prothoracicotropic hormone secretion by ecdysteroid – a mechanism that determines the timing of metamorphosis. *Insect Biochem. Mol. Biol.* 58:39–45.

Moerbitz, C. and S.K. Hetz. 2010. Tradeoffs between metabolic rate and spiracular conductance in discontinuous gas exchange of *Samia cynthia* (Lepidpoptera, Saturniidae). *J. Insect Physiol.* 56:536–542.

Mohamed, A.A.M., Q. Wang, J. Bembenek, N. Ichihara, S. Hiragaki, T. Suzuki, and M. Takeda. 2014. N-acetyltransferase (*nat*) is a critical conjunct of photoperiodism between the circadian system and endocrine axis in *Antheraea pernyi*. *PLoS One* 9:e92680.

Molleman, F., B.J. Zwann, P.M. Brakefeld, and J.R. Carey. 2007. Extraordinary long life spans in fruit-feeding butterflies can provide window on evolution of life span and aging. *Exp. Gerontol.* 42:472–482.

Mondy, N. and M.-F. Corio-Costet. 2004. Feeding insects with a phytopathogenic fungus influences their diapause and population dynamics. *Ecol. Ent.* 29:711–717.

Monro, J. 1972. Terminal diapause in geometrid pupae – a new link in the endocrine chain? In *Insect Endocrines*, ed. V.J.A. Novak and K. Sláma, Vol. 3, Prague: Academia, pp. 89–97.

Monteith, G.B. 1982. Dry season aggregations of insects in Australian monsoon forests. *Mem. Queensl. Mus.* 20:533–543.

Moore, A.J., D.A. Devine, and M.C. Bibby. 1994. Preliminary experimental anticancer activity of cecropins. *Peptide Res.* 7:265–269.

Moraiti, C.A., C.T. Nakas, and N.T. Papadopoulos. 2012. Prolonged pupal dormancy is associated with significant fitness cost for adults of *Rhagoletis cerasi* (Diptera: Tephritidae). *J. Insect Physiol.* 58:1128–1135.

Moraiti, C.A., C.T. Nakas, and N.T. Papadopoulos. 2013. Diapause termination of *Rhagoletis cerasi* pupae is regulated by local adaptation and phenotypic plasticity: escape in time through bet-hedging strategies. *J. Evol. Biol.* 27:43–54.

Moraiti, C.A. and N.T. Papadopoulos. 2017. Obligate annual and successive facultative diapause establish a bet-hedging strategy of *Rhagoletis cerasi* (Diptera: Tephritidae) in seasonally unpredictable environments. *Physiol. Ent.* 42:225–231.

Moraiti, C.A.,K. Köppler, H. Vogt, and N.T. Papadopoulos. 2020. Effects of photoperiod and relative humidity on diapause termination and post-winter development of *Rhagoletis cerasi* pupae. *Bull. Ent. Res.* 110:588–596.

Moran, N.A. and H.E. Dunbar. 2006. Sexual acquisition of beneficial symbionts in aphids. *Proc. Nat'l. Acad. Sci., USA* 103:12803–12806.

Morehouse, N.I., N. Mandon, J.-P. Christides, M. Body, G. Bimbard, and J. Casas. 2013. Seasonal selection and resource dynamics in a seasonally polyphonic butterfly. *J. Evol. Biol.* 26:175–185.

Moreira, D.C., D.P. Paula, and M. Hermes-Lima. 2012. Redox metabolism during tropical diapause in a Lepidoptera larva. In *Living in a Seasonal World*, ed. T. Ruf, C. Bieber, W. Arnold, and E. Millesi, Heidelberg: Springer-Verlag, pp. 399–409.

Moreira, D.C., M.F. Oliveira, L. Liz-Guimaraes, N. Diniz-Rojas, E.G. Campos, and M. Hermes-Lima. 2017. Current trends and research challenges regarding "preparation for oxidative stress." *Front. Physiol.* 8:702.

Morewood, W.D. and R.A. Ring. 1998. Revision of the life history of the high Arctic moth *Gynaephora groenlandica* (Wocke) (Lepidoptera: Lymantriidae). *Can. J. Zool.* 76:1371–1381.

Mori, A., J. Romero-Severson, and D.W. Severson. 2007. Genetic basis for reproductive diapause is correlated with life history traits within the *Culex pipiens* complex. *Insect Mol. Biol.* 16:515–524.

Moribayashi, A., H. Kurahashi, and T. Ohtaki. 1992. Physiological differentiation of the ring glands in mature larvae of the flesh fly, *Boettcherisca peregrina*, programmed for diapause or non-diapause. *J. Insect Physiol.* 38:177–183.

Moribayashi, A., H. Kurahashi, and T. Ohtaki. 1988. Different profiles of ecdysone secretion and its metabolism between diapause- and nondiapause-destined cultures of the fleshfly, *Boettcherisca peregrina. Comp. Biochem. Physiol. A* 91:157–164.

Moribayashi, A., T. Hayashi, D. Taylor, H. Kurahashi, and M. Kobayashi. 2008. Different responses to photoperiod in non-diapausing colonies of the flesh fly, *Boettcherisca peregrina. Physiol. Ent.* 33:31–36.

Moribe, Y., T. Niimi, O. Yamashita, and T. Yaganuma. 2001. *Samui,* a novel cold-inducible gene, encoding a protein with a BAG domain similar to silencer of death domains (SODD/BAG-4), isolated from *Bombyx* diapause eggs. *Eur. J. Biochem.* 268:3432–3442.

Moribe,Y., K. Oka, T. Niimi, O. Yamashita, and T. Yaginuma. 2010. Expression of heat shock protein 70a mRNA in *Bombyx mori* diapause eggs. *J. Insect Physiol.* 56:1246–1252.

Morita, A. and H. Numata. 1997. Distribution of the photoperiodic receptor in the compound eyes of the bean bug, *Riptortus clavatus. J. Comp. Physiol. A* 180:181–185.

Morita, A., T. Niimi, and O. Yamashita. 2003. Physiological differentiation of DH-PBAN-producing neurosecretory cells in the silkworm embryo. *J. Insect Physiol.* 49:1093–1102.

Moriyama, M. and H. Numata. 2008. Diapause and prolonged development in the embryo and their ecological significance in two cicadas, *Cryptotympana facialis* and *Graptopsaltria nigrofuscata. J. Insect Physiol.* 54:1487–1494.

Mousseau, T.A. and C.W. Fox, eds. 1998. *Maternal Effects as Adaptations.* Oxford: Oxford University Press.

Mukai, A. and S.G. Goto. 2016. The clock gene *period* is essential for the photoperiodic response in the jewel wasp *Nasonia vitripennis* (Hymenoptera: Pteromalidae). *App. Ent. Zool.* 51:185–194.

Mukherjee, K., A. Baudach, H. Vogel, and A. Vilcinskas. 2020. Seasonal phenotype-specific expression of microRNAs during metamorphosis in the European map butterfly *Araschnia levana. Arch. Insect Biochem. Physiol.* 2020:e21657.

Müller, H.J. 1970. Formen der Dormanz bei Insekten. *Nova Acta Leopoldina* 35:1–27.

Müller, H.J. 1976. Formen der Dormanz bei Insekten als Mechanismen ökologischer Anpassung. *Verh. Dtsch. Zool. Ges.* 1976:46–58.

Munyiri, F. and Y. Ishikawa. 2004. Endocrine changes associated with metamorphosis and diapause induction in the yellow-spotted longicorn beetle, *Psacothea hilaris. J. Insect Physiol.* 50:1075–1081.

Munyiri, F., Y. Shintani, and Y. Ishikawa. 2004. Evidence for the presence of a threshold weight for entering diapause in the yellow-spotted longicorn beetle, *Psacothea hilaris. J. Insect Physiol.* 50:295–301.

Murphy, S.M. and J.T. Lill. 2010. Winter predation of diapausing cocoons of slug caterpillars (Lepidoptera: Limacodidae). *Environ. Ent.* 39:1893–1902.

Mushegian, A.A. and K. Tougeron. 2019. Animal-microbe interactions in the context of diapause. *Biol. Bull.* 237:180–191.

Mushegian, A.A., N. Neupane, Z. Batz, M. Mogi, N. Tuno, T. Toma, I. Miyagi, L. Ries, and P.A. Armbruster. 2021. Ecological mechanism of climate-mediated selection in a rapidly evolving invasive species. *Ecol. Lett.* 24:698–707. doi: 10.1111/ele13686.

Musolin, D.L. 2012. Surviving winter: diapause syndrome in the southern green stink bug *Nezara viridula* in the laboratory, in the field, and under climate change conditions. *Physiol. Ent.* 37:309–322.

Musolin, D.L. and A.H. Saulich. 2000. Summer dormancy ensures univoltinism in the predatory bug *Picromerus bidens. Ent. Exp. Appl.* 95:259–267.

Musolin, D.L., H. Numata, and A.H. Saulich. 2001. Timing of diapause induction outside the natural distribution range of a species: an outdoor experiment with the bean bug *Riptortus clavatus. Ent. Exp. Appl.* 100:211–219.

Musolin, D.L. and H. Numata. 2003a. Photoperiodic and temperature control of diapause induction and colour change in the southern green stink bug *Nezara viridula. Physiol. Ent.* 28:65–74.

Musolin, D.L. and H. Numata. 2003b. Timing of diapause induction and its life-history consequences in *Nezara viridula*: is it costly to expand the distribution range? *Ecol. Ent.* 28:694–703.

Musolin, D.L. and H. Numata. 2004. Late-season induction of diapause in *Nezara viridula* and its effect on adult colouration and post-diapause reproductive performance. *Ent. Exp. Appl.* 111:1–6.

Musolin, D.L., K. Fujisaki, and H. Numata. 2007. Photoperiodic control of diapause termination, colour change and postdiapause reproduction in the southern green stinkbug, *Nezara viridula. Physiol. Ent.* 32:64–72.

Musolin, D.L. and K. Ito. 2008. Photoperiodic and temperature control of nymphal development and induction of reproductive diapause in two predatory *Orius* bugs: interspecific and geographic differences. *Physiol. Ent.* 33:291–301.

Musolin, D.L., D. Tougou, and K. Fujisaki. 2011. Photoperiodic response in the subtropical and warm-temperate zone populations of the southern green stink bug *Nezara viridula*: why does it not fit the common latitudinal trend? *Physiol. Ent.* 36:379–384.

Musolin D.L. and A.K. Saulich. 2018. Diapause in Pentatomoidea. In *Invasive Stink Bugs and Related Species (Pentatomoidea).* ed. J.E. McPherson, Boca Raton: CRC Press, pp. 497–564.

Musolin, D.L., M.Y. Dolgovskaya, V.Y. Protsenko, N.N. Karpun, S.Y. Reznik, and A.K. Saulich. 2019. Photoperiodic and temperature control of nymphal growth and adult diapause induction in the invasive Caucasian population of the brown marmorated stink bug, *Halyomorpha halys*. *J. Pest Sci.* 92:621–631.

Muszyńska-Pytel, M., R. Trzcińska, M. Aubry, M.A. Pszczółkowski, and B. Cymborowski. 1993. Regulation of prothoracic gland activity in diapausing larvae of the wax moth, *Galleria mellonella* L. (Lepidoptera). *Insect Biochem. Mol. Biol.* 23:33–41.

Nadal-Jimenez, P., J.S. Griffin, L. Davies, C.L. Frost, M. Marcello, and G.D.D. Hurst. 2019. Genetic manipulation allows in vivo tracking of the life cycle of the son-killer symbiont, *Arsenophonus nasoniae*, and reveals patterns of host invasion, tropism and pathology. *Environ. Microbiol.* 21:3172–3182.

Nagy, D., G. Andreatta, S. Bastianello, A.M. Anduaga, G. Mazzotta, C.P. Kyriaou, and R. Costa. 2018. A semi-natural approach for studying seasonal diapause in *Drosophila melanogaster* reveals robust photoperiodicity. *J. Biol. Rhyth.* 33:117–125.

Nakagaki, M., R. Takei, E. Nagashima, and T. Yaginuma. 1991. Cell cycles in embryos of the silkworm, *Bombyx mori*: G2-arrest at diapause stage. *Roux's Arch. Dev. Biol.* 200:223–229.

Nakamura, A., K. Miyado, Y. Takezawa, N. Ohnami, M. Sato, C. Ono Y. Harada, K. Yoshida, N. Kawano, S. Kanai, M. Miyado, and A. Umezawa. 2011. Innate immune system still works at diapause, a physiological state of dormancy in insects. *Biochem. Biophy. Res. Comm.* 410:351–357.

Nakamura, K. and H. Numata. 2000. Photoperiodic control of the intensity of diapause and diapause development in the bean bug, *Riptortus clavatus* (Heteroptera: Alydidae). *Eur. J. Ent.* 97:19–23.

Nakashima, Y. and Y. Hirose. 1997. Winter reproduction and photoperiodic effects on diapause induction of *Orius tantillus* (Motschulsky) (Heteroptera: Anthocoridae), a predator of *Thrips palmi*. *Appl. Ent. Zool.* 32:403–405.

Nanda, K.K. and K.C. Hamner. 1958. Studies on the nature of the endogenous rhythm affecting photoperiodic response of Biloxi soy bean. *Bot. Gaz.* 120:14-25.

Nasci, R.S., H.M. Savage, D.J. White, J.R. Miller, B.C. Cropp, M.S. Godsey, A.J. Kerst, P. Bennett, K.Gottfried, and R.S. Lanciotti. 2001. West Nile virus in overwintering *Culex* mosquitoes, New York City, 2000. *Emerg. Infect. Diseases* 7:742–744.

Nässel, D.R., O.I. Kubrak, Y. Liu, J. Luo, and O.V. Lushchak. 2013. Factors that regulate insulin producing cells and their output in *Drosophila*. *Front. Physiol.* 4:252.

Nealis, V.G. 2005. Diapause and voltinism in western and 2-year-cycle spruce budworms (Lepidoptera: Tortricidae) and their hybrid progeny. *Can. Ent.* 137: 584–597.

Nealis, V.G. and J. Regniere. 2016. Why western spruce budworms travel so far for winter. *Ecol. Ent.* 41:633–641.

Nelms, B.M., P.A. Macedo, L. Kothera, H.M. Savage, and W.K. Reisen. 2013. Overwintering biology of *Culex* (Diptera: Culicidae) mosquitoes in the Sacramento Valley of California. *J. Med. Ent.* 50:773–790.

Nelson, R.J., D.L. Denlinger, and D.E. Somers, eds. 2010. *Photoperiodism, the Biological Calendar*. Oxford: Oxford University Press.

Nesin, A.P., N.P. Simonenko, H. Numata, and S.I. Chernysh. 1995. Effects of photoperiod and parental age on the maternal induction of larval diapause in the blowfly *Calliphora vicina* Robineau-Desvoidy (Diptera Calliphoridae). *Appl. Ent. Zool.* 30:351–356.

Neven, L.G. and W.L. Yee. 2017. Impact of prolonged absence of low temperature on adult eclosion patterns of western cherry fruit fly (Diptera: Tephritidae). *Environ. Ent.* 46:708–713.

Nicholls, C.N. and A.S. Pullin. 2003. The effects of flooding on survivalship in overwintering larvae of the large copper butterfly *Lycaena dispar batavus* (Lepidoptera: Lycaenidae), and it possible implications for restoration management. *Eur. J. Ent.* 100:65–72.

Nijhout, H.F. and V. Callier. 2013. A new mathematical approach for qualitative modeling of the insulin-TOR-MAPK network. *Front. Physiol.* 4:245.

Nishida, T. 1955. The phenomenon of arrested insect development in the Hawaiian Islands. *Proc. Hawaii. Ent. Soc.* 15:575–582.

Nisimura, T. and H. Numata. 2001. Endogenous timing mechanism controlling the circannual pupation rhythm of the varied carpet beetle *Anthrenus verbasci*. *J. Comp. Physiol. A* 187:433–440.

Nisimura, T., M. Kon, and H. Numata. 2002. Bimodal life cycle of the burying beetle *Nicrophorus quadripunctus* in relation to its summer reproductive diapause. *Ecol. Ent.* 27:220–228.

Nisimura, T. and H. Numata. 2003. Circannual control of the life cycle in the varied carpet beetle *Anthrenus verbasci*. *Funct. Ecol.* 17:489–495.

Nnko, H. J., A. Ngonyoka, L. Salekwa, A.B. Estes, P.J. Hudson, P.S. Gwakisa, and I.M. Cattadori. 2017. Seasonal variation of tsetse fly species abundance and prevalence of trypanosomes in the Maasai Steppe, Tanzania. *J. Vector Ecol.* 42:24–33.

Noguchi, H. and Y. Hayakawa. 1997. Role of dopamine at the onset of pupal diapause in the cabbage armyworm, *Mamestra brassicae*. *FEBS Lett.* 413:157–161.

Noguchi, H. and Y. Hayakawa. 2001. Dopamine is a key factor for the induction of egg diapause of the silkworm *Bombyx mori*. *Eur. J. Biochem.* 268:774–780.

Noh, M.Y., S.H. Kim, M.J. Gorman, K.J. Kramer, S. Muthukrishnan, and Y.

Arakane. 2020. Yellow-g and yellow-g2 proteins are required for egg desiccation resistance and temporal pigmentation in the Asian tiger mosquito, *Aedes albopictus*. *Insect Biochem. Mol. Biol.* 122:103386.

Nomura, M. and Y. Ishikawa. 2000. Biphasic effect of low temperature on completion of winter diapause in the onion maggot, *Delia antiqua*. *J. Insect Physiol.* 46:373–377.

Noronha, C. and C. Cloutier. 2006. Effects of potato foliage age and temperature regime on prediapause Colorado potato beetle *Leptinotarsa decemlineata* (Coleoptera: Chrysomelidae). *Environ. Ent.* 35:590–599.

Numata, H. 2004. Environmental factors that determine the seasonal onset and termination of reproduction in seed-sucking bugs (Heteroptera) in Japan. *Appl. Ent. Zool.* 39:565–573.

Numata, H., S. Shiga, and A. Morita. 1997. Photoperiodic receptors in arthropods. *Zool. Sci.* 14:187–197.

Numata, H. and K. Nakamura. 2002. Photoperiodism and seasonal adaptations in some seed-sucking bugs (Heteroptera) in central Japan. *Eur. J. Ent.* 99:155–161.

Numata, H. and H. Udaka. 2010. Photoperiodism in Mollusks. In *Photoperiodism, the Biological Calendar*, ed. R.J. Nelson, D.L. Denlinger, and D.E. Somers, Oxford: Oxford University Press, pp. 173–192.

Numata, H., Y. Miyazaki, and T. Ikeno. 2015. Common features in diverse insect clocks. *Zool. Letters* 1:10.

Nylin, S. 2013. Induction of diapause and seasonal morphs in butterflies and other insects: knowns, unknowns and the challenge of integration. *Physiol. Ent.* 38:96–104.

Oberhauser, K., R. Wiederholt, J.E. Diffendorfer, D. Semmens, L. Ries W.E. Thogmartin, L. Lopez-Hoffman, and B. Semmens. 2017. A trans-national monarch butterfly population model and implications for regional conservation. *Ecol. Ent.* 42:51–60.

Obregon, R., J. Fernandez Haeger, and D. Jordano. 2017. Adaptive significance of the prolonged diapause in the western Mediterranean lycaenid butterfly *Tomares ballus* (Lepidoptera: Lycaenidae). *Eur. J. Ent.* 114:133–139.

Oh, S.W., A. Mukhopadhyay, B.I. Dixit, T. Raha, M.R. Green, and H.A. Tissenbaum. 2006. Identification of direct DAF-16 targets controlling longevity, metabolism and diapause by chromatin immunoprecipitation. *Nature Gen.* 38:251–257.

Ohashi, K, S.-E. Kawauchi, and Y. Sakuratani. 2003. Geographic and annual variation of summer-diapause expression in the ladybird beetle, *Coccinella septempunctata* (Coleoptera: Coccinellidae), in Japan. *Appl. Ent. Zool.* 38:187–196.

Ohtaki, J.M. 2008. Physiologically induced color-pattern changes in butterfly wings: mechanistic and evolutionary implications. *J. Insect Physiol.* 54:1099–1112.

Oikarinen, A. and J. Lumme. 1979. Selection against photoperiodic reproductive diapause in *Drosophila littoralis*. *Hereditas* 90:119–125.

Ojima, N., S. Ishiguro, Y. An, T. Kadosawa, and K. Suzuki. 2015. Male reproductive maturity and diapause termination in the leaf beetle *Gastrophysa atrocyanea*. *Physiol. Ent.* 40:277–283.

Ojima, N. ,Y. Hara, H. Ito, and D. Yamamoto. 2018. Genetic dissection of stress-induced reproductive arrest in *Drosophila melanogaster* females. *PLoS Genet.* 14:e1007434.

Oku, T. 1983. Aestivation and migration in noctuid moths. In *Diapause and Life Cycle Strategies in Insects*, ed. V.K. Brown and I. Hodek, the Hague: Dr. W. Junk, pp. 219–231.

Okuda, T. 1991. Factors inducing and terminating larval diapause ina stem borer, *Busseola fusca* in western Kenya. *Jap. Agr. Res. Quart.* 25:47–54.

Okuda, T., S. Tanaka, T. Kotaki, and H.-J. Ferenz. 1996. Role of the corpora allata and juvenile hormone in the control of imaginal diapause and reproduction in three species of locusts. *J. Insect Physiol.* 42:943–951.

Okuda, T. and S. Tanaka. 1997. An allostatic factor and juvenile hormone synthesis by corpora allata in *Locusta migratoria*. *J. Insect Physiol.* 43:635–641.

Olademehin, O.P., C. Liu, B. Rimal, N.F. Adegboyega, F. Chen, C. Sim, and S.J. Kim. 2020. Dsi-RNA knockdown of genes regulated by Foxo reduces glycogen and lipid accumulations in diapausing *Culex pipiens*. *Sci. Rep.* 10:17201.

Oliveira, C.M., J.R.S. Lopes, and L.R. Nault. 2013. Survival strategies of *Dalbulus maidis* during maize off-season in Brazil. *Ent. Exp. Appl.* 147:141–153.

O'Riordan, V.B. and A.M. Burnell. 1989. Intermediary metabolism in the dauer larva of the nematode *Caenorhabditis elegans* – I. Glycolysis, gluconeogenesis, oxidative phosphorylation and the tricarboxylic acid cycle. *Comp. Biochem. Physiol. B* 92:233–238.

Orshan, L. and M.P. Pener. 1976. Termination and reinduction of reproductive diapause by photoperiod and temperature in males of the grasshopper, *Oedipoda miniata*. *Physiol. Ent.* 4:55–61.

Osir, E.O., L.V. Labongo, and G.C. Unnithan. 1989. A high molecular weight diapause-associated protein from the stem-borer Busseola fusca: purification and properties. *Arch. Insect Biochem. Physiol.* 11:173–187.

Ovchinnikova, A.A., A.N. Ovchinnikov, M.Y. Dolgovskaya, S.Y. Reznik, and N.A. Belyakova. 2016. Trophic induction of diapause in native and invasive populations of *Harmonia axyridis* (Coleoptera: Coccinellidae). *Eur. J. Ent.* 113:469–475.

Overgaard, J. and H.A. MacMillan. 2017. The integrative physiology of insect chill tolerance. *Ann. Rev. Physiol.* 79:187–208.

Owen, D.F. 1971. *Tropical Butterflies*. Oxford: Clarendon.

Palli, S.R. 2017. New roles for old actors, ROS and PRMT1. *Proc. Nat'l. Acad. Sci., USA* 114:10810–10812.

Palli, S.R., T.R. Ladd, A.R. Ricci, M. Primavera, I.N. Mungrue, A.S.D. Pang, and A. Retnakaran. 1998. Synthesis of the same two proteins prior to larval diapause and pupation in the spruce budworm, *Choristoneura fumiferana*. *J. Insect Physiol.* 44:509–524.

Palli, S.R., R. Kothapalli, Q. Feng, T. Ladd, S.C. Perera, S.-C. Zheng, K. Gojtan, A.S.D. Pang, M. Primavera, W. Tomkins, and A. Retnakaran. 2001. Molecular analysis of overwintering diapause. In *Insect Timing: Circadian Rhythmicity to Seasonality*, ed. D.L. Denlinger, J.M. Giebultowicz, and D.S. Saunders, Amsterdam: Elsevier, pp. 133–144.

Panfilio, K.A. 2008. Extraembryonic development in insects and the acrobatics of blastokinesis. *Dev. Biol.* 313:471–491.

Paolucci, S., L. van de Zande, and L.W. Beukeboom. 2013. Adaptive latitudinal cline of photoperiodic diapause induction in the parasitoid *Nasonia vitripennis* in Europe. *J. Evol. Biol.* 26:705–718.

Paolucci, S., L. Salis, C.J. Vermeulen, L.W. Beukeboom, and L. van der Zande. 2016. QTL analysis of the photoperiodic response and clinal distribution of period alleles in *Nasonia vitripennis*. *Mol. Ecol.* 25:4805–4817.

Papa, G. and I. Negri. 2020. Cannibalism in the brown marmorated stink bug *Halyomorpha halys* (Stål). *Insects* 11:643.

Papanastasiou, S.A., D. Nestel, A.D. Diamantidis, C.T. Nakas, and N.T. Papadopoulos. 2011. Physiological and biological patterns of highland and coastal populations of the European cherry fruit fly during diapause. *J. Insect Physiol.* 57:83–93.

Parkash, R., S. Singh, and S. Ramniwas. 2009. Seasonal changes in humidity level in the tropics impact body color polymorphism and desiccation resistance in *Drosophila jambulina* – evidence for melanism-desiccation hypothesis. *J. Insect Physiol.* 55:358–368.

Parker, D.J., M.G. Ritchie, and M. Kankare. 2016. Preparing for winter: the transcriptomic response associated with different day lengths in *Drosophila montana*. *G3* 6:1373–1381.

Pavelka, J., K. Shimada, and V. Koštál. 2003. TIMELESS: a link between fly's circadian and photoperiodic clocks? *Eur. J. Ent.* 100:255–265.

Pavlides, S.C., S.A. Pavlides, and S.P. Tammariello. 2011. Proteomic and phosphoproteomic profiling during diapause entrance in the flesh fly, *Sarcophaga crassipalpis*. *J. Insect Physiol.* 57:635–644.

Pazyuk, I.M., M.Y. Dolgovskaya, S.Y. Reznik, and D.L. Musolin. 2018. Photoperiodic control of pre-adult development and adult diapause induction in zoophytophagous bug *Dicyphus errans* (Wolff) (Heteroptera, Miridae). *Ent. Rev.* 98:956–962.

Pegoraro, M., A. Bafna, N.J. Davies, D.M. Shuker, and E. Tauber. 2016. DNA methylation changes induced by long and short photoperiods in *Nasonia*. *Genome Res.* 26:203–210.

Pegoraro, M., V. Zonato, E.R. Tyler, G. Fedele, C.P. Kyriacou, and E. Tauber. 2017. Geographic analysis of diapause inducibility in European *Drosophila melanogaster* populations. *J. Insect Physiol.* 98:238–244.

Pener, M.P. 1992. Environmental cues, endocrine factors, and reproductive diapause in male insects. *Chronobiol. Int.* 9:102–113.

Pener, M.P. and M. Broza. 1971. The effect of implanted, active corpora allata on reproductive diapause in adult females of the grasshopper *Oedipoda miniata*. *Ent. Exp. Appl.* 14:190–202.

Pener, M.P. and L. Orshan. 1983. The reversibility and flexibility of the reproductive diapause in males of a "short day" grasshopper, *Oedipoda miniata*. In *Diapause and Life Cycle Strategies in Insects*, ed. V.K. Brown and I Hodek, the Hague: Dr. W. Junk, pp. 67–85.

Pengelley, E.T. and K.C. Fisher. 1961. Rhythmic arousal from hibernation in the golden-mantled ground squirrel, *Citellus lateralis tescorum. Can. J. Zool.* 39:105–120.

Penkov, S., B.K. Raghuram, C. Erkut, J. Oertel, R. Galli, E.J.M. Ackerman, D. Vorkel, J.-M. Verbavatz, E. Koch, K. Fahmy, A. Shevchenko, and T.V. Kurzchalia. 2020. A metabolic switch regulates the transition between growth and diapause in *C. elegans. BMC Biol.* 18:31.

Pepper, J.H. 1973. Breaking the dormancy in the sugar beet webworm, *Loxostege sticticalcis* L., by means of chemicals. *J. Econ. Ent.* 30:380.

Pérez-Hedo, M., M. Eizaguirre, and F. Sehnal. 2010. Brain-independent development in the moth *Sesamia nonagrioides. J. Insect Physiol.* 56:594–602.

Pérez-Hedo, M., W.C. Goodman, C. Schafellner, A. Martini, F. Sehnal, and M. Eizaguirre. 2011. Control of larval-pupal-adult molt in the moth *Sesamia nonagrioides* by juvenile hormone and ecdysteroids. *J. Insect Physiol.* 57:602–607.

Perrot-Minnot, M.J., L.R. Guo, and J.H. Werren. 1996. Single and double infections with *Wolbachia* in the parasitic wasp *Nasonia vitripennis*: effects on compatibility. *Genetics* 143:961–972.

Peschel, N. and C. Helfrich-Förster. 2011. Setting the clock by nature: circadian rhythm in the fruitfly *Drosophila melanogaster. FEBS Lett.* 585:1435–1442.

Petrice, T.R., D.L. Miller, L.S. Bauer, T.M. Poland, and F.W. Ravlin. 2019. Photoperiodic modulation of diapause induction and termination in *Oobius agrili* Zhang and Huang (Hymenoptera: Encyrtidae), an egg parasitoid of the invasive emerald ash borer. *Biol. Cont.* 138:104047.

Peypelut, L., P. Beydon, and L. Lavenseau. 1990. 20-hydroxyecdysone triggers the resumption of imaginal wing disc development after diapause in the European corn borer, *Ostrinia nubilalis. Arch. Insect Biochem. Physiol.* 15:1–19.

Pfenning, B., S. Gerstner, and H.J. Poethke. 2008. Alternative life histories in the water strider *Gerris lacustris*: time constraint on wing morph and voltinism. *Ent. Exp. Appl.* 129:235–242.

Piekarski, P.K., J.M. Carpenter, A.R. Lemmon, E.M. Lemmon, and B.J. Sharanowski. 2018. Phylogenetic evidence overturns current conceptions of social evolution in wasps (Vespidae). *Mol. Biol. Evol.* 35:2097–2109.

Pieloor, M.J. and J.E. Seymour. 2001. Factors affecting adult diapause initiation in the tropical butterfly *Hypolimnas bolina* L. (Lepidoptera: Nymphalidae). *Aust. J. Ent.* 40:376–379.

Piiroinen, S., T. Ketola, A. Lyytinen, and L. Lindstrom. 2011. Energy use, diapause behavior and northern range expansion potential in the invasive Colorado potato beetle. *Funct. Ecol.* 25:527–536.

Pinones-Tapia, D., R.S. Rios, and E. Gianoli. 2017. Pupal colour dimorphism in a desert swallowtail (Lepidoptera: Papilionidae) is driven by changes in food availability, not photoperiod. *Ecol. Ent.* 42:636–644.

Pintureau, B., J. Pizzol, and P. Bolland. 2003. Effects of endosymbiotic *Wolbachia* on the diapause in *Trichogramma* hosts and effects of the diapause on Wolbachia. *Ent. Exp. Appl.* 106:193–200.

Pinyarat, W., T. Shimada, W.-H. Xu, Y. Sato, O. Yamashita, and M. Kobayashi. 1995. Linkage analysis of the gene encoding precursor protein of diapause hormone and pheromone biosynthesis-activating neuropeptide in the silkworm, *Bombyx mori. Genet. Res.* 65:105–111.

Pires, C.S.S., E.R. Suji, E.M.G. Fontes, C.A. Tauber, and M.J. Tauber. 2000. Dry-season embryonic dormancy in *Deois flavopicta* (Homoptera: Cercopidae): role of temperature and moisture in nature. *Environ. Ent.* 29:714–720.

Pittendrigh, C.S. 1966. The circadian oscillation in *Drosophila pseudoobscura* pupae: a model for the photoperiodic clock. *Z. Pflanzenphysiol.* 54:275–307.

Pittendrigh, C.S. and D.H. Minis. 1964. The entrainment of circadian oscillations by light and their role as photoperiodic clocks. *Am. Nat.* 98:261–299.

Pittendrigh, C.S., J.H. Eichhorn, D.H. Minis, and V.G. Bruce. 1970. Circadian systems. VI. Photoperiodic time measurement in *Pectinophora gossypiella*. *Proc. Nat'l. Acad. Sci., USA* 66:758–764.

Pizzol, J. and B. Pintureau. 2008. Effect of photoperiod experienced by parents on diapause induction in *Trichogramma cacoeciae*. *Ent. Exp. Appl.* 127:72–77.

Ploomi, A., A. Kuusik, K. Jögar, L. Metspalu, K. Hiiesaar, R. Karise, I.H. Williams, I. Sibul, and M. Mänd. 2018. Variability in metabolic rate and gas exchange patterns of the Colorado potato beetle of winter and prolonged diapauses. *Physiol. Ent.* 43:251–258.

Poelchau, M.F., J.A. Reynolds, D.L. Denlinger, C.G. Elsik, and P.A. Armbruster. 2011. A *de novo* transcriptome of the Asian tiger mosquito, *Aedes albopictus*, to identify candidate transcripts for diapause preparation. *BMC Genomics* 12:619.

Poelchau, M.F., J.A. Reynolds, D.L. Denlinger, C.G. Elsik, and P.A. Armbruster. 2013a. Deep sequencing reveals complex mechanisms of diapause preparation in the invasive mosquito, *Aedes albopictus*. *Proc. R. Soc. B* 280:20130143.

Poelchau, M.F., J.A. Reynolds, C.G. Elsik, D.L. Denlinger, and P.A. Armbruster. 2013b. RNA-Seq reveals early distinctions and late convergence of gene expression between diapause and quiescence in the Asian tiger mosquito, *Aedes albopictus*. *J. Exp. Biol.* 216:4082–4090.

Poikela, N., V. Tyukmaeva, A. Hoikkala, and M. Kankare. 2021. Multiple paths to cold tolerance: the role of environmental cues, morphological traits and circadian clock gene *vrille*. *BMC Ecol. Evol.* 21:117.

Polgar, L.A., M. Mackauer, and W. Volkl. 1991. Diapause induction in two species of aphid parasitoids: the influence of aphid morph. *J. Insect Physiol.* 37:699–702.

Polgar, L.A., B. Darvas, and W. Volkl. 1995. Induction of dormancy in aphid parasitoids: implications for enhancing their field effectiveness. *Agric. Ecosyst. Environ.* 52:19–23.

Polgar, L.A. and J. Hardie. 2000. Diapause induction in aphid parasitoids. *Ent. Exp. Appl.* 97:21–27.

Posledovich, D., T. Toftegaard, C. Wiklund, J. Ehrlén, and K. Gotthard. 2015. Latitudinal variation in diapause duration and post-winter development in two pierid butterflies in relation to phenological specialization. *Oecologia* 177:181–190.

Poulton, E. 1936. Assemblies of coccinellid beetles observed in N. Uganda (1927) by Prof. Hale Carpenter and in Beuchuanaland (1935) by Dr. W. A. Lamborn. *Proc. Roy. Ent. Soc. London A* 11:99–100.

Poupardin, R., K. Schöttner, J. Korbelová, J. Provazník, D. Doležel, D. Pavlinic, V. Beneš, and V. Koštál. 2015. Early transcriptional events linked to induction of diapause revealed by RNAseq in larvae of drosophilid fly, *Chymomyza costata*. *BMC Genomics* 16:720.

Powell, J.A. 1987. Records of prolonged diapause in Lepidoptera. *J. Res. Lepid.* 25:83–109.

Powell, J.A. 2001. Longest insect dormancy: yucca moth larvae (Lepidoptera: Prodoxidae) metamorphose after 20, 25, and 30 years in diapause. *Ann. Ent. Soc. Am.* 94:677–680.

Powell, T.H.Q., A. Nguyen, Q. Xia, J.L. Feder, G.J. Ragland, and D.A. Hahn. 2020. A rapidly evolved shift in life history timing during ecological speciation is driven by the transition between developmental phases. *J. Evol. Biol.* 33:1371–1386.

Prasai, K. and B. Karlsson. 2011. Variation in immune defence in relation to developmental pathway in the green-veined white butterfly, *Pieris napi*. *Evol. Ecol. Res.* 13:295–305.

Pruisscher, P., H. Larsdotter-Mellström, C. Stefanescu, S. Nylin, C.W. Wheat, and K. Gotthard. 2017. Sex-linked inheritance of diapause induction in the butterfly *Pieris napi*. *Physiol. Ent.* 42:257–265.

Pruisscher, P., S. Nylin, K. Gotthard, and C.W. Wheat. 2018. Genetic variation underlying local adaptation of diapause induction along a cline in a butterfly. *Mol. Ecol.* 27:3613–3626.

Ptashne, M. 2013. Epigenetics: core miscon-cept. *Proc. Nat'l. Acad. Sci., USA* 110:7101–7103.

Pullin, A.S. 1987. Adult feeding time, lipid accumulation, and overwintering in *Aglais urticae* and *Inachis io* (Lepidoptera: Nymphalidae). *J. Zool., Lond.* 211: 631–641.

Pullin, A.S. and J.S. Bale. 1989. Effects of ecdysone, juvenile hormone and haemo-lymph transfer on cryoprotectant metabol-ism in diapausing and non-diapausing pupae of *Pieris brassicae*. *J. Insect Physiol.* 35:911–918.

Pullin, A.S. and H. Wolda. 1993. Glycerol and glucose accumulation during diapause in a tropical beetle. *Physiol. Ent.* 18:75–78.

Quan, W.-L., X.-L. Zheng, X.-X. Li, X-M. Zhou, W.-H. Ma, and X.-P. Wang. 2013. Overwintering strategy of endoparasitois in *Chilo suppressalis*: a perspective from the cold hardiness of a host. *Ent. Exp. Appl.* 146:398–403.

Raak-van den Berg, C.L., P.W. de Jong, L. Hemerik, and J.C. van Lenteren. 2013. Diapause and post-diapause quiescence demonstrated in overwintering *Harmonia axyridis* (Coleoptera: Coccinellidae) in northwestern Europe. *Eur. J. Ent.* 110:585–591.

Rabb, R.L. 1966. Diapause in *Protoparce sexta* (Lepidoptera: Sphingidae). *Ann. Ent. Soc. Am.* 59:160–165.

Rafferty, A.R. and R.D. Reina. 2012. Arrested embryonic development: a review of strat-egies to delay hatching in egg-laying rep-tiles. *Proc. Biol. Sci.* 279:2299 2308.

Ragland, G.J., J. Fuller, J.L. Feder, and D.A. Hahn. 2009. Biphasic metabolic rate trajec-tory of pupal diapause termination and post-diapause development in a tephritid fly. *J. Insect Physiol.* 55:344–350.

Ragland, G.J., D.L. Denlinger, and D.A. Hahn. 2010. Mechanisms of suspended ani-mation are revealed by transcript profiling of diapause in the flesh fly. *Proc. Nat'l. Acad. Sci., USA* 107:14909–14914.

Ragland, G.J., S.P. Egan, J.L. Feder, S.H. Berlocher, and D.A. Hahn. 2011. Developmental trajectories of gene expres-sion reveal candidates for diapause termin-ation: a key life-history transition in the apple maggot *Rhagoletis pomonella*. *J. Exp. Biol.* 214:3948–3959.

Ragland, G.J., M.M. Doellman, P.J. Meyers, G.R. Hood, S.P. Egan, T.H.Q. Powell, D.A. Hahn, P. Nosil, and J.L. Feder. 2017. A test of genomic modularity among life-history adaptations promoting speciation with gene flow. *Mol. Ecol.* 26:3926–3942.

Ragland, G.J. and E. Keep. 2017. Comparative transcriptomics support evolutionary conver-gence of diapause responses across Insecta. *Physiol. Ent.* 42:246–256.

Ragland, G.J., P.A. Armbruster, and M.E. Meuti. 2019. Evolutionary and functional genetics of insect diapause: a call for inte-gration. *Curr. Opin. Insect Sci.* 36:74–81.

Ramírez-Soria, M.J., F. Wäckers, and J.A. Sanchez. 2019. When natural enemies go to sleep: diapause induction and termin-ation in the pear psyllid predator *Pilophorus gallicus* (Hemiptera: Miridae). *Pest Manag. Sci.* 75:3293–3301.

Rankin, M.A. and S. Rankin. 1980. Some factors affecting presumed migratory flight activity of the convergent ladybeetle, *Hippodamia convergens* (Coccinellidae: Coleoptera). *Biol. Bull.* 158: 356–369.

Ravn, M.V., J.B. Campbell, L. Gerber, J.F. Harrison, and J. Overgaard. 2019. Effects of anoxia on ATP, water, ion and pH bal-ance in an insect (*Locusta migratoria*). *J. Exp. Biol.* 222.190850.

Ray, S., U.K. Valekunja, A. Stangherlin, S.A. Howell, A.P. Snijders, G. Damodaran, and A.B. Reddy. 2020. Circadian rhythms in the absence of the clock gene Bmal1. *Science* 367:800–806.

Readio, J., M.H. Chen, and R. Meola. 1999. Juvenile hormone biosynthesis in diapausing and nondiapausing *Culex pipiens* (Diptera: Culicidae). *J. Med. Ent.* 36:355–360.

Reierth, E. and K.-A. Stokkan. 1998. Activity rhythm in high Arctic Svalbard ptarmigan (*Lagopus mutus hyperboreus*). *Can. J. Zool.* 76:2031–2039.

Ren, S., Y.-J. Hao, B. Chen, and Y.-P. Yin. 2018. Global transcriptome sequencing reveals molecular profiles of summer diapause induction stage of onion maggot, *Delia antiqua* (Diptera: Anthomyiidae). *G3* 8:207–217.

Renfree, M.B. and J.C. Fenelon. 2017. The enigma of embryonic diapause. *Development* 144:3199–3210.

Reppert, S.M., R.J. Gegear, and C. Merlin. 2010. Navigational mechanisms of migrating monarch butterflies. *Trends Neurosci.* 33:399–406.

Reppert, S.M., P.A. Guerra, and C. Merlin. 2016. Neurobiology of monarch butterfly migration. *Ann. Rev. Ent.* 61:25–42.

Revel, F.G., A. Herwig, M.L. Garidou, H. Dardente, J.S. Menet, M. Masson-Pevet, V. Simonneaux, M. Saboureau, and P. Pevet. 2007. The circadian clock stops ticking during deep hibernation in the European hamster. *Proc. Nat'l. Acad. Sci., USA* 104:13816–13820.

Rewitz, K.F., N. Yamanaka, L.I. Gilbert, and M.B. O'Conner. 2009. The insect neuropeptide PTTH activates receptor tyrosine kinase torso to initiate metamorphosis. *Science* 326:1403–1405.

Reynolds, J.A. 2017. Epigenetic influences on diapause. *Adv. Insect Physiol.* 53:115–144.

Reynolds, J.A. 2019. Noncoding RNA regulation of dormant states in evolutionarily diverse animals. *Biol. Bull.* 237:192–209.

Reynolds, J.A. and S.C. Hand. 2009a. Decoupling development and energy flow during embryonic diapause in the cricket, *Allonemobius socius*. *J. Exp. Biol.* 212:2065–2074.

Reynolds, J.A. and S.C. Hand. 2009b. Embryonic diapause highlighted by differential expression of mRNAs for ecdysteroidogenesis, transcription and lipid sparing in the cricket *Allonemobius socius*. *J. Exp. Biol.* 212:2075–2084.

Reynolds, J.A., M.F. Poelchau, Z. Rahman, P. Armbruster, and D.L. Denlinger. 2012. Transcript profiling reveals mechanisms for lipid conservation during diapause in the mosquito, *Aedes albopictus*. *J. Insect Physiol.* 58:966–973.

Reynolds, J.A., J. Clark, S.J. Diakoff, and D.L. Denlinger. 2013. Transcriptional evidence for small RNA regulation of pupal diapause in the flesh fly, *Sarcophaga bullata*. *Insect Biochem. Mol. Biol.* 76:29–37.

Reynolds, J.A., R. Bautista-Jimenez, and D.L. Denlinger. 2016. Histone acetylation as potential mediators of pupal diapause in the flesh fly, *Sarcophaga bullata*. *Insect Biochem. Mol. Biol.* 76:29–37.

Reynolds, J.A., J.T. Peyton, and D.L. Denlinger. 2017. Changes in microRNA abundance may regulate diapause in the flesh fly *Sarcophaga bullata*. *Insect Biochem. Mol. Biol.* 84:1–14.

Reynolds, J.A., R.J. Nachman, and D.L. Denlinger. 2019. Distinct microRNA and mRNA responses elicited by ecdysone, diapause hormone and a diapause hormone analog at diapause termination in pupae of the corn earworm, *Helicoverpa zea*. *Gen. Comp. Endocrin.* 278:68–78.

Reznik, S.Y., T.S. Kats, T.Y. Umarova, and N.D. Voinovich. 2002. Maternal age and endogenous variation in maternal influence on photoperiodic response in the progeny diapause in *Trichogramma embryophgum* (Hymenoptera: Trichogrammatidae). *Eur. J. Ent.* 99:175–179.

Reznik, S.Y. and N.P. Vaghina. 2011. Photoperiodic control of development and reproduction in *Harmonia axyridis*

(Coleoptera: Coccinellidae). *Eur. J. Ent.* 108:385–390.

Reznik, S.Y., N.D. Voinovich, and N.P. Vaghina. 2011. Maternal regulation of *Trichogramma embryophagum* Htg. (Hymenoptera: Trichogrammatidae) diapause: photoperiodic sensitivity of adult females. *Biol. Cont.* 57:158–162.

Reznik, S.Y. and K.G. Samartsev. 2015. Multigenerational maternal inhibition of pre-pupal diapause in two *Trichogramma* species (Hymenoptera: Trichogrammatidae). *J. Insect Physiol.* 81:14–20.

Reznik, S.Y., M.Y. Dolgovskaya, and A.N. Ovchinnikov. 2015. Effect of photoperiod on adult size and weight in *Harmonia axyridis* (Coleoptera: Coccinellidae). *Eur. J. Ent.* 112:642–647.

Reznik, S.Y. and N. Voinovich. 2016. Diapause induction in *Trichogramma telengai*: the dynamics of maternal thermosensitivity. *Physiol. Ent.* 41:335–343.

Reznik, S.Y., A.A. Ovchinnikova, A.N. Ovchinnikov, L.V. Barabanova, and N.A. Belyakova. 2017. Inheritance of diapause regulation in the multicolored Asian ladybird *Harmonia axyridis* (Coleoptera: Coccinellidae). *Eur. J. Ent.* 114:416–421.

Reznik, S.Y., K.G. Samartsev, and M.Y. Dolgovskaya. 2020. Intrapopulation variability of the qualitative and quantitative photoperiodic responses in *Habrobracon hebetor* (Say)(Hymenoptera, Braconidae). *Ent. Rev.* 100:277–286.

Ricci, C., L. Ponti, and A. Pires. 2005. Migratory flight and pre-diapause feeding of *Coccinella septempunctata* (Coleoptera) adults in agricultural and mountain ecosystems of Central Italy. *Eur. J. Ent.* 102:531–538.

Rich, C. and T. Longcore, eds. 2006. *Ecological Consequences of Artificial Night Lighting*, Washington, DC: Island Press.

Richard, D.S. and D.S. Saunders. 1987. Prothoracic gland function in diapause and non-diapause *Sarcophaga argyrostoma* and *Calliphora vicina*. *J. Insect Physiol.* 33:385–392.

Richard, D.S., N.L. Watkins, R.B. Serfin, and L.I. Gilbert. 1998. Ecdysteroids regulate yolk protein uptake by *Drosophila melanogaster* oocytes. *J. Insect Physiol.* 44:637–644.

Richard, D.S., M. Gilbert, B. Crum, D.M. Hollinshead, S. Schelble, and D. Scheswohl. 2001a. Yolk protein endocytosis by oocytes in *Drosophila melanogaster*: immunofluorescent localization of clathrin, adaptin and yolk protein receptor. *J. Insect Physiol.* 47:715–723.

Richard, D.S., J.M. Jones, M.R. Barbarito, S. Cerula, J.P. Detweiler, S.J. Fisher, D.M. Brannigan, and D.M. Scheswohl. 2001b. Vitellogenesis in diapausing and mutant *Drosophila melanogaster*: further evidence for the relative roles of ecdysteroids and juvenile hormones. *J. Insect Physiol.* 47:905–913.

Riddle, D.L. 1997. The dauer larva. In *C. elegans II*, ed. D.L. Riddle, T. Blumenthal, B.J. Meyer, and J.R. Priess. Cold Spring Harbor: CSHL Press, pp. 739–768.

Rider, M.H., N. Hussain, S.M. Dilworth, J.M. Storey, and K.B. Storey. 2011. AMP-activated protein kinase and metabolic regulation in cold-hardy insects. *J. Insect Physiol.* 57:1453–1462.

Ridsdill-Smith, J., C. Pavri, E. de Boer and D. Kriticos. 2005. Predictions of summer diapause in the redlegged earth mite, *Halotydeus destructor* (Acari: Penthaleidae), in Australia. *J. Insect Physiol.* 51:717–726.

Riihimaa, A.J. and M.T. Kimura. 1988. A mutant strain of *Chymomyza costata* (Diptera: Drosophilidae) insensitive to diapause-inducing action of photoperiod. *Physiol. Ent.* 13:441–445.

Rinehart, J.P. and D.L. Denlinger. 2000. Heat-shock protein 90 is down-regulated during pupal diapause in the flesh fly, *Sarcophaga crassipalpis*, but remains responsive to thermal stress. *Insect Mol. Biol.* 9:641–645.

Rinehart, J.P., G.D. Yocum, and D.L. Denlinger. 2000. Developmental upregulation of inducible hsp70 transcripts, but not the cognate form, during pupal diapause in the flesh fly, *Sarcophaga crassipalpis*. *Insect Biochem. Mol. Biol.* 30:515–521.

Rinehart, J.P., R.A. Cikra-Ireland, R.D. Flannagan, and D.L. Denlinger. 2001. Expression of ecdysone receptor is unaffectedby pupal diapause in the flesh fly, *Sarcophaga crassipalpis*, while its dimerization partner, USP, is downregulated. *J. Insect Physiol.* 47:915–921.

Rinehart, J.P., S.J. Diakoff, and D.L. Denlinger. 2003. *Sarcotoxin II* from the flesh fly *Sarcophaga crassipalpis* (Diptera): a comparison of transcript expression in diapausing and nondiapausing pupae. *Eur. J. Ent.* 100:251–254.

Rinehart, J.P., R.M. Robich, and D.L. Denlinger. 2006. Enhanced cold and desiccation tolerance in diapausing adults of *Culex pipiens*, and a role for *Hsp70* in response to cold shock but not as a component of the diapause program. *J. Med. Ent.* 43:713–722.

Rinehart, J.P., A. Li, G.D. Yocum, R.M. Robich, S.A.L. Hayward, and D.L. Denlinger. 2007. Up-regulation of heat shock proteins is essential for cold survival during insect diapause. *Proc. Nat'l. Acad. Sci., USA* 104:11130–11137.

Rinehart, J.P., R.M. Robich, and D.L. Denlinger. 2010. Isolation of diapause-regulated genes from the flesh fly, *Sarcophaga crassipalpis* by suppressive subtractive hybridization. *J. Insect Physiol.* 56:603–609.

Ring, R.A. 1967. Maternal induction of diapause in larvae of *Lucilia Caesar* L. (Diptera: Calliphoridae). *J. Exp. Biol.* 46:123–136.

Ritdachyeng, E., M. Manaboon, S.S. Tobe, and T. Singtripop. 2013. Possible roles of juvenile hormone and juvenile hormone binding protein in changes in the integument during termination of larval diapause in the bamboo borer *Omphisa fuscidentalis*. *Physiol. Ent.* 38:183–191.

Rivers, D.B. and D.L. Denlinger. 1994. Redirection of metabolism in the flesh fly, *Sarcophaga bullata,* following envenomation by the ectoparasitoid *Nasonia vitripennis* and correlation of metabolic effects with the diapause status of the host. *J. Insect Physiol.* 40:207–215.

Rivers, D.B. and D.L. Denlinger. 1995. Fecundity and development of the ectoparasitic wasp *Nasonia vitripennis* are dependent on host quality. *Ent. Exp. Appl.* 76:15–24.

Rivers, D.B., R.E. Lee, Jr., and D.L. Denlinger. 2000. Cold hardiness of the fly pupal parasitoid *Nasonia vitripennis* is enhanced by its host, *Sarcophaga crassipalpis.* *J. Insect Physiol.* 46:99–106.

Robbins, H. M., G. Van Stappen, P. Sorgeloos, Y.Y. Sung, T.H. MacRae, and P. Bossier. 2010. Diapause termination and development of encysted Artemia embryos: roles for nitric oxide and hydrogen peroxide. *J. Exp. Biol.* 213:1464–1470.

Robich, R.M. and D.L. Denlinger. 2005. Diapause in the mosquito *Culex pipiens* evokes a metabolic switch from blood feeding to sugar gluttony. *Proc. Nat'l. Acad. Sci., USA* 102:15912–15917.

Robich, R.M., J.P. Rinehart, L.J. Kitchen, and D.L. Denlinger. 2007. Diapause-specific gene expression in the northern house mosquito, *Culex pipiens* L., identified by suppressive subtractive hybridization. *J. Insect Physiol.* 53:235–245.

Rochefort, S., R. Berthiaume, C. Hebert, M. Charest, and E. Bauce. 2011. Effect of temperature and host tree on cold hardiness of hemlock looper eggs along a latitudinal gradient. *J. Insect Physiol.* 57:751–759.

Rock, G.C. 1983. Thermoperiodic effects on the regulation of larval diapause in the tufted apple budworm (Lepidoptera: Tortricidae). *Environ. Ent.* 12:1500–1503.

Rockey, S.J. and D.L. Denlinger. 1983. Deuterium oxide alters pupal diapause response in the flesh fly, *Sarcophaga crassipalpis*. *Physiol. Ent.* 8:445–449.

Rockey, S.J. and D.L. Denlinger. 1986. Influence of maternal age on incidence of pupal diapause in the flesh fly, *Sarcophaga bullata*. *Physiol. Ent.* 11:199–203.

Rockey, S.J., J.H. Hainze, and J.M. Scriber. 1987. Evidence of a sex-linked diapause response in *Papilio glaucus* subspecies and their hybrids. *Physiol. Ent.* 12:181–184.

Rockey, S.J., B.B. Miller, and D.L. Denlinger. 1989. A diapause maternal effect in the flesh fly *Sarcophaga bullata*: transfer of information from mother to progeny. *J. Insect Physiol.* 35:553–558.

Roditakis, N.E. and M.G. Karandinos. 2001. Effects of photoperiod and temperature on pupal diapause induction of grape berry moth *Lobesia botrana*. *Physiol. Ent.* 26:329–340.

Romney A.L. and J.E. Podrabsky. 2017. Transcriptomic analysis of maternally provisioned cues for phenotypic plasticity in the annual killifish, *Austrofundulus limnaeus*. *EvoDevo* 8:6.

Roncalli, V., S.A. Sommer, M.C. Cieslak, C. Clarke, R.R. Hopcroft, and P.H. Lenz. 2018. Physiological characterization of the emergence from diapause: a transcriptomic approach. *Sci. Rep.* 8:12577.

Roseler, P.-F. 1985. A technique for year-round rearing of *Bombus terrestris* (Apidae, Bombini) colonies in captivity. *Apidologie* 16:165–170.

Roubaud, E. 1930. Suspension évolutive et hibernation larvaire obligatoire, provoquées par la chaleur, chez le moustique commun, *Culex pipiens* L. Les diapauses vraies et les pseudo-diapauses chez les insectes. *C.R. Acad. Sci., Paris* 190:324–326.

Roubik, D.W. and C.D. Michener. 1980. The seasonal cycle and nests of *Epicharis zonata*, a bee whose cells are below the wet-season water table (Hymenoptera, Anthophoridae). *Biotropica* 12:56–60.

Roulin, A.C., Y. Bourgeois, U. Stiefel, J.-C. Walser, and D. Ebert. 2016. A photoreceptor contributes to the natural variation of diapause induction in *Daphnia magna*. *Mol. Biol. Evol.* 33:3194–3204.

Rowarth, N.M. and T.H. MacRae. 2018a. Post-diapause synthesis of ArHsp40-2, a type 2J-domain protein from *Artemia franciscana*, is developmentally regulated and induced by stress. *PLoS One* 13:0201477.

Rowarth, N.M. and T.H. MacRae. 2018b. ArHsp40 and ArHsp40-2 contribute to stress tolerance and longevity in *Artemia franciscana*, but only ArHp40 influences diapause entry. *J. Exp. Biol.* 221:189001.

Roy, D.B. and T.H. Sparks. 2000. Phenology of British butterflies and climate change. *Glob. Change Biol.* 6:407–416.

Rozsypal, J. and V. Koštál. 2018. Supercooling and freezing as eco-physiological alternatives rather than mutually exclusive strategies: a case study in *Pyrrhocoris apterus*. *J. Insect Physiol.* 111:53–62.

Rozsypal, J., M. Moos, I. Rudolf, and V. Koštál. 2021. Do energy reserves and cold hardiness limit winter survival of *Culex pipiens*? *Comp. Biochem. Physiol. A* 255:110912.

Ruan C.-C., W.-M. Du, X.-M. Wang, J.-J. Zhang, and L.-S. Zang. 2012. Effect of long-term cold storage on the fitness of pre-wintering *Harmonia axyridis* (Pallas). *BioControl* 57:95–102.

Rull, J., E. Tadeo, R.L. Lasa, and M. Aluja. 2017. The effect of winter length on duration of dormancy and survival of specialized herbivorous *Rhagoletis* fruit flies from high elevation environments with acyclic climatic variability. *Bull. Ent. Res.* 108:461–470.

Rull, J., R. Lasa, and M. Aluja. 2019. The effect of seasonal humidity on survival and duration of dormancy on diverging Mexican *Rhagoletis pomonella* (Diptera:

Tephritidae) populations inhabiting different environments. *Environ. Ent.* 48:1121–1128.

Rundle, B.J. and A.A. Hoffmann. 2003. Overwintering of *Trichogramma funiculatum* Carver (Hymenoptera: Trichogrammatidae) under semi-natural conditions. *Environ. Ent.* 32:290–298.

Ryan, S.F., M.C. Fontaine, J.M. Scriber, M.E. Pfrender, S.T. O'Neil, and J.J. Hellman. 2016. Patterns of divergence across the geographic and genomic landscape of a butterfly hybrid zone associated with a climatic gradient. *Mol. Ecol.* 26:4725–4742.

Ryan, S.F., P. Valella, G. Thivierge, M.L. Aardema, and J.M. Scriber. 2018. The role of latitudinal, genetic and temperature variation in the induction of diapause of *Papilio glaucus* (Lepidoptera: Papilionidae). *Insect Sci.* 25:328–336.

Sadakiyo, S. and M. Ishihara. 2011. Rapid seasonal adaptation of an alien bruchid after introduction: geographic variation in life cycle synchronization and critical photoperiod for diapause induction. *Ent. Exp. Appl.* 140:69–76.

Sadakiyo, S. and M. Ishihara. 2012. Cost of male diapause indirectly affects female reproductive performance. *Ent. Exp. Appl.* 143:42–46.

Safranek, L. and C.M. Williams. 1980. Studies on the prothoracicotropic hormone in the tobacco hornworm, *Manduca sexta*. *Biol. Bull.* 158:141–153.

Sahoo, A., J. Dandapat, and L. Samanta. 2015. Oxidative damaged products, level of hydrogen peroxide, and antioxidant protection in diapausing pupae of the Tasar silk worm, *Antheraea mylitta*: a comparative study in two voltine groups. *Int'l. J. Insect Sci.* 7:11–17.

Sahota, T.S., D.S. Ruth, A. Ibaraki, S.H. Farris, and F.G. Peet. 1982. Diapause in the pharate adult stage of insect development. *Can. Ent.* 114:1179–1183.

Sahota, T.S., A. Ibaraki, and S.H. Farris. 1985. Pharate-adult diapause of *Barbara*

colfaxiana (Kft.): differentiation of 1- and 2-year dormancy. *Can. Ent.* 117:873–876.

Sakagami, S.F., K. Tanno, H. Tsutsui, and K. Honma. 1985. The role of cocoons in overwintering of the soybean pod borer *Leguminivora glycinivorella* (Lepidoptera: Tortricidae). *J. Kans. Ent. Soc.* 58:240–247.

Sakamoto, T. and K. Tomioka. 2007. Effects of unilateral compound eye removal on the photoperiodic response of nymphal development in the cricket *Modicogryllus siamensis. Zool. Sci.* 24:604–610.

Sakamoto, T., O. Uryu, and K. Tomioka. 2009. The clock gene *period* plays an essential role in photoperiodic control of nymphal development in the cricket *Modicogryllus siamensis. J. Biol. Rhyth.* 24:379–390.

Salama, M. and T.A. Miller. 1992. A diapause-associated protein of the pink bollworm *Pectinophora gossypiella. Arch. Insect Biochem. Physiol.* 21:1–11.

Salavert, V., C. Zamora-Munoz, M. Ruiz-Rodriguez, and J.J. Soler. 2011. Female-biased size dimorphism in a diapausing caddisfly, *Mesophylax aspersus*: effect of fecundity and natural and sexual selection. *Ecol. Ent.* 36:389–395.

Salman, M.H.R., K. Hellrigl, S. Minerbi, and A. Battisti. 2016. Prolonged pupal diapause drives population dynamics of the pine processionary moth (*Thaumetopoea pityocampa*) in an outbreak expansion area. *Forest Ecol. Manag.* 361:375–381.

Salman, M.H.R., C.P. Bonsignore, A.E.A. El Fels, F. Giomi, J.A. Hodar, M. Laparie, L. Marini, C. Merel, M.P. Zalucki, M. Zamoum, and A. Battisti. 2019a. Winter temperature predicts prolonged diapause in pine processionary moth species across geographic range. *PeerJ* 7:e6530.

Salman, M.H.R., F. Giomi, M. Laparie, P. Lehmann, A. Pitacco, and A. Battisti. 2019b. Termination of pupal diapause in the pine processionary moth *Thaumetopoea pityocampa. Physiol. Ent.* 44:53–59.

Salminen, T.S., L. Vesala, and A. Hoikkala. 2012. Photoperiodic regulation of life-history traits before and after eclosion: egg-to-adult development time, juvenile body mass and reproductive diapause in *Drosophila montana*. *J. Insect Physiol.* 58:1541–1547.

Salminen, T.S. and A. Hoikkala. 2013. Effect of temperature on the duration of sensitive period and on the number of photoperiodic cycles required for the induction of reproductive diapause in *Drosophila montana*. *J. Insect Physiol.* 59:450–457.

Salminen, T.S., L. Vesala, A. Laiho, M. Merisalo, A. Hoikkala, and M. Kankare. 2015. Seasonal gene expression kinetics between diapause phases in *Drosophila virilis* group species and overwintering differences between diapausing and non-diapausing females. *Sci. Rep.* 5:11197.

Salom, S.M., A.A. Sharov, W.T. Mays, and J.W. Neal, Jr. 2001. Evaluation of aestival diapause in hemlock woolly adelgid (Homoptera: Adelgidae). *Environ. Ent.* 30:877–882.

Samayoa, A.C. and S.-Y. Hwang. 2018. Degradation capacity and diapause effects on oviposition of *Hermetia illucens* (Diptera: Stratiomyidae). *J. Econ. Ent.* 111:1682–1690.

Sandrelli, F., E. Tauber, M. Pegoraro, G. Mazzotta, P. Cisotto et al. 2007. A molecular basis for natural selection at the *timeless* locus in *Drosophila melanogaster*. *Science* 316:1898–1900.

Santos, P.K.F., N. de Souza Araujo, E. Francoso, A.R. Zuntini, and M.C. Arias. 2018. Diapause in a tropical oil-collecting bee: molecular basis unveiled by RNA-Seq. *BMC Genomics* 19:305.

Santos, P.K.F., M.C. Arias, and K.M. Kapheim. 2019. Loss of developmental diapause as prerequisite for social evolution in bees. *Biol. Lett.* 15:20190398.

Saravanakumar, R., K.M. Ponnuvel, and S.M.H. Qadri. 2010. Genetic stability analysis of diapause-induced multivoltine silkworm *Bombyx mori* germplasm using inter simple sequence repeat markers. *Ent. Exp. Appl.* 135:170–176.

Sasibhushan, S., C.G.P. Rao, and K.M. Ponnuvel. 2013. Genome wide microarray based expression profiles during early embryogenesis in diapause induced and non-diapause eggs of polyvoltine silkworm *Bombyx mori*. *Genomics* 102:379–387.

Sato, A., T. Sokabe, M. Kashio, Y. Yasukochi, M. Tominaga, and K. Shiomi. 2014. Embryonic thermosensitive TRPA1 determines transgenerational diapause phenotype of the silkworm, *Bombyx mori*. *Proc. Nat'l. Acad. Sci., USA* 111: E1249–E1255.

Sato, K., H. Tanaka, Y. Saito, and K. Suzuki. 2002. Baculovirus-mediated production and antifungal activity of a diapause-specific peptide, diapausin, of the adult leaf beetle, *Gastrophysa atrocyanea* (Coleoptera: Chrysomelidae). *J. Insect Biotech. Ser.* 71:69–77.

Sato, Y., M. Oguchi, N. Menjo, K. Imai, H. Saito, M. Ikeda, M. Isobe, and O. Yamashita. 1993. Precursor polyprotein for multiple neuropeptides secreted from the subesophageal ganglion of the silkworm *Bombyx mori*: characterization of the cDNA encoding the diapause hormone precursor and identification of additional peptides. *Proc. Nat'l. Acad. Sci., USA* 90:3251–3255.

Sato, Y., K. Shiomi, H. Saito, K. Imai, and O. Yamashita. 1998. Phe-X-Pro-Arg-Leu-NH$_2$ peptide producing cells in the central nervous system of the silkworm, *Bombyx mori*. *J. Insect Physiol.* 44:333–342.

Sato, Y., P. Yang, Y. An, K. Matsukawa, K. Ito, S. Imanishi, H. Matsuda, Y. Uchiyama, K. Imai, S. Ito, Y. Ishida, and K. Suzuki. 2010. A palmitoyl conjugate of insect pentapeptide Yamamarin arrests cell proliferation and respiration. *Peptides* 31:827–833.

Saulich, A.K. 2010. Long life cycles in insects. *Ent. Rev.* 90:1127–1152.

Saulich, A.K. and D.L. Musolin. 2018. Summer diapause as a special seasonal adaptation in insects: diversity of forms, control mechanisms, and ecological importance. *Ent. Rev.* 97:1183–1212.

Saulich, A.K. and D.L. Musolin. 2019. Seasonal development of plant bugs (Heteroptera, Miridae): subfamily Bryocorinae. *Ent. Rev.* 99:275–291.

Saulich, A.K. and D.L. Musolin. 2020. Seasonal development of plant bugs (Heteroptera, Miridae): subfamily Mirinae, tribe Mirini. *Ent. Rev.* 100:7–38.

Sauman, I., A.D. Briscoe, H. Zhu, D. Shi, O. Froy, J. Stalleicken, Q. Yuan, A. Casselman, and S.M. Reppert. 2005. Connecting the navigational clock to sun compass input in monarch butterfly brain. *Neuron* 46:457–467.

Sauman, I. and S.M. Reppert. 1996. Molecular characterization of prothoracicotropic hormone (PTTH) from the giant silkmoth *Antheraea pernyi*: developmental appearance of PTTH-expressing cells and relationship to circadian clock cells in central brain. *Dev. Biol.* 178:418–429.

Sauman, I. and F. Sehnal. 1997. Immunohistochemistry of the products of male accessory glands in several hemimetabolous insects and the control of their secretion in *Pyrrhocoris apterus* (Heteroptera: Pyrrhocoridae). *Eur. J. Ent.* 94:349–360.

Saunders, D.S. 1965. Larval diapause of maternal origin: induction of diapause in *Nasonia vitripennis* (Walk.) (Hymenoptera: Pteromalidae). *J. Exp. Biol.* 42:395–508.

Saunders, D.S. 1970. Circadian clock in insect photoperiodism. *Science* 169:601–603.

Saunders, D.S. 1971. The temperature-compensated photoperiodic clock "programming" development and pupal diapause in the flesh-fly, *Sarcophaga argyrostoma*. *J. Insect Physiol.* 17:801–812.

Saunders, D.S. 1973. Thermoperiodic control of diapause in an insect: theory of internal coincidence. *Science* 181:358–360.

Saunders, D.S. 1974. Evidence for "dawn" and "dusk" oscillators in the *Nasonia* photoperiodic clock. *J. Insect Physiol.* 20:77–88.

Saunders, D.S. 1975a. Spectral sensitivity and intensity thresholds in *Nasonia* photoperiodic clock. *Nature* 253:732–734.

Saunders, D.S. 1975b. Skeleton photoperiods and the control of diapause and development in the flesh fly *Sarcophaga argyrostoma*. *J. Comp. Physiol. A* 97:97–112.

Saunders, D.S. 1981. Insect photoperiodism – the clock and counter: a review. *Physiol. Ent.* 6:99–116.

Saunders, D.S. 1984. Photoperiodic time measurement in *Sarcophaga argyrostoma*: an attempt to use daily temperature cycles to distinguish external from internal coincidence. *J. Comp. Physiol.* 154:789–794.

Saunders, D.S. 1990. The circadian basis of ovarian diapause regulation in *Drosophila melanogaster*: is the *period* gene causally involved in photoperiodic time measurement? *J. Biol. Rhyth.* 5:315–331.

Saunders, D.S. 1997. Under-sized larvae from short-day adults of the blow fly, *Calliphora vicina*, side-step the diapause programme. *Physiol. Ent.* 22:249–255.

Saunders, D.S. 2000. Larval diapause duration and fat metabolism in three geographic strains of the blow fly, *Calliphora vicina*. *J. Insect Physiol.* 46:509–517.

Saunders, D.S. 2002. *Insect Clocks*, 3rd ed., Amsterdam: Elsevier.

Saunders, D.S. 2005. Erwin Bünning and Tony Lees, two giants of chronobiology, and the problem of time measurement in insect photoperiodism. *J. Insect Physiol.* 51:599–608.

Saunders, D.S. 2010a. Photoperiodism in insects: migration and diapause responses. In *Photoperiodism, the Biological Calendar*, ed. R.J. Nelson, D.L. Denlinger, and D.E. Somers, Oxford: Oxford University Press, pp. 218–257.

Saunders, D.S. 2010b. Controversial aspects of photoperiodism in insects and mites. *J. Insect Physiol.* 56:1491–1502.

Saunders, D.S. 2011. Unity and diversity in the insect photoperiodic mechanism. *Ent. Sci.* 14:235–244.

Saunders, D.S. 2012. Insect photoperiodism: seeing the light. *Physiol. Ent.* 37:207–218.

Saunders, D.S. 2013. Insect photoperiodism: measuring the night. *J. Insect Physiol.* 59:1–10.

Saunders, D.S. 2016. The temporal "structure" and function of the insect photoperiodic clock: a tribute to Colin S. Pittendrigh. *Physiol. Ent.* 41:1–18.

Saunders, D.S. 2020a. Dormancy, diapause, and the role of the circadian system in insect photoperiodism. *Ann. Rev. Ent.* 65:373–389.

Saunders, D.S. 2020b. Insect photoperiodism: seasonal development on a revolving planet. *Eur. J. Ent.* 117:328–342.

Saunders, D.S., D. Sutton, and R.A. Jarvis. 1970. The effect of host species on diapause induction in *Nasonia vitripennis*. *J. Insect Physiol.* 16:405–416.

Saunders, D.S., V.C. Henrich, and L.I. Gilbert. 1989. Induction of diapause in *Drosophila melanogaster*: photoperiodic regulation and the impact of arrhythmic clock mutations on time measurement. *Proc. Nat'l. Acad. Sci., USA* 86:3748–3752.

Saunders, D.S., D.S. Richard, S.W. Applebaum, M. Ma, and L.I. Gilbert. 1990. Photoperiodic diapause in *Drosophila melanogaster* involves a block to juvenile hormone regulation of ovarian maturation. *Gen. Comp. Endocrinol.* 79:174–184.

Saunders, D.S. and B. Cymborowski. 1996. Removal of optic lobes of adult blow flies (*Calliphora vicina*) leaves photoperiodic induction of larval diapause intact. *J. Insect Physiol.* 42:807–811.

Saunders, D.S. and B. Cymborowski. 2003. Selection for high diapause incidence in blow flies (*Calliphora vicina*) maintained under long days increases the maternal critical daylength: some consequences for the photoperiodic clock. *J. Insect Physiol.* 49:777–784.

Saunders, D.S., R.D. Lewis, and G.R. Warman. 2004. Photoperiodic induction of

diapause: opening the black box. *Physiol. Ent.* 29:1–15.

Saunders, D.S. and R.C. Bertossa. 2011. Deciphering time measurement: the role of "circadian clock" genes and formal experimentation in insect photoperiodism. *J. Insect Physiol.* 57:557–566.

Scarbrough, A.G., J.G. Sternberg, and G.P. Waldbauer. 1977. Selection of the cocoon spinning site by the larvae of *Hyalophora cecropia* (Saturniidae). *J. Lep. Soc.* 31:153–166.

Schafellner, C., M. Eizaguirre, C. López, and F. Sehnal. 2008. Juvenile hormone esterase activity in the pupating and diapausing larvae of *Sesamia nonagrioides*. *J. Insect Physiol.* 54:916–921.

Scharf, I., S.S. Bauerfeind, W.U. Blanckenhorn, and M.A. Schafer. 2010. Effects of maternal and offspring environmental conditions on growth, development and diapause in latitudinal yellow dung fly populations. *Climate Res.* 43:115–125.

Schebeck, M., E.M. Hansen, A. Schopf, G.J. Ragland, C. Stauffer, and B.J. Benz. 2017. Diapause and overwintering of two spruce bark beetle species. *Physiol. Entomol.* 42:200–210.

Schechter, M.S., D.K. Hayes, and W.N. Sullivan. 1971. Manipulation of photoperiod to control insects. *Israel J. Ent.* 6:143–166.

Scheltes, P. 1976. The role of graminaceous host-plants in the induction of aestivation-diapause in the larvae of *Chilo zonellus* Swinhoe and *Chilo argyrolepia* Hamps. *Symp. Biol. Hung.* 16:247–253.

Scheltes, P. 1978. The condition of the host plant during aestivation-diapause of the stalk borers *Chilo partellus* and *Chilo orichalcociliella* (Lepidoptera, Pyralidae) in Kenya. *Ent. Exp. Appl.* 24:479–488.

Schiesari, L., C.P. Kyriacou, and R. Costa. 2011. The hormonal and circadian basis for insect photoperiodic timing. *FEBS Lett.* 585:1450–1460.

Schiesari, L., G. Andreatta, C.P. Kyriacou, M.B. O'Connor, and R. Costa. 2016. The insulin-like proteins dILPs-2/5 determine diapause inducibility in *Drosophila*. *PLoS One* 11:e0163680.

Schill, R.O., M. Pfannkuchen, G. Fritz, H.-R. Kohler, and F. Brummer. 2006. Quiescent gemmules of the freshwater sponge, *Spongilla lacustris* (Linnaeus, 1759), contains remarkably high levels of Hsp70 stress protein and hsp70 stress gene mRNA. *J. Exp. Zool.* 305:449–457.

Schmidt, P.S., L. Matzkin, M. Ippolito, and W.F. Eanes. 2005. Geographic variation in diapause incidence, life-history traits, and climatic adaptation in *Drosophila melanogaster*. *Evolution* 59:1721–1732.

Schmidt, P.S., C.T. Zhu, J. Das, M. Batavia, L. Yang, and W.F. Eanes. 2008. An amino acid polymorphism in the couch potato gene forms the basis for climatic adaptation in *Drosophila melanogaster*. *Proc. Nat'l. Acad. Sci., USA* 105:16207–16211.

Schneider, J.C. 2003. Overwintering of *Heliothis virescens* (F.) and *Helicoverpa zea* (Boddie) (Lepidoptera: Noctuidae) in cotton fields of northeast Mississippi. *J. Econ. Ent.* 96:1433–1447.

Schneiderman, H.A. and C.M. Williams. 1953. The respiratory metabolism of the Cecropia silkworm during diapause and metamorphosis. *Biol. Bull.* 105:320–334.

Schneiderman, H.A. and C.M. Williams. 1954. The physiology of insect diapause. IX. The cytochrome oxidase system in relation to the diapause and development of the Cecropia silkworm. *Biol. Bull.* 106:238–252.

Schneiderman, H.A. and J. Horwitz. 1958. Induction and termination of facultative diapause in the chalcid wasps *Mormoniella vitripennis* (Walker) and *Tritneptis klugii* (Ratzeburg). *J. Exp. Biol.* 35:520–551.

Schoonhoven, L.M. 1963. Spontaneous electrical activity in the brains of diapausing insects. *Science* 141:173–174.

Schwarzenberger, A., L. Chen, and L.C. Weiss. 2020. The expression of circadian clock genes in *Daphnia magna* diapause. *Sci. Rep.* 10:19928.

Scriber, J.M., E. Maher, and M.L. Aardema. 2012. Differential effects of short term winter thermal stress on diapausing tiger swallowtail butterflies (*Papilio* spp.). *Insect Sci.* 19:277–285.

Seuge, J. and K. Veith. 1976. Diapause de *Pieris brassicae*: role des photorecepteurs cephaliques, etude des carotenodes cerebraux. *J. Insect Physiol.* 22:1229–1235.

Seymour, J. and R.E. Jones. 2000. Humidity terminated diapause in the tropical braconid *Microplitis demolitor*. *Ecol. Ent.* 25:481–485.

Sgolastra, F., J. Bosch, R. Molowny-Horas, S. Maini, and W.P. Kemp. 2010. Effect of temperature regime on diapause intensity in an adult-wintering hymenopteran with obligate diapause. *J. Insect Phyiol.* 56:185–194.

Sgolastra, F., W.P. Kemp, J.S. Buckner, T.L. Pitts-Singer, and S. Maini. 2011. The long summer: pre-winter temperatures affect metabolic expenditure and winter survival in a solitary bee. *J. Insect Physiol.* 57:1651–1659.

Sgolastra, F., W.P. Kemp, S. Maini, and J. Bosch. 2012. Duration of prepupal summer dormancy regulates synchronization of adult diapause with winter temperatures in bees of the genus *Osmia*. *J. Insect Physiol.* 58:924–933.

Sgolastra, F., X. Arnan, T.L. Pitt-Singer, S. Maini, W.P. Kemp, and J. Bosch. 2016. Pre-winter conditions and post-winter performance in a solitary bee: does diapause impose an energetic cost on reproductive success? *Ecol. Ent.* 41:201–210.

Sgrò, C.M., J.S. Terblanche, and A.A. Hoffmann. 2016. What can plasticity contribute to insect responses to climate change? *Ann. Rev. Ent.* 61:433–451.

Shah, M. T. Suzuki, N A. Ghazy, H. Amano, and K. Ohyama. 2011. Night-interruption light inhibits diapause induction in the Kanzawa spider mite, *Tetranychus kanzawai* Kishida (Acari: Tetranychidae). *J. Insect Physiol.* 57:1185–1189.

Sharma, A., A.B. Nuss, and M. Gulia-Nuss. 2019. Insulin-like peptide signaling in mosquitoes: the road behind and the road ahead. *Front. Physiol.* 10:166.

Shaub, L.P., F.W. Ravlin, D.R. Gray, and J.A. Logan. 1995. Landscape framework to predict phenological events for gypsy moth (Lepidoptera: Lymantriidae) management programs. *Environ. Ent.* 24:10–18.

Shelford, V.E. 1929. *Laboratory and Field Ecology.* Baltimore: Willimas & Wekins.

Sherr, C.J. 1996. Cancer cell cycles. *Science* 274:1672–1677.

Shiga, S. and H. Numata. 1997. The adult blow fly (*Protophormia terraenovae*) perceives photoperiod through the compound eyes for induction of reproductive diapause. *J. Comp. Physiol. A* 181:35–40.

Shiga, S. and H. Numata. 2000. The role of neurosecretory neurons in the pars intercerebralis and pars lateralis in reproductive diapause of the blowfly, *Protophormia terraenovae.* *Naturwissenschaften* 87:125–128.

Shiga, S., Y. Hamanaka, Y. Tatsu, T. Okuda, and H. Numata. 2003a. Juvenile hormone biosynthesis in diapause and nondiapause females of the adult blow fly *Protophormia terraenovae. Zool. Sci.* 20:1199–1206.

Shiga, S., N.T. Davis, and J.G. Hildebrand. 2003b. Role of neurosecretory cells in the photoperiodic induction of pupal diapause of the tobacco hornworm *Manduca sexta. J. Comp. Neur.* 462:275–285.

Shiga, S. and H. Numata. 2007. Neuroanatomical approaches to the study of insect photoperiodism. *Phytochem. Photobiol.* 83:76–86.

Shiga, S. and H. Numata. 2009. Role of PER immunoreactive neurons in circadian rhythms and photoperiodism in the blow fly, *Protophormia terraenovae. J. Exp. Biol.* 212:867–877.

Shim, J.-K., H.-S. Bang, and K.-Y Lee. 2013. Differential regulation of heat shock protein genes by temperature in relation to initial diapause in the egg of the katydid *Paratlanticus ussuriensis. Physiol. Ent.* 38:140–149.

Shimada, K. 2005. Photoperiod-sensitive developmental delay in *facet* mutants of the drosophilid fly, *Chymomyza costata* and the genetic interaction with *timeless. J. Insect Physiol.* 51:649–653.

Shimada, K. and A. Riihimaa. 1990. Cold-induced freezing tolerance in diapausing and non-diapausing larvae of *Chymomyza costata* (Diptera: Drosophilidae) with accumulation of trehalose and proline. *Cryo Letters* 11:243–250.

Shimizu, I., S. Aoki, and T. Ichikawa. 1997. Neuroendocrine control of diapause hormone secretion in the silkworm, *Bombyx mori. J. Insect Phyisol.* 43:1101–1109.

Shimizu, K. and M. Kato. 1984. Carotenoid functions in photoperiodic induction in the silkworm, *Bombyx mori. Photochem. Photobiophys.* 7:47–52.

Shimizu, K., Y. Yamakawa, Y. Shimazaki, and T. Iwasa. 2001. Molecular cloning of *Bombyx* cerebral opsin (Boceropsin) and cellular localization of its expression in the silkworm brain. *Biochem. Biophys. Res. Com.* 287:27–34.

Shimizu, K. and K. Fjuisaki. 2002. Sexual differences in diapause induction of the cotton bollworm, *Helicoverpa armigera* (Hb.)(Lepidoptera: Noctuidae). *Appl. Ent. Zool.* 37:527–533.

Shimizu, K. and K. Fujisaki. 2006. Geographic variation in diapause induction under constant and changing conditions in *Helicoverpa armigera. Ent. Exp. Appl.* 121:253–260.

Shimizu, T. and K. Kawasaki. 2001. Geographic variability in diapause response of Japanese *Orius* species. *Ent. Exp. Appl.* 98:303–316.

Shimizu, Y., A. Mukai, and S.G. Goto. 2018a. Cell cycle arrest in the jewel wasp *Nasonia vitripennis* in larval diapause. *J. Insect Physiol.* 106:147–152.

Shimizu, Y., T. Tamai, and S.G. Goto. 2018b. Cell cycle regulator, small silencing RNA, and segmentation patterning gene expression in relation to embryonic diapause in the band-legged cricket. *Insect Biochem. Mol. Biol.* 102:75–83.

Shingleton, A.W., G.C. Sisk, and D.L. Stern. 2003. Diapause in the pea aphid (*Acyrthosiphon pisum*) is a slowing but not a cessation of development. *BMC Dev. Biol.* 3:7.

Shintani, Y. 2009a. Effect of seasonal variation in host-plant quality on the rice leaf bug, *Trigonotylus caelestialium*. *Ent. Exp. Appl.* 133:128–135.

Shintani, Y. 2009b. Artificial selection for responsiveness to photoperiodic change alters the response to stationary photoperiods in maternal induction of egg diapause in the rice leaf bug *Trigonotylus caelestialium*. *J. Insect Physiol.* 55:818–824.

Shintani, Y., S. Shiga, and H. Numata. 2009. Different photoreceptor organs are used for photoperiodism in the larval and adult stages of the carabid beetle, *Leptocarabus kumagaii*. *J. Exp. Biol.* 212:3651–3655.

Shintani, Y. and H. Numata. 2010. Photoperiodic response of larvae of the yellow-spotted longicorn beetle, *Psacothea hilaris* after removal of the stemmata. *J. Insect Physiol.* 56:1125–1129.

Shintani, Y., Y. Hirose, and M. Terao. 2011. Effects of temperature, photoperiod and soil humidity on induction of pseudopupal diapause in the bean blister beetle *Epicauta gorhami*. *Physiol. Ent.* 36:14–20.

Shintani, Y. and K. Nagamine. 2020. Microhabitat variation in egg diapause incidence in summer within a local population: an adaptation to decline in host-plant suitability in *Trigonotylus caelestialium* (Hemiptera: Miridae). *Environ. Ent.* 49:912–917.

Shiomi, K., Y. Fujiwara, Y. Yasukochi, Z. Kajiura, M. Nakagaki, and T. Yaginuma. 2007. The Pitx homeobox gene in *Bombyx mori*: regulation of DH-PBAN neuropeptide hormone gene expression. *Mol. Cell Neurosci.* 34:209–218.

Shiomi, K., Y. Takasu, M. Kunii, R. Tsuchiya, M. Mukaida, M. Kobayashi, H. Sezutsu, M. I. Takahama, and A. Mizoguchi. 2015. Disruption of diapause induction by TALEN-based gene mutagenesis in relation to a unique neuropeptide signaling pathway in *Bombyx*. *Sci. Rep.* 5:15566.

Short, C.A., M.E. Meuti, Q. Zhang, and D.L. Denlinger. 2016. Entrainment of eclosion and preliminary ontogeny of circadian clock gene expression in the flesh fly, *Sarcophaga crassipalpis*. *J. Insect Physiol.* 93–94:28–35.

Showers, W.B. 1979. Effect of diapause on the migration of the European corn borer into the southeastern United States. In *Movement of Highly Mobile Insects: Concepts and Methodology in Research*, ed. R.L. Rabb and G.G. Kennedy, Raleigh: North Carolina State University Press, pp. 420–432.

Shroyer, D.A. and G.B. Craig, Jr. 1983. Egg diapause in *Aedes triseriatus* (Diptera: Culicidae): geographic variationin photoperiodic response and factors influencing diapause termination. *J. Med. Ent.* 20:601–607.

Siegert, K.J. 1986. The effects of chilling and integumentary injury on carbohydrate and lipid metabolism in diapause and non-diapause pupae of *Manduca sexta*. *Comp. Biochem. Physiol.* 85A:257–262.

Sim, C. and D.L. Denlinger. 2008. Insulin signaling and FOXO regulate the overwintering diapause of the mosquito *Culex pipiens*. *Proc. Nat'l. Acad. Sci., USA* 105:6777–6781.

Sim, C. and D.L. Denlinger. 2009a. Transcription profiling and regulation of fat metabolism genes in diapausing adults of the mosquito *Culex pipiens*. *Physiol. Genomics* 39:202–209.

Sim, C. and D.L. Denlinger. 2009b. A shutdown in expression of an insulin-like peptide, ILP-1, halts ovarian maturation during the overwintering diapause of the mosquito *Culex pipiens*. *Insect Mol. Biol.* 18:325–332.

Sim, C. and D.L. Denlinger. 2011. Catalase and superoxide dismutase-2 enhance survival and protect ovaries during overwintering diapause in the mosquito *Culex pipiens*. *J. Insect Physiol.* 57:628–634.

Sim, C. and D.L. Denlinger. 2013a. Insulin signaling and the regulation of insect diapause. *Front. Physiol.* 4:189.

Sim, C. and D.L. Denlinger. 2013b. Juvenile hormone III suppresses forkhead of transcription factor in the fat body and reduces fat accumulation in the diapausing mosquito, *Culex pipiens*. *Insect Mol. Biol.* 22:1–11.

Sim, C., D.S. Kang, S. Kim, X. Bai, and D.L. Denlinger. 2015. Identification of FOXO targets that generate the diverse features of the diapause phenotype in the mosquito *Culex pipiens*. *Proc. Nat'l. Acad. Sci., USA* 112:3811–3816.

Simard, F., T. Lehmann, J.-J. Lemasson, M. Diatta and D. Fontenille. 2000. Persistence of *Anopheles arabiensis* during the severe dry season conditions in Senegal: an indirect approach using microsatellite loci. *Insect Mol. Biol.* 9:467–479.

Sims, S.R. 2007. Diapause dynamics, seasonal phenology, and pupal color dimorphism of *Papilio polyxenes* in southern Florida, USA. *Ent. Exp. Appl.* 123:239–245.

Sinclair, B.J. and K.E. Marshall. 2018. The many roles of fat in overwintering insects. *J. Exp. Biol.* 221:161836.

Singh, S.P. 1993. Species composition and diapause in citrus butterflies. *J. Insect Sci.* 6:48–52.

Singtripop, T., S. Wanichacheewa, S. Tsuzuki, and S. Sakurai. 1999. Larval growth and diapause in a tropical moth, *Omphisa fuscidentalis* Hampson. *Zool. Sci.* 16:725–733.

Singtripop, T., S. Wanichacheewa, and S. Sakuri. 2000. Juvenile hormone-mediated termination of larval diapause in the bamboo borer, *Omphisa fuscidentalis*. *Insect Biochem. Mol. Biol.* 30:847–854.

Singtripop, T., Y. Oda, S. Wanichacheewa, and S. Sakurai. 2002. Sensitivities to juvenile hormone and ecdysteroid in diapause larvae of *Omphisa fuscidentalis* based on the hemolymph trehalose dynamics index. *J. Insect Physiol.* 48:817–824.

Singtripop, T., M. Saeangsakda, N. Tatun, Y. Kaneko, and S. Sakurai. 2007. Correlation of oxygen consumption, cytochrome *c* oxidase, and cytochrome *c* oxidase subunit I gene expression in the termination of larval diapause in the bamboo borer, *Omphisa fuscidentalis*. *J. Insect Physiol.* 53:933–939.

Singtripop, T., M. Manaboon, N. Tatun, Y. Kaneko, and S. Sakurai. 2008. Hormonal mechanisms underlying termination of larval diapause by juvenilehormone in the bamboo borer, *Omphisa fuscidentalis*. *J. Insect Physiol.* 54:137–145.

Skillman, V.P., N.G. Wilman, and J.C. Lee. 2018. Nutrient declines in overwintering *Halyomorpha halys* populations. *Ent. Exp. Appl.* 166:778–789.

Sláma, K. 1980. Homeostatic functions of ecdysteroids in ecdysis and oviposition. *Acta Entomol. Bohem.* 77:145–168.

Sláma, K. 1988. A new look at insect respiration. *Biol. Bull.* 175:289–300.

Sláma, K. 2000. Extracardiac versus cardiac haemocoelic pulsations in pupae of the mealworm (*Tenebrio molitor* L.). *J. Insect Physiol* 46:977–992.

Sláma, K. and D.L. Denlinger. 1992. Infradian cycles of oxygen consumption in diapausing pupae of the flesh fly, *Sarcophaga crassipalpis*, monitored by a scanning microrespirographic method. *Arch. Insect Biochem. Physiol.* 20:135–143.

Sláma, K. and T.A. Miller. 2001. Physiology of heartbeat reversal in diapausing pupae of the tobacco hornworm, *Manduca sexta* (Lepidoptera: Sphingidae). *Eur. J. Ent.* 98:415–431.

Sláma, K. and D.L. Denlinger. 2013. Transitions in the heartbeat pattern during pupal diapause and adult development in the flesh fly, *Sarcophaga crassipalpis*. *J. Insect Physiol.* 59:767–780.

Slifer, E.H. 1946. The effects of xylol and other solvents on diapause in the grasshopper eggs; together with a possible explanation for the action of these agents. *J. Exp. Zool.* 102:333–356.

Slusarczyk, M. and B. Rybicka. 2011. Role of temperature in diapause response to fish kairomones in crustacean *Daphnia*. *J. Insect Physiol.* 57:676–680.

Smith, W.A., M.F. Bowen, W.E. Bollenbacher, and L.I. Gilbert. 1986. Cellular changes in the prothoracic glands of diapausing pupae of *Manduca sexta*. *J. Exp. Biol.* 120:131–142.

Smith, W.A., A. Lamattina, and M. Collins. 2014. Insulin signaling pathways in lepidopteran ecdysone secretion. *Front. Physiol.* 5:19.

Smits, S.A., J. Leach, E.D. Sonnenburg, C.G. Gonzalez, J.S. Lichtman, G. Reid, R. Knight, A. Manjurano, J. Changaluch, J.E. Elias, M.G. Dominguez-Bello, and J.L. Sonnenburg. 2017. Seasonal cycling in the gut microbiome of the Hadza hunter-gatherers of Tanzania. *Science* 357:802–806.

Smykal, V., A. Bajgar, J. Provazník, S. Fexova, M. Buricova, K. Takaki, M. Hodková, M. Jindra, and D. Doležel. 2014. Juvenile hormone signaling during reproduction and development of the linden bug, *Pyrrhocoris apterus. Insect Biochem. Mol. Biol.* 45:69–76.

Smykal, V., M. Pivarč, J. Provazník, O. Bazalová, P. Jedlička, O. Lukšan A. Horák, H. Vanečková, V. Beneš, I. Fiala, R. Hanus, and D.Doležel. 2020. Complex evolution of insect insulin receptors and homologous decoy receptors, and functional significance of their multiplicity. *Mol. Biol. Evol.* 37:1775–1789.

Sniegula, S. and F. Johansson. 2010. Photoperiod affects compensating developmental rate across latitudes in the damselfly *Lestes sponsa*. *Ecol. Ent.* 35:149–157.

Snodgrass, G.L., R.E. Jackson, O.P. Perera, K.C. Allen, and R.G. Luttrell. 2012. Effect of food and temperature on emergence from diapause in the tarnished plant bug (Hemiptera: Miridae). *Environ. Ent.* 41:1302–1310.

Socha, R. 2007. Factors terminating ovarian arrest in long-winged females of a flightless bug, *Pyrrhocoris apterus* (Heteroptera: Pyrrhocoridae). *Eur. J. Ent.* 104:15–22.

Socha, R. 2010. Pre-diapause mating and overwintering of fertilized adult females: new aspects of the life cycle of the wing-polymorphic bug *Pyrrhocoris apterus* (Heteroptera: Pyrrhocoridae). *Eur. J. Ent.* 107:521–525.

Socha, R. and D. Kodrik. 1999. Differences in adipokinetic response of *Pyrrhocoris apterus* (Heteroptera) in relation to wing dimorphism and diapause. *Physiol. Ent.* 24:278–284.

Socías, M.G., G. van Nieuwenhove, A.S. Casmuz, E. Willink, and G.G. Liljeström. 2016. The role of rainfall in *Sternechus subsignatus* (Coleoptera: Curculionidae) adult emergence from the soil after its winter dormant period. *Environ. Ent.* 45:1049–1057.

Söderlind, L. and S. Nylin. 2011. Genetics of diapause in the comma butterfly *Polygonia c-album*. *Physiol. Ent.* 36:8–13.

Soderstrom, E.L., D.G. Brandl, and B. Mackey. 1990. Responses of codling moth (Lepidoptera: Tortricidae) life stages to high carbon dioxide or low oxygen atmospheres. *J. Econ. Ent.* 83:472–475.

Solbreck, C. and O. Widenfalk. 2012. Very long diapause and extreme resistance to population disturbance in a galling insect. *Ecol. Ent.* 37:51–55.

Sonobe, H. and R. Yamada. 2004. Ecdysteroids during early embryonic development in silkworm *Bombyx mori*: metabolism and functions. *Zool. Sci.* 21:503–516.

Sonoda, S., K. Fukumoto, Y. Izumi, M. Ashfaq, H. Yoshida, and H. Tsumuki.

2006. Methionine-rich storage protein gene in the rice stem borer, *Chilo suppressalis*, is expressed during diapause in response to cold. *Insect Mol. Biol.* 15:853–859.

Soula, B. and F. Menu. 2005. Extended life cycle in the chestnut weevil: prolonged or repeated diapause? *Ent. Exp. Appl.* 115:333–340.

Spacht, D.E., N.M. Teets, and D.L. Denlinger. 2018. Two isoforms of *Pepck* in *Sarcophaga bullata* and their distinct expression profiles through development, diapause, and in response to stresses of cold and starvation. *J. Insect Physiol.* 111:41–46.

Spacht, D.E., J.D. Ganz, R.E. Lee, Jr., and D.L. Denlinger. 2020. Onset of seasonal metabolic depression in the Antarctic midge, *Belgica antarctica*, appears to be independent of environmental cues. *Physiol. Ent.* 45:16–21.

Spielman, A. 1964. Studies on autogeny in *Culex pipiens* populations in nature. I. Reproductive isolation between autogenous and anautogenous populations. *Am. J. Hygiene* 80:175–183.

Spielman, A. and J. Wong. 1973. Environmental control of ovarian diapause in *Culex pipiens*. *Ann. Ent. Soc. Amer.* 66:905–907.

Spieth, H.R. 1995. Change in photoperiodic sensitivity during larval development of *Pieris brassicae*. *J. Insect Physiol.* 41:77–83.

Spieth, H.R. 2002. Estivation and hibernation of *Pieris brassicae* (L.) in southern Spain: synchronization of two complex behavioral patterns. *Pop. Ecol.* 44:273–280.

Spieth, H.R. and E. Schwarzer. 2001. Aestivation in *Pieris brassicae* (Lepidoptera: Pieridae): implications for parasitism. *Eur. J. Ent.* 98:171–176.

Spieth, H.R., U. Porschmann, and C. Teiwes. 2011. The occurrence of summer diapause in the large white butterfly *Pieris brassicae* (Lepidoptera: Pieridae): a geographical perspective. *Eur. J. Ent.* 108:377–384.

Spieth, H.R. and R. Cordes. 2012. Geographic comparison of seasonal migration events of the large white butterfly, *Pieris brassicae*. *Ecol. Ent.* 37:439–445.

Squire, F.A. 1939. Observations on the pink bollworm, *Platyedra gossypiella*, Saund. *Bull. Ent. Res.* 30:475–481.

Srygley, R. 2019. Coping with drought: diapause plasticity in katydid eggs at high, mid, and low elevations. *Ecol. Ent.* 45:485–492.

Srygley, R. 2020. Parental photoperiod affects egg diapause in a montane population of Mormon crickets (Orthoptera: Tettigoniidae). *Environ. Ent.* 49:895–901.

Srygley, R. and L. Senior. 2018. The laboratory curse: variation in temperature stimulates embryonic development and shortens diapause. *Environ. Ent.* 47:725–733.

Stålhandske, S., P. Lehmann, P. Pruisscher, and O. Leimar. 2015. Effect of winter cold duration on spring phenology of the orange tip butterfly, *Anthocharis cardamines*. *Ecol. Evol.* 5:5509–5520.

Stanewsky, R., M. Kaneko, P. Emery, B. Beretta, K. Wager-Smith et al. 1998. The cry(b) mutation identifies cryptochrome as a circadian photoreceptor in *Drosophila*. *Cell* 95:681–692.

Stefanescu, C. 2001. The nature of migration in the red admiral butterfly *Vanessa atalanta*: evidence from the population ecology in its southern range. *Ecol. Ent.* 26:525–536.

Stefanescu, C., R.R. Askew, J. Corbera, and M.R. Shaw. 2012. Parasitism and migration in southern Palaearctic populations of the painted lady butterfly, *Vanessa cardui* (Lepidoptera: Nymphalidae). *Eur. J. Ent.* 109:85–94.

Stegwee, D., E.C. Kimmel, J.A. de Boer, and S. Henstra. 1963. Hormonal control of reversible degeneration of flight muscle in the Colorado potato beetle, *Leptinotarsa decemlineata* Say (Coleoptera). *J. Cell Biol.* 19:519–527.

Stehlik, J., R. Zavodska, K. Shimada, I. Sauman, and V. Kostal. 2008.

Photoperiodic induction of diapause requires regulated transcription of *timeless* in the larval brain of *Chymomyza costata*. *J. Biol. Rhyth.* 23:129–139.

Štetina, T.,P. Hula, M. Moos, P. Šimek, P. Šmilauer, and V. Koštál. 2018. Recovery from supercooling, freezing, and cryopreservation stress in larvae of the drosophilid fly, *Chymomyza costata*. *Sci. Rep.* 8:4414.

Stevenson, T.J. and B.J. Prendergast. 2013. Reversible DNA methylation regulates seasonal photoperiodic time measurement. *Proc. Nat'l. Acad. Sci., USA* 110:16651–16656.

Stevenson, T.J., M.E. Visser, W. Arnold, P. Barrett, S. Biello, A. Dawson, D.L. Denlinger, D. Dominoni, F.J. Ebling, S. Elton, N. Evans, H.M. Ferguson, R.G. Foster, M. Hau, D.T. Haydon, D.G. Halerigg, P. Heideman, J.G.C. Hopcraft, N.N. Jonsson, N. Kronfeld-Schor, V. Kumar, G.A. Lincoln, R. MacLeod, S.A.M. Martin, M. Martinez-Bakker, R.J. Nelson, T. Reed, J.E. Robinson, D. Rock, W.J. Schwartz, I. Steffan-Dewenter, E. Tauber, S.J. Thackeray, C. Umstatter, T. Yoshimura, and B. Helms. 2015. Disrupted seasonal biology impacts health, food security, and ecosystems. *Proc. Roy. Soc. B* 282:20151453.

Stoeckli, S., M. Hirschi, C. Spirig, P. Calanca, M.W. Rotach, and J. Samietz. 2012. Impact of climate change on voltinism and prospective diapause induction of a global pest insect – *Cydia pomonella* (L.). *PLoS One* 7:e35723.

Stoffolano, J.G., Jr. 1967. The synchronization of the life cycle of diapausing face flies, *Musca autumnalis*, and of the nematode, *Heterotylenchus autumnalis. J. Invert. Path.* 9:395–397.

Stoffolano, J.G., Jr. 1968. The effect of diapause and age on the tarsal acceptance threshold of the fly, *Musca autumnalis. J. Insect Physiol.* 14:1205–1214.

Stoffolano, J.G., Jr. 1975. Central control of feeding in the diapausing adult blowfly *Phormia regina. J. Exp. Biol.* 63:265–271.

Storey, K.B. and J.M. Storey. 1991. Biochemistry of cryoprotectants. In *Insects at Low Temperature*, ed. R.E. Lee, Jr. and D.L. Denlinger, New York: Chapman & Hall, pp. 64–93.

Storey, K.B. and J.M. Storey. 2004. Metabolic rate depression in animals: transcriptional and translational controls. *Biol. Rev. Camb. Philos. Soc.* 79:207–233.

Storey, K.B. and J.M. Storey. 2007. Putting life on "pause" – a molecular regulation of hypometabolism. *J. Exp. Biol.* 210:1700–1714.

Storey, K.B. and J.M. Storey. 2010. Oxygen: stress and adaptation in cold-hardy insects. In *Low Temperature Biology of Insects*, ed. D.L. Denlinger and R.E. Lee, Jr., Cambridge: Cambridge University Press, pp. 141–165.

Storey, K.B. and J.M. Storey. 2012. Insect cold hardiness: metabolic, gene, and protein adaptation. *Can. J. Zool.* 90:456–475.

Stuckas, H., M.B. Mende, and A.K. Hundsdoerfer. 2014. Response to cold acclimation in diapausing pupae of *Hyles euphorbiae* (Lepidoptera: Sphingidae): candidate biomarker identification using proteomics. *Insect Mol. Biol.* 23:444–456.

Su, Y., L. Wei, H. Tan, J. Li, W. Li, L. Fu, T. Wang, L. Kang, and X.S. Yao. 2019. OCT as a non-invasive 3D real time imaging tool for the rapid evaluation of phenotypic variations in insect embryonic development. *J. Biophotonics* 13:e201960047.

Su, Z.-H., M. Ikeda, Y. Sato, H. Saito, K. Imai, M. Isobe, and O. Yamashita. 1994. Molecular characterization of ovary trehalase of the silkworm, *Bombyx mori* and its transcriptional activation by diapause hormone. *Biochem. Biophys. Acta* 1218:366–374.

Suang, S., M. Manaboon, P. Chantawannakul, T. Singtripop, K. Hiruma, and Y. Kaneko. 2015. Molecular cloning, developmental expression and tissue distribution of diapause hormone and pheromone biosynthesis activating neuropeptide in the bamboo borer *Omphisa fuscidentalis*. *Physiol. Ent.* 40:247–256.

Sula, J., D. Kodrik, and R. Socha. 1995. Hexameric haemolymph protein related to adult diapause in the red firebug, *Pyrrhocoris apterus* (L.) (Heteroptera). *J. Insect Physiol.* 41:793–800.

Suman, D.S., Y. Wang, and R. Gaugler. 2015. The insect growth regulator pyriproxyfen terminates egg diapause in the Asian tiger mosquito, *Aedes albopictus PLoS One* 10: e0130499.

Sunose, T. 1983. Insect prolonged diapause and its ecological significance. *Bull. Jap. Soc. Pop. Ecol.* (Kotaigun Seitai Gakkai Kaiho) 37:35–48. (in Japanese)

Susset, E.C., A. Magro, and J.-L. Hemptinne. 2017a. Using species distribution models to locate animal aggregations: a case study with *Hippodamia undecimnotata* (Schneider) overwintering aggregation sites. *Ecol. Ent.* 42:345–354.

Susset, E.C., J.-L. Hemptinne, and A. Magro. 2017b. Overwintering sites might not be safe haven for *Hippodamia undecimnotata* (Schneider) (Coleoptera: Coccinellidae). *Coleop. Bull.* 71:556–564.

Susset, E.C., J.-L. Hemptinne, E. Danchin, and A. Magro. 2018. Overwintering aggregations are part of *Hippodamia undecimnotata*'s (Coleoptera: Coccinellidae) mating system. *PLoS One* 13:e0197108.

Sussky, E.M. and J.S. Elkinton. 2015. Survival and near extinction of hemlock woolly adelgid (Hemiptera: Adelgidae) during summer aestivation in a hemlock plantation. *Ecol. Ent.* 44:153–159.

Suwa, A. and T. Gotoh. 2006. Geographic variation in diapause induction and mode of diapause inheritance in *Tetranychus pueraricola*. *J. Appl. Ent.* 130:329–335.

Suzuki, K., T. Minagawa, T. Kumagai, S.-I. Naya, Y. Endo, M. Osanai, and E. Kuwano. 1990. Control mechanism of diapause of the pharate first-instar larva of the silkmoth *Antheraea yamamai*. *J. Insect Physiol.* 36:855–860.

Suzuki, K., T. Nakamura, T. Yanbe, M. Kurihra, and E. Kuwano. 1993. Termination of diapause in pharate first-instar larvae of the gypsy moth *Lymantria dispar japonica* by an imidazole derivative KK-42. *J. Insect Physiol.* 39:107–110.

Suzuki, T. and M. Takeda. 2009. Diapause-inducing signals prolong nymphal development in the two-spotted spider mite *Tetranychus urticae*. *Physiol. Ent.* 34:278–283.

Szücs, M., U. Schaffner, W.J. Price, and M. Schwarzländer. 2012. Post-introduction evolution in the biological control agent *Longitarsus jacobaeae* (Coleoptera: Chrysomelidae). *Evol. Appl.* 5:858–868.

Tachibana, S.-I. and H. Numata. 2004. Parental and direct effects of photoperiod and temperature on the induction of larval diapause in the blow fly *Lucilia sericata*. *Physiol. Ent.* 29:39–44.

Tachibana, S.-I., H. Numata, and S.G. Goto. 2005. Gene expression of heat-shock proteins (*Hsp23*, *Hsp70* and *Hsp90*) during and after larval diapause in the blow fly *Lucilia sericata*. *J. Insect Physiol.* 51:641–647.

Tachibana, S.-I. and T. Watanabe. 2007. Sexual differences in the crucial environmental factors for the timing of postdiapause development in the rice bug *Leptocorisa chinensis*. *J. Insect Physiol.* 53:1000–1007.

Tachibana, S.-I. and T. Watanabe. 2008. Regulation of gonad development and respiratory metabolism associated with food availability and reproductive diapause in the rice bug *Leptocorisa chinensis*. *J. Insect Physiol.* 54:445–453.

Tachibana, S.-I., S. Matsuzaki, M. Tanaka, M. Shiota, D. Motooka, S. Nakamura, and S.G. Goto. 2019. The autophagy-related protein GABARAP is induced during overwintering in the bean bug (Hemiptera: Alydidae). *J. Econ. Ent.* 113:427–434.

Tagaya, J., H. Numata, and S. Goto. 2010. Sexual difference in the photoperiodic induction of pupal diapause in the flesh fly *Sarcophaga similis. Ent. Sci*.13:311–319.

Takagi, S. and T. Miyashita. 2008. Host plant quality influences diapause induction of *Byasa alcinous* (Lepidoptera: Papilionidae). *Ann. Ent. Soc. Am.* 101:392–396.

Takano, Y., M.S. Ullah, and T. Gotoh. 2017. Effect of temperature on diapause termination and post-diapause development in *Eotetranychus smithi* (Acari: Tetranychidae). *Exp. Appl. Acarol.* 73:353–363.

Takeda, K., D.L. Musolin, and K. Fujisaki. 2010. Dissecting insect responses to climate warming: overwintering and post-diapause performance in the southern green stink bug, *Nezara viridula*, under simulated climate-change conditions. *Physiol. Ent.* 35:343–353.

Takeda, M. 2005. Differentiation in life cycle of sympatric populations of two forms of *Hyphantria* moth in central Missouri. *Ent. Sci.* 8:211–218.

Takeda, M. and S. Masaki. 1979. Asymmetric perception of twilight affecting diapause induction by the fall webworm, *Hyphantria cunea. Ent. Exp. Appl.* 25:317–327.

Takeda, M. and G.M. Chippendale. 1982. Environemental and genetic control of the larval diapause of the southwestern corn borer, *Diatraea grandiosella. Physiol. Ent.* 7:99–110.

Takeda, S. and K. Hasegawa. 1975. Alteration of egg diapause in *Bombyx mori* by ouabain injected into diapause egg producers. *J. Insect Physiol.* 21:1995–2003.

Tamai, T., S. Shiga, and S.G. Goto. 2019. Roles of the circadian clock and endocrine regulator in the photoperiodic response of the brown-winger green bug *Plautia stali. Physiol. Ent.* 44:43–52.

Tamaki, S., S. Takemoto, O. Uryu, Y. Kamae, and K. Tomioka. 2013. Opsins are involved in nymphal photoperiodic responses in the cricket *Modicogryllus siamensis. Physiol. Ent.* 38:163–172.

Tammariello, S.P. 2001. Regulation of the cell cycle during diapause. In *Insect Timing: Circadian Rhythmicity to Seasonality*, ed. D.L. Denlinger, J. Giebultowicz, and D.S. Saunders, Amsterdam: Elsevier, pp. 173–183.

Tammariello, S.P. and D.L. Denlinger. 1998. G0/G1 cell cycle arrest in the brain of *Sarcophaga crassipalpis* during pupal diapause and the expression of the cell cycle regulator proliferating cell nuclear antigen. *Insect Biochem. Mol. Biol.* 28:83–89.

Tammariello, S.P., J.P. Rinehart, and D.L. Denlinger. 1999. Desiccation elicits heat shock protein transcription in the flesh fly, *Sarcophaga crassipalpis*, but does not enhance tolerance to high or low temperatures. *J. Insect Physiol.* 45:933–938.

Tan, Q.-Q., W. Liu, F. Zhu, C.-L. Lei, and X.-P. Wang. 2017a. *Fatty acid synthase 2* contributes to diapause preparation in a beetle by regulating lipid accumulation and stress tolerance genes expression. *Sci. Rep.* 7:40509.

Tan, Q.-Q., W. Liu, F. Zhu, C.-L. Lei, D.A. Hahn, and X.-P. Wang. 2017b. Describing the diapause-preparatory proteome of the beetle *Colaphellus bowringi* and identifying candidates affecting lipid accumulation using isobaric tags for mass spectrometry-based proteome quantification (iTRAQ). *Front. Physiol.* 8:251.

Tanaka, A., Y. Kuga, Y. Tanaka, S.G. Goto, H. Numata, and S. Shiga. 2013. Effects of ablation of the pars intercerbralis on ecdysteroid quantities and yolk protein expression in the blowfly *Protophormia terraenovae. Physiol. Ent.* 38:192–201.

Tanaka, H., K. Sato, Y. Saito, T. Yamashita, M. Agoh, J. Okunishi, E. Tachikawa, and K. Suzuki. 2003. Insect diapause-specific peptide from the leaf beetle has a consensus with a putative iridovirus peptide. *Peptides* 24:1327–1333.

Tanaka, K. 1997. Evolutionary relationship between diapause and cold hardiness in the house spider, *Achaearanea tepidariorum* (Araneae: Theridiidae). *J. Insect Physiol.* 43:271–274.

Tanaka, K. and K. Murata. 2016. Rapid evolution of photoperiodic response in a recently introduced pest *Ophraella communa* along geographic gradients. *Ent. Sci.* 19:207–214.

Tanaka, K. and K. Murata. 2017. Genetic basis underlying rapid evolution of an introduced insect *Ophraella communa* (Coleoptera: Chrysomelidae): heritability of photoperiodic response. *Environ. Ent.* 46:167–173.

Tanaka, M., S.-I. Tachibana, and H. Numata. 2008. Sensitive stages for photoperiodic induction of pupal diapause in the flesh fly *Sarcophaga similis* (Meade) (Diptera: Sarcophagidae). *Appl. Entomol. Sci.* 43:403–407.

Tanaka, S. 2001. Endocrine mechanisms controlling body-color polymorphism in locusts. *Arch. Insect Biochem. Physiol.* 47:139–149.

Tanaka, S. and V.J. Brookes. 1983. Altitudinal adaptation of the life cycle in *Allonemobius fasciatus* DeGeer (Orthoptera: Gryllidae). *Can. J. Zool.* 61:1986–1990.

Tanaka, S. and H. Wolda. 1987. Dormancy and aggregation in a tropical insect *Jadera obscura* (Hemiptera: Rhopalidae). *Proc. Kon. Nederl. Akad. Wetensch. Ser. C.* 90:351–366.

Tanaka, S., D.L. Denlinger, and H. Wolda. 1987a. Daylength and humidity as environmental cues for diapause termination in a tropical beetle. *Physiol. Ent.* 12:213–224.

Tanaka, S., H. Wolda, and D.L. Denlinger. 1987b. Abstinence from mating by sexually mature males of the fungus beetle, *Stenotarsus rotundus*, during a tropical dry season. *Biotropica* 19:252–254.

Tanaka, S., H. Wolda, and D.L. Denlinger. 1987c. Seasonality and its physiological regulation in three neotropical insect taxa from Barro Colorado Island, Panama. *Insect Sci. Applic.* 8:507–514.

Tanaka, S., H. Wolda, and D.L. Denlinger. 1988. Group size affects the metabolic rate of a tropical beetle. *Physiol. Ent.* 13:239–241.

Tanaka, S. and D.-H. Zhu. 2003. Presence of three diapauses in a subtropical cockroach: control mechanisms and adaptive significance. *Physiol. Ent.* 28:323–330.

Tanaka, S., F. Yukuhiro, H. Yasui, M. Fukay, T. Akino, and S. Wakamura. 2008. Presence of larval and adult diapauses in a subtropical scarab beetle: graded thermal response for synchronized sexual maturation and reproduction. *Physiol. Ent.* 33:334–345.

Tanaka, S., H. Sakamoto, T. Hata, and R. Sugahara. 2018. Hatching synchrony is controlled by a two-step mechanism in the migratory locust *Locusta migratoria* (Acrididae: Orthoptera): roles of vibrational stimuli. *J. Insect Physiol.* 107:125–135.

Tanaka, S.I., C. Imai, and H. Numata. 2002. Ecological significance of adult summer diapause after nymphal winter diapause in *Poecilocoris lewisi* (Distant)(Heteroptera: Scutelleridae). *Appl. Entomol. Zool.* 37:469–475.

Tani, N., G. Kamada, K. Ochiai, M. Isobe, S. Suwan, and H. Kai. 2001. Carbohydrate moiety of time-interval measuring enzyme regulates time measurement through its interaction with time-holding peptide PIN. *J. Biochem.* 129:221–227.

Tanigawa, N.A., S. Shiga, and H. Numata. 1999. Role of the corpus allatum in the control of reproductive diapause in the male blow fly, *Protophormia terraenovae*. *Zool. Sci.* 16:639–644.

Tanigawa, N., K. Matsumoto, K. Yasuyama, H. Numata, and S. Shiga. 2009. Early embryonic development and diapause stage in the band-legged ground cricket *Dianemobius nigrofasciatus*. *Dev. Genes Evol.* 219:589–596.

Taniguchi, N. and K. Tomioka. 2003. Duration of development and number of nymphal instars are differentially regulated by photoperiod in the cricket *Modicogryllus siamensis* (Orthoptera: Gryllidae). *Eur. J. Ent.* 100:275–281.

Tanzubil, P.B., G.W.K. Mensah, and A.R. McCaffery. 2002. Dry season survival, diapause duration and timing of diapause termination in the millet stem borer, *Coniesta ignefusalis* (Hampson) (Lepidoptera: Pyralidae) in northern Ghana. *Int'l J. Pest Managment* 48:33–37.

Tarazona, E., J.I. Lucas-Liedó, M.J. Carmona, and E.M. García-Roger. 2020. Gene expression in diapausing rotifer eggs in response to divergent environmental predictability regimes. *Sci. Rep.* 10:21366.

Tarrant, A.M., M.F. Baumgartner, T. Verslycke, and C.L. Johnson. 2008. Differential gene expression in diapausing and active *Calanus finmarchicus* (Copepoda). *Mar. Ecol. Prog. Ser.* 355:193–207.

Tatar, M. and C.-M. Yin. 2001. Slow aging during insect reproductive diapause: why butterflies, grasshoppers and flies are like worms. *Exp. Gerontol.* 36:723–738.

Tatar, M., S.A. Chien, and N.K. Priest. 2001. Negligible senescence during reproductive dormancy in *Drosophila melanogaster*. *Am. Nat.* 158:248–258.

Tatun, N., T. Singtripop, and S. Sakurai. 2008. Dual control of midgut trehalase activity by 20-hydroxyecdysone and an inhibitory factor in the bamboo borer *Omphisa fuscidentalis* Hampson. *J. Insect Physiol.* 54:351–357.

Tauber, C.A., M.J. Tauber, and J.R. Nechols. 1977. Two genes control seasonal isolation in sibling species. *Science* 197:592–593.

Tauber, E., M. Zordan, F. Sandrelli, M. Pegoraro, N. Osterwalder, C. Breda, A. Daga, A. Selmin, K. Monger, C. Benna, E. Rosata, C.P. Kyriacou, and R. Costa. 2007. Natural selection favors a newly derived *timeless* allele in *Drosophila melanogaster*. *Science* 316:1895–1898.

Tauber, M.J. and C.A. Tauber. 1970a. Adult diapause in *Chrysopa carnea*: stages sensitive to photoperiodic induction. *J. Insect Physiol.* 16:2075–2080.

Tauber, M.J. and C.A. Tauber. 1970b. Photoperiodic induction and termination of diapause in an insect: response to changing daylength. *Science* 167:170.

Tauber, M.J., C.A. Tauber, and C.J. Denys. 1970. Adult diapause in *Chrysopa carnea:* photoperiodic control of duration and colour. *J. Insect Physiol.* 16:949–955.

Tauber, M.J. and C.A. Tauber. 1973. Natural daylengths regulate insect seasonality by two mechanisms. *Nature* 258:711–712.

Tauber, M.J. and C.A. Tauber. 1976. Insect seasonality – diapause maintenance, termination, and postdiapause development. *Ann. Rev. Ent.* 21:81–107.

Tauber, M.J. and C.A. Tauber. 1981. Insect seasonal cycles: genetics and evolution. *Ann. Rev. Ecol. Syst.* 12:281–308.

Tauber, M.J., C.A. Tauber, and S. Masaki. 1986. *Seasonal Adaptations of Insects*. Oxford: Oxford University Press.

Tauber, M.J., C.A. Tauber, J.J. Obrycki, B. Gollands, and R.J. Wright. 1988. Voltinism and the induction of aestival diapause in the Colorado potato beetle, *Leptinotarsa decemlineata* (Coleoptera: Chrysomelidae). *Ann. Ent. Soc. Amer.* 81:748–754.

Tauber, M.J., C.A. Tauber, and S. Gardescu. 1993. Prolonged storage of *Chrysoperla carnea* (Neuroptera: Chrysopidae). *Environ. Ent.* 22:843–848.

Tauber, M.J., G.S. Albuquerque, and C.A. Tauber. 1997. Storage of nondiapausing *Chrysoperla externa* adults: influence on survival and reproduction. *Biol. Cont.* 10:69–72.

Tauber, M.J. and C.A. Tauber. 2002. Prolonged dormancy in *Leptinotarsa decemlineata* (Coleoptera: Chrysomelidae): a ten-year field study with implications for crop rotation. *Environ. Ent.* 31:499–504.

Tawfik, A.I., Y. Tanaka, and S. Tanaka. 2002a. Possible involvement of ecdysteroids in photoperiodically induced suppression of ovarian development in a Japanese strain of the migratory locust, *Locusta migratoria*. *J. Insect Physiol.* 48:411–418.

Tawfik, A.I., Y. Tanaka, and S. Tanaka. 2002b. Possible involvement of ecdysteroids in embryonic diapause of *Locusta migratoria. J. Insect Physiol.* 48:743–749.

Tazima, Y. 1964. *The Genetics of the Silkworm.* London: Logos Press.

Teder, T., T. Esperk, T. Remmel, A. Sang, and T. Tammaru. 2010. Counterintuitive size patterns in bivoltine moths: late-season larvae grow larger despite lower food quality. *Oecologia* 162:117–125.

Teets, N.M., J.T. Peyton, H. Colinet, D. Renault, J. Kelley, Y. Kawarasaki, R.E. Lee, and D.L. Denlinger. 2012. Gene expression changes governing extreme dehydration tolerance in an Antarctic insect. *Proc. Nat'l. Acad. Sci., USA* 109:20744–20749.

Teets, N.M. and D.L. Denlinger. 2013. Physiological mechanisms of seasonal and rapid cold-hardening in insects. *Physiol. Ent.* 38:105–116.

Teixeira, L.A.F. and S. Polavarapu. 2005. Diapause development in the blueberry maggot *Rhagoletis mendax* (Diptera: Tephritidae). *Environ. Ent.* 34:47–53.

Terao, M., Y. Hirose, and Y. Shintani. 2012. Effects of temperature and photoperiod on termination of pseudopupal diapause in the bean blister beetle, *Epicauta gorhami. J. Insect Physiol.* 58:737–742.

Terblanche, J.S., E. Marais, S.K. Hetz, and S.L. Chown. 2008. Control of discontinuous gas exchange in *Samia cynthia*: effects of atmospheric oxygen, carbon dioxide and moisture. *J. Exp. Biol.* 211:3272–3280.

Ti, X., N. Tuzuki, N. Tani, E. Morigami, M. Isobe, and H. Kai. 2004. Demarcation of diapause development by cold and its relations to time-interval activation of TIME-ATPase in eggs of the silkworm, *Bombyx mori. J. Insect Physiol.* 50:1053–1064.

Tian, K. and W.-H. Xu. 2013. High expression of PP2A-Aa is associated with diapause induction during the photoperiod-sensitive stage of the cotton bollworm *Helicoverpa armigera. J. Insect Physiol.* 59:588–594.

Tikkanen, O.-P. and P. Lyytikäinen-Saarenmaa. 2002. Adaptation of a generalist moth, *Operophtera brumata*, to variable budburst phenology of host plants. *Ent. Exp. Appl.* 103:123–133.

Tobin, P.C., S. Nagarkatti, G. Loeb, and M.C. Saunders. 2008. Historical and projected interaction between climate change and insect voltinism in a multivoltine species. *Global Change Biol.* 14:951–957.

Toepfer, S., H. Gu, and S. Dorn. 2000. Selection of hibernation sites by *Anthonomus pomorum*: preferences and ecological consequences. *Ent. Exp. Appl.* 95:241–249.

Toepfer, S., H. Gu, and S. Dorn. 2002. Phenological analysis of spring colonization of apple trees by *Anthonomus pomorum. Ent. Exp. Appl.* 103:151–159.

Togashi, K. 2014. Effects of larval food shortage on diapause induction and adult traits in Taiwanese *Monochamus alternatus alternatus. Ent. Exp. Appl.* 151:34–42.

Tojo, S., Y. Nagase, and L. Filippi. 2005. Reduction of respiration rates by forming aggregations in diapausing adults of the shield bug, *Parastrachia japonensis. J. Insect Physiol.* 51:1075–1082.

Tomioka, K., N. Agui, and W.E. Bollenbacher. 1995. Electrical properties of the cerebral prothoracicotropic hormone cells in diapausing and non-diapausing pupae of the tobacco hornworm, *Manduca sexta. Zool. Sci.* 12:165–173.

Toni, L.S. and P.A. Padilla. 2016. Developmentally arrested *Austrofundulus limnaeus* embryos have changes in post-translational modifications of histone H3. *J. Exp. Biol.* 219:544–552.

Toprak, U. 2020. The role of peptide hormones in insect lipid metabolism. *Front. Physiol.* 11:434.

Torson, A.S., G.D. Yocum, J.P. Rinehart, W.P. Kemp, and J.H. Bowsher. 2015. Transcriptional responses to fluctuating thermal regimes underpinning differences in survival in the solitary bee *Megachile rotundata. J. Exp. Biol.* 218:1060–1068.

Tougeron, K., G. Hraoui, C. Le Lann, J. van Baaren, and J. Brodeur. 2018. Intraspecific maternal competition induces summer diapause in insect parasitoids. *Insect Sci.* 25:1080–1088.

Tougeron, K., J. Brodeur, J. van Baaren, D. Renault, and C. Le Lann. 2019. Sex makes them sleepy: host reproductive status induces diapause in a parasitoid population experiencing harsh winters. bioRxiv 371385, ver 6 recommended by *PCI Ecology*. doi: 10.1101/37138.

Tougeron, K., M. Devogel, J. van Baaren, C. Le Lann, and T. Hance. 2020. Trans-generational effects on diapause and life-history-traits of an aphid parasitoid. *J. Insect Physiol.* 121:104001.

Toxopeus, J., R. Jakobs, L.V. Ferguson, T.D. Gariepy, and B.J. Sinclair. 2016. Reproductive arrest and stress resistance in winter-acclimated *Drosophila suzukii*. *J. Insect Physiol.* 89:37–51.

Toxopeus, J. and B.J. Sinclair. 2018. Mechanisms underlying insect freeze tolerance. *Biol. Rev.* 93:1891–1914.

Tschinkel, W.R. 1993. Sociometry and sociogenesis of colonies of the fire ant *Solenopsis invicta* during one annual cycle. *Ecol. Monogr.* 63:425–457.

Tsuchiya, R., A. Kaneshima, M. Kobayashi, M. Yamazaki, Y. Takasu, H. Sezutsu, Y. Tanaka, A. Mizoguchi, and K. Shiomi. 2021. Maternal GABAergic and GnRH/corazonin pathway modulates egg diapause phenotype of the silkworm *Bombyx mori*. *Proc. Nat'l. Acad. Sci., USA* 118: e2020028118.

Tsumuki, H. and K. Kanehisa. 1980. Changes in enzyme activities related to glycerol synthesis in hibernating larvae of the rice stem borer *Chilo suppressalis*. *Appl. Ent. Zool.* 15:285–292.

Tsunoda, T., L.F. Chaves, G.T.T. Nguyen, Y.T. Nguyen, and M. Takagi. 2015. Winter activity and diapause of *Aedes albopictus* (Diptera: Culicidae) in Hanoi, North Vietnam. *J. Med. Ent.* 52:1203–1212.

Tsurumaki, J., J. Ishiguro, A. Yamanaka, and K. Endo. 1999. Effects of photoperiod and temperature on seasonal morph development and diapause egg oviposition in a bivoltine race (Diazo) of the silkmoth, *Bombyx mori* (L.). *J. Insect Physiol.* 45:101–106.

Tsurumaru, S., A. Kawamori, K. Mitsumasu, T. Niimi, K. Imai, O. Yamashita, and T. Yaginuma. 2010. Disappearance of chorion proteins from *Bombyx mori* eggs treated with HCl solution to prevent diapause. *J. Insect Physiol.* 56:1721–1727.

Tu, M.-P., C.-M. Yin, and M. Tatar. 2002. Impaired ovarian ecdysone synthesis of *Drosophila melanogaster* insulin receptor mutants. *Aging Cell* 1:158–160.

Tu, M.-P., C.-M. Yin, and M. Tatar. 2005. Mutations in insulin signaling pathway alter juvenile hormone synthesis in *Drosophila melanogaster*. *Gen. Comp. Endocrin.* 142:347–356.

Tu, X., J. Wang, K. Hao, D.W. Whitman, Y. Fan, G. Cao, and Z. Zhang. 2015. Transcriptomic and proteomic analysis of pre-diapause and nondiapause eggs of migratory locust, *Locusta migratoria* L. (Orthoptera: Acridoidea). *Sci. Rep.* 5:11402.St.

Tungjitwitayakul, J., T. Singtripop, A. Nettagul, Y. Oda, N. Tatun, T. Sekimoto, and S. Sakurai. 2008. Identification, characterization, and developmental regulation of two storage proteins in the bamboo borer *Omphis fuscidentalis*. *J. Insect Phyisol.* 54:62–76.

Turnock, W.J. and P.G. Fields. 2005. Winter climates and cold hardiness in terrestrial insects. *Eur. J. Ent.* 102:561–576.

Tyukmaeva, V., P. Lankinen, J. Kinnunen, H. Kauranen, and A. Hoikkala. 2020. Latitudinal clines in timing and temperature-sensitivity of photoperiodic reproductive diapause in *Drosophila montana*. *Ecography* 43:1–11.

Tyukmaeva, V.I., T.S. Salminen, M. Kankare, K.E. Knott, and A. Hoikkala. 2011. Adaptation to a seasonally varying environment: a strong latitudinal cline in reproductive diapause combined with high gene flow in *Drosophila montana. Ecol. Evol.* 1:160–168.

Tyukmaeva, V.I., P. Veltsos, J. Slate, E. Gregson, H. Kauranen, M. Kankare, M.G. Ritchie, R.K. Butlin, and A. Hoikkala. 2015. Localization of quantitative trait loci for diapause and other photoperiodically regulated life history traits important in adaptation to seasonally varying environments. *Mol. Ecol.* 24:2809–2819.

Uehara, H., Y. Senoh, K. Yoneda, Y. Kato, and K. Shiomi. 2011. An FXPRL neuropeptide induces seasonal reproductive polyphenism underlying a life-history tradeoff in the tussock moth. *PLoS One* 6(8):e24213.

Uno, T., A. Nakasuji, M. Shimoda, and Y. Aizono. 2004. Expression of cytochrome *c* oxidase subunit I gene in the brain at any early stage in the termination of pupal diapause in the sweet potato hornworm, *Agrius convolvuli. J. Insect Physiol.* 50:35–42.

Urbanová, V., O. Bazalová, H. Vaněčková, and D. Doležel. 2016. Photoperiod regulates growth of male accessory glands through juvenile hormone signaling in the linden bug, *Pyrrhocoris apterus. Insect Biochem. Mol. Biol.* 70:184–190.

Urbanski, J.M., A. Aruda, and P. Armbruster. 2010a. A transcriptional element of the diapause program in the Asian tiger mosquito, *Aedes albopictus*, identified by suppressive subtractive hybridization. *J. Insect Physiol.* 56:1147–1154.

Urbanski, J.M., J.B. Benoit, M.R. Michaud, D.L. Denlinger, and P. Armbruster. 2010b. The molecular physiology of increased egg desiccation resistance during diapause in the invasive mosquito, *Aedes albopictus. Proc. Roy. Soc. B* 277:2683–2692.

Urbanski, J.M., M. Mogi, D. O'Donnell, M. DeCotiis, T. Toma, and P. Armbruster. 2012. Rapid adaptive evolution of photoperiodic response during invasion and range expansion across a climatic gradient. *Am. Nat.* 179:490–500.

Urquhart, F.A. and N.R. Urquhart. 1976. The overwintering site of the eastern population of the monarch butterfly (*Danaus p. plexippus*; Danaidae) in southern Mexico. *J. Lep. Soc.* 30:153–158.

Uryu, M., Y. Ninomiya, T. Yokoi, S. Tsuzuki, and Y. Hayakawa. 2003. Enhanced expression of genes in the brains of larvae of *Mamestra brassicae* (Lepidoptera; Noctuidae) exposed to short daylengths or fed Dopa. *Eur. J. Ent.* 100:245–250.

Ushatinskaya, R.S. 1976a. Insect dormancy and classification. *Zool. Jb. Syst.* 103:76–97.

Ushatinskaya, R.S. 1976b. Prolonged diapause in Colorado beetle and conditions of its formation. In *Ecology and Physiology of Diapause in the Colorado Beetle*, ed. K.V. Arnoldi, New Dehli: Indian National Scientific Documentation Centre, (translation from Russian), pp. 168–200.

Ushatinskaya, R.S. 1984. A critical review of the superdiapause in insects. *Ann. Zool.* 21:3–30.

Usua, E.J. 1970. Diapause in the maize stemborer. *J. Econ. Ent.* 63:1605–1610.

Uzelac, I., M. Avramov, T. Celić, E. Vukašinović, S. Gošić-Dondo, J. Purać, D. Kojić, D. Blagojević, and Z.D. Popović. 2020. Effect of cold acclimation on selected metabolic enzymes during diapause in the European corn borer *Ostrinia nubilalis* (Hbn.). *Sci. Rep.* 10:9085.

Vanatoa, A., A. Kuusik, U. Tartes, L. Metspalu, and K. Hiiesaar. 2006. Respiration rhythms and heartbeats of diapausing Colorado potato beetles, *Leptinotarsa decemlineata*, at low temperatures. *Ent. Exp. Appl.* 118:21–31.

Varpe, Ø. and M.J. Ejsmond. 2018. Tradeoffs between storage and survival affect diapause timing in capital breeders. *Evol. Ecol.* 32:623–641.

Vaz Nunes, M. and A. Veerman. 1982. Photoperiodic time measurement in the spider mite *Tetranychus urticae*: a novel concept. *J. Insect Physiol.* 28:1041–1053.

Vaz Nunes, M. and J. Hardie. 1993. Circadian rhythmicity is involved in photoperiodic time measurement in the aphid *Megoura viciae*. *Cell. Mol. Life Sci.* 49:711–713.

Vaz Nunes, M. and D.S. Saunders. 1999. Photoperiodic time measurement in insects: a review of clock models. *J. Biol. Rhyth.* 14:84–104.

Vaze, K.M. and C. Helfrich-Förster. 2016. *Drosophila ezoana* uses an hour-glass or highly damped circadian clock for measuring night length and inducing diapause. *Physiol. Ent.* 41:378–389.

Veerman, A. 1985. Diapause. In *Spider Mites: Their Biology, Natural Enemies and Control*, ed. W. Helle and M.W. Sabelis, Vol. 1A, Amsterdam: Elsevier, pp. 279–316.

Veerman, A. 2001. Photoperiodic time measurement in insects and mites: a critical evaluation of the oscillator-clock hypothesis. *J. Insect Physiol.* 47:1097–1109.

Veerman, A., W.P.J. Overmeer, M.F.J. Alderliest, and R.L. Veenendaal. 1985. Photoperiodic induction of diapause in an insect is vitamin A dependent. *Experientia* 41:1194–1195.

Veerman, A. and R.L. Veenendaal. 2003. Experimental evidence for a non-clock role of the circadian system in spider mite photoperiodism. *J. Insect Physiol.* 49:727–732.

Ventura Garcia, P., E. Wajnberg, J. Pizzol, and M.L.M. Oliveira. 2002. Diapause in the egg parasitoid *Trichogramma cordubensis*: role of temperature. *J. Insect Physiol.* 48:349–355.

Vermeulen, A., S. Engels, L. Engqvist, and K.P. Sauer. 2009. Phenotypic plasticity in sperm traits in scorpionflies (Mecoptera: Panorpidae): Consequences of larval history and seasonality on sperm length and sperm transfer. *Eur. J. Ent.* 106:347–352.

Vermunt, A.M.W., A.B. Koopmanschap, J.M. Vlak, and C.A.D. de Kort. 1999. Expression of the juvenile hormone esterase gene in the Colorado potato beetle, *Leptinotarsa decemlineata*: photoperiodic and juvenile hormone analog response. *J. Insect Physiol.* 45:135–142.

Verspagen, N., S. Ikonen, M. Saastamoinen, and E. van Bergen. 2020. Multidimensional plasticity in the Glanville fritillary butterfly: larval performance is temperature, host and family specific. *Proc. R. Soc. B* 287:20202577.

Vesala, L., T.S. Salminen, M. Kankare, and A. Hoikkala. 2012a. Photoperiodic regulation of cold tolerance and expression levels of *regucalcin* gene in *Drosophila montana*. *J. Insect Physiol.* 58:704–709.

Vesala, L., T.S. Salminen, V. Koštál, H. Zahradníčková, and A. Hoikkala. 2012b. *Myo*-inositol as a main metabolite in overwintering flies: seasonal metabolomic profiles and cold stress tolerance in a northern drosohilid fly. *J. Exp. Biol.* 215:2891–2897.

Vinogradova, E.B. 1974. The pattern of reactivation of diapausing larvae in the blowfly, *Calliphora vicina*. *J. Insect Physiol.* 20:2487–2496.

Vinogradova, E.B. 2000. *Culex pipiens pipiens Mosquitoes: Taxonomy, Distribution, Ecology, Physiology, Genetics, Applied Importance and Control*. Sofia: Pensoft.

Vinogradova, E.B. and K.B. Zinovjeva. 1972. Maternal induction of larval diapause in the blowfly, *Calliphora vicina*. *J. Insect Physiol.* 18:2401–2409.

Vinogradova, E.B. and S.Ya. Reznik. 2013. Induction of larval diapause in the blowfly, *Calliphora vicina* R.D. (Diptera, Calliphoridae) under field and laboratory conditions. *Ent. Rev.* 93:935–941.

Vinogradova, E.B. and S.Ya. Reznik. 2015. Photothermal control of larval diapause in the blowfly, *Calliphora vicina* R.D. (Diptera, Calliphoridae) from the Lofoten Islands (Northern Norway). *Ent. Rev.* 95:296–304.

Visser, B., C.M. Williams, D.A. Hahn, C.A. Short, and G. López-Martinez. 2018.

Hormetic benefits of prior anoxia exposure in buffering anoxia stress in a soil-pupating insect. *J. Exp. Biol.* 221:jeb167825.

Voinovich, N.D., N.P. Vaghina, and S.Y. Reznik. 2013. Comparative analysis of maternal and grand-mother photoperiodic responses of *Trichogramma* species (Hymenoptera: Trichogrammatidae). *Eur. J. Ent.* 110:451–460.

Voinovich, N.D. and S.Y. Reznik. 2017. On the factors inducing the inhibition of diapause in progeny of diapause females of *Trichogramma telengai*. *Physiol. Ent.* 42:274–281.

Voinovich, N.D. and S.Y. Reznik. 2020. Grandmaternal temperature effect on diapause induction in *Trichogramma telengai* (Hymenoptera: Trichogrammatidae). *J. Insect Physiol.* 124:104072.

Vu, H.M., J.E. Pennoyer, K.R. Ruiz, P. Portmann, and J.G. Duman. 2019. Beetle, *Dendroides canadensis*, antifreeze proteins increased high temperature survivorship in transgenic fruit flies, *Drosophila melanogaster*. *J. Insect Physiol.* 112:68–72.

Vuillaume, M. and A. Berkaloff. 1974. LSD treatment of *Pieris brassicae* and consequences on the progeny. *Nature* 251:314–315.

Vukašinović, E., D.W. Pond, G. Grubor-Lajšić, M.R. Worland, D. Kojić, J. Purać, Z.D. Popović, and D.P. Blagojević. 2017. Temperature adaptation of lipids in diapausing *Ostrinia nubilalis*: an experimental study to distinguish environmental versus endogenous controls. *J. Comp. Physiol. B* 188:27–36.

Waddington, C.H. 1952. *The Epigenetics of Birds*, Cambridge: Cambridge University Press.

Wade, F.A. and S.R. Leather. 2002. Overwintering of the sycamore aphid, *Drepanosiphum platanoidis*. *Ent. Exp. Appl.* 104:241–253.

Wadsworth, C.B., W.A. Woods, D.A. Hahn, and E.B. Dopman. 2013. One phase of the dormancy developmental pathway is critical for the evolution of insect seasonality. *J. Evol. Biol.* 26:2359–2368.

Wadsworth, C.B. and E.B. Dopman. 2015. Transcriptome profiling reveals mechanisms for the evolution of insect seasonality. *J. Exp. Biol.* 218:3611–3622.

Wadsworth, C.B., Y. Okada, and E.B. Dopman. 2020. Phenology-dependent cold exposure and thermal performance of *Ostrinia nubilalis* ecotypes. *BMC Evol. Biol.* 20:34.

Wagner, D., P. Doak, T. Sformo, P.M. Steiner, and B. Carlson. 2012. Overwintering physiology and microhabitat use of *Phyllocnistis populiella* (Lepidoptera: Gracilliariidae) in interior Alaska. *Environ. Ent.* 41:180–187.

Waldbauer, G.P. 1978. Phenological adaptation and the polymodal emergence patterns of insects. In *Evolution of Insect Migration and Diapause*, ed. H. Dingle, New York: Springer-Verlag, pp. 127–144.

Waldbauer, G. 1996. *Insects through the Seasons*. Cambridge: Harvard University Press.

Waldbauer, G., J.G. Sternburg, and G.R. Wilson. 1978. The effect of injections of β-ecdysterone on the bimodal emergence of *Hyalophora cecropia*. *J. Insect Physiol.* 24:623–628.

Waldbauer, G. and J.S. Sternburg. 1986. The bimodal emergence of adult *Hyalophora cecropia*: conditions required for initiation of development by second mode pupae. *Ent. Exp. Appl.* 41:315–317.

Walker, T.J. 1980. Mixed oviposition in individual females of *Gryllus firmus:* graded proportions of fast-developing and diapause eggs. *Oecologia* 47:291–298.

Wallingford, A.K. and G.M. Loeb. 2016. Developmental acclimation of *Drosophila suzukii* (Diptera: Drosophilidae) and its effect on diapause and winter stress tolerance. *Eviron. Ent.* 45:1081–1089.

Walsh, B.D. 1867. The apple-worm and the apple-maggot. *Am. J. Hortic.* 2:338–343.

Waltrick, D., C. Awruch, and C. Simpfendorfer. 2012. Embryonic diapause in the elasmobranchs. *Rev. Fish Biol. Fisheries* 22:849–859.

Wang, F., H. Gong, and J. Qin. 1999. Termination of pupal diapause in the bollworm *Helicoverpa armigera* by an imidazole derivative KK-42. *Ent. Exp. Appl.* 93:331–333.

Wang, J. and S.K. Kim. 2003. Global analysis of dauer gene expression in *Caenorhabditis elegans*. *Development* 130:1621–1634.

Wang, J., H. Fan, P. Wang, and Y.-H. Liu. 2019a. Expression analysis reveals the association of several genes with pupal diapause in *Bactrocera minax* (Diptera: Tephritidae). *Insects* 10:169.

Wang, J., L.-L. Ran, Y. Li, and Y.-H. Liu. 2020. Comparative proteomics provides insights into diapause program of *Bactrocera minax* (Diptera: Tephritidae). *PLoS One* 15:e0244493.

Wang, J.D. and P.A. Levin. 2009. Metabolism, cell growth and the bacterial cell cycle. *Nature Rev. Microbiol.* 7:822–827.

Wang, M.C., E.J. O'Rourke, and G. Ruvkun. 2008. Fat metabolism links germline stem cells and longevity in *C. elegans*. *Science* 322:957–960.

Wang, Q., A.A.M. Mohamed, and M. Takeda. 2013. Serotonin receptor B may lock the gate of PTTH release/synthesis in the Chinese silk moth, *Antheraea pernyi*; a diapause initiation/maintenance mechanism? *PLoS One* 8:e79381.

Wang, Q., Y. Egi, M. Takeda, K. Oishi, and K. Sakamoto. 2015a. Melatonin pathway transmits information to terminate pupal diapause in the Chinese oak silkmoth *Antheraea pernyi* and through reciprocated inhibition of dopamine pathway functions as a photoperiodic counter. *Ent. Sci.* 18:74–84.

Wang, Q., I. Hanatani, M. Takeda, K. Oishi, and K. Sakamoto. 2015b. D2-like dopamine receptors mediate regulation of pupal diapause in Chinese oak silkmoth *Antheraea pernyi*. *Ent. Sci.* 18:193–198.

Wang, T., S.-L. Geng, Y.-M. Guan, and W.-H. Xu. 2018. Deacetylation of metabolic enzymes by Sirt2 modulates pyruvate

homeostasis to extend insect lifespan. *Aging* 10:1053–1072.

Wang, X., J. Fan, M. Zhou, G. Gao, L. Wei, and L. Kang. 2021. Interactive effect of photoperiod and temperature on the inductionand termination of embryonic diapause in the migratory locust. *Pest Manag. Sci.* 77:2854–2862.

Wang, X.-P., F.-S. Xue, F. Ge, C.-A. Zhou, and L.-S. You. 2004. Effects of thermoperiods on diapause induction in the cabbage beetle, *Colaphellus bowringi* (Coleoptera: Chrysomelidae). *Physiol. Ent.* 29:419–425.

Wang, X.-P., F.-S. Xue, A. Hua, and F. Ge. 2006a. Effects of diapause duration on future reproduction in the cabbage beetle, *Colaphellus bowringi*: positive or negative? *Physiol. Ent.* 31:190–196.

Wang, X.-P., F. Ge, and F.-S. Xue. 2006b. Host plant mediation of diapause induction in the cabbage beetle, *Colaphellus bowringi* Baly (Coleoptera: Chrysomelidae). *Insect Sci.* 13:189–193.

Wang, X.-P., F.-S. Xue, Y.-Q. Tan, and C.-L. Lei. 2007a. The role of temperature and photoperiod in diapause induction in the brassica leaf beetle, *Phaedon brassicae* (Coleoptera: Chrysomelidae). *Eur. J. Ent.* 104:693–697.

Wang, X.-P., F.-S. Xue, X.-M. Zhou, and C.-L. Lei. 2007b. Thermoperiodic response and effect of photoperiod on thermoperiodic induction of diapause in *Colaphellus bowringi*. *Ent. Exp. Appl.* 124:299–304.

Wang, X.-P., Q.-S. Yang, X.-M. Zhou, S. Xu, and C.-L. Lei. 2009. Effects of photoperiod and temperature on diapause induction and termination in the swallowtail, *Sericinus montelus*. *Physiol. Ent.* 34:158–162.

Wang, Z., Y. Chen, R. Diaz, and R.A. Laine. 2019b. Physiology of crapemyrtle bark scale, *Acanthococcus lagerstroemiae* (Kuwana), associated with seasonally altered cold tolerance. *J. Insect Physiol.* 112:1–8.

Wasserthal, L.T. 1996. Interaction of circulation and tracheal ventilation in holometabolous insects. *Adv. Insect Physiol.* 26:297–351.

Wassielewski, O., T. Wojciechowicz, K. Giejdasz, and N. Krishnan. 2011. Influence of methoprene and temperature on diapause termination in adult females of the over-wintering solitary bee, *Osmia rufa* L. *J. Insect Physiol.* 57:1682–1688.

Watanabe, K., J.J. Hull, T. Niimi, K. Imai, S. Matsumoto, T. Yaginuma, and H. Kataoka. 2007. FXPRL-amide peptides induce ecdysteroidogenesis through a G-protein coupled receptor expressed in the prothoracic gland of *Bombyx mori*. *Mol. Cell. Endocrin.* 273:51–58.

Watanabe, M. and K. Tanaka. 1998. Adult diapause and cold hardiness in *Aulacophora nigripennis* (Coleoptera: Chrysomelidae). *J. Insect Physiol.* 44:1103–1110.

Watari, Y. 2005. Comparison of the circadian eclosion rhythm between non-diapause and diapause pupae in the onion fly, *Delia antiqua*: the change in rhythmicity. *J. Insect Physiol.* 51:11–16.

Webb, M.-L. and D.L. Denlinger. 1998. GABA and picrotoxin alter expression of a maternal effect that influences pupal diapause in the flesh fly, *Sarcophaga bullata*. *Physiol. Ent.* 23:184–191.

Weed, A.S., J.S. Elkinton, and N.K. Lany. 2016. Density-dependent recruitment and diapause in the spring-feeding generation of hemlock woolly adelgid (Hemiptera: Adelgidae) in western North America. *Environ. Ent.* 45:1352–1359.

Wei, L.-H. and J.U. Guo. 2020. Coding functions of "noncoding" RNAs. *Science* 367:1074–1075.

Wei, X., F. Xue, and A. Li. 2001. Photoperiodic clock of diapause induction in *Pseudopidorus fasciata* (Lepidoptera: Zygaenidae). *J. Insect Physiol.* 47:1367–1375.

Wei, X.-T., Y.-C. Zhou, H.-J. Xiao, X.-P. Wang, Z.-M. Bao, and F.-S. Xue. 2010. Relationship between the natural duration of diapause and the post-diapause reproduction in the cabbage beetle, *Colaphellus bowringi* (Coleoptera: Chrysomelidae). *Eur. J. Ent.* 107:337–340.

Wei, Y., J. Zhang, S. Xu, X. Peng, X. Yan, X. Li, H. Wang, H. Chang, and Y. Gao. 2018. Controllable oxidative stress and tissue specificity in major tissues during the torpor-arousal cyle in hibernating Daurian ground squirrels. *Open Biol.* 8:180068.

Wei, Z.-J., Q.-R. Zhang, L. Kang, W.-H. Xu, and D.L. Denlinger. 2005. Molecular characterization and expression of prothoracicotropic hormone during development and pupal diapause in the cotton bollworm, *Helicoverpa armigera*. *J. Insect Physiol.* 51:691–700.

Wellso, S.G. and P.L. Adkisson. 1966. A long-day short-day effect in the photoperiodic control of the pupal diapause of the bollworm, *Heliothis zea* (Boddie). *J. Insect Physiol.* 12:1455–1465.

West-Eberhard, M.J. 2003. *Developmental Plasticity and Evolution*, Oxford: Oxford University Press.

Westby, K.M. and K.A. Medley. 2020. Cold nights, city lights: artificial light at night reduces photoperiodically induced diapause in urban and rural populations of *Aedes albopictus* (Diptera: Culicidae). *J. Med. Ent.* 57:1694–1699.

Weyda, F., J. Pflegerova, T. Staskova, A. Tomala, E. Prenerova, R. Zemek, L. Volter, and D. Kodrik. 2015. Ultrastructural and biochemical comparison of summer active and summer diapausing pupae of the horse chestnut leaf miner, *Cameraria ohridella* (Lepidoptera: Gracillariidae). *Eur. J. Ent.* 112:197–203.

Wheeler, W.M. 1893. A contribution to insect embryology. *J. Morph.* 8:141–160.

White, W.H. 2008. Utilizing diapause in a sugarcane borer (Lepidoptera: Crambidae) laboratory colony as a cost saving measure. *Midsouth Ent.* 1:76–80.

Whitman, D.W. and A.A. Agrawal. 2009. What is phenotypic plasticity and why is it important?. In *Phenotypic Plasticity in Insects. Mechanisms and Consequences*, ed. D.W. Whitman and T.N. Ananthakrishnan, Einfield: Science Publishers, pp. 1–63.

Whitney, T.D., B.N. Philip, and J.D. Harwood. 2014. Tradeoff in two winter-active wolf spiders: increased mortality for increased growth. *Ent. Exp. Appl.* 153:191–198.

Wijayaratne, L.K.W. and P.G. Fields. 2012. Effects of rearing conditions, geographic origin, and selection on larval diapause in the Indianmeal moth, *Plodia interpunctella*. *J. Insect Sci.* 12:119.

Wiklund, C., P.-O. Wickman, and S. Nylin. 1992. A sex difference in the propensity to enter direct/diapause development: a result of selection for protandry. *Evolution* 46:519–528.

Wiklund, C. and B.S. Tullberg. 2004. Seasonal polyphenism and leaf mimicry in the comma butterfly. *Animal Behav.* 68:621–627.

Wiklund, C., A. Vallin, M. Friberg, and S. Jakobsson. 2008. Rodent predation on hibernating peacock and small tortoiseshell butterflies. *Behav. Ecol. Sociobiol.* 62:379–389.

Wiklund, C., C. Stefnecu, and M. Friberg. 2017. Host plant exodus and larval wandering behavior in a butterfly: diapause generation larvae wander for longer periods than do non-diapause generation larvae. *Ecol. Ent.* 42:531–534.

Wiklund, C., P. Lehmann, and M. Friberg. 2019. Diapause decision in the small tortoiseshell butterfly, *Aglais urticae*. *Ent. Exp. Appl.* 167:433–441.

Wilches, D.M., R.A. Laird, K.D. Floate, and P.G. Fields. 2017. Effects of acclimation and diapause on the cold tolerance of *Trogoderma granarium*. *Ent. Exp. Appl.* 165:169–178.

de Wilde, J. and H. Bonga. 1958. Observations on threshold intensity and sensitivity of different wavelengths of photoperiodic response in the Colorado beetle (*Leptinotarsa decemlineata* Say). *Ent. Exp. Appl.* 1:301–307.

de Wilde, J., C.S. Duintjer, and L. Mook. 1959. Physiology of diapause in the adult Colorado potato beetle (*Leptinotarsa decemlineata* Say). I. The photoperiod as a controlling factor. *J. Insect Physiol.* 3:75–85.

de Wilde, J. and J.A. de Boer. 1961. Physiology of diapause in the adult Colorado beetle. II. Diapause as a case of pseudoallatectomy. *J. Insect Physiol.* 6:152–161.

de Wilde, J., W. Bongers, and H. Schooneveld. 1969. Effects of hostplant age on phytophagous insects. *Ent. Exp. Appl.* 12:714–720.

Williams, C.M. 1946. Physiology of insect diapause: the role of the brain in the production and termination of pupal dormancy in the giant silkworm *Platysamia cecropia*. *Biol. Bull.* 90:234–243.

Williams, C.M. 1952. Physiology of insect diapause. IV. The brain and prothoracic glands as an endocrine system in the cecropia silkworm. *Biol. Bull.* 103:120–138.

Williams, C.M. 1969. Photoperiodism and the endocrine aspects of insect diapause. *Symp. Soc. Exp. Biol.* 23:285–300.

Williams, C.M. and P.L. Adkisson. 1964. Physiology of insect diapause. XIV. An endocrine mechanism for the photoperiodic control of pupal diapause in the oak silkworm *Antheraea pernyi*. *Biol. Bull.* 127:511–525.

Williams, C.M., P.L. Adkisson, and C. Walcott. 1965. Physiology of insect diapause. XV. The transmission of photoperiod signals to the brain of the oak silkworm, *Antheraea pernyi*. *Biol. Bull.* 128:497–507.

Williams, C.M., K.E. Marshall, H.A. MacMillan, J.D.K. Dzurisin, J.J. Hellmann, and B.J. Sinclair. 2012. Thermal variability increases the impact of autumnal warming and drives metabolic depression in an overwintering butterfly. *PLoS One* 7:e34470.

Williams, C.M., H.A.L. Henry, and B.J. Sinclair. 2015. Cold truths: how winter drives responses of terrestrial organisms to climate change. *Biol. Rev.* 90:214–235.

Williams, D.R., L.E. Epperson, W. Li, M.A. Hughes, R. Taylor, J. Rogers, S.L. Martin, A.R. Cossins, and A.Y. Gracey. 2005. Seasonally hibernating phenotype assessed through transcript screening. *Physiol. Genomics* 24:13–22.

Williams, K.D., M. Busto, M.L. Suster, A.K.-C. So, Y. Ben-Shahar, S.J. Leevers, and M.B. Sokolowski. 2006. Natural variation in *Drosophila melanogaster* diapause due to the insulin-regulated PI3-kinase. *Proc. Nat'l. Acad. Sci., USA* 103:15911–15915.

Wilson, R.L. and C.A. Abel. 1996. Storage conditions for maintaining *Osmia cornifrons* (Hymenoptera: Megachilidae) for use in germplasm pollination. *J. Kansas Ent. Soc.* 69:270–272.

Win, A.T. and Y. Ishikawa. 2015. Effects of diapause on post-diapause reproductive investment in the moth *Ostrinia scapulalis*. *Ent. Exp. Appl.* 157:346–353.

Winder, D., W.H. Gunzburg, V. Erfle, and B. Salmons. 1998. Expression of antimicrobial peptides has an antitumor effect in human cells. *Biochem. Biophys. Res. Com.* 242:608–612.

Winston, M. 1987. *The Biology of the Honey Bee*, Cambridge: Harvard University Press.

Wipking, W. 1988. Repeated larval diapause and diapause-free development in geographic strains of the burnet moth *Zygaena trifolii* Esp. (Insecta, Lepidoptera). *Oecologia* 77:557–564.

Wipking, W., M. Viebahn, and D. Neumann. 1995. Oxygen consumption, water, lipid and glycogen content of early and late diapause and non-diapause larvae of the burnet moth *Zygaena trifolii*. *J. Insect Physiol.* 41:47–56.

Wipking, W. and J. Kurtz. 2000. Genetic variability in the diapause response of the

burnet moth *Zygaena trifolii* (Lepidoptera: Zygaenidae). *J. Insect Physiol.* 46:127–134.

Wobschall, A. and S.K. Hetz. 2004. Oxygen uptake by convection and diffusion in diapausing moth pupae (*Attacus atlas*). *Animals Environ.* 1275:157–164.

Wolda, H. 1978. Fluctuations in abundance of tropical insects. *Am. Nat.* 112:1017–1045.

Wolda, H. 1988. Insect seasonality: Why? *Ann. Rev. Ecol. Syst.* 19:1–18.

Wolda, H. and D.L. Denlinger. 1984. Diapause in a large aggregation of a tropical beetle. *Ecol. Ent.* 9:217–230.

Woll, S.C. and J.E. Podrabsky. 2017. Insulin-like growth factor signaling regulates developmental trajectory associated with diapause in embryos of the annual killifish *Austrofundulus limnaeus*. *J. Exp. Biol.* 220:2777–2786.

Wolschin, F. and J. Gadau. 2009. Deciphering proteomic signatures of early diapause in *Nasonia. PLoS One* 4:e6394.

Woods, H.A. and S.J. Lane. 2016. Metabolic recovery from drowning by insect pupae. *J. Exp. Biol.* 219:3126–3136.

Wu, S., O.S. Kostromytska, F. Xue, and A.M. Koppenhöfer. 2017. Chilling effect on termination of reproductive diapause in *Listronotus maculicollis* (Coleoptera: Curculionidae). *J. Insect Physiol.* 104:25–32.

Wu, S.-B., M.-Q. Wang, and G. Zhang. 2010. Effects of putrescine on diapause induction and intensity, and post-diapause development of *Helicoverpa armigera*. *Ent. Exp. Appl.* 136:199–205.

Wu, S.-H., D. Yang, X.-T. Lai, and F.-S. Xue. 2006. Induction and termination of prepupal summer diapause in *Pseudopidorus fasciata* (Lepidoptera: Zygaenidae). *J. Insect Physiol.* 52:1095–1104.

Wu, T., Y. Chen, W. Chen, S. Zou, Y. Zhang, Y. Lin, and D. Tang. 2013. Transgenic expression of an insect diapause-specific peptide (DSP) in *Arabidopsis* resists phytopathogenic fungal attacks. *Eur. J. Plant Path.* 137:93–101.

Wu, Q. and M.R. Brown. 2006. Signaling and function of insulin-like peptides in insects. *Ann. Rev. Ent.* 51:1–24.

Wylie, H.G. 1980. Factors affecting diapause of *Microtonus vittatae* (Hymenoptera: Braconidae). *Can. Ent.* 112:747–749.

Xiao, H.-J., D. Yang, and F.-S. Xue. 2006. Effect of photoperiod on the duration of summer and winter diapause in the cabbage butterfly, *Pieris melete* (Lepidoptera: Pieridae). *Eur. J. Ent.* 103:537–540.

Xiao, H.-J., F. Li, X.-T. Wei, and F.-S. Xue. 2008a. A comparison of photoperiodic control of diapause between aestivation and hibernation in the cabbage butterfly *Pieris melete*. *J. Insect Physiol.* 54:755–764.

Xiao, H.-J., H.-M. He, F. Li, and F.-S. Xue. 2008b. Influence of pre-diapause temperature on intensity of summer and winter diapause in the cabbage butterfly *Pieris melete* (Lepidoptera: Pieridae). *Eur. J. Ent.* 105:607–611.

Xiao, H.-J, F.-C. Mou, X.-F. Zhu, and F.-S. Xue. 2010. Diapause induction, maintenance and termination in the rice stem borer *Chilo suppressalis* (Walker). *J. Insect Physiol.* 56:1558–1564.

Xie, D., L. Luo, T.W. Sappington, X. Jiang, and L. Zhang. 2012. Comparison of reproductive and flight capacity of *Loxostege sticticalis* (Lepidoptera: Pyralidae), developing from diapause and non-diapause larvae. *Environ. Ent.* 41:1199–1207.

Xinxin, Z., Y. Shuang, Z. Xunming, W. Shang, Z. Juhong, and X. Jinghui. 2020. TMT-based quantitative proteomic profiling of overwintering *Lissorhoptrus oryzophilus*. *Front. Physiol.* 10:1623.

Xu, S., M.-L. Wang, N. Ding, W.-H. Ma, Y.-N. Li, C.-L. Lei, and X-P. Wang. 2011. Relationship between body weight of overwintering larvae and supercooling capacity; diapause intensity and post-diapause reproductive potential in *Chilo suppressalis* Walker. *J. Insect Physiol.* 57:653–659.

Xu, W.-H., Y. Sato, M. Ikeda, and O. Yamashita. 1995. Stage-dependent and temperature controlled expression of the gene encoding the precursor protein of diapause hormone and pheromone biosynthesis activating neuropeptide in the silkworm, *Bombyx mori*. *J. Biol. Chem.* 270:3804–3808.

Xu, W.-H. and D.L. Denlinger. 2003. Molecular characterization of prothoracicotropic hormone and diapause hormone in *Heliothis virescens* during diapause, and a new role for diapause hormone. *Insect Mol. Biol.* 12:509–516.

Xu, W.-H. and D.L. Denlinger. 2004. Identification of a cDNA encoding DH, PBAN and other FXPRL neuropeptides from the tobacco hornworm, *Manduca sexta*, and expression associated with pupal diapause. *Peptides* 25:1099–1106.

Xu, W.-H., Y.-X. Lu, and D.L. Denlinger. 2012. Cross-talk between the fat body and brain regulates insect developmental arrest. *Proc. Nat'l. Acad. Sci., USA* 109:14687–14692.

Xue, F., H.G. Kalenborn, and H.-Y. Wei. 1997. Summer and winter diapause in pupae of the cabbage butterfly, *Pieris melete* Menetries. *J. Insect Physiol.* 43:701–707.

Xue, F., H.R. Spieth, A. Li and H. Ai. 2002. The role of photoperiod and temperature in determination of summer and winter diapause in the cabbage beetle, *Colaphellus bowringi* (Coleoptera: Chrysomelidae). *J. Insect Physiol.* 48:279–286.

Yagi, S. and M. Fukaya. 1974. Juvenile hormone as a key factor regulating larval diapause of the rice stem borer *Chilo suppressalis* (Lepidoptera: Pyralidae). *App. Ent. Zool.* 9:247–255.

Yaginuma, T. and O. Yamashita. 1999. Oxygen consumption in relation to sorbitol utilization at the termination of diapause in eggs of the silkworm, *Bombyx mori*. *J. Insect Physiol.* 45:621–627.

Yamada, N. and A. Mizoguchi. 2017. Endocrine changes during diapause development in the cabbage army moth

Mamestra brassicae. Physiol. Ent. 42:239–245.

Yamada, N., H. Kataoka, and A. Mizoguchi. 2017. Myosuppression is involved in the regulation of pupal diapause in the cabbage army moth *Mamestra brassicae. Sci. Rep.* 7:41651.

Yamaguchi, K. and K. Nakamura. 2015. Effects of environmental factors on the regulation of egg diapause in the walking-stick insect, *Ramulus irregulariterdentatus* (Phasmatodea: Phasmatidae). *Eur. J. Ent.* 111:35–40.

Yamaguchi, K. and S.G. Goto. 2019. Distinct physiological mechanisms induce latitudinal and sexual differences in the photoperiodic induction of diapause in a fly. *J. Biol. Rhyth.* 34:293–306.

Yamamoto, K., Y. Tsujimura, M. Kometani, C. Kitazawa, A.T.M.F. Islam, and A. Yamanaka. 2011. Diapause pupal color diphenism induced by temperature and humidity conditions in *Byasa alcinous* (Lepidoptera: Papilionidae). *J. Insect Physiol.* 57:930–934.

Yamamoto, T., K. Mase, and H. Sawada. 2013. Diapause prevention effect of *Bombyx mori* by dimethyl sulfoxide. *PLoS One* 8:64124.

Yamanaka, A., M. Adachi, H. Imai, T. Uchiyama, M. Inoue, A.T.M.F. Islam, C. Kitazawa, and K. Endo. 2006. Properties of orange-pupa-inducing factor (OPIF) in the swallowtail butterfly, *Papilio zuthus* L. *Peptides* 27:534–538.

Yamanaka, A., M. Kometani, K. Yamamoto, Y. Tsujimura, M. Motomura, C. Kitazawa, and K. Endo. 2009. Hormonal control of pupal coloration in the painted lady butterfly *Vanessa cardui. J. Insect Physiol.* 55:512–517.

Yamanaka, N., K.F. Rewitz, and M.B. O'Connor. 2013a. Ecdysone control of developmental transitions: lessons from *Drosophila* research. *Ann. Rev. Ent.* 58:497–516.

Yamanaka, N., N.M. Romero, F.A. Martin, K.F. Rewitz, M. Sun, M.B. O'Connor,

and P. Léopold. 2013b. Neuroendocrine control of *Drosophila* larval light preference. *Science* 3431:1113–1116.

Yamashita, O. 1996. Diapause hormone of the silkworm, *Bombyx mori*: structure, gene expression and function. *J. Insect Physiol.* 42:669–679.

Yamashita, O. and K. Hasegawa. 1985. Embryonic diapause. In *Comprehensive Insect Physiology, Biochemistry and Pharmacology*, ed. G.A. Kerkut and L.I. Gilbert, Vol. 1, Oxford: Pergamon Press, pp. 407–434.

Yang, D., X.-T. Lai, L. Sun, and F.-S. Xue. 2007a. Parental effects: physiological age, mating pattern, and diapause duration on incidence of progeny in the cabbage beetle, *Colaphellus bowringi* Baly (Coleoptera: Chrysomelidae). *J. Insect Physiol.* 53:900–908.

Yang, H., S. Berry, T.S.G. Olsson, M. Hartley, M. Howard, and C. Dean. 2017a. Distinct phases of Polycomb silencing to hold epigenetic memory of cold in *Arabidopsis. Science* 357:1142–1149.

Yang, H.-Z., X.-Y. Tu, Q.-W. Xia, H.-M. He, C. Chen, and F.-S. Xue. 2014. Photoperiodism of diapause induction and diapause termination in *Ostrinia furnacalis. Ent. Exp. Appl.* 153:34–46.

Yang, J., J. Zhu, and W.-H. Xu. 2010. Differential expression, phosphorylation of COX subunit 1 and COX activity during diapause phase in the cotton bollworm, *Helicoverpa amigera. J. Insect Physiol.* 56:1992–1998.

Yang, L., D.L. Denlinger, and P. Piermarini. 2017b. The diapause program impacts renal excretion and molecular expression of aquaporins in the northern house mosquito, *Culex pipiens. J. Insect Physiol.* 98:141–148.

Yang, P., S. Abe, Y.-P. Zhao, Y. An, and K. Suzuki. 2004. Growth suppression of rat hepatoma cells by a pentapeptide from *Antheraea yamamai. J. Insect Biotech. Ser.* 73:1–13.

Yang, P., S. Abe, Y. Sato, T. Yamashita, F. Matsuda, T. Hamayasu, K. Imai, and K. Suzuki. 2007b. A palmitoyl conjugate of an insect pentapeptide causes growth arrest in mammalian cells and mimics the action of diapause hormone. *J. Insect Biotech. Ser.* 76:63–69.

Yang, P., H. Tanaka, E. Kuwano, and K. Suzuki. 2008. A novel cytochrome P450 gene (*CYP4G25*) of the silkmoth *Antheraea yamamai*: cloning and expression pattern in pharate first instar larvae in relation to diapause. *J. Insect Physiol.* 54:636–643.

Yaro, A.S., A.I. Traore, D.L. Huestis, A. Adamou, S. Timbine, Y. Kassogue, M. Diallo, A. Dao, S.F. Traore, and T. Lehmann. 2012. Dry season reproductive depression of *Anopheles gambiae* in the Sahel. *J. Insect Physiol.* 58:1050–1059.

Ye, H.-L., D.-R Li, J.-S. Yang, D.-F. Chen, S. de Vos, M. Vuylsteke, P. Sorgeloos, G. van Stappen, P. Bossier, H. Nagasawa, and W.-J. Yang. 2017. Molecular characterization and functional analysis of a diapause hormone receptor-like gene in parthenogenetic *Artemia*. *Peptides* 90:100–110.

Yin, C.M. and G.M. Chippendale. 1973. Juvenile hormone regulation of the larval diapause of the southwestern corn borer, *Diatraea grandiosella*. *J. Insect Physiol.* 19:2403–2420.

Yin, C.M., Z.S. Wang, and W.D. Chaw. 1985. Brain neurosecretory cell and ecdysiotropin activity of the nondiapausing, pre-diapausing and diapausing southwestern cornborer, *Diatraea grandiosella Dyar*. *J. Insect Physiol.* 31:659–667.

Yin, Z.-J., X.-L. Dong, K. Kang, H. Chen, X.-Y. Dai, G.-A. Wu, L. Zheng Y. Yu, and Y.-F. Zhai. 2018. FoxO transcription factor regulates hormone mediated signaling on nymphal diapause. *Front. Physiol.* 9:1654.

Yocum, G.D. 2001. Differential expression of two *HSP70* transcripts in response to cold shock, thermoperiod, and adult diapause in the Colorado potato beetle. *J. Insect Physiol.* 47:1139–1145.

Yocum, G.D. 2003. Isolation and characterization of three diapause-associated transcripts from the Colorado potato beetle, *Leptinotarsa decemlineata*. *J. Insect Physiol.* 49:161–169.

Yocum, G.D., K.H. Joplin, and D.L. Denlinger. 1998. Upregulation of a 23 kDa small heat shock protein transcript during pupal diapause in the flesh fly, *Sarcophaga crassipalpis*. *Insect Biochem. Mol. Biol.* 28:677–682.

Yocum, G.D., W.P. Kemp, J. Bosch, and J.N. Knoblett. 2005. Temporal variation in overwintering gene expression and respiration in the solitary bee *Megachile rotundata*. *J. Insect Physiol.* 51:621–629.

Yocum, G.D., W.P. Kemp, J. Bosch, and J.N. Knoblett. 2006. Thermal history influences diapause development in the solitary bee *Megachile rotundata*. *J. Insect Physiol.* 52:1113–1120.

Yocum, G.D., J.P. Rinehart, A. Chirumamilla-Chapara, and M.L. Larson. 2009a. Characterization of gene expression patterns during the inititation and maintenance phases of diapause in the Colorado potato beetle, *Leptinotarsa decemlineata*. *J. Insect Physiol.* 55:32–39.

Yocum, G.D., J.P. Rinehart, and M.L. Larson. 2009b. Down-regulation of gene expression between the diapause initiation and maintenance phases of the Colorado potato beetle, *Leptinotarsa decemlineata* (Coleoptera: Chrysomelidae). *Eur. J. Ent.* 106:471–476.

Yocum, G.D., J.P. Rinehart, and M.L. Larson. 2011a. Monitoring diapause development in the Colorado potato beetle, *Leptinotarsa deemlineata*, under field conditions using molecular biomarkers. *J. Insect Physiol.* 57:645–652.

Yocum, G.D., J.S. Buckner, and C.L Fatland. 2011b. A comparison of internal and external lipids of nondiapausing and diapause initiation phase adult Colorado potato beetles, *Leptinotarsa decemlineata*. *Comp. Biochem. Physiol. B* 159:163–170.

Yocum, G.D., J.P. Rinehart, and M.A. Boetel. 2012. Water balance in the sugarbeet root maggot *Tetanops myopaeformis*, during long-term low-temperature storage and after freezing. *Physiol. Ent.* 37:340–344.

Yocum, G.D., J.P. Rinehart, and W.P. Kemp. 2014. Cell position during larval development affects postdiapause development in *Megachile rotundata* (Hymenoptera: Megachilidae). *Environ. Ent.* 43:1045–1052.

Yocum, G.D., J.P. Rinehart, D.P. Horvath, W.P. Kemp, J. Bosch, R. Alroobi, and S. Salem. 2015. Key molecular processes of the diapause to post-diapause quiescence transition in the alfalfa leafcutting bee *Megachile rotundata* identified by comparative transcriptome analysis. *Physiol. Ent.* 40:103–112.

Yocum, G.D., A.K. Childers, J.P. Rinehart, A. Rajamohan, T.L. Pitts-Singer, K.J. Greenlee, and J.H. Bowsher. 2018. Environmental history impacts gene expression during diapause development in the alfalfa leafcutter bee, *Megachile rotundata*. *J. Exp. Biol.* 221:jeb173443.

Yoder, J.A. and D.L. Denlinger. 1991. Water balance in flesh fly pupae and water vapor absorption associated with diapause. *J. Exp. Biol.* 157:273–286.

Yoder, J.A. and D.L. Denlinger. 1992. Water vapour uptake by diapausing eggs of a tropical walking stick. *Physiol. Ent.* 17:97–103.

Yoder, J.A., D.L. Denlinger, M.W. Dennis, and P.E. Kolattukudy. 1992a. Enhancement of diapausing flesh fly puparia with additional hydrocarbons and evidence for alkane biosynthesis by a decarbonylation mechanism. *Insect Biochem. Mol. Biol.* 22:237–243.

Yoder, J.A., D.L. Denlinger, and H. Wolda. 1992b. Aggregation promotes water conservation during diapause in the tropical fungus beetle, *Stenotarsus rotundus*. *Ent. Exp. Appl.* 63:203–205.

Yoder, J.A., D.B. Rivers, and D.L. Denlinger. 1994. Water relationships in the ectoparasitoid *Nasonia vitripennis* during larval diapause. *Physiol. Ent.* 19:373–378.

Yoder, J.A., G.J. Blomquist, and D.L. Denlinger. 1995. Hydrocarbon profiles from puparia of diapausing and nondiapausing flesh flies (*Sarcophaga crassipalpis*) reflect quantitative rather than qualitative differences. *Arch. Insect Biochem. Physiol.* 28:377–385.

Yoder, J.A., J.B. Benoit, D.L. Denlinger, and D.B. Rivers. 2006. Stress-induced accumulation of glycerol in the flesh fly, *Sarcophaga bullata*: evidence indicating anti-desiccant and cryoprotectant functions of this polyol and a role for the brain in coordinating the response. *J. Insect Physiol.* 52:202–214.

Yoshimura, H. and Y.Y. Yamada. 2018. Caste-fate determination primarily occurs after adult emergence in a primitively eusocial paper wasp: significance of the photoperiod during the adult stage. *Sci. Nature* 105:15.

Young, A.M. 1982. *Population Biology of Tropical Insects*. New York: Plenum.

Yuan, Q., D. Metterville, A.D. Briscoe, and S.M. Reppert. 2007. Insect cryptochromes: gene duplication and loss define diverse ways to construct insect circadian clocks. *Mol. Biol. Evol.* 24:948–955.

Zalucki, M.P. and A.R. Clarke. 2004. Monarchs across the Pacific: the Columbus hypothesis revisited. *Biol. J. Linn. Soc.* 81:111–121.

Zani, D., T.W. Crowther, L. Mo, S.S. Renner, and C.M. Zohner. 2020. Increased growing-season productivity drives earlier autumn leaf senescence in temperate trees. *Science* 370:1066–1071.

Ždárek, J. and D.L. Denlinger. 1975. Action of ecdysteroids, juvenoids, and non-hormonal agents on termination of pupal diapause in the flesh fly. *J. Insect Physiol.* 21:1193–1202.

Ždárek, J. and D.L. Denlinger. 1987. Pupal ecdysis in flies: the role of ecdysteroids in its regulation. *J. Insect Physiol.* 33:123–128.

Ždárek, J., R. Čtvrtečka, O. Hovorka, and V. Koštál. 2000. Activation of gonads and disruption of imaginal diapause in the apple blossom weevil, *Anthomomus pomorum* (Coleoptera: Curculionidae), with juvenoids in laboratory and field trials. *Eur. J. Ent.* 97:25–31.

Zeender, V., J. Roy, A. Wegmann, M.A. Schafer, N. Gourgoulianni, W.U. Blanckenhorn, and P.T. Rohner. 2019. Comparative reproductive dormancy differentiation in European black scavenger flies (Diptera: Sepsidae). *Oecologia* 189:905–917.

Zeng, J.P., Y. Wang, and X.-P. Liu. 2013. Influence of photoperiod on the development of diapause in larvae and its cost for individuals of a univoltine population of *Dendrolimus punctatus* (Lepidoptera: Lasiocampidae). *Eur. J. Ent.* 110:95–101.

Zhai, Y., X. Dong, H. Gao, H. Chen, P. Yang, P. Li, Z. Yin, L. Zheng, and Y. Yu. 2019. Quantitative proteomic and transcriptomic analyses of metabolic regulation of adult reproductive diapause in *Drosophila suzukii* (Diptera: Drosophilidae) females. *Front. Physiol.* 10:344.

Zhan, S., C. Merlin, J.L. Boore, and S.M. Reppert. 2011. The monarch butterfly genome yields insights into long-distance migration. *Cell* 147:1171–1185.

Zhan, S., W. Zhang, K. Niitepold, J. Hsu, J.F. Haeger, M.P. Zalucki, S. Altizer, J.C. de Roode, S.M. Reppert, and M.R. Kronforst. 2014. The genetics of monarch butterfly migration and warning colouration. *Nature* 514:317–321.

Zhang, B., Y. Peng, X.-J. Zhao, A.A. Hoffmann, R. Liu, and C.-S. Ma. 2016a. Emergence of the overwintering generation of peach fruit moth (*Carposina* sasakii) depends on diapause and spring soil temperatures. *J. Insect Physiol.* 86:32–39.

Zhang, B., F. Zhao, A. Hoffmann, G. Ma, H.-M. Ding, and C.-S. Ma. 2016b. Warming accelerates carbohydrate consumption in the diapausing overwintering peach fruit moth *Carposin sasakii* (Lepidoptera: Carposinidae). *Environ. Ent.* 45:1287–1293.

Zhang, B., L. Zhao, J. Ning, J.D. Wickham, H. Tian, X. Zhang, M. Yang, X. Wang, and J. Sun. 2020a. miR-31-5p regulates cold acclimation of the wood-boring beetle *Monochamus alternatus* via ascaroside signaling. *BMC Biol.* 18:184.

Zhang, C., D. Wei, G. Shi, X. Huang, P. Cheng, G. Liu, X. Guo, L. Liu, H. Wang, F. Miao, and M. Gong. 2019. Understanding the regulation of overwintering diapause molecular mechanisms in *Culex pipiens pallens* through comparative proteomics. *Sci. Rep.* 9:6485.

Zhang, C., I. Daubnerova, Y.-H. Jang, S. Kondo, D. Žitnan, and Y.-J. Kim. 2021. The neuropeptide allatostatin C from clock-associated DN1p neurons generates the circadian rhythm for oogenesis. *Proc. Nat'l. Acad. Sci., USA* 118:e2016878118.

Zhang, H., Y. Lin, G. Shen, X. Tan, C. Lei, W. Long, H. Liu, Y. Zhang, Y. Xu, J. Wu, J. Gu, Q. Xia, and P. Zhao. 2017a. Pigmentary analysis of eggs of the silkworm *Bombyx mori. J. Insect Physiol.* 101:142–150.

Zhang, H.-Z., Y.-Y. Li, T. An, F.-X. Huang, M.-Q. Wang, C.-X. Liu, J.-J. Mao, and L.-S. Zhang. 2018a. Comparative transcriptome and iTRAQ proteome analyses reveal the mechanisms of diapause in *Aphidius gifuensis* Ashmead (Hymenoptera: Aphidiidae). *Front. Physiol.* 9:1697.

Zhang, J.-J., X. Zhang, L.-S. Zang, W.-M. Du, Y.-Y. Hou, C.-C. Ruan, and N. Desneux. 2018b. Advantages of diapause in *Trichogramma dendrolimi* mass production on eggs of the Chinese silkworm, *Antheraea pernyi. Pest Manag. Sci.* 74:959–965.

Zhang, T.-Y., L. Kang, Z.-F. Zhang, and W.-H. Xu. 2004a. Identification of a POU factor involved in regulating the neuron-specific expression of the gene encoding diapause hormone and pheromone biosynthesis-activating neuropeptide in *Bombyx mori. Biochem. J.* 380:255–263.

Zhang, T.-Y., J.-S. Sun, L.-B. Zhang, J.-L. Shen, and W.-H. Xu. 2004b. Cloning and expression of the cDNA encoding the FXPRL family of peptides and a functional analysis of their effect on breaking pupal diapause in *Helicoverpa armigera. J. Insect Physiol.* 50:25–33.

Zhang, T.-Y., J.-S. Sun, Q.-R. Zhang, J. Xu, R.-J. Jiang, and W.-H. Xu. 2004c. The diapause hormone-pheromone biosynthesis activating neuropeptide gene of *Helicoverpa armigera* encodes multiple peptides that break, rather than induce, diapause. *J. Insect Physiol.* 50:547–554.

Zhang, Q., J. Ždárek, R.J. Nachman, and D.L. Denlinger. 2008. Diapause hormone in the corn earworm, *Helicoverpa zea*: optimum temperature for activity, structure-activity relationships, and efficacy in accelerating flesh fly pupariation. *Peptides* 29:196–205.

Zhang, Q., R.J. Nachman, P. Zubrzak, and D.L. Denlinger. 2009. Conformational aspects and hyperpotent agonists of diapause hormone for termination of pupal diapause in the corn earworm. *Peptides* 30:596–602.

Zhang, Q. and D.L. Denlinger. 2011a. Molecular structure of the prothoracic hormone gene in the northern house mosquito, *Culex pipiens*, and its expression analysis in association with diapause and blood feeding. *Insect Mol. Biol.* 20:201–213.

Zhang, Q. and D.L. Denlinger. 2011b. Elevated *couch potato* transcripts associated with adult diapause in the mosquito *Culex pipiens. J. Insect Physiol.* 57:620–627.

Zhang, Q., R.J. Nachman, K. Kaczmarek, J. Zabrocki, and D.L. Denlinger. 2011a. Disruption of insect diapause using agonists and an antagonist of diapause hormone. *Proc. Nat'l. Acad. Sci., USA* 108:16922–16926.

Zhang, Q. and D.L. Denlinger. 2012. Dynamics of diapause hormone and prothoracicotropic hormone transcript expression at diapause termination in pupae of the corn earworm, *Helicoverpa zea. Peptides* 34:120–126.

Zhang, Q., Y.-X. Lu, and W.-H. Xu. 2013. Proteomic and metabolomic profiles of larval hemolymph associated with diapause in the cotton bollworm, *Helicoverpa armigera. BMC Genomics* 14:751.

Zhang, Q., P.M. Piermarini, R.J. Nachman, and D.L. Denlinger. 2014. Molecular identification and expression analysis of a diapause hormone receptor in the corn earworm, *Helicoverpa zea. Peptides* 53:250–257.

Zhang, Q., R.J. Nachman, and D.L. Denlinger. 2015a. Diapause hormone in the *Helicoverpa/Heliothis* complex: a review of gene expression, peptide structure and activity, analog and antagonist development, and the receptor. *Peptides* 72:196–201.

Zhang, Q., R.J. Nachman, K. Kaczmarek, K. Kierus, J. Zabrocki, and D.L. Denlinger. 2015b. Development of neuropeptide analogs capable of traversing the integument: a case study using diapause hormone analogs in *Helicoverpa zea. Insect Biochem. Mol. Biol.* 67:87–93.

Zhang, X., R. Zabinsky, Y. Teng, M. Cui, and M. Han. 2011b. microRNAs play critical roles in the survival and recovery of *Caenorhabditis elegans* from starvation-induced L1 diapause. *Proc. Nat'l. Acad. Sci., USA* 108:17997–18002.

Zhang, X., W. Du, J. Zhang, Z. Zou, and C. Ruan. 2020b. High-throughput profiling of diapause regulated genes from *Trichogramma dendrolimi*, an important egg parasitoid. *BMC Genomics* 21:664.

Zhang, X.-S., T. Wang, X.-W. Lin, D.L. Denlinger, and W.-H. Xu. 2017b. Reactive oxygen species extend insect life span using components of the insulin-signaling pathway. *Proc. Nat'l. Acad. Sci., USA* 114: e7832–7840.

Zhang, Y., M.J. Markert, S.C. Groves, P.E. Hardin, and C. Merlin. 2017c. Vertebrate-like CRYPTOCHROME 2 from monarch regulates circadian transcription via independent repression of CLOCK and BMAL1 activity. *Proc. Nat'l. Acad. Sci., USA* 114:e7516–e7525.

Zhao, J.-Y., X.-T. Zhao, J.-T. Sun, L.-F. Zou, S.-X. Yang, X. Han, W.-C. Zhu, Q. Yin, and X.-Y. Hong. 2017. Transcriptome and proteome analysis reveal complex mechanisms of reproductive diapause in the two-spotted spider mite, *Tetranychus urticae*. *Insect Mol. Biol.* 26:215–232.

Zhao, L., J. Wit, N. Svetic, and D.J. Begun. 2015. Parallel gene expression differences between low and high latitude populations of *Drosophila melanogaster* and *D. simulans*. *PLoS Genet.* 11:e1005184.

Zhao, L., W. Wang, Y. Qiu, and A.S. Torson. 2021. Physiological mechanisms of variable nutrient accumulation patterns between diapausing and non-diapausing fall webworm (Lepidoptera: Arctiidae) pupae. *Environ. Ent.* (in press).

Zhao, L.-L., F. Jin, X. Ye, L. Zhu, J.-S. Yang, and W.-J. Yang. 2015. Expression profiles of miRNAs and involvement of miR-100 and miR-34 in regulation of cell cycle arrest in *Artemia*. *Biochem. J.* 470:223–231.

Zhao, X., A.O. Bergland, E.L. Behrman, B.D. Gregory, D.A. Petrov, and P.S. Schmidt. 2016. Global transcriptional profiling of diapause and climatic adaptation in *Drosophila melanogaster*. *Mol. Biol. Evol.* 33:707–720.

Zhou, C.-M., T.-Q. Zhang, Xi Wang, S. Yu, H. Lian, H. Tang, Z.-Y. Feng, J. Zozomova-Lihová, and J.-W. Wang. 2013. Molecular basis of age-dependent vernalization in *Cardamine flexuosa*. *Science* 340: 1097–1100.

Zhou, G. and R.L. Miesfeld. 2009. Energy metabolism during diapause in *Culex pipiens* mosquitoes. *J. Insect Physiol.* 55:40–46.

Zhou, H.-Z. and W. Topp. 2000. Diapause and polyphenism of life-history of *Lagria hirta*. *Ent. Exp. Appl.* 94:201–210.

Zhou, J., J. Li, R. Wang, X. Sheng, S. Zhong, Q. Weng, and Y. Luo. 2016. Ecdysteroid titers and expression of *Halloween* genes and ecdysteroid receptor in relation to overwintering and the long larval phase in the seabuckthorn carpenterworm, *Holcocerus hippophaecolus*. *Ent. Exp. Appl.* 160:133–146.

Zhou, X., F.M. Oi, and M.E. Scharf. 2006. Social exploitation of hexamerin: RNAi reveals a major caste-regulatory factor in termites. *Proc. Nat'l. Acad. Sci., USA* 103:4499–4504.

Zhou, Y., D. Sun, W.-L. Quan, N. Ding, W. Liu, W.-H. Ma, and X.-P. Wang. 2018. Divergence in larval diapause induction between the rice and water-oat populations of the striped stem borer, *Chilo suppressalis* (Walker) (Lepidoptera: Crambidae). *Environ. Sci. Pollut. Res.* 25:29715–29724.

Zhou, Z., Y. Li, C. Yuan, D. Doucet, Y. Zhang, and L. Qu. 2016. Overexpression of TAT-PTD-diapause hormone fusion protein in tobacco and its effect on the larval development of *Helicoverpa armigera* (Lepidoptera: Noctuidae). *Pest Manag. Sci.* 73:1197–1203.

Zhou, Z., X. Dong, Q. Su, Z. Xia, Z. Wang, J. Yuan, and C. Li. 2020. Effects of pre-diapause temperature and body weight on the diapause intensity of the overwintering generation of *Bactrocera minax* (Diptera: Tephritidae). *J. Insect Sci.* 20(12):1–6.

Zhu, D.-H. and S. Tanaka. 2004. Photoperiod and temperature affect the life cycle of a subtropical cockroach, *Opisoplatia orientalis*: seasonal pattern shaped by winter mortality. *Physiol. Ent.* 29:16–23.

Zhu, D.-H., Y.-P. Yang, and Z. Liu. 2009. Reversible change in embryonic diapause intensity by mild temperature in the Chinese rice grasshopper, *Oxya chinensis*. *Ent. Exp. Appl.* 133:1–8.

Zhu, F., F. Xue, and C. Lei. 2006. The effect of environmental conditions on diapause in the blister beetle, *Mylabris phalerata* (Coleoptera: Meloidae). *Eur. J. Ent.* 103:531–535.

Zhu, H., Q. Yuan, O. Froy, A. Casselman, and S.M. Reppert. 2005. The two CRYs of the butterfly. *Curr. Biol.* 15:953–954.

Zhu, H., R.J. Gegear, A. Casselman, S. Kanginakudru, and S.M. Reppert. 2009. Defining behavioral and molecular differences between summer and migratory monarch butterflies. *BMC Biol.* 7:14.

Zhu, L., Z. Tian, S.Guo, W. Liu, F. Zhu, and X.-P. Wang. 2019. Circadian clock genes link photoperiodic signals to lipid accumulation during diapause preparation in the diapause-destined female cabbage beetles *Colaphellus bowringi. Insect Biochem. Mol. Biol.* 104:1–10.

van Zon, A.C., W.P.J. Overmeer, and A. Veerman. 1981. Carotenoids function in photoperiodic induction of diapause in a predacious mite. *Science* 213:1131–1133.

Zonato, V., G. Fedele, and C.P. Kyricou. 2016. An intonic polymorphism in *couch potato* is not distributed clinally in European *Drosophila melanogaster* populations nor does it affect diapause inducibility. *PLoS One* 11:e0162370.

Zonato, V., L. Collins, M. Pegoraro, E. Tauber, and C.P. Kyriacou. 2017. Is diapause an ancient adaptation in *Drosophila*? *J. Insect Physiol.* 98:267–274.

Zonato, V., S. Vanin, R. Costa, E. Tauber, and C.P. Kyriacou. 2018. Inverse european latitudinal cline at the *timeless* locus of *Drosophila melanogaster* reveals selection on a clock gene: population genetics of ls-tim. *J. Biol. Rhyth.* 33:15–23.

Zverev, V., M.V. Kozlov, A. Forsman, and E.L. Zvereva. 2018. Ambient temperatures differently influence colour morphs of the leaf beetle *Chrysomela lapponica*: roles of thermal melanism and developmental plasticity. *J. Therm. Biol.* 74:100–109.

Zvereva, E.L. 2002. Effects of host plant quality on overwintering success of the leaf beetle *Chrysomela lapponica* (Coleoptera: Chrysomelidae). *Eur. J. Ent.* 99:189–195.

Species Index

Subject Index

Printed in the United States
by Baker & Taylor Publisher Services